Contemporary Mathematics in Context

2nd Edition

Christian R. Hirsch • James T. Fey • Eric W. Hart
Harold L. Schoen • Ann E. Watkins
with
Beth E. Ritsema • Rebecca K. Walker • Sabrina Keller
Robin Marcus • Arthur F. Coxford • Gail Burrill

New York, New York Columbus, Ohio Chicago, Illinois Peoria, Illinois Woodland Hills, California

The McGraw·Hill Companies

This material is based upon work supported, in part, by the National Science Foundation under grant no. ESI 0137718. Opinions expressed are those of the authors and not necessarily those of the Foundation.

Send all inquiries to:
Glencoe/McGraw-Hill
8787 Orion Place
Columbus, OH 43240-4027

ISBN-13: 978-0-07-861521-4 (Student Edition)
ISBN-10: 0-07-861521-6 (Student Edition)

Core-Plus Mathematics
Contemporary Mathematics in Context
Course 1 Student Edition

Printed in the United States of America.

5 6 7 8 9 10 WDQ 15 14 13 12 11 10

Core-Plus Mathematics 2 Development Team

Senior Curriculum Developers
Christian R. Hirsch (Director)
Western Michigan University

James T. Fey
University of Maryland

Eric W. Hart
Maharishi University of Management

Harold L. Schoen
University of Iowa

Ann E. Watkins
California State University, Northridge

Contributing Curriculum Developers
Beth E. Ritsema
Western Michigan University

Rebecca K. Walker
Grand Valley State University

Sabrina Keller
Michigan State University

Robin Marcus
University of Maryland

Arthur F. Coxford (deceased)
University of Michigan

Gail Burrill
Michigan State University (First edition only)

Principal Evaluator
Steven W. Ziebarth
Western Michigan University

Advisory Board
Diane Briars
Pittsburgh Public Schools

Jeremy Kilpatrick
University of Georgia

Robert E. Megginson
University of Michigan

Kenneth Ruthven
University of Cambridge

David A. Smith
Duke University

Mathematical Consultants
Deborah Hughes-Hallett
University of Arizona / Harvard University

Stephen B. Maurer
Swarthmore College

William McCallum
University of Arizona

Doris Schattschneider
Moravian College

Richard Scheaffer
University of Florida

Evaluation Consultant
Norman L. Webb
University of Wisconsin-Madison

Collaborating Teachers
Mary Jo Messenger
Howard Country Public Schools, Maryland

Jacqueline Stewart
Okemos, Michigan

Technical Coordinator
James Laser
Western Michigan University

Production and Support Staff
Angie Reiter
Teri Ziebarth
Western Michigan University

Graduate Assistants
Allison BrckaLorenz
Christopher Hlas
University of Iowa

Michael Conklin
University of Maryland

Jodi Edington
Karen Fonkert
Dana Grosser
Anna Kruizenga
Diane Moore
Western Michigan University

Undergraduate Assistants
Cassie Durgin
University of Maryland

Rachael Kaluzny
Jessica Tucker
Ashley Wiersma
Western Michigan University

Core-Plus Mathematics 2 Field-Test Sites

Core-Plus Mathematics 2 builds on the strengths of the 1st edition, which was shaped by multi-year field tests in 36 high schools in Alaska, California, Colorado, Georgia, Idaho, Iowa, Kentucky, Michigan, Ohio, South Carolina, and Texas. Each revised text is the product of a three-year cycle of research and development, pilot testing and refinement, and field testing and further refinement. Special thanks are extended to the following teachers and their students who participated in the testing and evaluation of 2nd Edition Course 1.

Hickman High School
Columbia, Missouri
Peter Doll

Holland Christian High School
Holland, Michigan
Jeff Goorhouse
Tim Laverell
Brian Lemmen
Mike Verkaik

Jefferson Junior High School
Columbia, Missouri
Marla Clowe
Lori Kilfoil
Martha McCabe
Paul Rahmoeller
Evan Schilling

Malcolm Price Lab School
Cedar Falls, Iowa
James Maltas
Josh Wilkinson

North Shore Middle School
Holland, Michigan
Sheila Schippers
Brenda Katerberg

Oakland Junior High School
Columbia, Missouri
Teresa Barry
Erin Little
Christine Sedgwick
Dana Sleeth

Riverside University High School
Milwaukee, Wisconsin
Cheryl Brenner
Alice Lanphier
Ela Kiblawi

Rock Bridge High School
Columbia, Missouri
Nancy Hanson

Sauk Prairie High School
Prairie du Sac, Wisconsin
Joel Amidon
Shane Been
Kent Jensen
Scott Schutt
Dan Tess
Mary Walz

Sauk Prairie Middle School
Sauk City, Wisconsin
Julie Dahlman
Janine Jorgensen

South Shore Middle School
Holland, Michigan
Lynn Schipper

Washington High School
Milwaukee, Wisconsin
Anthony Amoroso
Debbie French

West Junior High School
Columbia, Missouri
Josephus Johnson
Rachel Lowery
Mike Rowson
Amanda Schoenfeld
Patrick Troup

Overview of Course 1

Unit 1 Patterns of Change

Patterns of Change develops student ability to recognize and describe important patterns that relate quantitative variables, to use data tables, graphs, words, and symbols to represent the relationships, and to use reasoning and calculating tools to answer questions and solve problems.

Topics include variables and functions, algebraic expressions and recurrence relations, coordinate graphs, data tables and spreadsheets, and equations and inequalities.

Lesson 1 Cause and Effect
Lesson 2 Change Over Time
Lesson 3 Tools for Studying Patterns of Change
Lesson 4 Looking Back

Unit 2 Patterns in Data

Patterns in Data develops student ability to make sense of real-world data through use of graphical displays, measures of center, and measures of variability.

Topics include distributions of data and their shapes, as displayed in dot plots, histograms, and box plots; measures of center including mean and median, and their properties; measures of variability including interquartile range and standard deviation, and their properties; and percentiles and outliers.

Lesson 1 Exploring Distributions
Lesson 2 Variability
Lesson 3 Looking Back

Unit 3 Linear Functions

Linear Functions develops student ability to recognize and represent linear relationships between variables and to use tables, graphs, and algebraic expressions for linear functions to solve problems in situations that involve constant rate of change or slope.

Topics include linear functions, slope of a line, rate of change, modeling linear data patterns, solving linear equations and inequalities, equivalent linear expressions.

Lesson 1 Modeling Linear Relationships
Lesson 2 Linear Equations and Inequalities
Lesson 3 Equivalent Expressions
Lesson 4 Looking Back

Overview of Course 1

Vertex-Edge Graphs

Vertex-Edge Graphs develops student understanding of vertex-edge graphs and ability to use these graphs to represent and solve problems involving paths, networks, and relationships among a finite number of elements, including finding efficient routes and avoiding conflicts.

Topics include vertex-edge graphs, mathematical modeling, optimization, algorithmic problem solving, Euler circuits and paths, matrix representation of graphs, vertex coloring and chromatic number.

Lesson 1 Euler Circuits: Finding the Best Path
Lesson 2 Vertex Coloring: Avoiding Conflict
Lesson 3 Looking Back

UNIT 5 Exponential Functions

Exponential Functions develops student ability to recognize and represent exponential growth and decay patterns, to express those patterns in symbolic forms, to solve problems that involve exponential change, and to use properties of exponents to write expressions in equivalent forms.

Topics include exponential growth and decay functions, data modeling, growth and decay rates, half-life and doubling time, compound interest, and properties of exponents.

Lesson 1 Exponential Growth
Lesson 2 Exponential Decay
Lesson 3 Looking Back

UNIT 6 Patterns in Shape

Patterns in Shape develops student ability to visualize and describe two- and three-dimensional shapes, to represent them with drawings, to examine shape properties through both experimentation and careful reasoning, and to use those properties to solve problems.

Topics include Triangle Inequality, congruence conditions for triangles, special quadrilaterals and quadrilateral linkages, Pythagorean Theorem, properties of polygons, tilings of the plane, properties of polyhedra, and the Platonic solids.

Lesson 1 Two-Dimensional Shapes
Lesson 2 Polygons and Their Properties
Lesson 3 Three-Dimensional Shapes
Lesson 4 Looking Back

Overview of Course 1

Unit 7 Quadratic Functions

Quadratic Functions develops student ability to recognize and represent quadratic relations between variables using data tables, graphs, and symbolic formulas, to solve problems involving quadratic functions, and to express quadratic polynomials in equivalent factored and expanded forms.

Topics include quadratic functions and their graphs, applications to projectile motion and economic problems, expanding and factoring quadratic expressions, and solving quadratic equations by the quadratic formula and calculator approximation.

Unit 8 Patterns in Chance

Patterns in Chance develops student ability to solve problems involving chance by constructing sample spaces of equally-likely outcomes and to solve more complex probability problems by using simulation or geometric models.

Topics include sample spaces, equally-likely outcomes, probability distributions, mutually exclusive events, Addition Rule, simulation, Law of Large Numbers, and geometric probability.

Contents

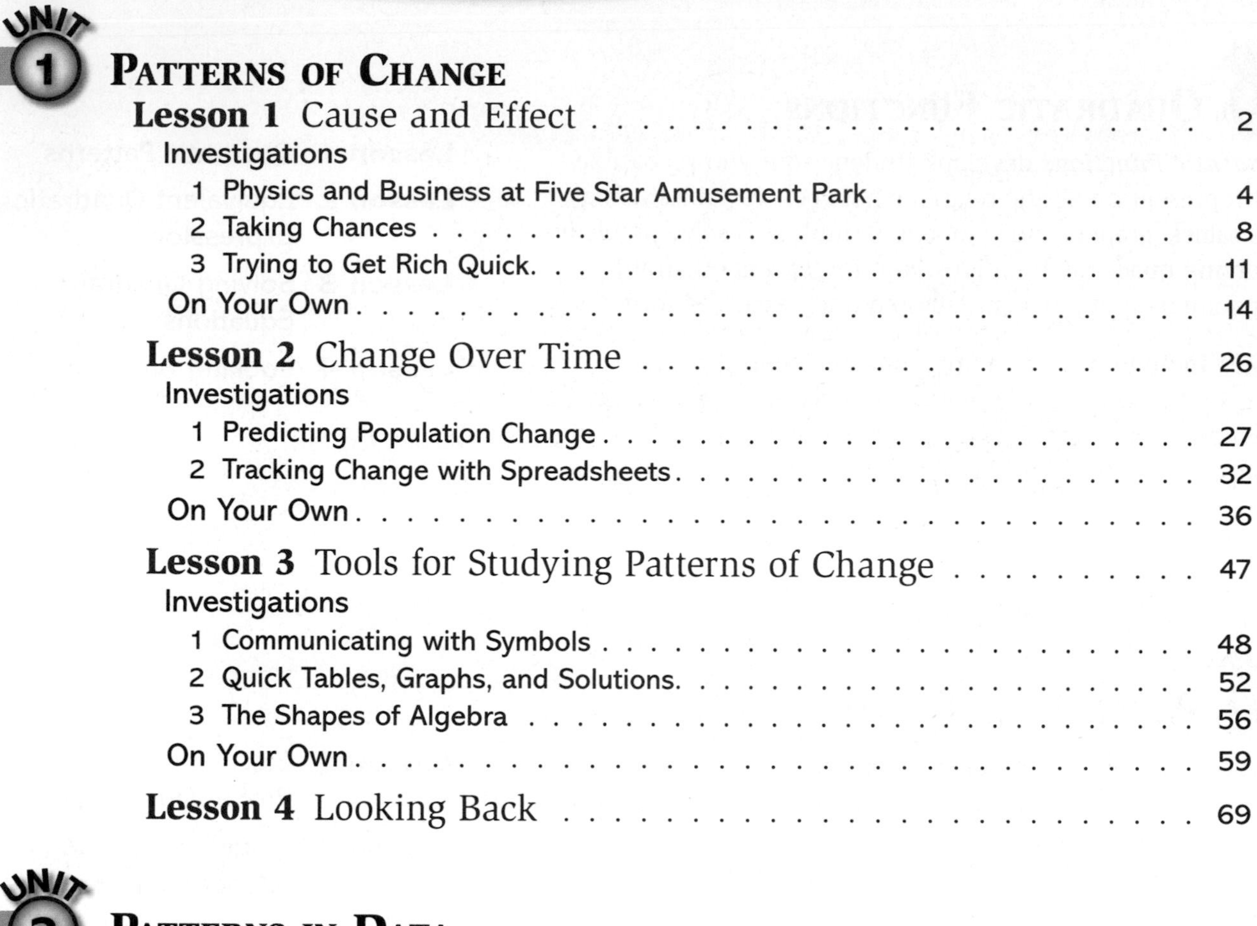

Unit 1 Patterns of Change

Unit 2 Patterns in Data

Contents

UNIT 3 Linear Functions

UNIT 4 Vertex-Edge Graphs

Contents

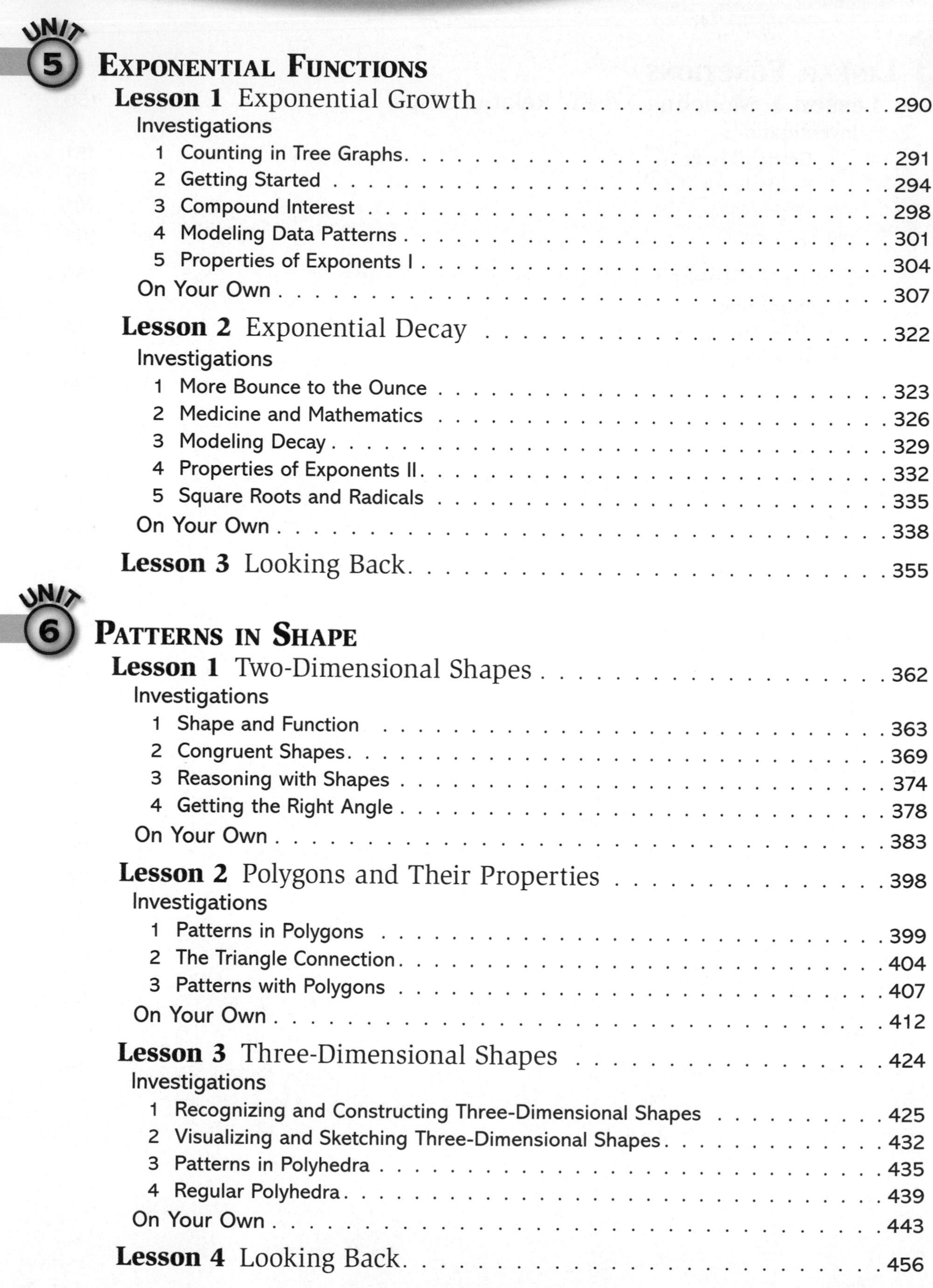

Contents

Preface

The first three courses in *Core-Plus Mathematics* provide a significant common core of broadly useful mathematics for all students. They were developed to prepare students for success in college, in careers, and in daily life in contemporary society. Course 4 continues the preparation of students for success in college mathematics and statistics courses. The program builds upon the theme of mathematics as sense-making. Through investigations of real-life contexts, students develop a rich understanding of important mathematics that makes sense to them and which, in turn, enables them to make sense out of new situations and problems.

Each course in *Core-Plus Mathematics* shares the following mathematical and instructional features.

- **Integrated Content** Each year the curriculum advances students' understanding of mathematics along interwoven strands of algebra and functions, statistics and probability, geometry and trigonometry, and discrete mathematics. These strands are unified by fundamental themes, by common topics, and by mathematical habits of mind or ways of thinking. Developing mathematics each year along multiple strands helps students develop diverse mathematical insights and nurtures their differing strengths and talents.
- **Mathematical Modeling** The curriculum emphasizes mathematical modeling including the processes of data collection, representation, interpretation, prediction, and simulation. The modeling perspective permits students to experience mathematics as a means of making sense of data and problems that arise in diverse contexts within and across cultures.
- **Access and Challenge** The curriculum is designed to make mathematics accessible to more students while at the same time challenging the most able students. Differences in student performance and interest can be accommodated by the depth and level of abstraction to which core topics are pursued, by the nature and degree of difficulty of applications, and by providing opportunities for student choice on homework tasks and projects.
- **Technology** Numeric, graphic, and symbolic manipulation capabilities such as those found on many graphing calculators are assumed and appropriately used throughout the curriculum. The curriculum materials also include a suite of computer software called *CPMP-Tools* that provide powerful aids to learning mathematics and solving mathematical problems. (See page xviii for further details.) This use of technology permits the curriculum and instruction to emphasize multiple representations (verbal, numerical, graphical, and symbolic) and to focus on goals in which mathematical thinking and problem solving are central.
- **Active Learning** Instructional materials promote active learning and teaching centered around collaborative investigations of problem situations followed by teacher-led whole-class summarizing activities that lead to analysis, abstraction, and further application of underlying mathematical ideas and principles. Students are actively engaged in exploring, conjecturing, verifying, generalizing, applying, proving, evaluating, and communicating mathematical ideas.
- **Multi-dimensional Assessment** Comprehensive assessment of student understanding and progress through both curriculum-embedded assessment opportunities and supplementary assessment tasks supports instruction and enables monitoring and evaluation of each student's performance in terms of mathematical processes, content, and dispositions.

Integrated Mathematics

Core-Plus Mathematics replaces the traditional Algebra-Geometry-Advanced Algebra/Trigonometry-Precalculus sequence of high school mathematics courses with a sequence of courses that features concurrent and connected development of important mathematics drawn from four strands.

The Algebra and Functions strand develops student ability to recognize, represent, and solve problems involving relations among quantitative variables. Central to the development is the use of functions as mathematical models. The key algebraic models in the curriculum are linear, exponential, power, polynomial, logarithmic, rational, and trigonometric functions. Modeling with systems of equations, both linear and nonlinear, is developed. Attention is also given to symbolic reasoning and manipulation.

The primary goal of the Geometry and Trigonometry strand is to develop visual thinking and ability to construct, reason with, interpret, and apply mathematical models of patterns in visual and physical contexts. The focus is on describing patterns in shape, size, and location; representing patterns with drawings, coordinates, or vectors; predicting changes and invariants in shapes under transformations; and organizing geometric facts and relationships through deductive reasoning.

The primary role of the Statistics and Probability strand is to develop student ability to analyze data intelligently, to recognize and measure variation, and to understand the patterns that underlie probabilistic situations. The ultimate goal is for students to understand how inferences can be made about a population by looking at a sample from that population. Graphical methods of data analysis, simulations, sampling, and experience with the collection and interpretation of real data are featured.

The Discrete Mathematics strand develops student ability to solve problems using vertex-edge graphs, recursion, matrices, systematic counting methods (combinatorics), and voting methods. Key themes are discrete mathematical modeling, optimization, and algorithmic problem-solving.

Each of these strands of mathematics is developed within focused units connected by fundamental ideas such as symmetry, matrices, functions, data analysis, and curve-fitting. The strands also are connected across units by mathematical habits of mind such as visual thinking, recursive thinking, searching for and explaining patterns, making and checking conjectures, reasoning with multiple representations, inventing mathematics, and providing convincing arguments and proofs.

The strands are unified further by the fundamental themes of data, representation, shape, and change. Important mathematical ideas are frequently revisited through this attention to connections within and across strands, enabling students to develop a robust and connected understanding of mathematics.

Active Learning and Teaching

The manner in which students encounter mathematical ideas can contribute significantly to the quality of their learning and the depth of their understanding. *Core-Plus Mathematics* units are designed around multi-day lessons centered on big ideas. Each lesson includes 2–5 mathematical investigations that engage students in a four-phase cycle of classroom activities, described in the following paragraph—Launch, Explore, Share and Summarize, and Check Your Understanding. This cycle is designed to engage students in investigating and making sense of problem situations, in constructing important mathematical concepts and methods, in generalizing and proving mathematical relationships, and in communicating, both orally and in writing, their thinking and the results of their efforts. Most classroom activities are designed to be completed by students working collaboratively in groups of two to four students.

The launch phase of a lesson promotes a teacher-led class discussion of a problem situation and of related questions to think about, setting the context for the student work to follow. In the second or explore phase, students investigate more focused problems and questions related to the launch situation. This investigative work is followed by a teacher-led class discussion in which students summarize mathematical ideas developed in their groups, providing an opportunity to construct a shared understanding of important concepts, methods, and approaches. Finally, students are given tasks to complete on their own, to check their understanding of the concepts and methods.

Each lesson also includes homework tasks to engage students in applying, connecting, reflecting on, extending, and reviewing their mathematical understanding. These On Your Own tasks are central to the learning goals of each lesson and are intended primarily for individual work outside of class. Selection of tasks should be based on student performance and the availability of time and technology. Students can exercise some choice of tasks to pursue, and at times they should be given the opportunity to pose their own problems and questions to investigate.

Multiple Approaches to Assessment

Assessing what students know and are able to do is an integral part of *Core-Plus Mathematics*. There are opportunities for assessment in each phase of the instructional cycle. Initially, as students pursue the investigations that comprise the curriculum, the teacher is able to informally assess student understanding of mathematical processes and content and their disposition toward mathematics. At the end of each investigation, a class discussion to Summarize the Mathematics provides an opportunity for the teacher to assess levels of understanding that various groups of students have reached as they share and explain their findings. Finally, the Check Your Understanding tasks and the tasks in the On Your Own sets provide further opportunities to assess the level of understanding of each individual student. Quizzes, in-class tests, take-home assessment tasks, and extended projects are included in the teacher resource materials.

Acknowledgments

Development and evaluation of the student text materials, teacher materials, assessments, and computer software for *Core-Plus Mathematics 2nd Edition* was funded through a grant from the National Science Foundation to the Core-Plus Mathematics Project (CPMP). We express our appreciation to NSF and, in particular, to our program officer John Bradley for his long-term trust, support, and input.

We are also grateful to Texas Instruments and, in particular, Dave Santucci for collaborating with us by providing classroom sets of graphing calculators to field-test schools.

As seen on page v, CPMP has been a collaborative effort that has drawn on the talents and energies of teams of mathematics educators at several institutions. This diversity of experiences and ideas has been a particular strength of the project. Special thanks is owed to the exceptionally capable support staff at these institutions, particularly to Angela Reiter, Matthew Tuley, and Teresa Ziebarth at Western Michigan University.

We are grateful to our Advisory Board, Diane Briars (Pittsburgh Public Schools), Jeremy Kilpatrick (University of Georgia), Robert E. Megginson (University of Michigan), Kenneth Ruthven (University of Cambridge), and David A. Smith (Duke University) for their ongoing guidance and advice. We also acknowledge and thank Norman L. Webb (University of Wisconsin-Madison) for his advice on the design and conduct of our field-test evaluations.

Special thanks are owed to the following mathematicians: Deborah Hughes-Hallett (University of Arizona/Harvard University), Stephen B. Maurer (Swarthmore College), William McCallum (University of Arizona), Doris Schattschneider (Moravian College), and to statistician Richard Scheaffer (University of Florida) who reviewed and commented on units as they were being developed, tested, and refined.

Our gratitude is expressed to the teachers and students in our 13 evaluation sites listed on page vi. Their experiences using the revised *Core-Plus Mathematics* units provided constructive feedback and suggested improvements that were immensely helpful.

Finally, we want to acknowledge Lisa Carmona, Heather Holliday, Rachel Norton, and their colleagues at Glencoe/McGraw-Hill who contributed to the publication of this program.

To the Student

Have you ever wondered ...

- How the ticket price for a concert is decided?
- How the size of a wildlife population 10 years from now can be predicted?
- How to decide on the best choice of long distance or cell phone plans for your family?
- Why honeycomb cells are the shape they are?
- How a town or city decides on routes that snowplows or sweepers use to clean the streets?
- On average, how many boxes of a particular cereal your family would need to buy to get one each of a complete set of prizes that are being offered?

The mathematics you will learn in *Core-Plus Mathematics* Course 1 will help you answer questions like these.

Because real-world situations and problems often involve data, shape, quantity, change, or chance, you will study concepts and methods from several interwoven strands of mathematics. In particular, you will develop an understanding of broadly useful ideas from algebra and functions, geometry, statistics and probability, and discrete mathematics. In the process, you will also see many connections among these strands.

In this course, you will learn important mathematics as you investigate and solve interesting problems. You will develop the ability to reason and communicate about mathematics as you are actively engaged in understanding and applying mathematics. You will often be learning mathematics in the same way that many people work in their jobs—by working in teams and using technology to solve problems.

In the 21st century, anyone who faces the challenge of learning mathematics or using mathematics to solve problems can draw on the resources of powerful information technology tools. Calculators and computers can help with calculations, drawing, and data analysis in mathematical explorations and solving mathematical problems.

Graphing calculators and computer software tools will be useful in work on many of the investigations in *Core-Plus Mathematics.*

To the Student

The curriculum materials include computer software called *CPMP-Tools* that will be of great help in learning and using the mathematical topics of each CPMP course.

The software toolkit includes four families of programs:

- Algebra—The software for work on algebra problems includes an electronic spreadsheet and a computer algebra system (CAS) that produces tables and graphs of functions, manipulates algebraic expressions, and solves equations and inequalities.
- Geometry—The software for work on geometry problems includes an interactive drawing program for constructing, measuring, and manipulating geometric figures and a set of custom tools for studying geometric models of physical mechanisms, tessellations, and special shapes.
- Statistics—The software for work on data analysis and probability problems provides tools for graphic display and analysis of data, simulation of probabilistic situations, and mathematical modeling of quantitative relationships.
- Discrete Mathematics—The software for work on graph theory problems provides tools for constructing, manipulating, and analyzing vertex-edge graphs.

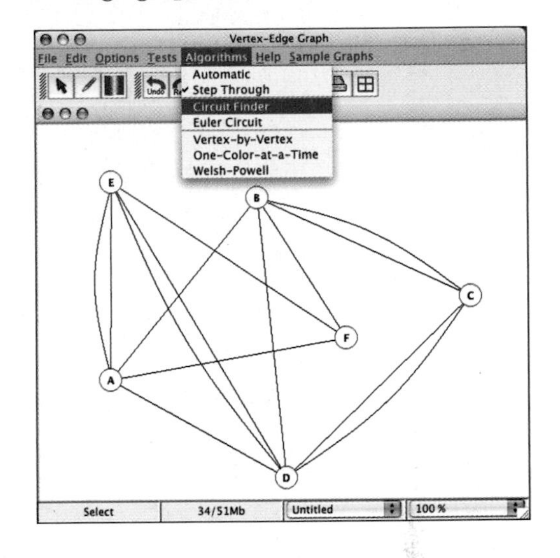

In addition to the general purpose tools provided for work on tasks in each strand of the curriculum, *CPMP-Tools* includes files of most data sets essential for work on problems in each *Core-Plus Mathematics* course. When you see an opportunity to use computer tools for work on a particular investigation, select the *CPMP-Tools* menu corresponding to the content involved in the problem. Then select the submenu items corresponding to the required mathematical operations and data sets.

In Course 1, you're going to learn a lot of useful mathematics. It will make sense to you and you can use it to make sense of your world. You're going to learn a lot about working collaboratively on problems and communicating with others as well. You're also going to learn how to use technological tools intelligently and effectively. Finally, you'll have plenty of opportunities to be creative and inventive. Enjoy!

PATTERNS OF CHANGE

Change is an important and often predictable aspect of the world in which we live. For example, in the thrill sport of bungee jumping, the stretch of the bungee cord is related to the weight of the jumper. A change in *jumper weight* causes change in *stretch of the cord*. Because bungee jumpers come in all sizes, it is important for jump operators to understand the connection between the key variables.

In this first unit of *Core-Plus Mathematics*, you will study ideas and reasoning methods of algebra that can be used to describe and predict patterns of change in quantitative variables. You will develop understanding and skill in use of algebra through work on problems in three lessons.

Lessons

1 *Cause and Effect*

Use tables, graphs, and algebraic rules to represent relationships between independent and dependent variables. Describe and predict the patterns of change in those cause-and-effect relationships.

2 *Change Over Time*

Use tables, graphs, and algebraic rules to describe, represent, and analyze patterns in variables that change with the passage of time. Use calculators and computer spreadsheets to study growth of populations and investments.

3 *Tools for Studying Patterns of Change*

Use calculator and computer tools to study relationships between variables that can be represented by algebraic rules. Explore connections between function rules and patterns of change in tables and graphs.

LESSON 1

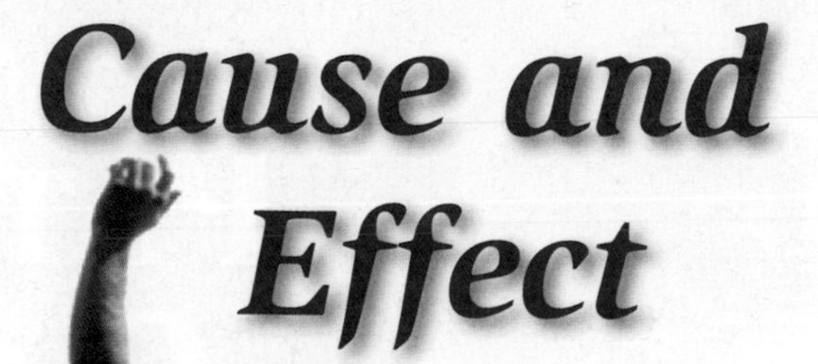

Cause and Effect

Popular sports like baseball, basketball, football, soccer, tennis, and golf have been played around the world for many years. But creative athletes are always looking for new thrills and challenges.

One new sport began when some young daredevils in New Zealand found a bridge over a deep river gorge and invented bungee jumping. They tied one end of a strong elastic cord to the bridge and the other end around their waists or feet. Then they jumped off the bridge and bounced up and down at the end of the cord to entertain tourists.

Soon word of this new sport got back to the United States. It wasn't long before some Americans tried bungee jumping on their own. Now amusement parks around the world have installed bungee jumps to attract customers who want a thrilling experience. Those parks had important planning to do before opening their bungee jumps for business.

As you can imagine, bungee jumping is risky, especially if the jump operator doesn't plan ahead carefully. If the apparatus isn't designed correctly, the consequences can be fatal. So, the first planning task is to make sure that the bungee apparatus is safe. Once the bungee jump is ready, it is time to consider business problems like setting prices to attract customers and maximize profit.

Think About This Situation

Suppose that operators of Five Star Amusement Park are considering installation of a bungee jump.

a How could they design and operate the bungee jump attraction so that people of different weights could have safe but exciting jumps?

b Suppose one test with a 50-pound jumper stretched a 60-foot bungee cord to a length of 70 feet. What patterns would you expect in a table or graph showing the stretched length of the 60-foot bungee cord for jumpers of different weights?

Jumper Weight (in pounds)	50	100	150	200	250	300
Stretched Cord Length (in feet)	70	?	?	?	?	?

Stretched Cord Length (in feet): 0, 50, 100, 150

Jumper Weight (in pounds): 0, 100, 200, 300

c How could the Five Star Amusement Park find the price to charge each customer so that daily income from the bungee jump attraction is maximized?

d What other safety and business problems would Five Star Amusement Park have to consider in order to set up and operate the bungee attraction safely and profitably?

As you complete the investigations in this lesson, you will learn how data tables, graphs, and algebraic rules express relations among variables. You will also learn how they can help in solving problems and making decisions like those involved in design and operation of the Five Star bungee jump.

Investigation 1 Physics and Business at Five Star Amusement Park

The distance that a bungee jumper falls before bouncing back upward *depends on* the jumper's weight. In designing the bungee apparatus, it is essential to know how far the elastic cord will stretch for jumpers of different weights.

The number of customers attracted to an amusement park bungee jump depends on the price charged per jump. Market research by the park staff can help in setting a price that will lead to maximum income from the attraction.

As you work on the problems of this investigation, look for answers to these questions:

How is the stretch of a bungee cord related to the weight of the bungee jumper?

How are number of customers and income for a bungee jump related to price charged for a jump?

How can data tables, graphs, and rules relating variables be used to answer questions about such relationships between variables?

Bungee Physics In design of any amusement park attraction like a bungee jump, it makes sense to do some testing before opening to the public. You can get an idea about what real testing will show by experimenting with a model bungee apparatus made from rubber bands and small weights. The pattern relating jumper weight and cord stretch will be similar to that in a real jump.

When scientists or engineers tackle problems like design of a safe but exciting bungee jump, they often work in research teams. Different team members take responsibility for parts of the design-and-test process. That kind of team problem solving is also effective in work on classroom mathematical investigations.

As you collect and analyze data from a bungee simulation, you may find it helpful to work in groups of about four, with members taking specific roles like these:

Experimenters	Perform the actual experiment and make measurements.
Recorders	Record measurements taken and prepare reports.
Quality Controllers	Observe the experiment and measurement techniques and recommend retests when there are doubts about accuracy of work.

Different mathematical or experimental tasks require different role assignments. But, whatever the task, it is important to have confidence in your partners, to share ideas, and to help others.

1. Make a model bungee jump by attaching a weight to an elastic cord or to a chain of rubber bands.

   a. Use your model to collect test data about bungee cord stretch for at least five weights. Record the data in a table and display it as a *scatterplot* on a graph.

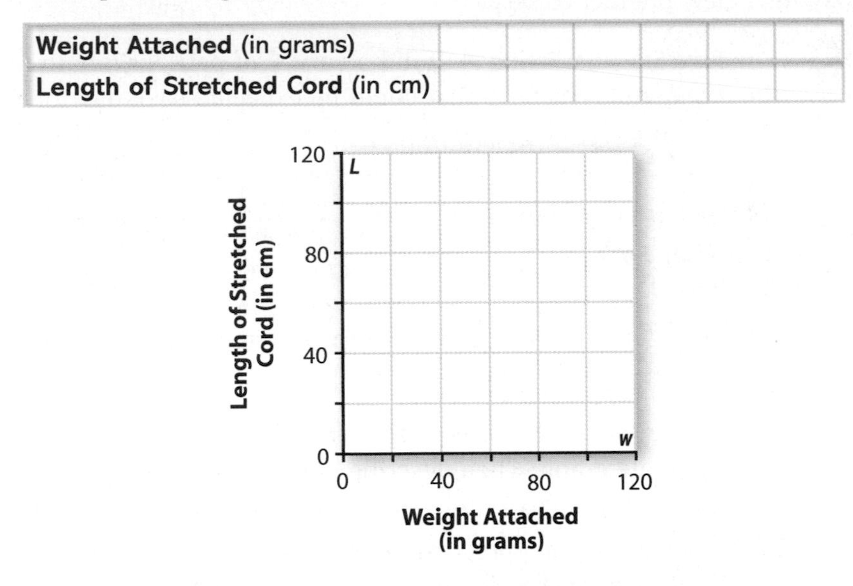

Weight Attached (in grams)						
Length of Stretched Cord (in cm)						

   b. Use the pattern in your experimental data to predict length of the stretched bungee cord for weights different from those already tested. Then test the accuracy of your predictions.

   c. Compare your results to those of others doing the same experiment. What might explain any differences in results?

2. When a group of students in Iowa did the bungee jump experiment, they proposed an algebraic rule relating the length of the stretched bungee cord L (in centimeters) to the attached weight w (in grams). They said that the rule $L = 30 + 0.5w$ could be used to predict the stretched cord length for any reasonable weight.

   a. Use the Iowa students' rule to make a table and a graph of sample (w, L) values for $w = 0$ to 120 in steps of 20 grams.

   b. How, if at all, do the numbers 30 and 0.5 in the Iowa students' rule relate to the pattern of (w, L) values shown in the table and graph? What do they tell about the way the length of the cord changes as the attached weight changes?

   c. Is the pattern of change in the rule-based (w, L) values in Part a different from the pattern of change in your experimental data? If so, what differences in experimental conditions might have caused the differences in results?

Bungee Business Designing the bungee jump apparatus is only part of the task in adding the attraction to Five Star Amusement Park. It is also important to set a *price per jump* that will make the operation profitable. When businesses face decisions like these, they get helpful information from market research. They ask people how much they would be willing to pay for a new product or service.

3 The Five Star marketing staff did a survey of park visitors to find out the *number of customers* that could be expected each day for the bungee jump at various possible *price per jump* values. Their survey produced data that they rounded off and presented in this table.

Market Survey Data

Price per Jump (in dollars)	0	5	10	15	20	25	30
Likely Number of Customers	50	45	40	35	30	25	20

a. Plot the (*price per jump, number of customers*) data on a coordinate graph. Then describe how the predicted *number of customers* changes as *price per jump* increases from \$0 to \$30.

b. The Five Star data processing department proposed the rule $N = 50 - p$ for the relationship between *number of customers N* and *price per jump p*. Does this rule represent the pattern in the market research data? Explain your reasoning.

4 The Five Star staff also wanted to know about *daily income* earned from the bungee jump attraction.

a. If the price per jump is set at \$5, the park can expect 45 bungee jump customers per day. In this case, what is the daily income?

b. Use the market survey data from Problem 3 to estimate the *daily income* earned by the bungee jump for prices from \$0 to \$30 in steps of \$5. Display the (*price per jump, daily income*) data in a table and in a graph. Then describe the pattern relating those variables.

c. What do the results of the Five Star market research survey and the income estimates suggest as the best price to charge for the bungee jump attraction? How is your answer supported by data in the table and the graph of (*price per jump, daily income*) values?

5 In situations where values of one variable depend on values of another, it is common to label one variable the **independent variable** and the other the **dependent variable.** Values of the dependent variable are a function of, or depend on, values of the independent variable. What choices of independent and dependent variables make sense in:

a. studying design of a bungee jump apparatus?

b. searching for the price per jump that will lead to maximum income?

Summarize the Mathematics

To describe relationships among variables, it is often helpful to explain how one variable *is a function of* the other or how the value of one variable *depends on* the value of the other.

a How would you describe the way that:

i. the stretch of a bungee cord depends on the weight of the jumper?

ii. the number of customers for a bungee jump attraction depends on the price per customer?

iii. income from the jump depends on price per customer?

b What similarities and what differences do you see in the relationships of variables in the physics and business questions about bungee jumping at Five Star Amusement Park?

c In a problem situation involving two related variables, how do you decide which should be considered the independent variable? The dependent variable?

d What are the advantages and disadvantages of using tables, graphs, algebraic rules, or descriptions in words to express the way variables are related?

e In this investigation, you were asked to use patterns in data plots and algebraic rules to make predictions of bungee jump stretch, numbers of customers, and income. How much confidence or concern would you have about the accuracy of those predictions?

Be prepared to share your thinking with the whole class.

✓ Check Your Understanding

The design staff at Five Star Amusement Park had another idea—selling raffle tickets for chances to win prizes. The prize-winning tickets would be drawn at random each day.

a. Suppose that a market research study produced the following estimates of raffle ticket sales at various prices.

Price per Ticket (in dollars)	1	2	3	4	5	10	15
Number of Tickets Sold	900	850	800	750	700	450	200

i. Plot the (*price per ticket, number of tickets sold*) estimates on a graph. Because *price per ticket* is the independent variable in this situation, its values are used as x-coordinates of the graph. Because *number of tickets sold* is the dependent variable, its values are used as the y-coordinates of the graph.

ii. Describe the pattern relating values of those variables and the way that the relationship is shown in the table and the graph.

iii. Does the rule $N = 950 - 50p$ produce the same pairs of (*price per ticket p, number of tickets sold N*) values as the market research study?

b. Use the data in Part a relating *price per ticket* to *number of tickets sold* to estimate the *income* from raffle ticket sales at each of the proposed ticket prices.

i. Record those *income* estimates in a table and plot the (*price per ticket, income*) estimates on a graph.

ii. Describe the relationship between raffle ticket price and income from ticket sales. Explain how the relationship is shown in the table and the graph of (*price per ticket, income*) estimates.

iii. What do your results in parts i and ii suggest about the ticket price that will lead to maximum income from raffle ticket sales? How is your answer shown in the table and graph of part i?

Investigation 2 Taking Chances

Students at Banneker High School hold an annual *Take a Chance* carnival to raise funds for special class projects. The planning committee is often puzzled about ways to predict profit from games of chance.

In one popular game, a fair die is rolled to find out whether you win a prize. Rules of the game are:

- You win a $4 prize if the top face of the die is a 4.
- You donate $1 to the school special project fund if the top face of the die is 1, 2, 3, 5, or 6.

As you work on the problems of this investigation, look for answers to these questions:

What is the pattern of change relating profit to number of players in the die-tossing game?

How is that pattern of change illustrated in tables and graphs of data from plays of the game?

How is the pattern of change in profit similar to and different from the patterns of change in bungee jump cord length and number of bungee jump customers?

1 Use a fair die to play the die-tossing game at least 20 times. Record your results in a table like this:

Play Number	1	2	3	4	5	6	7	...
Outcome ($ won or lost for school)								...
Cumulative Profit ($ won or lost by school)								...

2 Plot the data from your test of the game on a graph that shows how *cumulative profit* for the school changes as the *number of plays* increases. Since the school can lose money on this game, you will probably need a graph (like the one below) showing points below the horizontal axis. Connecting the plotted points will probably make patterns of change in fund-raiser profit clearer. Use the graph to answer the questions that follow.

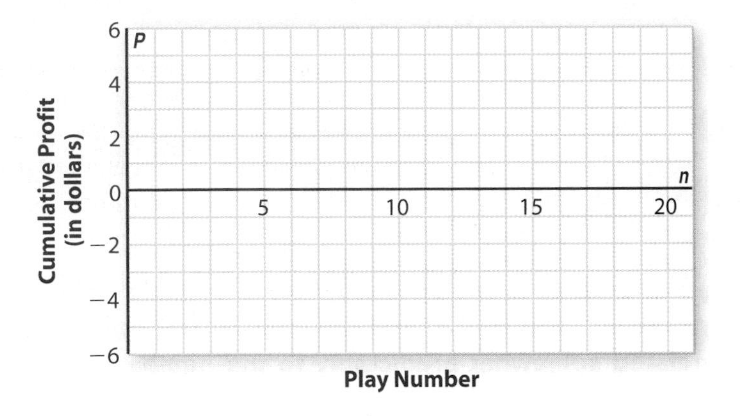

a. Describe the pattern of change in profit or loss for the school as clearly and precisely as you can. Explain how the pattern is shown in the table and the graph.

b. See if you can express the pattern as a rule relating *cumulative profit* P to *number of plays* n.

3 Combine your results from the die-tossing experiment with those of other students to produce a table showing results of many more plays. If each student or group contributes cumulative results for 20 plays, you could build a table like this:

Number of Plays	20	40	60	80	100	120	140	160
Cumulative Profit (in $)								

Plot the resulting (*number of plays, cumulative profit*) data.

a. How is the pattern in this experiment with many plays similar to or different from the patterns of your experiment with fewer plays?

b. See if you can express the pattern as a rule relating *cumulative profit* P to *number of plays* n.

4 Suppose that the game operators change the prize payoff from $4 to $6.

a. What similarities and differences would you expect in the way *cumulative profit* for the school changes as the *number of plays* increases in the new game compared to the original game? How will those patterns appear in a data table and a graph of results?

b. Repeat the die-tossing experiment to test profit prospects for the fund-raiser with the new payoff scheme. Try to explain differences between what you predicted would happen in Part a and what actually did happen.

5 What payoff amounts (for winning and losing) might make this a fair game—that is, a game in which profit for the school is expected to be zero?

Summarize the Mathematics

In this investigation, you explored patterns of change for a variable with outcomes subject to the laws of probability. You probably discovered in the die-tossing game that *cumulative profit* is related somewhat predictably to the *number of plays* of the game.

a After many plays of the two games with payoffs of $4 or $6, who seemed to come out ahead in the long run—the players or the school fund-raiser? Why do you think those results occurred?

b How is the pattern of change in *cumulative profit* for the school fund-raiser similar to, or different from, patterns you discovered in the investigation of bungee physics and business?

Be prepared to share your ideas and reasoning with the class.

Check Your Understanding

Suppose that another game at the *Take a Chance* carnival has these rules:

Three coins—a nickel, a dime, and a quarter—are flipped.

If all three turn up heads or all three turn up tails, the player wins a $5 prize.

For any other result, the player has to contribute $2 to the school fund.

The school fund-raiser is most likely to win $2 on any individual play of the game, but there is also a risk of losing $5 to some players. The challenge is to predict change in fund-raiser profit as more and more customers play this game.

a. If you keep a tally of your *cumulative profit* (or loss) for many plays of this game:

  i. What pattern would you expect to find in your cumulative profit as the number of plays increases?

  ii. How would the pattern you described in part i appear in a graph of the recorded (*play number, cumulative profit*) data?

b. Use three coins to play the game at least 20 times. In a table, record the results of each play and the cumulative profit (or loss) after each play. Make a plot of your (*play number, cumulative profit*) data and describe the pattern shown by that graph.

c. How are the results of your actual plays similar to what you predicted in Part a? If there are differences, how can they be explained?

Investigation 3 Trying to Get Rich Quick

Relationships between independent and dependent variables occur in a wide variety of problem situations. Tables, graphs, and algebraic rules are informative ways to express those relationships. The problems of this investigation illustrate two other common patterns of change. As you work on the next problems—about NASCAR racing and pay-for-work schemes—look for answers to these questions:

Why are the relationships involved in these problems called nonlinear patterns of change?

How do the dependent variables change as the independent variables increase?

NASCAR Racing Automobile racing is one of the most popular spectator sports in the United States. One of the most important races is the NASCAR Daytona 500, a 500-mile race for cars similar to those driven every day on American streets and highways. The prize for the winner is over $1 million. Winners also get lots of advertising endorsement income.

1. The average speed and time of the Daytona 500 winner varies from year to year.

   a. In 1960, Junior Johnson won with an average speed of 125 miles per hour. The next year Marvin Panch won with an average speed of 150 miles per hour. What was the difference in race time between 1960 and 1961 (in hours)?

   b. In 1997, Jeff Gordon won with an average speed of 148 miles per hour. The next year the winner was Dale Earnhardt with an average speed of 173 miles per hour. What was the difference in race time between 1997 and 1998 (in hours)? (Source: www.nascar.com/races/)

2. Complete a table like that shown here to display sample pairs of (*average speed, race time*) values for completion of a 500-mile race.

Average Speed (in mph)	50	75	100	125	150	175	200
Race Time (in hours)							

a. Plot the sample (*average speed, race time*) data on a graph. Describe the relationship between those two variables.

b. Write a symbolic rule that shows how to calculate *race time t* as a function of *average speed s* in the Daytona 500 race. Show with specific examples that your rule produces correct race time for given average speed.

3 In the 1960–61 and 1997–98 comparisons of winning speed and time for the Daytona 500 race, the differences in average speed are both 25 miles per hour. The time differences are not the same. At first, this might seem like a surprising result.

How is the fact that equal changes in average speed don't imply equal changes in race time illustrated in the shape of the graph of sample (*average speed, race time*) data?

4 How are the table, graph, and algebraic rule relating *average speed* and *race time* similar to or different from those you have seen in work on earlier problems?

Part-Time Work ... Big-Time Dollars When Devon and Kevin went looking for part-time work to earn spending money, their first stop was at the Fresh Fare Market. They asked the manager if they could work helping customers carry groceries to their cars. When the manager asked how much they wanted to earn, Devon and Kevin proposed \$2 per hour plus tips from customers.

The Fresh Fare Market manager proposed a different deal, to encourage Devon and Kevin to work more than a few hours each week. The manager's weekly plan would pay each of them \$0.10 for the first hour of work, \$0.20 for the second hour, \$0.40 for the third hour, \$0.80 for the fourth hour, and so on.

5 Which pay plan do you think would be best for Devon and Kevin to choose? To provide evidence supporting your ideas, complete a table showing the earnings (without tips) for each student from each plan for work hours from 1 to 10. Plot graphs showing the patterns of growth in earnings for the two plans.

Hours Worked in a Week	1	2	3	4	5	6	7	8	9	10
Earnings in \$ Plan 1	2	4	...							
Earnings in \$ Plan 2	0.10	0.30	...							

Based on the pattern of earnings, which of the two pay plans would you recommend to Devon and Kevin?

6 Would you change your choice of pay plan if the manager's offer was:

a. Plan 3: Only $0.05 for the first hour of work, $0.10 for the second hour, $0.20 for the third hour, $0.40 for the fourth hour, and so on? Why or why not?

b. Plan 4: Only $0.01 for the first hour of work, but $0.03 for the second hour, $0.09 for the third, $0.27 for the fourth, and so on? Why or why not?

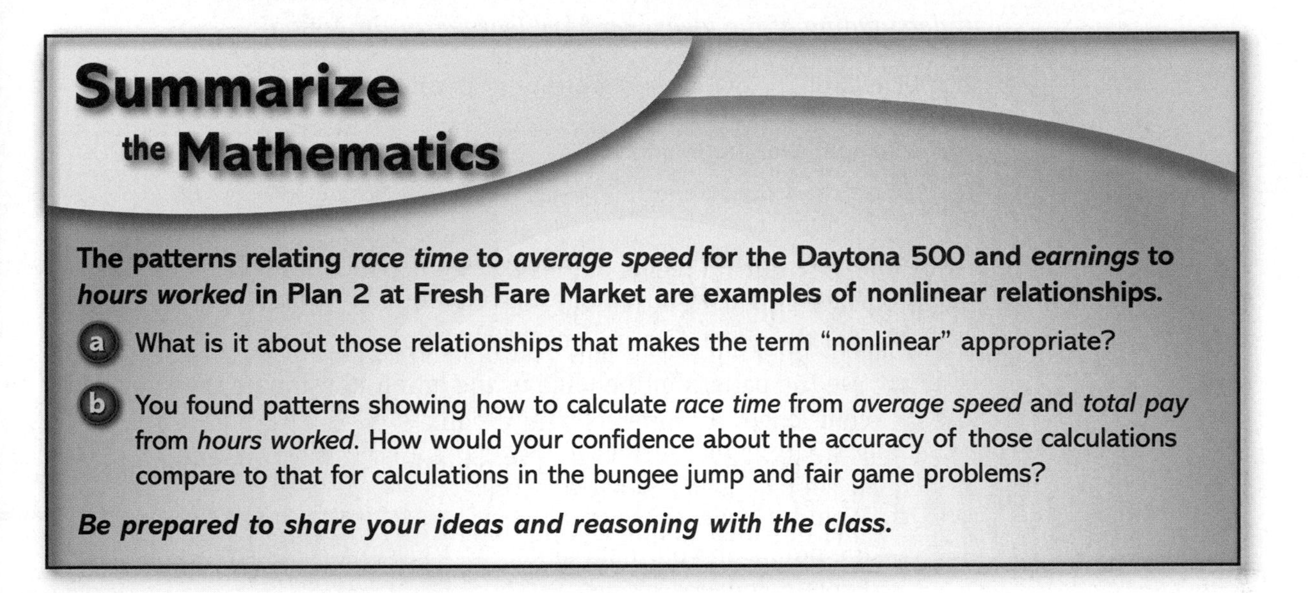

Summarize the Mathematics

The patterns relating *race time* to *average speed* for the Daytona 500 and *earnings* to *hours worked* in Plan 2 at Fresh Fare Market are examples of nonlinear relationships.

a What is it about those relationships that makes the term "nonlinear" appropriate?

b You found patterns showing how to calculate *race time* from *average speed* and *total pay* from *hours worked.* How would your confidence about the accuracy of those calculations compare to that for calculations in the bungee jump and fair game problems?

Be prepared to share your ideas and reasoning with the class.

Check Your Understanding

Use these problems to test your skill in analyzing nonlinear relationships like those in the NASCAR and Fresh Fare Market problems.

a. The Iditarod Trail Sled Dog Race goes 1,100 miles from Anchorage to Nome, Alaska, in March of each year. The winner usually takes about 10 days to complete the race.

i. What is a typical average speed (in miles per day) for Iditarod winners?

ii. Make a table and sketch a graph showing how *average speed* for the Iditarod race depends on *race time.* Use times ranging from 2 (not really possible for this race) to 20 days in steps of 2 days.

iii. What rule shows how to calculate *average speed s* for any Iditarod *race time t*?

iv. Compare the table, graph, and rule showing Iditarod *average speed* as a function of *race time* to that showing Daytona 500 *race time* as a function of *average speed* in Problem 2 of this investigation. Explain how relationships in the two situations are similar and how they are different.

b. Ethan and Anna tried to get a monthly allowance of spending money from their parents. They said, "You only have to pay us 1 penny for the first day of the month, 2 pennies for the second day of the month, 4 pennies for the third day, and so on." According to Ethan and Anna's idea, how much would the parents have to pay on days 10, 20, and 30 of each month?

On Your Own

Applications

These tasks provide opportunities for you to use and strengthen your understanding of the ideas you have learned in the lesson.

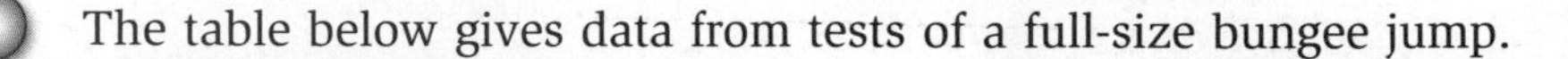

1 The table below gives data from tests of a full-size bungee jump.

Jumper Weight (in pounds)	100	125	150	175	200
Stretched Cord Length (in feet)	50	55	60	65	70

a. Which variable does it make sense to consider independent and which dependent?

b. Plot the given data on a coordinate graph.

c. Use the pattern in the table or the graph to estimate the stretched cord length for jumpers who weigh:

i. 85 pounds　　**ii.** 135 pounds　　**iii.** 225 pounds

d. Would it make sense to connect the points on your data plot? Explain your reasoning.

e. Describe the overall pattern relating *stretched cord length L* to *jumper weight w*.

f. The technician who did the tests suggested that the pattern could be summarized with a symbolic rule $L = 30 + 0.2w$. Does that rule give estimates of stretched cord length that match the experimental data? Explain.

2 To help in estimating the number of customers for an amusement park bungee jump, the operators hired a market research group to visit several similar parks that had bungee jumps. They recorded the number of customers on a weekend day. Since the parks charged different prices for their jumps, the collected data looked like this:

Price per Jump (in dollars)	15	20	25	28	30
Number of Customers	25	22	18	15	14

a. In this situation, which variable makes sense as the independent variable and which as the dependent variable?

b. Plot these data on a coordinate graph.

c. Does it make sense to connect the points on your data plot? Explain your reasoning.

d. Use the pattern in the table or the graph to estimate the *number of customers* if the *price per jump* is:

i. \$18 **ii.** \$23 **iii.** \$35

e. Describe the overall pattern of change relating *price per jump* to *number of customers.*

f. The market research staff suggested that the pattern could be summarized with a rule $n = 35 - 0.7p$. Does that rule produce estimates of number of customers n at various prices p like those in the survey data?

3 Use the data in Applications Task 2 to study the relationship between price per bungee jump and income from one day's operation at the five parks that were visited.

a. Complete a table showing sample (*price per jump, daily income*) values.

Price per Jump (in dollars)	15	20	25	28	30
Daily Income (in dollars)					

b. In this situation, what choice of independent and dependent variables makes most sense?

c. Plot the data relating *price per jump* and *daily income* on a coordinate graph.

d. Would it make sense to connect the points on the graph? Explain your reasoning.

e. Describe the overall pattern in the relationship between *price per jump* and *daily income.*

f. Use the data table and graph pattern to estimate the *price per jump* that seems likely to yield maximum *daily income.*

4 Suppose that you go to a school carnival night and play a game in which two fair coins are tossed to find out whether you win a prize. The game has these rules:

- Two heads or two tails showing—you win \$1.
- One head and one tail showing—you lose \$1.

a. If you keep score for yourself in 20 plays of this game:

i. What pattern would you expect in your cumulative score as the plays occur?

ii. How would the pattern you described in part i appear in a graph of (*play number, cumulative score*) data?

b. Use two coins to play the game 20 times. Record the results of each play and the cumulative score after each play in a table. Make a scatterplot of your (*play number, cumulative score*) data and describe the pattern shown by that graph.

c. How are the results of your actual plays similar to what you predicted in Part a? How are they different?

5. The postage cost for U.S. first-class mail is related to the weight of the letter or package being shipped. The following table gives the regulations in 2006 for relatively small letters or packages.

Weight (in ounces)	up to 1	up to 2	up to 3	up to 4	up to 5
Postage Cost (in dollars)	0.39	0.63	0.87	1.11	1.35

a. Make a coordinate graph showing (*weight, postage cost*) values for letters or packages weighing 1, 2, 3, 4, and 5 ounces.

b. What postage costs would you expect for letters or small packages sent by first-class mail, if those items weighed:

i. 1.5 ounces **ii.** 4.25 ounces **iii.** 7 ounces

c. Add the (*weight, postage cost*) values from Part b to your graph. How should the points on your graph be connected (if at all)?

6. The Olympic record for the men's 400-meter hurdle race is 46.78 seconds. It was set by Kevin Young in 1992. His average running speed was $400 \div 46.78 \approx 8.55$ meters per second.

a. Make a table and a graph showing how 400-meter *race time* changes as *average speed* increases from 2 meters per second to 10 meters per second in steps of 1 meter per second.

b. Describe the pattern of change shown in your table and graph.

c. Write a rule showing how to calculate *race time t* for any *average speed s*.

d. Which change in *average speed* will reduce the *race time* most: an increase from 2 to 4 meters per second or an increase from 8 to 10 meters per second? Explain how your answer is illustrated in the shape of your graph.

7. The Olympic record in the women's 100-meter freestyle swim race is 53.52 seconds. It was set by Australian Jodie Henry in 2004. She swam at an average speed of $100 \div 53.52 \approx 1.87$ meters per second.

a. Make a table and a graph showing the way *average speed* for the 100-meter race changes as *time* increases from 40 seconds to 120 seconds (2 minutes) in steps of 10 seconds.

b. Describe the pattern of change shown in your table and graph.

c. Write a rule showing how to calculate *average speed s* for any *race time t.*

d. Which change in *race time* will cause the greatest change in *average speed*: an increase from 50 to 60 seconds or an increase from 110 to 120 seconds? Explain how your answer is illustrated in the shape of your graph.

8 The Water World Amusement Park has a huge swimming pool with a wave machine that makes you feel like you are swimming in an ocean. Unfortunately, the pool is uncovered and unheated, so the temperature forecast for a day affects the number of people who come to Water World.

On a summer day when the forecast called for a high temperature of 90°F, about 3,000 people visited the park. On another day, when the forecast called for a high temperature of 70°F, only 250 people came for the ocean-wave swimming.

a. Complete this table of (*temperature forecast, number of swimmers*) data in a way that you think shows the likely pattern relating *temperature forecast* to *number of swimmers.*

Temperature Forecast (in °F)	70	75	80	85	90	95
Number of Swimmers	250				3,000	

b. Graph the data in Part a. Then draw a line or curve that seems to match the pattern in your data points and could be used to predict *number of swimmers* at other temperatures.

c. Describe the pattern of change in *number of swimmers* as *temperature forecast* increases and explain how much confidence you would have in using that pattern to predict *number of swimmers* on any particular day.

d. Use your table and/or graph to estimate the number of swimmers for temperatures of:

i. 77°F   ii. 83°F   iii. 98°F

e. Suppose that Water World charges $15 for admission. Use this information and your estimates for number of swimmers at various forecast temperatures to make a table and graph showing the relationship between *forecast high temperature* and *park income.*

f. Use the information from Part e to estimate the park income when the high temperature is forecast to be:

i. 87°F   ii. 92°F

g. Why would you have limited confidence in using the data patterns of Parts a and e to predict park income when the forecast high temperature is 40°F or 110°F?

Connections

These tasks will help you to build links between mathematical topics you have studied in the lesson and to connect those topics with other mathematics that you know.

9 The table below shows latitude of some major northern hemisphere cities and the average high temperatures in those cities in mid-summer and in mid-winter. Use that data and the scatterplots on page 19 to answer Parts a–c about the relationship between latitude and typical temperatures.

a. Does the pattern of points relating *mid-summer average high temperature* to *geographic latitude* suggest a close relationship between those variables? Explain your reasons for saying yes or no.

b. Does the pattern of points relating *mid-winter average high temperature* to *geographic latitude* suggest a close relationship between those variables? Explain your reasons for saying yes or no.

c. What factors other than latitude might influence summer and winter temperatures?

City	Latitude Degrees N	Mid-Summer Temperature °F	Mid-Winter Temperature °F
Athens, Greece	40	89	56
Bangkok, Thailand	14	90	90
Barrow, Alaska	72	45	−7
Berlin, Germany	53	74	35
Bombay, India	19	86	85
Cairo, Egypt	30	94	66
Chicago, Illinois	42	84	29
Jerusalem, Israel	32	84	53
Lagos, Nigeria	7	83	90
London, England	51	71	44
Los Angeles, California	34	84	68
Mexico City, Mexico	19	74	70
Miami, Florida	26	89	75
Manila, Philippines	14	89	86
New York City, New York	41	84	37
Reykjavik, Iceland	64	56	35
Seattle, Washington	48	75	45
Tokyo, Japan	36	84	49

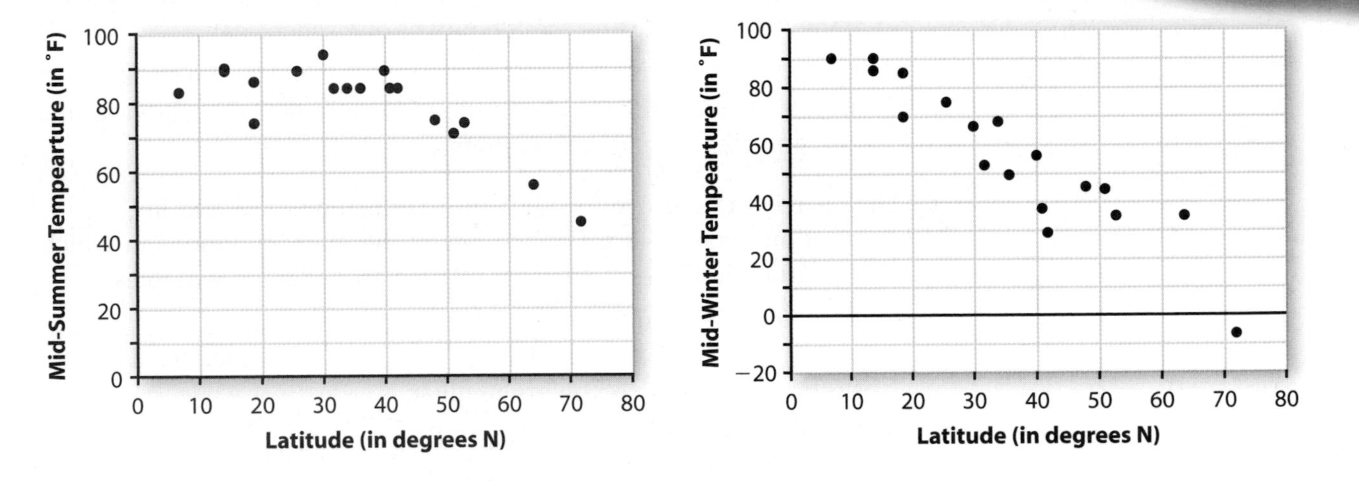

10 Random events such as the outcomes of flipping a fair coin often have predictable patterns.

a. What is the probability of flipping a coin once and getting a head?

b. What is the probability of flipping a coin two times and getting two heads?

c. What is the probability of flipping a coin three times and getting three heads?

d. What is the probability of flipping a coin four times and getting four heads?

e. How would you describe the pattern in the probabilities of getting all heads as the number of coin tosses increases?

11 Jamal's average on history quizzes changed throughout the first quarter.

a. After the first two quizzes, his average was 7, but he earned a 9 on the third quiz. What was his average for the first three quizzes?

b. After the first eight quizzes, his average had slipped again to 7, but he earned 9 on the ninth quiz. What was his average for all nine quizzes of the quarter?

c. Why did Jamal's 9 on the third quiz improve his overall average more than his 9 on the ninth quiz?

12 When the value of a quantity changes, there are several standard ways to describe *how much* it has changed. For example, if a boy who is 60 inches tall at the start of grade 8 grows to 66 inches twelve months later, we could say his height has increased:

- by 6 inches (the *difference* between original and new height)
- by 10% (the *relative* or *percent change* in his height)
- by 0.5 inches per month (an *average rate of change*)

Express each of the following quantitative changes in three ways similar to those above:

a. The enrollment of Wayzata High School increased from 1,000 to 1,250 in the five-year period from 1998 to 2003.

b. The balance in a student's bank savings account increased from $150 to $225 while she worked during the three-month summer break from school.

c. The supply of soft drinks in a school vending machine decreased from 200 to 140 during the 8 hours of one school day.

d. From the start of practice in March until the end of the track season in June, Mike's time in the 800-meter race decreased from 2 minutes 30 seconds to 2 minutes.

13 The related variables you studied in Investigations 1–3 and in Applications Tasks 1–8 are only a few of the many situations in which it helps to understand the pattern relating two or more variables.

a. Write a sentence in the form "__________ depends on __________" or "__________ is a function of __________" that describes a situation with which you are familiar.

b. For the situation you described in Part a:

i. Explain how change in one variable relates to or causes change in the other.

ii. Make a table showing at least 5 sample pairs of values that you would expect for the related variables.

iii. Plot a graph of the sample data in part ii and connect the points in a way that makes sense (if at all).

Reflections

These tasks provide opportunities for you to re-examine your thinking about ideas in the lesson.

14 Experimentation with one bungee cord suggested that the rule $L = 30 + 0.2w$ would be a good predictor of the stretched cord length as a function of jumper weight. The operators of the bungee jump decided to adjust the jump-off point for each jumper to the height L calculated from the rule. What reasons can you think of to question that plan?

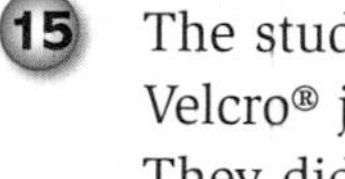

15 The student government at Banneker High School decided to set up a Velcro® jump (pictured at the left) as a fund-raiser for a school trip. They did a survey to see how many students would try the Velcro jump at various prices.

The data were as follows:

Price per Jump (in dollars)	0.50	1.00	2.00	3.00	5.00
Expected Number of Jumps	95	80	65	45	15

When several groups of Banneker mathematics students were asked to study the survey data about profit prospects of the rented Velcro jump, they produced different kinds of reports.

How would you rate each of the following reports, on a scale of 5 (excellent) to 0 (poor)? Explain why you gave each report the rating you did.

Report a: *Making Money for Banneker*

The survey shows that a price of $0.50 will lead to the most customers, so that will bring in the biggest profit.

Report b: *Sticking it to the Velcro Customers*

The survey shows that the more you charge, the fewer customers you will have. If you multiply each price by the expected number of customers, you get a prediction of the income from the Velcro jump.

When we did that, we found that a price of $3.00 leads to the greatest income, so that is what should be charged. If you want to let the most kids have fun, you should charge only $0.50. If the operators don't want to work very hard, they should charge $10, because then no one will want to pay to jump.

Report c: *Velcro Profit Prospects*

Data from our market survey suggest a pattern in which the number of customers will decrease as the price increases. Each increase of $1 in the price will lead to a decrease of 15–20 customers. This pattern is shown in a plot of the survey data.

The trend in the data is matched well by the line drawn on the graph that follows. That line also helps in predicting the number of customers for prices not included in the survey.

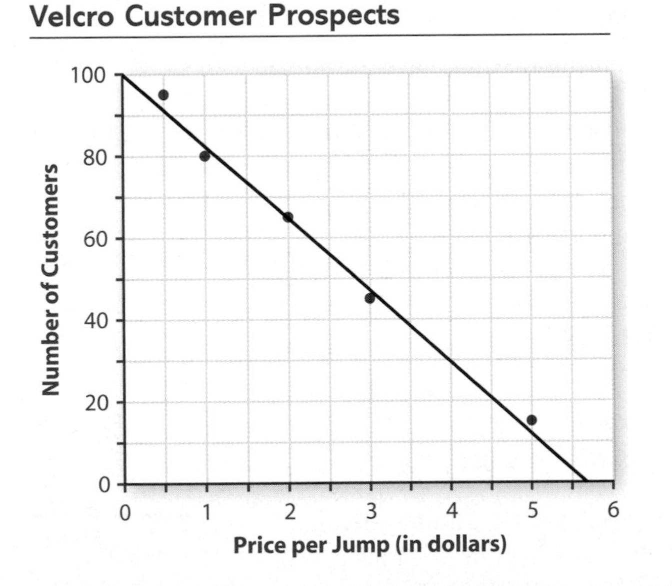

To see how the amount of money earned by the Velcro game would be related to the price per jump, we added another row to the table, showing income. For example, 95 customers at $0.50 per jump will bring income of $47.50.

Price per Jump (in dollars)	0.50	1.00	2.00	3.00	5.00
Expected Number of Jumps	95	80	65	45	15
Expected Income (in dollars)	47.50	80.00	130.00	135.00	75.00

A graph of the *price per jump* and *income* data is shown at the right. It suggests that a price between \$2 and \$3 per jump will lead to the greatest income. Since the rental charge is a fixed dollar amount, greatest income means greatest profit.

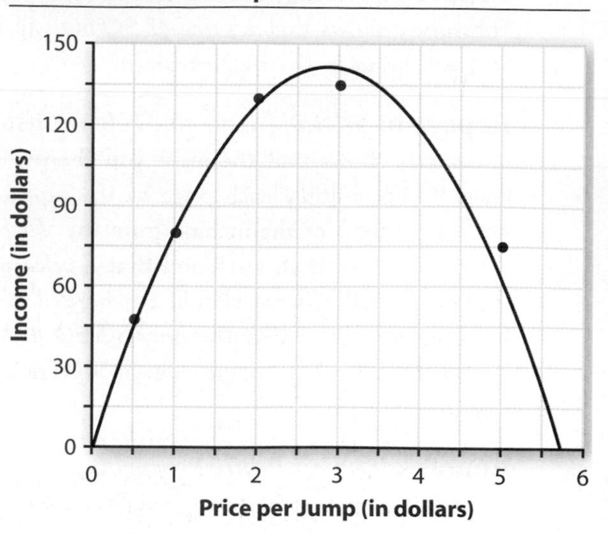

16 If you were asked to look for a pattern relating the values of two variables in a problem, would you prefer to have:

- a table of (x, y) data,
- a plot of points with coordinates (x, y), or
- a symbolic rule showing how values of y could be calculated from values of x?

Explain the reasons for your choice.

17 When there appears to be a relationship between values of two variables, how do you decide which should be considered the *independent variable* and which should be considered the *dependent variable*?

Extensions

These tasks provide opportunities for you to explore further or more deeply the mathematics you are learning.

18 Suppose that for a fund-raising event, your school can rent a climbing wall for \$275. Complete the following tasks to help find the likely profit from using the climbing wall at the event.

a. Do a survey of your class to find out how many customers you might expect for various possible prices. Then use your data to estimate the number of students from your whole school who would try the climbing wall at various possible prices.

Climb Price (in dollars)	1	2	3	4	5	6	7	8	9	10
Number of Customers										

b. Plot a graph of the survey data and explain how it shows the pattern of change relating *number of customers* to *climb price*. Be sure to explain which number it makes sense to consider the independent variable and which the dependent variable in this situation.

c. Display the data relating *number of customers* to *climb price* in a table and a graph. Then use the pattern in the data to estimate the *income* that would be earned at various possible prices.

d. What do you recommend as the price that will maximize *profit* from the climbing wall rental? Explain how your decision is based on patterns in the data tables and graphs you've displayed.

19 One of the most important principles of physics is at work when two kids play on a teeter-totter. You probably know that for two weights on opposite sides of the *fulcrum* to balance, those weights need to be placed at just the right distances from the fulcrum.

a. Suppose that a 50-pound weight is placed at one end of a teeter-totter, 6 feet from the fulcrum. How far from the fulcrum should a person sit to balance the 50-pound weight if the person weighs:

i. 50 pounds

ii. 100 pounds

iii. 150 pounds

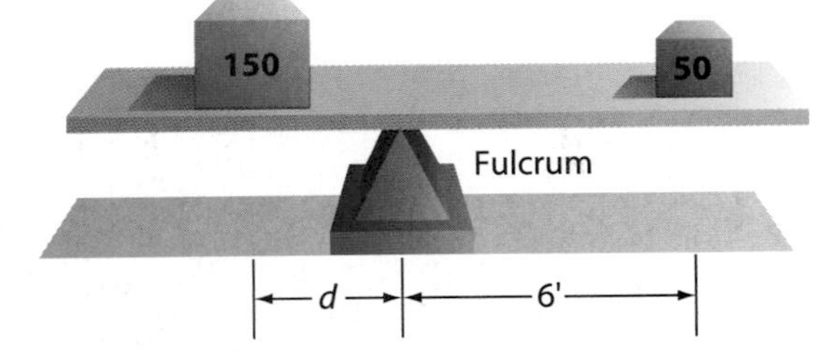

(If you are unsure of the physical relationship required to make a balance, do some experiments with a meter stick as the teeter-totter and stacks of pennies as the weights.)

b. Sketch a graph showing the distance from the fulcrum required for various weights to balance a 50-pound weight that has been placed on the opposite side, 6 feet from the fulcrum. Describe the pattern relating distance from the fulcrum to the counter-balancing weight.

c. What rule relates distance from the fulcrum d (in feet) to weight w (in pounds) when the weight balances a 50-pound weight on the opposite side and 6 feet from the fulcrum?

20 In many problems, it is helpful to express the relationship between dependent and independent variables with a symbolic rule that shows how values of one variable can be calculated from the values of the other.

a. If number of customers n at a bungee jump is related to price per jump p by the rule $n = 50 - p$, what rule shows how to calculate income I from values of n and p? What rule shows how to calculate values of I from the value of p alone?

b. What rule shows how to calculate Ethan and Anna's allowance on day n of a month if they receive 1 penny on day one, 2 pennies on day two, 4 pennies on day three, 8 pennies on day four, and so on?

c. Devon and Kevin were offered a pay scheme for work at Fresh Fare Market that would earn each of them $0.10 for the first hour in a week, $0.20 for the second hour, $0.40 for the third hour, $0.80 for the fourth hour, and so on. What rule shows how to calculate their pay for the nth hour in a week?

Review

These tasks provide opportunities for you to review previously learned mathematics and to refine your skills in using that mathematics.

21 Suppose a fair die is rolled.

a. What is the probability that the top face is a 6?

b. What is the probability that the top face is a 3?

c. What is the probability that the top face is an even number?

d. What is the probability that the top face is *not* a 6?

22 Micah and Keisha are renting a boat. The charge for the boat is $25 for the first hour and $12 for every hour (or portion of an hour) after the first.

a. How much will it cost if they rent the boat at 1:00 P.M. and return it at 3:50 P.M.?

b. They have been saving money all summer and have $80. What is the maximum amount of time that they can keep the boat?

23 The speed at which you travel, the length of time you travel, and the distance you travel are related in predictable ways. In particular, *speed* • *time* = *distance*. Use this relationship to help you answer the following questions.

a. Dave rides his bike for 2 hours with an average speed of 8.6 miles per hour. How far does he travel?

b. Kristen lives 4 miles from her friend's house. It is 2:30 P.M. and she needs to meet her friend at 3:00 P.M. How fast must she ride her bike in order to get to her friend's house on time?

c. Jessie leaves home at 7:30 A.M. and rides his bike to school at a speed of 9 miles per hour. If his school is 3 miles from his house, what time will he get to school?

24 Consider this scale drawing of Mongoose Lake. Using the given scale, estimate the perimeter to the nearest 10 meters and the area to the nearest 100 square meters.

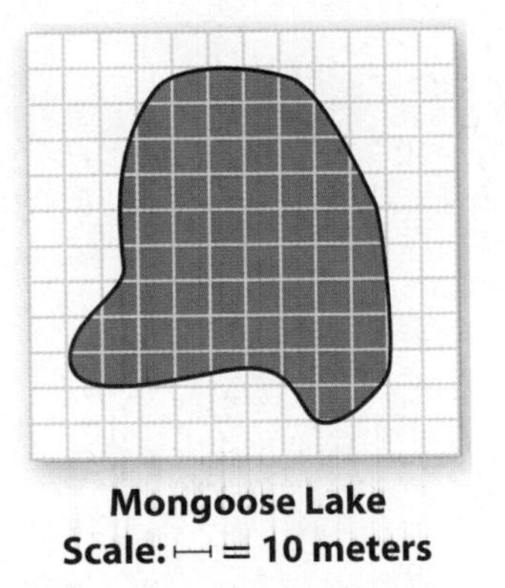

Mongoose Lake
Scale: ⊢⊣ = 10 meters

25 Convert each of these percents into equivalent decimals.

a. 75%

b. 5.4%

c. 0.8%

d. 0.93%

26 The table below gives some measurements associated with four different rectangles. Use the relationships between the lengths of the sides of a rectangle and the area and perimeter of a rectangle to complete the table.

Length (in cm)	**Width** (in cm)	**Perimeter** (in cm)	**Area** (in cm^2)
25	10	?	?
15	?	42	?
?	25	?	150
?	?	28	40

27 Convert each of these decimals into equivalent percents.

a. 0.8

b. 0.25

c. 2.45

d. 0.075

28 Suppose that a student has $150 in a bank savings account at the start of the school year. Calculate the change in that savings account during the following year in case it

a. earns 5% interest over that year.

b. grows from monthly deposits of $10 throughout the year.

c. earns 6.5% interest over that year.

d. declines by 8.7% over the year because withdrawals exceed interest earned.

e. declines at an average rate of $2.50 per month.

LESSON 2

Change Over Time

Every 10 years, the U.S. Census Bureau counts every American citizen and permanent resident. The 2000 census reported the U.S. population to be 281 million, with growth at a rate of about 1% each year. The world population is over 6 billion and growing at a rate that will cause it to exceed 9 billion by the year 2050.

National, state, and local governments and international agencies provide many services to people across our country and around the world. To match resources to needs, it is important to have accurate population counts more often than once every 10 years. However, complete and accurate census counts are very expensive.

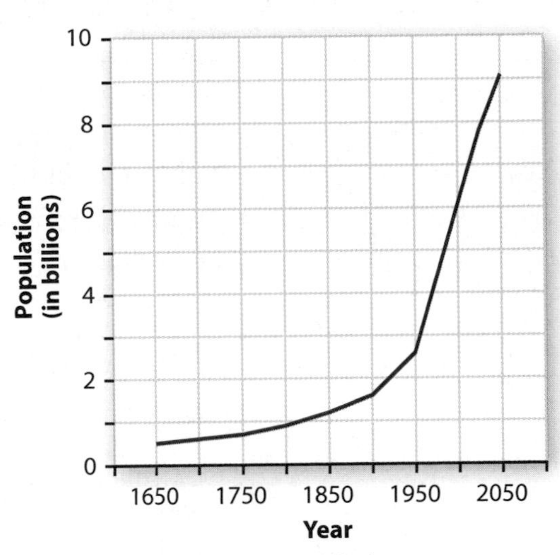

Source: www.census.gov/ipc/www/world.html

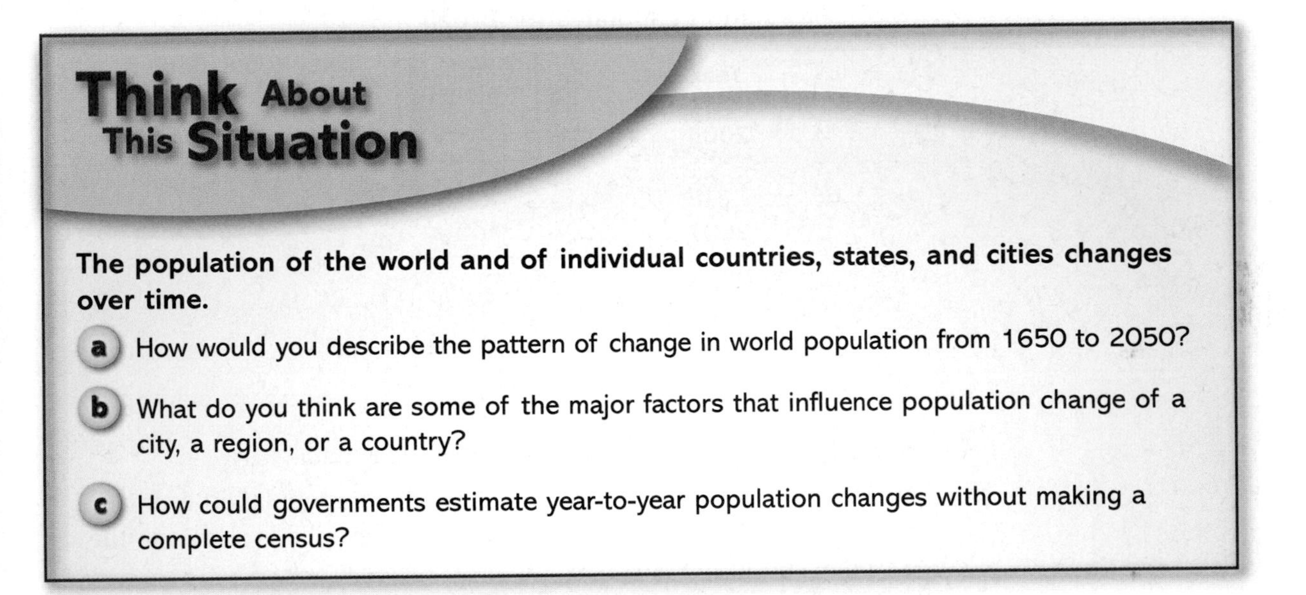

Think About This Situation

The population of the world and of individual countries, states, and cities changes over time.

a. How would you describe the pattern of change in world population from 1650 to 2050?

b. What do you think are some of the major factors that influence population change of a city, a region, or a country?

c. How could governments estimate year-to-year population changes without making a complete census?

Your work on investigations of this lesson will develop your understanding and skill in using algebra to solve problems involving variables like populations that change as time passes.

Investigation 1 Predicting Population Change

If you study trends in population data over time, you will often find patterns that suggest ways to predict change in the future. There are several ways that algebraic rules can be used to explain and extend such patterns of change over time. As you work on the problems of this investigation, look for an answer to this question:

What data and calculations are needed to predict human and animal populations into the future?

Population Change in Brazil Brazil is the largest country in South America. Its population in the year 2005 was about 186 million.

Census statisticians in Brazil can estimate change in that country's population from one year to the next using small surveys and these facts:

- Based on recent trends, births every year equal about 1.7% of the total population of the country.
- Deaths every year equal about 0.6% of the total population.

Source: *CIA—The World Factbook 2005*

1. How much of the estimated change in Brazil's population from 2005 to 2006 was due to:

   a. births? **b.** deaths? **c.** both causes combined?

2. Calculate estimates for the population of Brazil in 2006, 2007, 2008, 2009, and 2010. Record those estimates and the year-to-year changes in a table like the one below.

Population Estimates for Brazil

Year	Change (in millions)	Total Population (in millions)
2005	—	186
2006	?	?
2007	?	?
2008	?	?
2009	?	?
2010	?	?

   a. Make a plot of the (*year, total population*) data.

   b. Describe the pattern of change over time in population estimates for Brazil. Explain how the pattern you describe is shown in the table and in the plot.

3. Which of these strategies for estimating *change* in Brazil's population from one year to the next uses the growth rate data correctly? Be prepared to justify your answer in each case.

   a. 0.017(*current population*) − 0.006(*current population*) = *change in population*

   b. 0.011(*current population*) = *change in population*

   c. 0.17(*current population*) − 0.06(*current population*) = *change in population*

   d. 1.7%(*current population*) − 0.6% = *change in population*

4. Which of the following strategies correctly use the given growth rate data to estimate the *total* population of Brazil one year from now? Be prepared to justify your answer in each case.

   a. (*current population*) + 0.011(*current population*) = *next year's population*

b. (*current population*) + 0.017(*current population*) − 0.006(*current population*) = *next year's population*

c. 1.011(*current population*) = *next year's population*

d. 186 million + 1.7 million − 0.6 million = *next year's population*

5 Use the word *NOW* to stand for the population of Brazil in any year and the word *NEXT* to stand for the population in the next year to write a rule that shows how to calculate *NEXT* from *NOW*. Your rule should begin "*NEXT* = ..." and then give directions for using *NOW* to calculate the value of *NEXT*.

The Whale Tale In 1986, the International Whaling Commission declared a ban on commercial whale hunting to protect the small remaining stocks of several whale types that had come close to extinction.

Scientists make census counts of whale populations to see if the numbers are increasing. While it's not easy to count whales accurately, research reports have suggested that one population, the bowhead whales of Alaska, was probably between 7,700 and 12,600 in 2001.

The difference between whale births and natural deaths leads to a natural increase of about 3% per year. However, Alaskan native people are allowed to hunt and kill about 50 bowhead whales each year for food, oil, and other whale products used in their daily lives.

6 Assume that the 2001 bowhead whale population in Alaska was the low estimate of 7,700.

a. What one-year change in that population would be due to the difference between births and natural deaths?

b. What one-year change in that population would be due to hunting?

c. What is the estimate of the 2002 population that results from the combination of birth, death, and hunting rates?

7 Use the word *NOW* to stand for the Alaskan bowhead whale population in any given year and write a rule that shows how to estimate the population in the *NEXT* year.

8 Which of the following changes in conditions would have the greater effect on the whale population over the next few years?

- decrease in the natural growth rate from 3% to 2%, or
- increase in the Alaskan hunting quota from 50 to 100 per year

In studies of population increase and decrease, it is often important to predict change over many years, not simply from one year to the next. It is also interesting to see how changes in growth factors affect changes in populations. Calculators and computers can be very helpful in those kinds of studies.

For example, the following calculator procedure gives future estimates of the bowhead whale population with only a few keystrokes:

Calculator commands	Expected display
7700 ENTER .03 × Answer + Answer − 50 ENTER ENTER ENTER ENTER ENTER ENTER ENTER ENTER ENTER	7700 7700 .03*Ans+Ans-50 7881 8067 8259 8457 8259 8457 8661 8871 9087 9310 9539

9 Examine the calculator procedure above.

a. What seem to be the purposes of the various keystrokes and commands?

b. How do the instructions implement a *NOW-NEXT* rule for predicting population change?

10 Modify the given calculator steps to find whale population predictions starting from the 2001 high figure of 12,600 and a natural increase of 3% per year.

a. Find the predicted population for 2015 if the annual hunt takes 50 whales each year.

b. Suppose that the hunt takes 200 whales each year instead of 50. What is the predicted population for 2015 in this case?

c. Experiment to find a hunt number that will keep the whale population stable at 12,600.

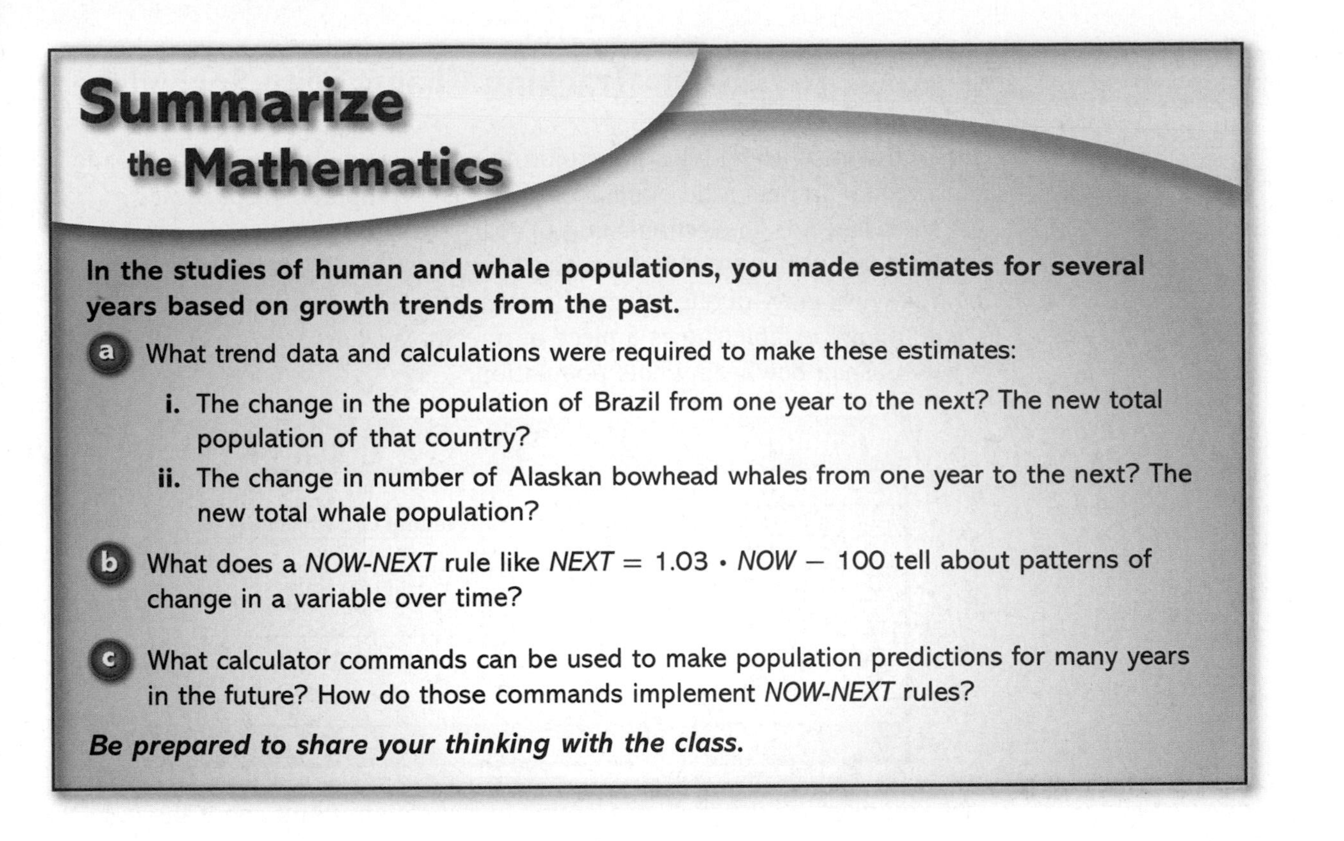

Summarize the Mathematics

In the studies of human and whale populations, you made estimates for several years based on growth trends from the past.

a What trend data and calculations were required to make these estimates:

i. The change in the population of Brazil from one year to the next? The new total population of that country?

ii. The change in number of Alaskan bowhead whales from one year to the next? The new total whale population?

b What does a *NOW-NEXT* rule like $NEXT = 1.03 \cdot NOW - 100$ tell about patterns of change in a variable over time?

c What calculator commands can be used to make population predictions for many years in the future? How do those commands implement *NOW-NEXT* rules?

Be prepared to share your thinking with the class.

Check Your Understanding

The 2000 United States Census reported a national population of about 281 million, with a birth rate of 1.4%, a death rate of 0.9%, and net migration of about 1.1 million people per year. The net migration of 1.1 million people is a result of about 1.3 million immigrants entering and about 0.2 million emigrants leaving each year.

a. Use the given data to estimate the U.S. population for years 2001, 2005, 2010, 2015, 2020.

b. Use the words *NOW* and *NEXT* to write a rule that shows how to use the U.S. population in one year to estimate the population in the next year.

c. Write calculator commands that automate calculations required by your rule in Part b to get the U.S. population estimates.

d. Modify the rule in Part b and the calculator procedure in Part c to estimate U.S. population for 2015 in case:

i. The net migration rate increased to 1.5 million per year.

ii. The net migration rate changed to −1.0 million people per year. That is, if the number of emigrants (people leaving the country) exceeded the number of immigrants (people entering the country) by 1 million per year.

Investigation 2 Tracking Change with Spreadsheets

One of the most useful tools for exploring relations among birth rates, death rates, migration rates, and population totals is a computer *spreadsheet.*

A spreadsheet is an electronic grid of cells in which numerical data or labels can be stored. The cells of a spreadsheet can be related by formulas, so that the numerical entry of one cell can be calculated from data in other cells.

The following table shows a piece of one spreadsheet that predicts growth of the Alaskan bowhead whale population.

Whale Population.xls

◇	A	B	C
1	Year	Population	Natural Growth Rate
2	2001	7700	1.03
3	2002	7881	Hunting Rate
4	2003	8067	50
5	2004	8259	
6	2005	8457	
7			

As you work on the problems in this investigation, think about the following question:

How do basic spreadsheet methods use the NOW-NEXT way of thinking to help solve problems about change over time?

1 From your earlier work with calculators, the numbers in column B of the preceding spreadsheet probably look familiar. However, you can't see how the spreadsheet actually produced those numbers. The next table shows the formulas used to calculate entries in columns A and B of the first display.

Whale Population.xls

A6 =A5+1

◇	A	B	C
1	Year	Population	Natural Growth Rate
2	2001	7700	1.03
3	=A2+1	=1.03*B2−50	Hunting Rate
4	=A3+1	=1.03*B3−50	50
5	=A4+1	=1.03*B4−50	
6	=A5+1	=1.03*B5−50	
7			

Compare the formula cell entries to the numerical cell values in the display above to help answer the next questions about how spreadsheets actually work.

a. How do you think the formulas in cells A3, A4, A5, and A6 produce the pattern of entries 2002, 2003, 2004, and 2005 in the numerical form of the spreadsheet?

b. How do you think the formulas in cells B3, B4, B5, and B6 produce the pattern of entries 7881, 8067, 8259, and 8457?

c. Why would it make sense to call the formulas in cells A3–A6 and B3–B6 *NOW-NEXT* formulas?

d. What are the starting values for the formulas in columns A and B?

The real power of a spreadsheet comes from a feature not shown in this table of formulas. After entering the starting values in cells A2 and B2 and the *NOW-NEXT* formulas in cells A3 and B3, the spreadsheet command "fill down" will automatically produce formulas for the cells below, changing the cell reference A2 to A3, B2 to B3, and so on.

Suppose that you were interested in studying population growth of the United States in 10-year intervals corresponding to the national census counts. With the 2000 population of 281 million, a natural 10-year growth rate of about 5%, and 10-year migration of about 11 million, a spreadsheet to make predictions for several decades might begin like the one below.

U.S. Population.xls

B3 =1.05*B2+11

◇	A	B	C
1	Year	Population	Natural Growth Rate
2	2000	281	1.05
3	=A2+10	=1.05*B2+11	Migration Rate
4			11
5			

a. What formula and numerical entries would you expect in cells A3, A4, A5, and A6 if you use a fill down command in that column?

b. What formula and numerical entries would you expect in cells B3, B4, B5, and B6 if you use a fill down command in that column?

A second feature of spreadsheets makes exploratory work even more efficient. If you mark column and/or row labels with a dollar sign symbol, they will not change in response to fill down or fill across commands.

Suppose that you want to study the effects of change in both natural growth and migration rates for the U.S. population.

a. What numerical value do you think will result from the formula "=C$2*B2+C$4" in cell B3 of the spreadsheet below?

U.S. Population.xls

B3 =C$2*B2+C$4

◇	A	B	C
1	Year	Population	Natural Growth Rate
2	2000	281	1.05
3	=A2+10	=C$2*B2+C$4	Migration Rate
4	=A3+10		11
5	=A4+10		

b. What formulas and numerical values will appear in cells **B4** and **B5** following a fill down command?

c. What formulas and numerical values will appear in cells **B4** and **B5** if the entry in cell **C2** is changed to 1.06 and the entry in cell **C4** is changed to 12?

d. What changes in natural growth and migration rates are implied by those changes in the spreadsheet?

When Robin got a summer job, she decided she could save $25 from her pay every week.

a. Construct a spreadsheet that will display Robin's total savings at the end of each week during the 10-week summer job.

b. If necessary, modify your spreadsheet so that the amount saved each week can be found by changing only one cell entry. Then use the new spreadsheet to display Robin's total savings at the end of each week if she actually saves only $17.50 per week.

Suppose that in September, Robin invests her summer savings of $250 in a bank account that pays interest at the rate of 0.5% per month (an annual rate of 6%).

a. Construct a spreadsheet that will display Robin's bank balance at the end of each month for the next year.

b. Modify your spreadsheet to account for Robin's habit of withdrawing $20 at the beginning of every month for extra spending money.

6 Modify the spreadsheet in Problem 5 to compare two possible savings plans.

Plan 1: Deposit $100 in September and add $10 per month thereafter.
Plan 2: Deposit $0 in September and add $20 per month thereafter.

a. How long will it take before Plan 2 gives a greater balance than Plan 1?

b. How will the answer to Part a change if the monthly interest rate decreases to 0.4%, 0.3%, 0.2%, or 0.1%?

7 When José was considering purchase of a $199 portable music player and ear phones, he was told that resale value of the gear would decline by about 5% per month after he bought it.

a. Construct a spreadsheet that will display the value of José's music gear at the start of each month over two years from its purchase.

b. Modify your spreadsheet to analyze the changing value of a PDA that would cost $499 to purchase and decline in value at about the same percent rate.

c. Explain why your spreadsheets in Parts a and b do not show loss of all value for the music player or the PDA in 20 months, even though $20 \cdot 5\% = 100\%$.

Summarize the Mathematics

In this investigation, you learned basic spreadsheet techniques for studying patterns of change.

a. How are cells in a spreadsheet grid labeled and referenced by formulas?

b. How are formulas used in spreadsheets to produce numbers from data in other cells?

c. How is the "fill" command used to produce cell formulas rapidly?

d. How are the cell formulas in a spreadsheet similar to the *NOW-NEXT* rules you used to predict population change?

Be prepared to share your ideas with other students.

Check Your Understanding

The number pattern that begins 1, 1, 2, 3, 5, 8, 13, 21, 34, 55, ... is known as the **Fibonacci sequence**. The pattern appears many places in nature. It also has been the subject of many mathematical investigations.

a. Study the pattern. What are the next five numbers in the sequence?

b. Write spreadsheet formulas that will produce columns A and B in the next table (and could be extended down to continue the pattern).

Fibonacci Sequence.xls

◇	A	B
1	1	1
2	2	1
3	3	2
4	4	3
5	5	5
6	6	8
⋮	⋮	⋮

c. Modify the spreadsheet of Part b to produce terms in the number pattern that begins 5, 5, 10, 15, 25, ... and grows in the Fibonacci way. Use the spreadsheet to find the next 10 numbers in the pattern.

d. Compare the number patterns in Parts a and c. What explains the way the patterns are related?

On Your Own

Applications

1 The People's Republic of China is the country with the largest population in the world. The population of China in 2005 was approximately 1.3 billion. Although families are encouraged to have only one child, the population is still growing at a rate of about 0.6% per year.

a. Estimate the population of China for each of the next 5 years and record your estimates in a data table.

b. When is it likely that the population of China will reach 1.5 billion?

c. How would your prediction in Part b change if the growth rate were 1.2%, double the current rate?

d. Using the word *NOW* to stand for the population in any year, write rules that show how to calculate the population in the *NEXT* year:

i. if the growth rate stays at 0.6%.

ii. if the growth rate doubles to 1.2%.

2 The country with the second largest population in the world is India, with about 1.1 billion people in 2005. The birth rate in India is about 2.2% per year and the death rate is about 0.8% per year.

a. Estimate the population of India for each of the next 5 years and record your estimates in a data table.

b. When is it likely that the population of India will reach 1.5 billion?

c. How would your prediction in Part b change if the birth rate slows to 2.0%?

d. Using the word *NOW* to stand for the population in any year, write rules that show how to calculate the population in the *NEXT* year:

i. if the birth rate stays at 2.2%.

ii. if the birth rate slows to 2.0%.

3 Timber wolves were once very common in wild land across the northern United States. However, when the Endangered Species Act was passed in 1973, wolves were placed on the endangered list.

Thirty years later, the wolf populations have recovered in the northern Rockies and in the forests of Minnesota, Wisconsin, and Michigan.

In 2003, estimates placed the Midwest wolf population at more than 3,100 with an annual growth rate of 25% to 30%. (Source: "Timber Wolves Resurgent in Upper Midwest," *The Washington Post,* Monday, February 10, 2003.)

a. Use the given wolf population estimate and the 25% growth rate to predict populations for 10 years (from 2003 to 2013). Record your results in a data table.

b. Estimate the time when the Midwest wolf population will reach 30,000 (the number believed to have lived in the Great Lakes region 500 years ago).

c. How does your answer to Part b change if you use the higher growth rate estimate of 30%?

d. Using the word *NOW* to stand for the Midwest wolf population in any year, write rules that show how to calculate the population in the *NEXT* year:

i. if the growth rate stays at 25%.

ii. if the growth rate increases to 30%.

4 Midwestern farmers who raise dairy cattle are concerned that growing wolf populations described in Task 3 above threaten the safety of their herds. They want permission to eliminate wolves that kill livestock.

a. Make a table showing how the Midwest wolf population of 3,100 would change over the next 10 years if an annual harvest of 250 animals were allowed, but the natural growth rate continued at 25% per year.

b. When is it likely that the Midwest wolf population would reach 30,000 if the annual harvest of 250 animals were permitted?

c. How would your answer to Part b change if the annual harvest were increased to 500?

d. Using the word *NOW* to stand for the population in any year, write rules that show how to calculate the population in the *NEXT* year:

i. if the natural growth rate stays at 25% and 250 wolves are killed each year.

ii. if the growth rate stays at 25% but the annual harvest increases to 500 wolves per year.

5. China experiences annual negative net migration of its population. People leave for other countries of the world in large numbers.

   a. How would the current 1.3 billion population of China change in 10 years in case of natural growth rate of 0.6% and net migration of about −500,000 people per year? (Remember to use uniform units.)

   b. What net migration would have to occur for China to reach *zero population growth,* assuming that the natural growth rate remained at 0.6% per year?

   c. Using *NOW* to stand for the population of China in any year, write a rule that shows how to calculate the population in the *NEXT* year if the natural growth rate is 0.6% and the net migration is about −500,000 people per year.

6. India has an annual negative migration to somewhat offset its natural population growth.

   a. How would the current 1.1 billion population of India change in 10 years in case of a natural growth rate of 1.4% and net migration of about only −80,000 people per year? (Use uniform units.)

   b. What net migration would have to occur for India to reach *zero population growth,* assuming that the natural growth rate remained at 1.4% per year?

   c. Using *NOW* to stand for the population of India in any year, write a rule that shows how to calculate the population in the *NEXT* year if the natural growth rate is 1.4%, and the net migration is about −80,000 people per year.

7. If money is invested in a savings account, a business, or real estate, its value usually increases each year by some percent. For example, investment in common stocks yields growth in value of about 10% per year in the long run. Suppose that when a child is born, the parents invest $1,000 in a mutual fund account.

   a. If that fund actually grows in value at a rate of 10% per year, what will its value be after 1 year? After 2 years? After 5 years? After 18 years when the child is ready to go to college?

   b. Using *NOW* to stand for the investment value at the end of any year, write a rule showing how to calculate the value at the end of the *NEXT* year.

   c. How will your answers to Parts a and b change if:

      i. the initial investment is only $500?

      ii. the initial investment is $1,000, but the growth rate is only 5% per year?

   d. How will your answers to Parts a and b change if, in addition to the percent growth of the investment, the parents add $500 per year to the account?

8. Select one of Applications Tasks 1–3 and develop a spreadsheet that could be used to answer the population growth questions asked in those items. Use the spreadsheet to answer those questions.

9 Select one of Applications Tasks 4–7 and develop a spreadsheet that could be used to answer the questions about population or investment growth over time. Use the spreadsheet to answer those questions.

Connections

Data in the next table show population (in thousands) of some major U.S. cities in 1990 and in 2000. Use the data to complete Connections Tasks 10–13 that follow.

Major U.S. Cities: 1990 and 2000 Population (in 1,000s)

U.S. City	1990	2000	U.S. City	1990	2000
Atlanta, GA	394	416	Independence, MO	112	113
Aurora, CO	222	276	Milwaukee, WI	628	597
Berkeley, CA	103	103	Newark, NJ	275	274
Boise, ID	127	186	Portland, OR	437	529
El Paso, TX	515	564	St. Louis, MO	397	348
Hartford, CT	140	122	Washington, DC	607	572

10 The population of Berkeley, California, changed by fewer than 1,000 people. Among the remaining cities in the list, which cities had:

a. the greatest absolute decrease in population?

b. the greatest absolute increase in population?

c. the greatest percent decrease in population?

d. the greatest percent increase in population?

11 What were the mean and median population change for the listed cities?

12 Suppose that the population of Aurora, Colorado, continues increasing at the rate it changed between 1990 and 2000.

a. What population for Aurora would be predicted for 2010, 2020, 2030, 2040, and 2050 if population increases by the same number of people in each decade?

b. What *NOW-NEXT* rule describes the pattern of change in Part a?

c. What was the percent change in the Aurora population between 1990 and 2000?

d. What *NOW-NEXT* rule describes the pattern of change in Aurora's population each decade, if the percent rate of change from Part c is used?

e. What population is predicted for Aurora in 2010, 2020, 2030, 2040, and 2050 if growth occurs at the percent rate of Part c?

f. How are the predicted change patterns in Parts a and e similar, and how are they different? Why are they not the same?

g. What reasons could you have to doubt the predictions of Parts a or e?

13 Suppose that the population of Washington, D.C., continues decreasing at the rate it changed between 1990 and 2000.

a. What population for Washington, D.C., would be predicted for 2010, 2020, 2030, 2040, and 2050 if population decreases by the same number of people in each decade?

b. What *NOW-NEXT* rule describes the pattern of change in Part a?

c. What was the percent change in the Washington, D.C., population between 1990 and 2000?

d. What *NOW-NEXT* rule describes the pattern of change in the Washington, D.C., population each decade, if the percent rate of change from Part c is used?

e. What population is predicted for Washington, D.C., in 2010, 2020, 2030, 2040, and 2050 if growth occurs at the percent rate of Part c?

f. How are the predicted change patterns in Parts a and e similar, and how are they different? Why are they not the same?

g. What reasons could you have to doubt the predictions of Parts a or e?

14 Sketch graphs that match each of the following stories about quantities changing over time. On each graph, label the axes to indicate reasonable scale units for the independent variable and the dependent variable. For example, use "time in hours" and "temperature in degrees Fahrenheit" for Part a.

a. On a typical summer day where you live, how does the temperature change from midnight to midnight?

b. When a popular movie first appears in video rental stores, demand for rentals changes as time passes.

c. The temperature of a cold drink in a glass placed on a kitchen counter changes as time passes.

d. The number of people in the school gymnasium changes before, during, and after a basketball game.

15 Each part below gives a pair of *NOW-NEXT* rules. For each rule in each pair, produce a table of values showing how the quantities change from the start through 5 stages of change. Then compare the patterns of change produced by each rule in the pair and explain how differences are related to differences in the *NOW-NEXT* rules.

a. Rule 1: $NEXT = NOW + 10$, starting at 5
Rule 2: $NEXT = NOW + 8$, starting at 5

Sample Table:

Stage	0	1	2	3	4	5
Rule 1	5	15	25			
Rule 2	5	13	...			

b. Rule 1: $NEXT = 2 \cdot NOW$, starting at 5
Rule 2: $NEXT = 1.5 \cdot NOW$, starting at 10

c. Rule 1: $NEXT = 0.5 \cdot NOW$, starting at 100
Rule 2: $NEXT = 0.9 \cdot NOW$, starting at 50

d. Rule 1: $NEXT = 2 \cdot NOW + 10$, starting at 8
Rule 2: $NEXT = 3 \cdot NOW - 10$, starting at 8

16 The graph below shows how the amount of water in a city's reservoir changed during one recent year.

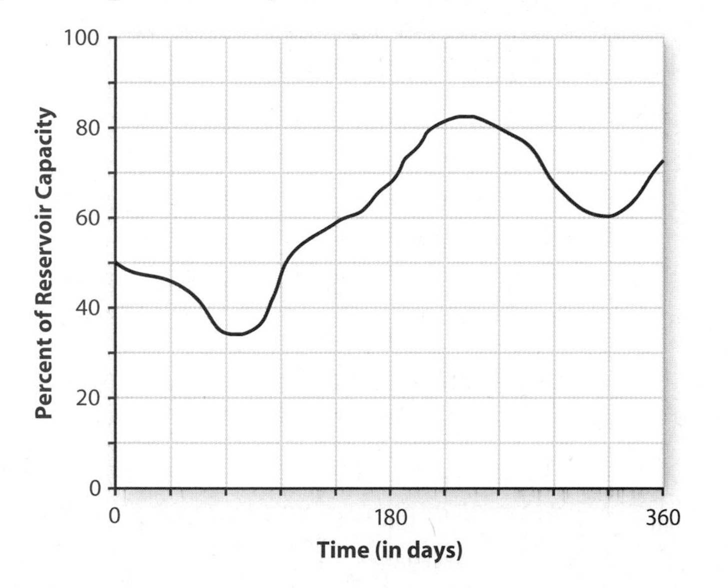

On a copy of the graph, mark points that show when the reservoir's water supply is:

a. increasing at the fastest rate—label the point(s) with the letter "A".

b. decreasing at the fastest rate—label the point(s) with the letter "B".

c. increasing at a constant rate—label the point(s) with the letter "C".

d. decreasing at a constant rate—label the point(s) with the letter "D".

e. neither increasing nor decreasing—label the point(s) with the letter "E".

Reflections

17 How are patterns in the data tables and graphs arising in the studies of human and whale populations similar to or different from those that related:

a. *weight* and *stretched length* of a bungee cord (page 5)?

b. *price per jump* and *number of customers* for a bungee jump (page 6)?

c. *price per jump* and *daily income* for operation of the bungee jump (page 6)?

d. *number of plays* and fund-raiser *cumulative profit* in the *Take a Chance* die-tossing game (pages 8–9)?

e. *average speed* and *race time* for a 500-mile NASCAR race (pages 11–12)?

f. *hours worked* and *earnings* at Fresh Fare Market (page 12)?

18 In what ways are the methods used to describe "change over time" patterns similar to or different from the methods used to study "cause and effect" patterns?

19 Consider the *NOW-NEXT* rules:

$$NEXT = NOW + 0.05 \cdot NOW \quad \text{and} \quad NEXT = 1.05 \cdot NOW$$

a. Find several values produced by these *NOW-NEXT* rules, starting from $NOW = 10$.

b. Then explore the patterns produced by each rule for some other common starting values.

c. Explain why the results of the explorations in Parts a and b are not surprising.

20 Do the ideas of independent and dependent variables have useful meaning in the study of "change over time" patterns? If so, how? If not, why not?

21 Both animal and human population growth rates commonly change as the years pass.

a. What factors might cause change in the percent growth rates of a population?

b. Why, if growth rates change, does it still make sense to use current growth rates for predictions of future populations?

Extensions

22 The amusement park ride test team took their radar gun for a ride on the Ferris wheel. They aimed the gun at the ground during two nonstop trips around on the wheel, giving a graph relating height above the ground to time into the trip.

a. The total time for the ride was 100 seconds. Sketch a graph showing how you think height will change over time during the ride. Then write an explanation of the pattern in your graph. (*Hint:* You might experiment with a bicycle wheel as a model of a Ferris wheel; as you turn the wheel, how does the height of the air valve stem change?)

b. Given next is the graph of (*time into ride, height of rider*) data for one Ferris wheel test ride. Write an explanation of what the graph tells about that test ride.

First Test Ride

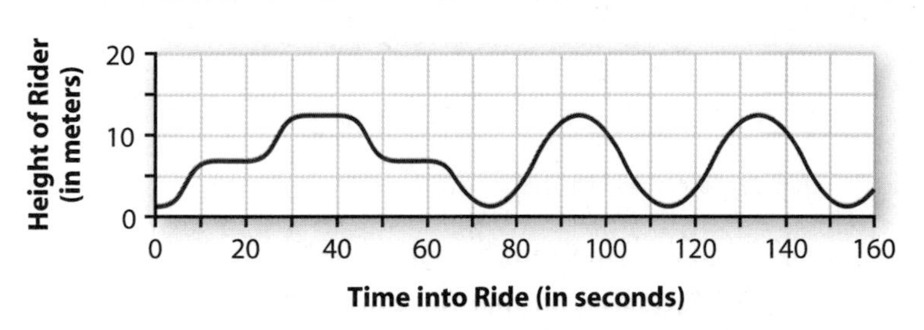

c. Given next are some (*time, height*) data from a second test ride. Write a short description of the pattern of change in height over time during this ride.

Second Test Ride

Time (in seconds)	0	2	5	10	15	20	22.5	25	30	35	40	42.5	45	50	55	60	62.5	65
Height (in meters)	1	1	3	3	11	11	13	11	11	3	3	1	3	3	11	11	13	11

d. Plot the data from the second test ride on a coordinate graph. Connect the points if it seems to make sense to do so. Then explain whether you think the graph or the table better shows the pattern of change in height during the ride.

23 A manatee is a large sea mammal native to Florida waters that is listed as endangered. The chart below gives the number of manatees killed in watercraft collisions near the Gulf Coast of Florida every year from 1985 through 2004.

Manatee/Watercraft Mortalities

Year	Number of Manatees Killed	Year	Number of Manatees Killed
1985	33	1995	42
1986	33	1996	60
1987	39	1997	54
1988	43	1998	66
1989	50	1999	82
1990	47	2000	78
1991	53	2001	81
1992	38	2002	95
1993	35	2003	73
1994	49	2004	69

Source: www.savethemanatee.org/mortalitychart.htm

a. Prepare a plot of the number of manatees killed in watercraft collisions between 1985 and 2004. Connect the points in that plot to help you study trends in manatee/watercraft mortalities.

b. Describe the pattern of change in mortalities that you see in the table and the *plot over time.*

c. During what one-year period was there the greatest change in manatee deaths if one measures change by:

   i. difference?

   ii. percent change?

d. How is the time of greatest change in manatee deaths shown in the table? In the graph?

e. What factors in the marine life and boating activity of Florida might be causing the increase in manatee deaths? What actions could be taken by government to reduce the number of deaths?

24 The Fibonacci sequence has many interesting and important properties. One of the most significant is revealed by studying the ratios of successive terms in the sequence. Consider the Fibonacci sequence 1, 1, 2, 3, 5, 8, 13, … .

a. Modify the spreadsheet you wrote to generate terms of the Fibonacci sequence to include a new column C. In cell C2, enter the formula "=B2/B1" and then repeat this formula (with cell references changing automatically down the column C). Record the sequence of terms generated in that column.

b. What pattern of numerical values do you notice as you look farther and farther down column C?

Leonardo Fibonacci
1180–1250

25 In an earlier problem, you explored the rate at which the allowance paid to Ethan and Anna would increase if it began with only 1 penny on the first day of the month but doubled each day thereafter.

a. Write a spreadsheet with three columns: "Day of the Month" in column A, "Daily Allowance" in column B, and "Cumulative Allowance" in column C, with rows for up to 31 days.

b. Use the spreadsheet to find the total allowance paid to Ethan and Anna in a month of 31 days.

Review

26 Find the value of each expression.

a. $4 \cdot 2 - 3$ **b.** $2(-5) + 2(3)$ **c.** $-5^2 - (3 - 5)$

d. $(-3)(2)(-5)$ **e.** $-5 + 2 + 10 + (-5)^2$ **f.** $|-5| + 15 - |5 - 3|$

27 In figures A and B, squares are built on each side of a right triangle.

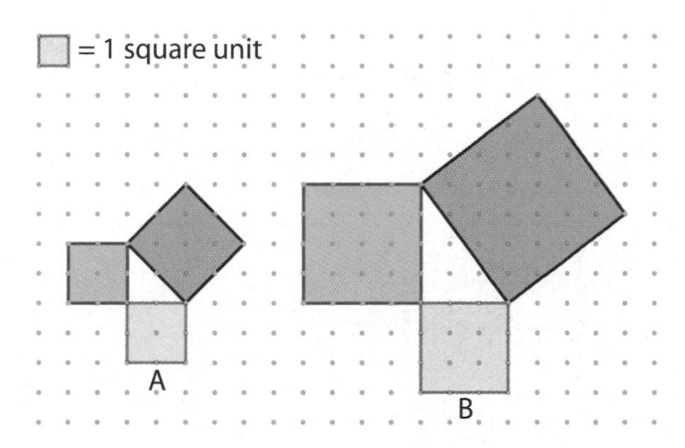

a. For figure A:

i. Find the area of the triangle.

ii. Find the area of each square. How are the areas related?

iii. Find the perimeter of the triangle.

b. For figure B:

i. Find the area of the triangle.

ii. Find the area of each of the squares. How are the areas of the squares related in this case?

iii. Find the perimeter of the triangle.

c. How is the work you did in Parts a and b related to the Pythagorean Theorem?

28 Consider the circle drawn below.

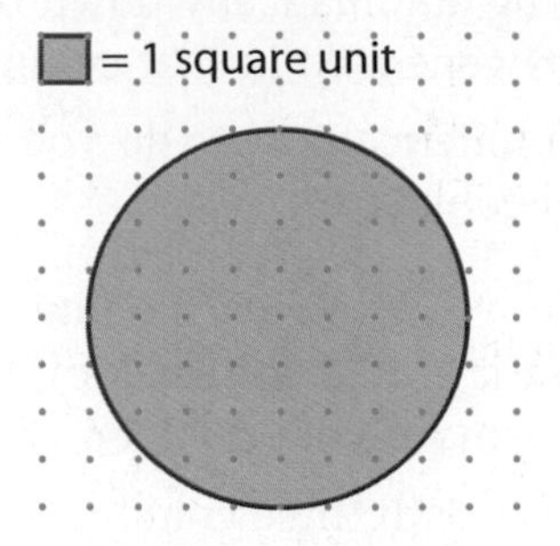

a. Use the dot grid to find the approximate area of the circle.

b. Use the formula $C = 2\pi r$ or $C = \pi d$ to find the circumference of the circle.

c. Use the formula $A = \pi r^2$ to find the area of the circle.

d. In what kind of unit is area measured? How can you use this fact to avoid confusing the formulas $2\pi r$ and πr^2 when computing the area of a circle?

29 The population of India is about 1.1 billion people. Suppose the population of country X is 1.1 million, and the population of country Y is 11 million.

a. How many times larger is the population of India than that of country Y? Than that of country X?

b. What percent of the Indian population is the population of country Y? Is the population of country X?

30 Consider the three parallelograms shown below.

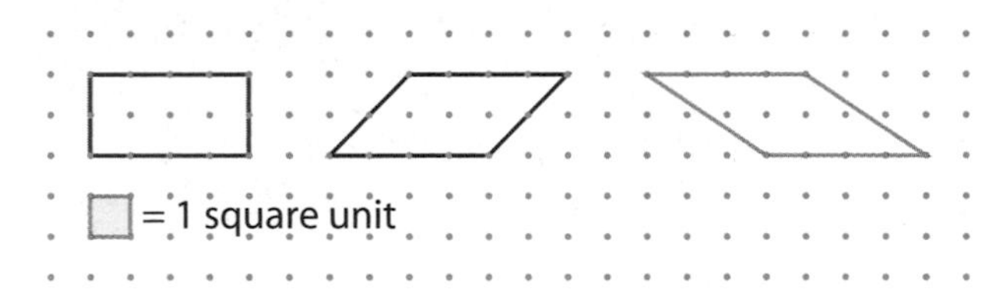

a. Rachel thinks that all three parallelograms have the same area. Is she correct? Explain your reasoning.

b. Sketch a parallelogram, different from those above, that has an area of 8 square units.

31 Place the following numbers in order from smallest to largest.

2.25 2.05 −2.35 −2.75 0 2.075

LESSON 3

Tools for Studying Patterns of Change

In your work on problems of Lesson 1, you studied a variety of relationships between dependent and independent variables. In many cases, those relationships can be expressed by calculating rules that use letter names for the variables. For example,

The *stretched length L* of a simulated bungee cord depends on the *attached weight w* in a way that is expressed by the rule $L = 30 + 0.5w$.

The *number of customers n* for a bungee jump depends on the *price per jump p* in a way that is expressed by the rule $n = 50 - p$.

The *time t* of a 500-mile NASCAR race depends on the *average speed s* of the winning car in a way that is expressed by the rule $t = \frac{500}{s}$.

These symbolic rules give directions for calculating values of dependent variables from given values of related independent variables. They also enable use of calculator and computer tools for solving problems involving the relationships.

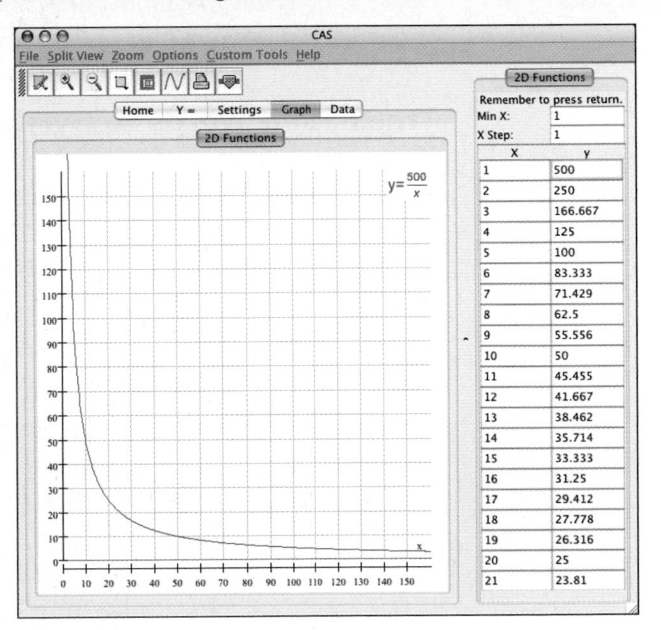

Think About This Situation

If you were asked to solve problems in situations similar to those described on the previous page:

a How would you go about finding algebraic rules to model the relationships between dependent and independent variables in any particular case?

b What ideas do you have about how the forms of algebraic rules are connected to patterns in the tables and graphs of the relationships that they produce?

c How could you use calculator or computer tools to answer questions about the variables and relationships expressed in rules?

Your work in this lesson will help you develop skills in writing algebraic rules to express relationships between variables. You will also use calculator and computer strategies to determine relationships expressed by those rules.

Investigation 1 Communicating with Symbols

The first challenge in using algebraic expressions and rules to study a relationship between variables is to write the relationship in symbolic form. There are several ways information about such relationships occur and several strategies for translating information into symbolic form. As you work on the problems of this investigation, look for answers to this question:

What are some effective strategies for finding symbolic expressions that represent relationships between variables?

Translating Words to Symbols In many problems, important information about a relationship between variables comes in the form of written words. Sometimes it is easy to translate those words directly into algebraic expressions.

For example, if a restaurant adds a 15% gratuity to every food bill, the *total bill T* is related to the *food charges F* by the rule:

$$T = F + 0.15F.$$

In other cases, you might need to calculate the value of the dependent variable for several specific values of the independent variable to see how the two are related in general. Suppose that a library loans books free for a week, but charges a fine of $0.50 each day the book is kept beyond the first week. To find a rule relating *library fines* for books to the *number of days* the book is kept, you might begin by calculating some specific fines, like these:

Book Kept 10 Days: Fine = $0.50(10 - 7)$
Book Kept 21 Days: Fine = $0.50(21 - 7)$

1. Can you create a rule relating *library fines* for new books to the *number of days* the book is kept? Write your rule in symbolic form, using F for the fine and d for number of days the book is kept.

2. Midwest Amusement Park charges $25 for each daily admission. The park has daily operating expenses of $35,000.

   a. What is the operating profit (or loss) of the park on a day when 1,000 admission tickets are sold? On a day when 2,000 admission tickets are sold?

   b. Write a symbolic rule showing how daily profit P for the park depends on the number of paying visitors n.

3. A large jet airplane carries 150,000 pounds of fuel at takeoff. It burns approximately 17,000 pounds of fuel per hour of flight.

   a. What is the approximate amount of fuel left in the airplane after 3 hours of flight? After 7 hours of flight?

   b. Write a rule showing how the amount of fuel F remaining in the plane's tanks depends on the elapsed time t in the flight.

4. The costs for a large family reunion party include $250 for renting the shelter at a local park and $15 per person for food and drink.

   a. Write a rule showing how the total cost C for the reunion party depends on the number of people n who will attend.

   b. Write another rule showing how the cost per person c (including food, drink, and a share of the shelter rent) depends on the number of people n who plan to attend.

Measurement Formulas Many of the most useful symbolic rules are those that give directions for calculating measurements of geometric figures. You probably know several such formulas from prior mathematics studies.

5. Figure $BCDE$ below is a rectangle.

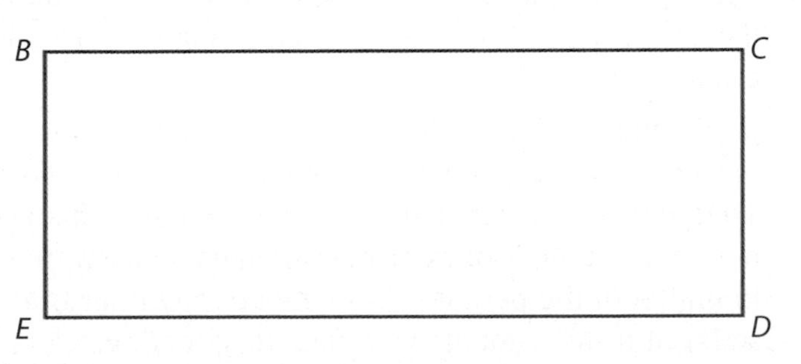

   a. Use a ruler to make measurements from which you can estimate the perimeter and area of rectangle $BCDE$. Then calculate those estimates.

b. For any given rectangle, what is the minimum number of ruler measurements you would need in order to find both its perimeter and area? What set(s) of measurements will meet that condition?

c. What formulas show how to calculate perimeter P and area A of a rectangle from the measurements described in Part b?

Figure *QRST* below is a parallelogram.

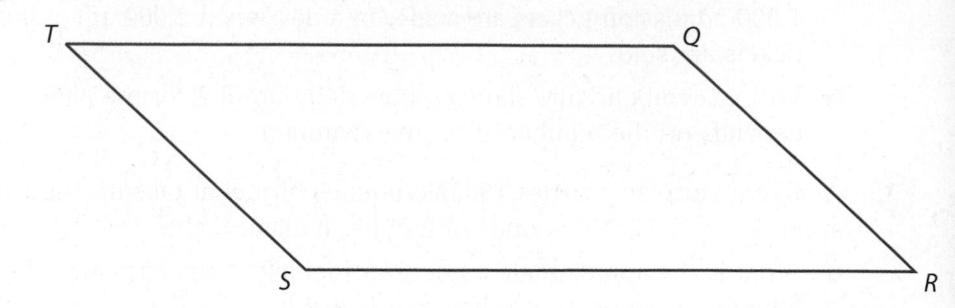

a. Use a ruler to make measurements from which you can estimate the perimeter and the area of *QRST*. Then calculate those estimates.

b. For any given parallelogram, what is the minimum number of measurements you would need in order to find both the perimeter and area? What measurements will meet that condition?

c. What formulas show how to calculate perimeter P and area A of a parallelogram from the measurements you described in Part b?

The two figures below are triangles—one a right triangle and one an obtuse triangle.

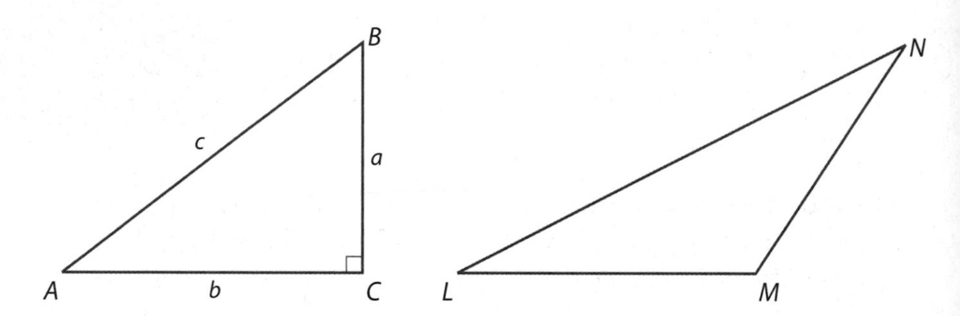

a. Use a ruler to make measurements from which you can estimate the perimeter and the area of each triangle. Then calculate those estimates.

b. If the lengths of the sides of a right triangle are a, b, and c, with the side of length c opposite the right angle, the **Pythagorean Theorem** guarantees that $a^2 + b^2 = c^2$. Using this fact, what is the minimum number of ruler measurements you would need in order to find both the perimeter and area of any right triangle? What set(s) of measurements will meet that condition?

c. What formulas show how to calculate perimeter P and area A of a right triangle from the measurements you described in Part b?

d. What is the minimum number of measurements you would need in order to find both the perimeter and the area of any nonright triangle? What measurements will meet that condition?

e. What formulas show how to calculate perimeter P and area A of a nonright triangle from the measurements you described in Part d?

8 The figure below is a circle with center O.

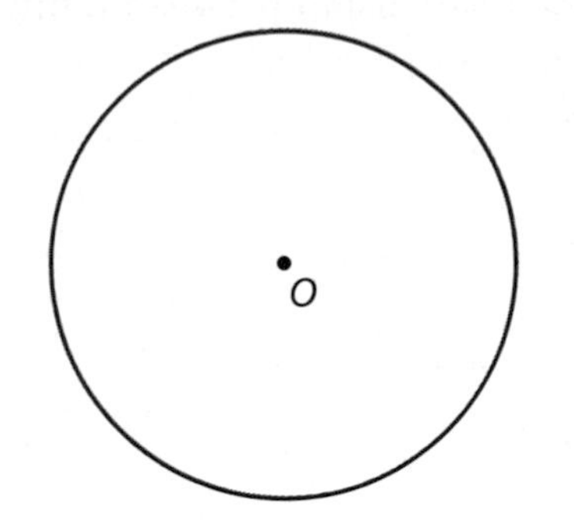

a. Use a ruler to make measurements from which you can estimate the circumference and area of the circle. Then calculate those estimates.

b. For any given circle, what is the minimum number of measurements you would need in order to find both the circumference and the area? What measurements will meet that condition?

c. What formulas show how to calculate circumference C and area A of a circle from the measurements you described in Part b?

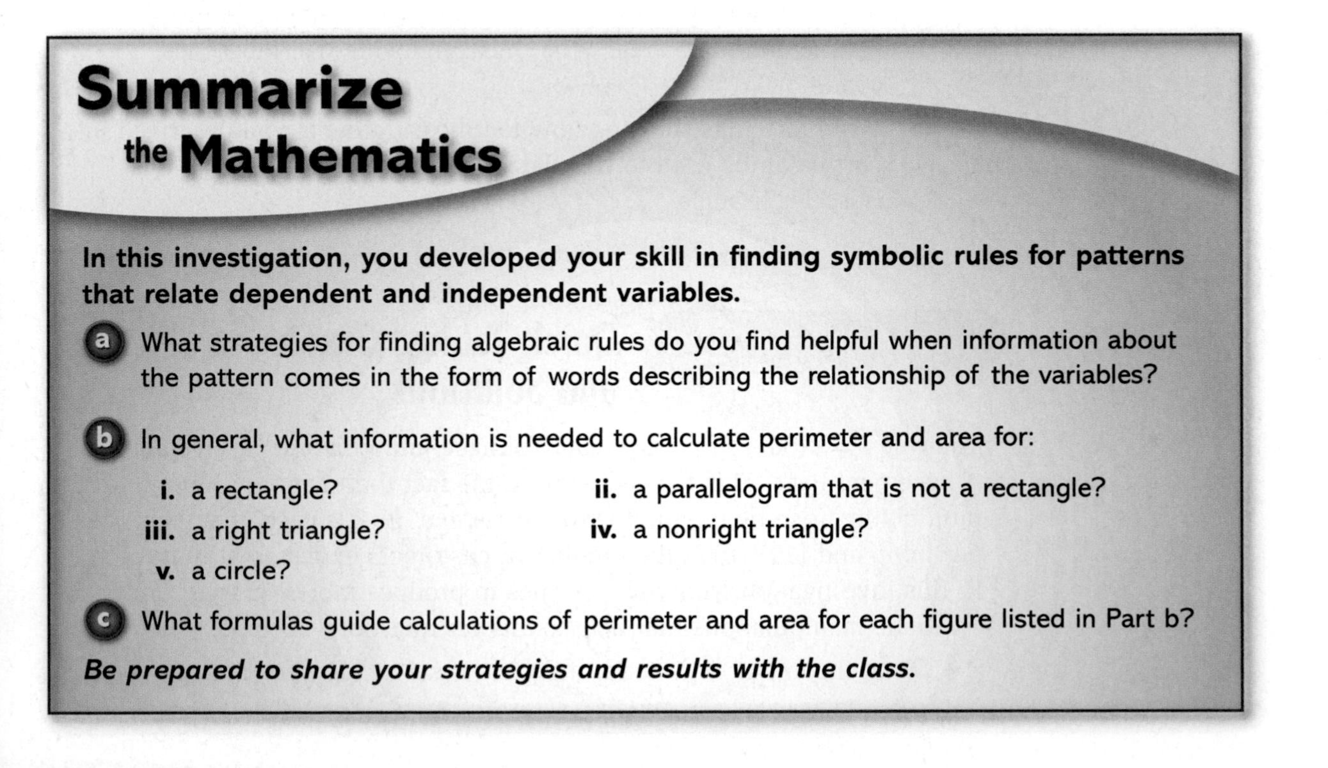

Summarize the Mathematics

In this investigation, you developed your skill in finding symbolic rules for patterns that relate dependent and independent variables.

a What strategies for finding algebraic rules do you find helpful when information about the pattern comes in the form of words describing the relationship of the variables?

b In general, what information is needed to calculate perimeter and area for:

i. a rectangle?
ii. a parallelogram that is not a rectangle?
iii. a right triangle?
iv. a nonright triangle?
v. a circle?

c What formulas guide calculations of perimeter and area for each figure listed in Part b?

Be prepared to share your strategies and results with the class.

Check Your Understanding

Write algebraic rules expressing the relationships in these situations.

a. Students and parents who attend the Banneker High School *Take a Chance* carnival night spend an average of $12 per person playing the various games. Operation of the event costs the student government $200. What is the relationship between profit P of the carnival night and number of people who attend n?

b. The figure drawn below is an isosceles triangle with $KL = LM$.

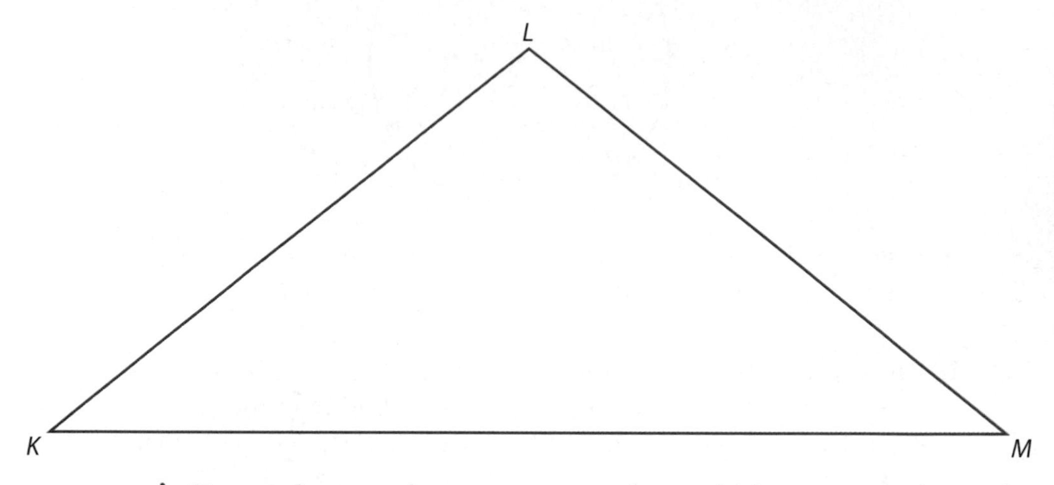

i. Use a ruler to make measurements from which you can estimate the perimeter and area of triangle KLM.

ii. For any isosceles triangle, what is the minimum number of measurements you would need in order to estimate both its perimeter and its area? What measurements will meet that condition?

iii. Write formulas showing how the measurements you described in part ii can be used to calculate perimeter P and area A of an isosceles triangle.

Investigation 2 Quick Tables, Graphs, and Solutions

The rule $I = p(50 - p)$ predicts daily bungee jump income at Five Star Amusement Park. This rule arises from the fact that income is computed by multiplying price by the number of customers. In this case, p is the price per jump and $(50 - p)$ is the number of customers expected at that price. In this investigation, you will use rules to produce tables, graphs, and symbolic manipulations that help to answer questions such as:

- What income is expected if the price is set at $10 per jump?
- What should the price be in order to get income of at least $500 per day?
- What price per jump will produce maximum daily income?

Sometimes you can answer questions like these by doing some simple arithmetic calculations. In other cases, calculators and computers provide useful tools for the work. As you work through the investigation, look for answers to this question:

How can you use calculator or computer tools to produce tables, graphs, and symbolic manipulations, which can help you to study relationships between variables?

1. **Using Tables** You can use computer software or a graphing calculator to produce tables of related values for the independent and dependent variables. For example, examine the table of sample (*price per jump, daily income*) data below.

Producing a Table

Enter Rule | **Set Up Table** | **Display Table**

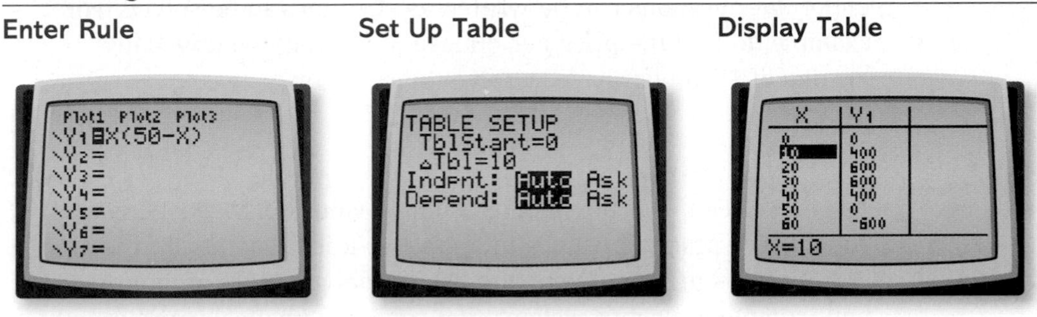

Scanning the table you can see, for example, that with the price set at \$10, Five Star expects a daily bungee jump income of \$400.

Use the software or calculator you have to produce and scan tables for the rule $I = p(50 - p)$ in order to estimate answers for these questions:

a. What daily income will result if the price is set at \$19?

b. To reach a daily income of at least \$500, why should the price be at least \$14, but not more than \$36?

c. What price(s) will yield a daily income of at least \$300?

d. What price will yield the maximum possible daily income?

e. How would you describe the pattern of change in income as price increases from \$0 to \$50 in steps of \$1?

2. **Using Graphs** Computer software and graphing calculators can also be used to produce graphs of relationships between variables. For example, see the graph below for $I = p(50 - p)$ relating *price per jump* and *daily income* for the Five Star bungee jump.

Producing a Graph

Enter Rule | **Set Viewing Window** | **Display Graph**

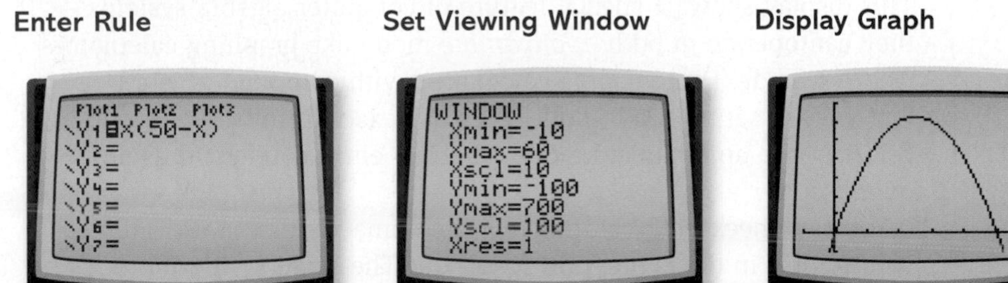

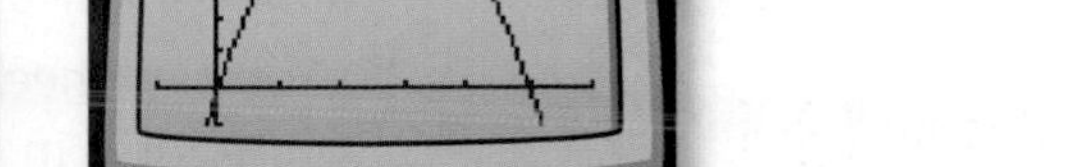

Use the software or calculator you have to produce and trace a graph for $I = p(50 - p)$ and estimate answers to the following questions. In each case, report your results with a sketch that shows how the answer is displayed on the calculator or computer screen.

a. What income is expected if the price is set at $17?

b. What price(s) will lead to a daily income of about $550?

c. How does the predicted income change as the price increases from $0 to $50?

d. What price will lead to maximum daily income from the bungee jump attraction?

3 **Using Computer Algebra Systems** When you can express the connection between two variables with a symbolic rule, many important questions can be written as equations to be solved. For example, to find the price per bungee jump that will give daily income of $500, you have to solve the equation

$$p(50 - p) = 500.$$

As you've seen, it is possible to estimate values of p satisfying this equation by scanning values in a table or tracing points on the graph of $I = p(50 - p)$. Computers and calculators are often programmed with computer algebra systems that solve automatically and exactly. One common form of the required instructions looks like this:

solve(p•(50–p)=500,p)

When you execute the command (often by simply pressing ENTER), you will see the solution(s) displayed in a form similar to that shown below.

Solving an Equation

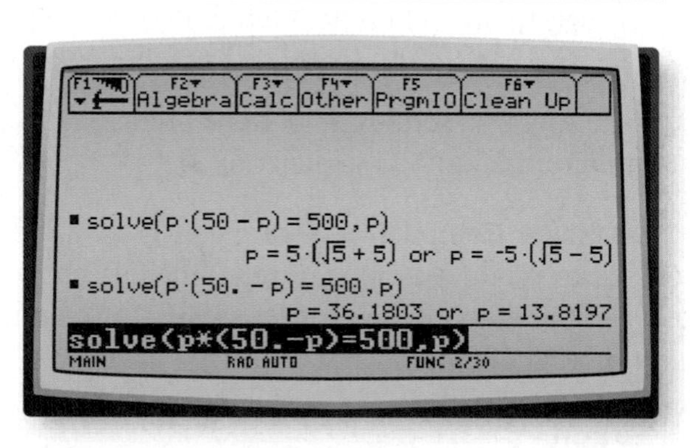

This display shows a special feature of computer algebra systems—they can operate in both *approximate* mode like graphing calculators or *exact* mode. (When some calculators with computer algebra systems are set in AUTO mode, they use exact form where possible. But they use approximate mode when an entry contains a decimal point.)

You can check both solutions with commands that substitute the values for p in the expression $p(50 - p)$. The screen will look something like the following display.

Evaluating an Expression

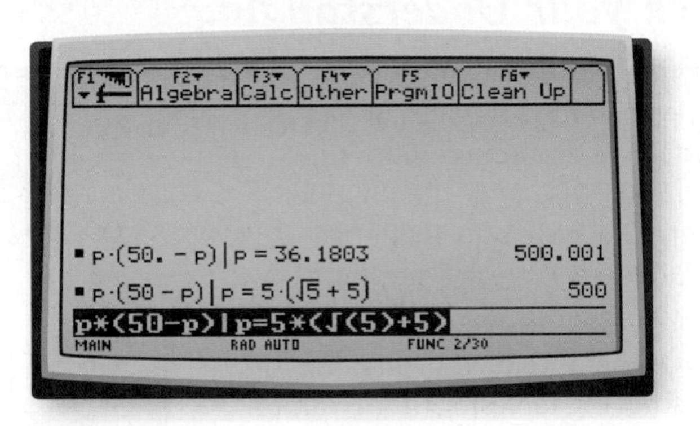

Computer algebra systems can do many other algebraic operations that you will learn about in future study. To get started, modify the instructions illustrated above to answer the following questions. In each case, check your results by using the same computer algebra system, scans of graphing calculator tables or traces of graphs, or arithmetic calculations.

a. What bungee jump price will give a predicted daily income of \$450? An income of \$0?

b. What daily income is predicted for a jump price of \$23? For a jump price of \$42?

c. What question will be answered by solving the equation $p(50 - p) = 225$? What is the answer?

d. How could you solve the equation $p(50 - p) = 0$ just by thinking about the question, "What values of p will make the expression $p(50 - p)$ equal to zero?"

Summarize the Mathematics

In this investigation, you developed skill in use of calculator or computer tools to study relations between variables. You learned how to construct tables and graphs of pairs of values and how to use a computer algebra system to solve equations.

a Suppose that you were given the algebraic rule $y = 5x + \frac{10}{x}$ relating two variables. How could you use that rule to find:

- the value of y when $x = 4$
- the value(s) of x that give $y = 15$

i. using a table of (x, y) values?

ii. using a graph of (x, y) values?

iii. using a computer algebra system?

b What seem to be the strengths and limitations of each tool—table, graph, and computer algebra system—in answering questions about related variables? What do these tools offer that makes problem solving easier than it would be without them?

Be prepared to share your thinking with the class.

Check Your Understanding

Weekly profit at the Starlight Cinema theater depends on the number of theater customers according to the rule $P = 6.5n - 2{,}500$. Use table, graph, and computer algebra system methods to complete Parts a–c. For each question:

- Report the setups you use to answer the questions by making and studying tables of (n, P) values. In each case, give the starting value and step size for the table that shows a satisfactory estimate of the answer.
- Report the window setups used to answer the questions by tracing a graph of the rule $P = 6.5n - 2{,}500$.
- Report the computer algebra system commands used to answer the questions and the results in approximate and exact modes.

a. To find the **break-even point** for the business, you need to find the value of n that produces a value of P equal to 0. That means you have to solve the equation $0 = 6.5n - 2{,}500$. What values of n satisfy that equation?

b. What profit is predicted if the theater has 750 customers in a week?

c. What number of customers will be needed to make a profit of \$1,000 in a week?

d. How could you answer the questions in Parts a–c if the only "tool" you had was your own arithmetic skills or a calculator that could only do the basic operations of arithmetic $(+, -, \times, \div)$?

Investigation 3 The Shapes of Algebra

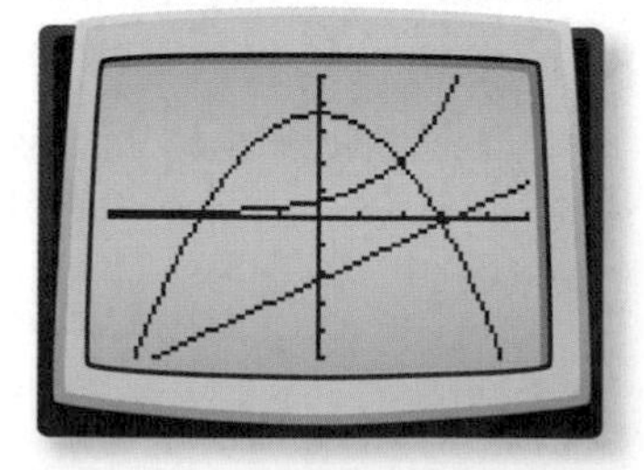

The patterns you discovered while working on problems of earlier investigations illustrate only a few of the many ways that tables, graphs, and algebraic rules are useful in studying relations among variables. To find and use rules that relate independent and dependent variables or that predict change in one variable over time, it helps to be familiar with the table and graph patterns associated with various symbolic forms.

As you work on the explorations of this investigation, look for answers to this question:

How do the forms of algebraic rules give useful information about the patterns in tables and graphs produced by those rules?

You can get ideas about connections between symbolic rules and table and graph patterns by exploration with a graphing calculator or a computer tool. You might find it efficient to share the following explorations among groups in your class and share examples within an exploration among individuals in a group.

In each exploration, you are given several symbolic rules to compare and contrast. To discover similarities and differences among the examples of each exploration:

- For each rule in the set, produce a table of (x, y) values with integer values of x and graphs of (x, y) values for x between –5 and 5.
- Record the table patterns and sketches of the graphs in your notes as shown here for the example $y = 0.5x - 1$.

x	−5	−4	−3	−2	−1	0	1	2	3	4	5
$y = 0.5x - 1$	−3.5	−3.0	...								

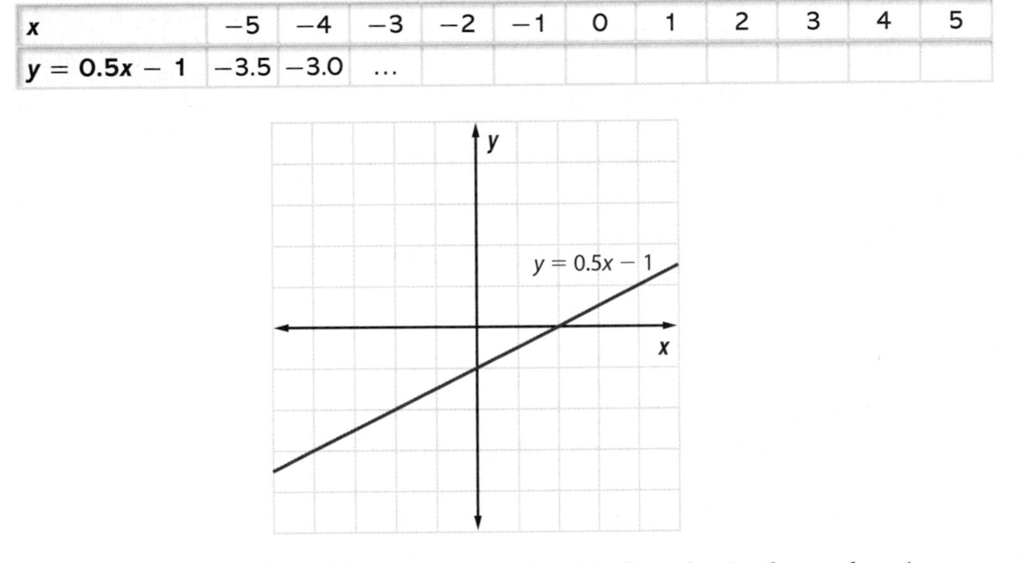

- Then compare the tables, graphs, and symbolic rules in the exploration. Note similarities, differences, and connections between the symbolic rules and the table and graph patterns. Explore some other similar rules to test your ideas.
- Try to explain why the observed connection between rules and table/graph patterns makes sense.

Exploration 1. Compare the patterns of change in tables and graphs for these rules.

a. $y = 2x - 4$ **b.** $y = -0.5x + 2$ **c.** $y = 0.5x + 2$ **d.** $y = \frac{2}{x} - 4$

Exploration 2. Compare the patterns of change in tables and graphs for these rules.

a. $y = x^2$ **b.** $y = 2^x$ **c.** $y = -x^2$ **d.** $y = -x^2 + 2$

Exploration 3. Compare the patterns of change in tables and graphs for these rules.

a. $y = \frac{1}{x}$ **b.** $y = \frac{x}{3}$ **c.** $y = \frac{3}{x}$ **d.** $y = -\frac{5}{x}$

Exploration 4. Compare the patterns of change in tables and graphs for these rules.

a. $y = 3^x$ **b.** $y = x^3$ **c.** $y = 1.5^x$ **d.** $y = 4^x$

Summarize the Mathematics

As a result of the explorations, you probably have some ideas about the patterns in tables of (x, y) values and the shapes of graphs that can be expected for various symbolic rules. Summarize your conjectures in statements like these:

a If we see a rule like … , we expect to get a table like … .

b If we see a rule like … , we expect to get a graph like … .

c If we see a graph pattern like … , we expect to get a table like … .

Be prepared to share your ideas with others in your class.

✓ *Check Your Understanding*

Each item here gives three algebraic rules—one of which will have quite different table and graph patterns than the other two. In each case, spot the "alien" rule and explain how and why its graph and/or table pattern will look different from the other two.

a. $y = \frac{10}{x}$

$y = 10x$

$y = x + 10$

b. $y = x^2 + 1$

$y = x + 1$

$y = 1 - x^2$

c. $y = 1.5x - 4$

$y = (1.5^x) - 4$

$y = 2^x$

d. $y = 1.5x - 4$

$y = 0.5x - 4$

$y = -1.5x - 4$

On Your Own

Applications

1. Members of the LaPorte High School football team have decided to hold a one-day car wash to raise money for trophies and helmet decals. They plan to charge $7.50 per car, but they need to pay $55 for water and cleaning supplies. Write a rule that shows how car wash profit is related to the number of car wash customers.

2. Juan and Tiffany work for their town's park department cutting grass in the summer. They can usually cut an acre of grass in about 2 hours. They have to allow 30 minutes for round-trip travel time from the department equipment shop to a job and back. What rule tells the time required by any job as a function of the number of acres of grass to be cut on that job?

3. When a summer thunder-and-lightning storm is within several miles of your home, you will see the lightning and then hear the thunder produced by that lightning. The lightning travels 300,000 kilometers per second, but the sound of the thunder travels only 330 meters per second. That means that the lightning arrives almost instantly, while the thunder takes measurable time to travel from where the lightning strikes to where you are when you hear it.

   You can estimate your distance from a storm center by counting the time between seeing the lightning and hearing the thunder. What formula calculates your distance from the lightning strike as a function of the time gap between lightning and thunder arrival?

4. Rush Computer Repair makes service calls to solve computer problems. They charge $40 for technician travel to the work site and $55 per hour for time spent working on the problem itself. What symbolic rule shows how the cost of a computer repair depends on actual time required to solve the problem?

5. The freshman class officers at Interlake High School ordered 1,200 fruit bars to sell as a fund-raising project. They paid $0.30 per bar at the time the order was placed. They plan to sell the fruit bars at school games and concerts for $0.75 apiece. No returns of unsold bars are possible. What rule shows how project profit depends on the number of bars sold?

On Your Own

6 Janitorial assistants at Woodward Mall start out earning \$6 per hour. However, the \$75 cost of uniforms is deducted from the pay that they earn.

a. Explain how the rule $E = 6.00h - 75$ shows how a new employee's earnings depend on the number of hours worked.

b. How many hours will a new employee have to work before receiving a paycheck for some positive amount?

c. How many hours will a new employee have to work to earn pay of \$100 before taxes and other withholdings?

d. Sketch a graph of the rule relating pay earned to hours worked, and label points with coordinates that provide answers to Parts b and c.

7 Experiments with a bungee jump suggested the rule $L = 30 + 0.2w$ relating stretched length of the cord (in feet) to weight of the jumper (in pounds).

a. What will be the stretched cord length for a jumper weighing 140 pounds?

b. What jumper weights will stretch the cord to a length of at most 65 feet?

c. Sketch a graph of the cord length relationship and label points with coordinates that give answers to Parts a and b.

d. Study entries in a table of (w, L) values for $w = 0$ to $w = 300$ in steps of 10. Try to figure out what the values 30 and 0.2 tell about the bungee jump experience.

8 When promoters of a special Bruce Springsteen Labor Day concert did some market research, they came up with a rule $N = 15{,}000 - 75p$ relating number of tickets that would be sold to the ticket price.

a. Income from ticket sales is found by multiplying the number of tickets sold by the price of each ticket. The rule $I = p(15{,}000 - 75p)$ shows how *income* depends on *ticket price.*

i. What do the terms p and $(15{,}000 - 75p)$ each tell about how ticket price affects the concert business?

ii. Why does the product give income as a function of ticket price?

b. What ticket price(s) is likely to produce concert income of at least \$550,000?

c. What is the predicted concert income if the ticket price is set at \$30?

d. What ticket price is likely to lead to the greatest concert income?

e. What ticket price(s) will lead to 0 income?

f. Sketch a graph of the relationship between concert income and ticket price. Then label the points with coordinates that provide answers to Parts b, c, and d.

9 When members of the LaPorte High School football team ran their fund-raising car wash, they expected profit to be related to number of cars washed by $P = 7.50n - 55$.

a. If their goal was to earn a $500 profit, how many cars would the team have to wash?

b. How many cars would the team need to wash to break even?

10 Without use of your graphing calculator or computer software, sketch graphs you would expect from these rules. Explain your reasoning in each case.

a. $y = 7x^2 + 4$

b. $y = 7 - \frac{1}{4}x$

c. $y = 4^x - 7$

11 Without use of your graphing calculator or computer software, match the following four rule types to the tables below. Explain your reasoning in each case.

a. $y = ax + b$ b. $y = ax^2 + b$ c. $y = \frac{a}{x}$ d. $y = a^x$

I

x	−4	−3	−2	−1	0	1	2	3	4	5
y	18	11	6	3	2	3	6	11	18	27

II

x	−4	−3	−2	−1	0	1	2	3	4	5
y	16	14	12	10	8	6	4	2	0	−2

III

x	−4	−3	−2	−1	0	1	2	3	4	5
y	0.0625	0.125	0.25	0.5	1	2	4	8	16	32

IV

x	−4	−3	−2	−1	0	1	2	3	4	5
y	−1.5	−2	−3	−6	error	6	3	2	1.5	1.2

12 Without use of your graphing calculator or computer software, match the following four rule types to the graph sketches below. Explain your reasoning in each case.

a. $y = ax + b$ b. $y = ax^2 + b$ c. $y = \frac{a}{x}$ d. $y = a^x$

I

II

III

IV

Connections

13 Three familiar formulas relate circumference and area of any circle to the radius or diameter of the circle. All three involve the number π, which is approximately 3.14.

Circumference: $C = \pi d$ and $C = 2\pi r$

Area: $A = \pi r^2$

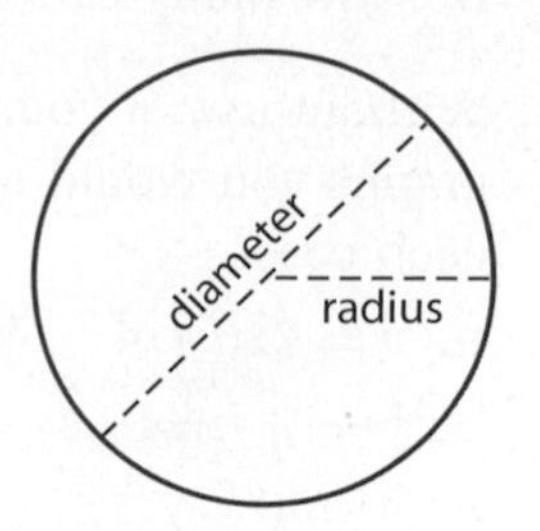

a. Complete a table like the one below to show the pattern of change in circumference and area of a circle as the radius increases.

Radius r	0	1	2	3	4	5	10	20
Circumference C								
Area A								

b. Compare the pattern of change in area to the pattern of change in circumference as radius increases. Explain differences in the patterns of change by comparing the formulas.

c. How will the area change if the radius is doubled? If it is tripled?

d. How will the circumference change if the radius is doubled? If it is tripled?

e. Which change in the size of a circle will cause the greater increase in circumference—doubling the radius or doubling the diameter? Which of those changes will cause the greater increase in area?

f. Tony's Pizza Place advertises 2-item, 10-inch pizzas for $7.95 and 2-item, 12-inch pizzas for $9.95. Which pizza is the better buy?

14 For polygons like triangles and rectangles, the formulas for perimeter and area often involve two variables—usually *base* and *height*.

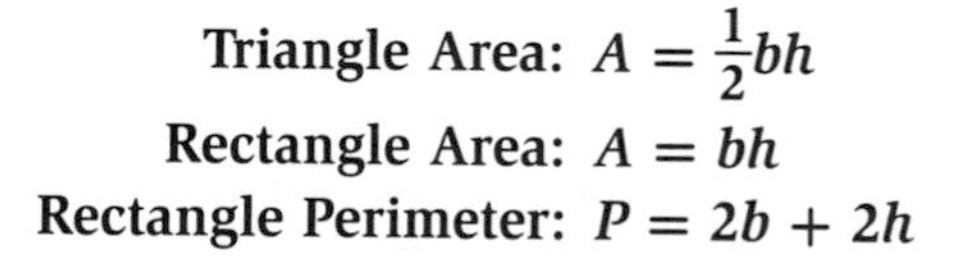

Triangle Area: $A = \frac{1}{2}bh$

Rectangle Area: $A = bh$

Rectangle Perimeter: $P = 2b + 2h$

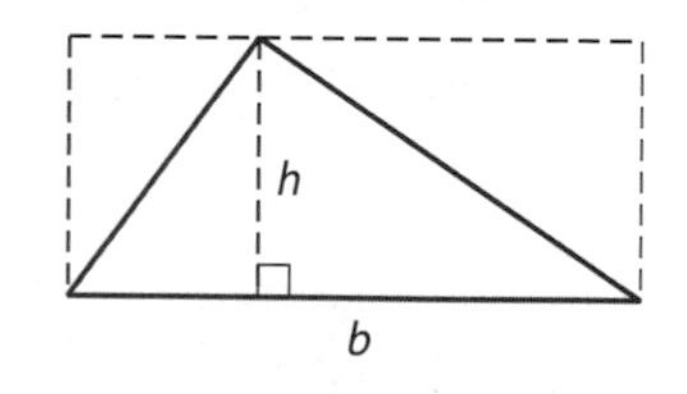

a. Complete entries in a table like the following to give a sample of triangle areas for different base and height values. The base values are in column A; the height values are in row 1; the areas go in cell B2 through J10. Then use the table patterns to answer questions in Parts b and c.

You might find it helpful to write a spreadsheet program to do the calculations.

Triangle Areas.xls

◇	A	B	C	D	E	F	G	H	I	J
1		1	2	3	4	5	6	7	8	9
2	1	0.5	1							
3	2	1	2							
4	3	1.5								
5	4									
6	5									
7	6									
8	7									
9	8									
10	9									
11										

b. Which change in the size of a triangle causes the greater increase in area—doubling the base or doubling the height?

c. How will the area of a triangle change if both the base and the height are doubled? What if both are tripled?

15 Create a table like that in Connections Task 14 to explore patterns of change in area of rectangles for base and height values from 1 to 9. Answer the same questions about the effects of doubling base and height for a rectangle.

16 Create a table like that in Connections Task 14 to explore patterns of change in perimeter of rectangles for base and height values from 1 to 9. Answer similar questions about the effects on perimeter of doubling base and height for a rectangle.

17 To answer Connections Task 14 with a spreadsheet, Mr. Conklin wrote some formulas for a few cells and then used "fill down" and "fill right" to get the rest of the sheet. Check his ideas in Parts a–c and explain why each is correct or not.

a. In cell B1, he entered "1". Then in cell C1, he wrote "=B1+1" and did "fill right" to complete row 1 of the table.

b. In cell A2, he entered "1". Then in cell A3, he wrote "=A2+1" and did "fill down."

c. In cell B2, he wrote "=0.5*$A2*B$1" and then did "fill down" and "fill right."

18 How could the instructions in Connections Task 17 be modified to produce the tables for

a. rectangle area?

b. rectangle perimeter?

19 For each algebraic rule below, use your calculator or computer software to produce a table and then write a rule relating *NOW* and *NEXT* values of the y variable.

a. $y = 5x + 2$ **b.** $y = 10x - 3$

c. $y = 2^x$ **d.** $y = 4^x$

On Your Own

20 A cube is a three-dimensional shape with square faces.

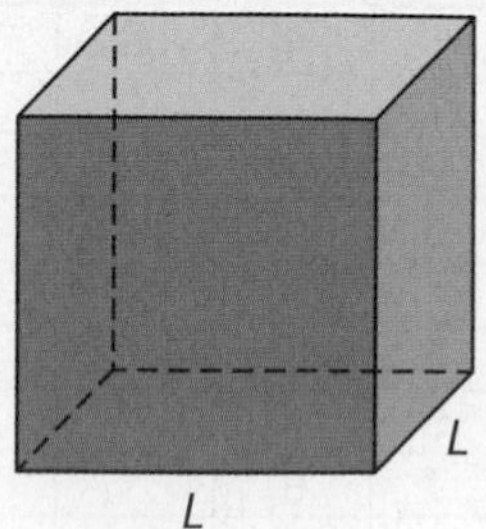

a. If the length of an edge of a cube is L, write an expression for the area of one of its faces.

b. Write a rule that gives the total surface area A of a cube as a function of the length L of an edge.

c. Suppose you wished to design a cube with surface area of 1,000 square centimeters. To the nearest 0.1 centimeter, what should be the length of the edge of the cube?

Reflections

21 If you are asked to write a rule or formula relating variables in a problem, how would you decide:

a. what the variables are?

b. which of the variables seems most natural to be considered the independent variable and which the dependent variable?

c. what symbols should be used as shorthand names for the variables?

d. whether to express the relationship with "$y = \ldots$" or *NOW-NEXT* form?

If you enter the rule $y = 5x + 100$ in your calculator and press the GRAPH key, you might at first find no part of the graph on your screen. The plotted points may not appear in your graphing window. Talk with others in your class about strategies for making good window choices. Write down good ideas as a reminder to yourself and as a help to others.

Look back at your work for Part c of Connections Task 20.

a. What technology tool, if any, did you use in answering that question? How did you decide to use that tool?

b. How could you answer Part c using only the arithmetic capabilities of your calculator?

Suppose that you were asked to answer the following questions about a relationship between variables given by $y = 3.4x + 5$. Explain the tool *you* would choose for answering each question—calculation in your head, arithmetic with a calculator, study of a calculator- or computer-produced table of (x, y) values, study of a calculator- or computer-produced graph, or use of a computer algebra system command. Also, explain how you would use the tool.

a. Do the values of y increase or decrease as values of x increase?

b. How rapidly do the values of y change as the values of x increase?

c. What is the value of y when $x = 7.5$?

d. What is the value of x when $y = 23.8$?

Extensions

25 The following sketches show the first four stages in a geometric pattern of rectangular grids made up of unit squares.

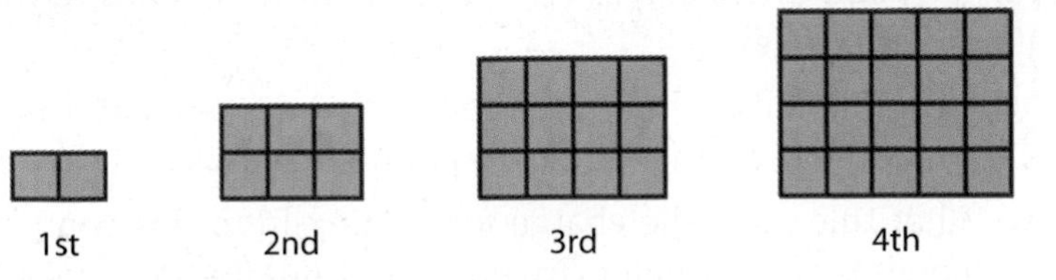

a. Describe geometrically how the grids change from one stage to the next.

b. What is the perimeter of the 5th rectangle?

c. What is the perimeter of the nth rectangle?

d. What is the area of the 5th rectangle?

e. What is the area of the nth rectangle?

26 Two different civic groups operate concession stands during games at the local minor-league baseball stadium. Group A sells hot dogs and soft drinks. Their profit P_A depends on the number of customers m and is given by $P_A = 3m - 100$. Group B sells ice cream. Their profit P_B depends on the number of customers n and is given by $P_B = 2n - 40$.

a. What are the break-even numbers of customers for each concession stand?

b. Is there any number of customers for which both stands make the same profit?

c. Which stand is likely to make the greater profit when the game draws a small crowd? When the game draws a large crowd?

27 Metro Cab Company charges a base price of \$1.50 plus 80¢ per mile. A competitor, Tack See Inc., charges a base price of \$2.50 plus 60¢ per mile.

a. What rules give the charge for a trip with each company as a function of the length of the trip?

b. If you need to travel 3 miles, which cab company is the least expensive?

c. If you need to travel 15 miles, which cab company is the least expensive?

d. For what trip length are the costs the same for the two cab companies?

28 Suppose, as part of an agreement with her father to do some work for him during the summer, Tanya will receive 2¢ for the first day of work, but every day after that her pay will double.

a. What rule shows how to calculate Tanya's daily pay p on work day n.

b. What rule using *NOW* and *NEXT* shows how Tanya's pay grows as each additional day of work passes?

c. If Tanya's pay for a day is $10.24, how many days has she worked?

d. Find Tanya's pay for a day after she has worked 20 days.

e. For how many days will she earn less than $20 per day?

29 One car rental company charges $35 per day, gives 100 free miles per day, and then charges 35¢ per mile for any miles beyond the first 100 miles per day.

a. What rule gives the charge for renting for one day from this company as a function of the number of miles m driven that day?

b. What rule gives the charge for renting from this company as a function of both miles driven m and number of days d?

c. A business person plans a trip of 300 miles that could be made in one day. However, she would arrive home late and is considering keeping the rental car until the next morning. What would you suggest? Explain your reasoning.

30 At the start of a match race for two late-model stock cars, one stalls and has to be pushed to the pits for repairs. The other car roars off at an average speed of 2.5 miles per minute. After 5 minutes of repair work, the second car hits the track and maintains an average speed of 2.8 miles per minute.

a. How far apart in the race are the two cars 5 minutes after the start? How far apart are they 10 minutes after the start?

b. What three rules can be used to calculate the distance traveled by each car and the distance between the two cars at any time after the start of the race?

c. On the same coordinate axes, make graphs displaying the distances traveled by each car as a function of time. Use a horizontal scale that allows you to see the first 60 minutes of the race.

d. On a different set of axes, make another graph showing the distance between the two cars as a function of time since the start of the race.

e. Explain what the patterns of the graphs in Parts c and d show about the progress of the race.

f. Write and solve equations that will answer each of these questions about the race:

i. How long after the start of the race will it take the first car to travel 75 miles? The second car?

ii. At what time after the start of the race will the second car catch up to the first car?

Review

31 Evaluate each expression if $x = 1$, $y = 3$, $a = -1$, and $b = 2$.

a. $a^2x + b^3y$

b. $a^2(x + by)$

c. $\frac{x^3(y + 1)}{by + a^3}$

32 A random sample of 100 students is chosen to survey about lunch preferences.

25 say their first choice is pizza.

30 say their first choice is chicken nuggets.

15 say their first choice is salad bar.

20 say their first choice is tacos.

10 say their first choice is subs.

If the entire school population is 1,500 students, how many students can you predict will have pizza as a first choice?

33 Estimate the measure of each angle. Check your estimates with a protractor.

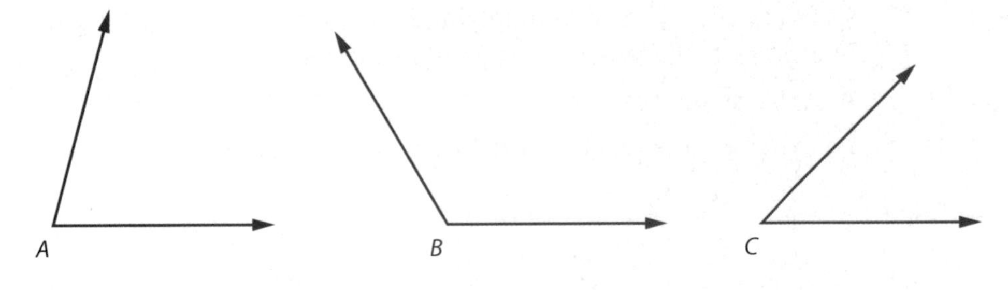

34 The dot plot below indicates the number of students in the 40 first-hour classes at Lincoln High School.

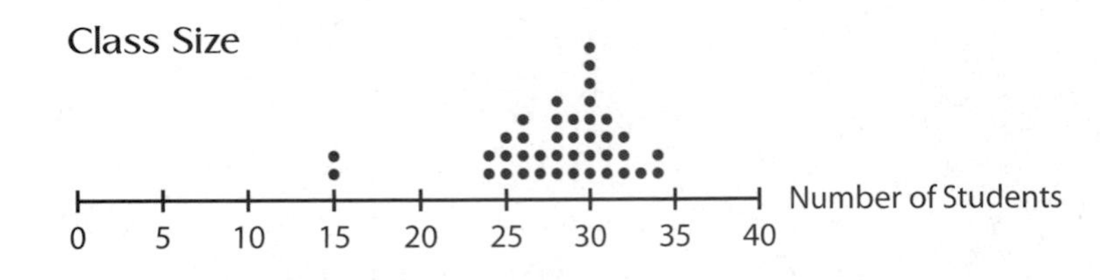

a. What was the smallest class size?

b. What was the largest class size?

c. What percent of the classes had 30 students in them?

d. What percent of the students had fewer than 25 students in them?

On Your Own

35 Consider the triangles drawn below. Assume that angles that look like right angles are right angles and that segments that appear to be the same length are the same length.

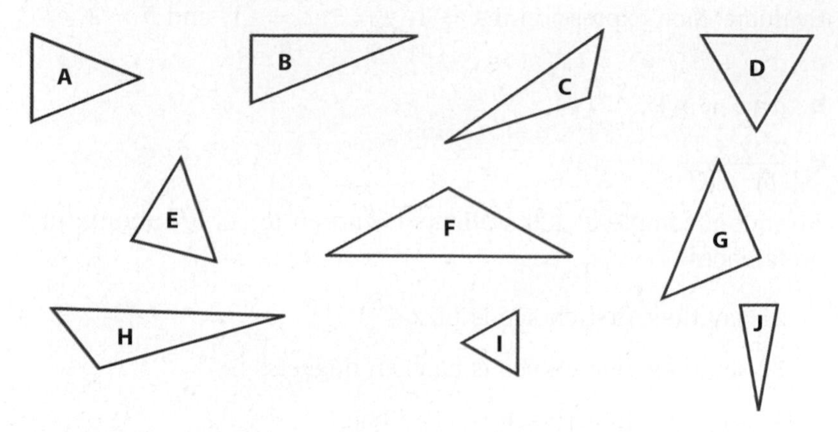

a. Identify all acute triangles.

b. Identify all obtuse triangles.

c. Identify all isosceles triangles.

d. Identify all scalene triangles.

e. Identify all equilateral triangles.

f. Identify all right triangles.

36 The lengths of the sides of a triangle are 4, 5, and 6 inches. These sides are scaled up by multiplying by a factor so that the length of the longest side of the new triangle is 10 inches.

a. What is the scale factor?

b. What are the lengths of the two shorter sides of the new triangle?

Looking Back

In your work on problems and explorations of this unit, you studied many different patterns of change in variables. In some cases, the aim was to describe and predict patterns of change in a dependent variable that are caused by change in values of an independent variable. In other cases, the goal was to describe and predict patterns of change in values of a single variable with the passage of time. For example:

For each weight attached to a bungee cord, there was a predicted stretched length for the cord.

For each year after the census in 2001, there was a predicted population of Alaskan bowhead whales.

For each possible price for a bungee jump at Five Star Amusement Park, there was a predicted daily income from the attraction.

In mathematics, relations like these—where each possible value of one variable is associated with exactly one value of another variable—are called **functions**. The use of the word "function" comes from the common English phrase that appears in statements like "cord stretch *is a function of* attached weight" or "average speed for the Iditarod Sled Race *is a function of* time to complete the race."

Many functions of interest to mathematicians have no particular cause-and-effect or change-over-time story attached. The only condition required for a relationship to be called a function is that each possible value of the independent variable is paired with one value of the dependent variable.

As a result of your work on Lessons 1–3, you should be better able to:

- recognize situations in which variables are related in predictable ways,
- use data tables and graphs to display patterns in those relationships,
- use symbolic rules to describe and reason about functions, and
- use spreadsheets, computer algebra systems, and graphing calculators to answer questions about functions.

The tasks in this final lesson of the *Patterns of Change* unit will help you review, pull together, and apply your new knowledge as you work to solve several new problems.

1 **Five Star Swimming** In addition to bungee jumping and rides like roller coasters and a Ferris wheel, Five Star Amusement Park has a large lake with a swimming beach and picnic tables.

Every spring when the park is preparing to open, lifeguards at the beach put out a rope with buoys outlining the swimming space in the lake. They have 1,000 feet of rope, and they generally outline a rectangular swimming space like that shown below.

When working on this task one year, the lifeguards wondered whether there was a way to choose dimensions of the rectangular swimming space that would provide maximum area for swimmers.

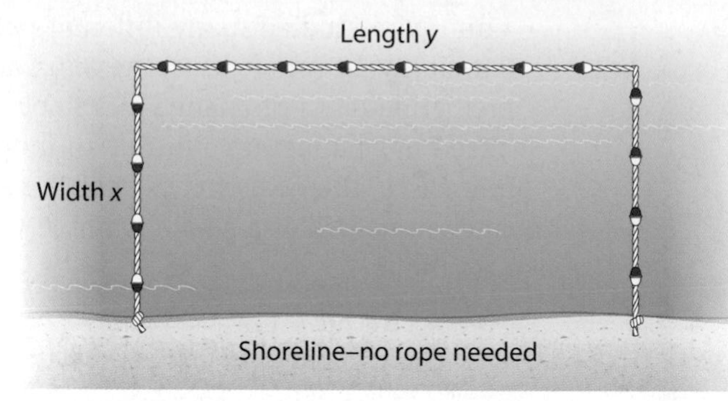

a. Complete entries in a table like this, showing how dimensions of the swimming space are related to each other. Then write a rule giving y as a function of x.

Width *x* (in feet)	50	100	150	200	250	300	350	400	450
Length *y* (in feet)									

b. One of the lifeguards claimed that the rule in Part a can be used to write another rule that shows how area A of the swimming space depends on choice of the width x. She said that $A = x(1{,}000 - 2x)$ would do the job. Is she right? How do you think she arrived at this area rule?

c. Use the area function in Part b and strategies you have for reasoning about such relationships to answer the following questions. To show what you've learned about using different tools for studying functions:

- Answer one question by producing and scanning entries in a table of values for the area function.
- Answer another question by producing and tracing coordinates of points on a graph of the function.
- Then answer another question using a computer algebra system equation solver.

When you report your results, explain your strategies as well.

i. What dimensions of the swimming space will give maximum area? What is that area?

ii. What dimensions will give a swimming area of 100,000 square feet?

iii. What dimensions will give a swimming area of 50,000 square feet?

2 **Borrowing to Expand** When the Five Star Amusement Park owners decided to expand park attractions by adding a new giant roller coaster, they borrowed \$600,000 from a local bank. Terms of the bank loan said that each month interest of 0.5% would be added to the outstanding balance, and the park would have to make monthly payments of \$10,000. For example, at the end of the first month of the loan period, the park would owe $600{,}000 + 0.005(600{,}000) - 10{,}000 = \$593{,}000$.

a. Make a table showing what the park owes the bank at the end of each of the first 12 months.

b. Write a *NOW-NEXT* rule that shows how the loan balance changes from one month to the next.

c. Use a calculator or spreadsheet strategy to find out how long it will take to pay off the loan.

d. Plot a graph showing the amount owed on the loan at the end of months 0, 6, 12, 18, ... until it is paid off. Describe the pattern of change in loan balance over that time.

3 **Setting the Price** Because Five Star managers expected the new roller coaster to be a big attraction, they planned to set a high price for riders. They were unsure about just what that price should be. They decided to do some market research to get data about the relationship between *price per ride* and *number of riders* that would be expected each day.

a. Complete a table that shows how you believe the *number of riders* will depend on the *price per ride*. Explain the pattern of entries you make and your reasons for choosing that pattern.

Price per Ride (in dollars)	0	5	10	15	20	25	30	35
Number of Riders								

b. Add a third row to the table in Part a to give the predicted *income* from the new roller coaster for each possible *price per ride*. Then plot the (*price per ride, income*) data and describe the pattern of change relating those variables.

c. Estimate the price per ride that will give maximum daily income.

d. What factors other than price are likely to affect daily income from the roller coaster ride? How do you think each factor will affect income?

4 Without using a graphing calculator or doing any calculation of (x, y) values, match each of the following functions with the graph that best represents it.

a. $y = -0.5x - 4$ **b.** $y = x^2 - 4$ **c.** $y = \frac{4}{x}$ **d.** $y = (1.5^x)$

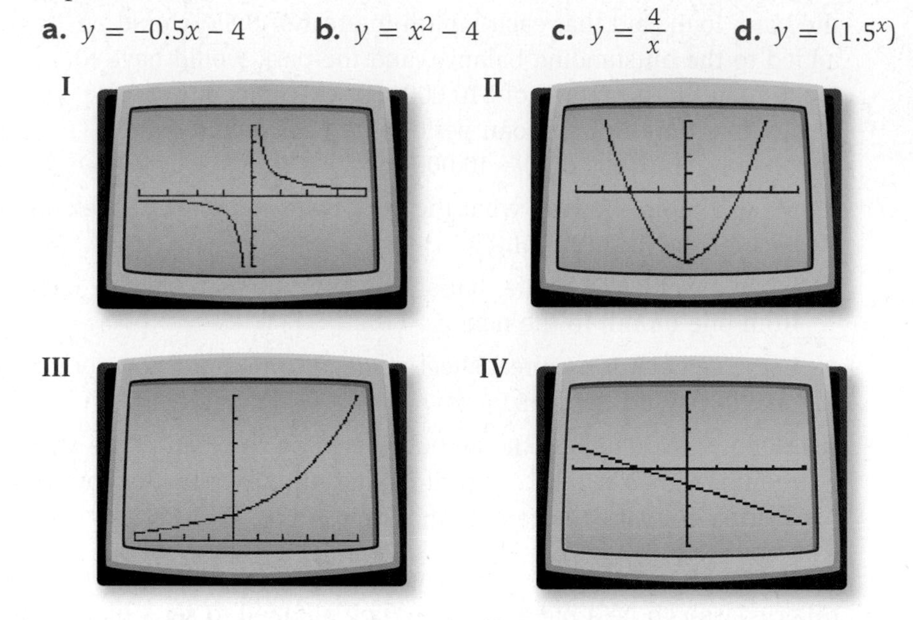

Summarize the Mathematics

When two variables change in relation to each other, the pattern of change often fits one of several common forms. These patterns can be recognized in tables and graphs of (x, y) data, in the rules that show how to calculate values of one variable from given values of the other, and in the conditions of problem situations.

a Sketch at least four graphs showing different patterns relating change in two variables or change in one variable over time. For each graph, write a brief explanation of the pattern shown in the graph and describe a problem situation that involves the pattern.

b Suppose that you develop or discover a rule that shows how a variable y is a function of another variable x. Describe the different strategies you could use to:

- **i.** Find the value of y associated with a specific given value of x.
- **ii.** Find the value of x that gives a specific target value of y.
- **iii.** Describe the way that the value of y changes as the value of x increases or decreases.
- **iv.** Find values of x that give maximum or minimum values of y.

Be prepared to share your ideas and reasoning with the class.

Check Your Understanding

Write, in outline form, a summary of the important mathematical concepts and methods developed in this unit. Organize your summary so that it can be used as a quick reference in future units and courses.

PATTERNS IN DATA

In this unit of *Core-Plus Mathematics*, you will explore principles and techniques for organizing and summarizing data. Whether the data are the results of a science experiment, from a test of a new medical procedure, from a political poll, or from a survey of what consumers prefer, the basic principles and techniques of analyzing data are much the same: make a plot of the data; describe its shape, center, and spread with numbers and words; and interpret your results in the context of the situation. If you have more than one distribution, compare them.

Key ideas of data analysis will be developed through your work in two lessons.

Lessons

1 *Exploring Distributions*

Plot single-variable data using dot plots, histograms, and relative frequency histograms. Describe the shape and center of distributions.

2 *Measuring Variability*

Calculate and interpret percentiles, quartiles, deviations from the mean, and standard deviation. Calculate and interpret the five-number summary and interquartile range and construct and interpret box plots. Predict the effect of linear transformations on the shape, center, and spread of a distribution.

Exploring Distributions

The statistical approach to problem solving includes refining the question you want to answer, designing a study, collecting the data, analyzing the data collected, and reporting your conclusions in the context of the original question. For example, consider the problem described below.

A Core-Plus Mathematics teacher in Traverse City, Michigan, was interested in whether eye-hand coordination is better when students use their dominant hand than when they use their nondominant hand. She refined this problem to the specific question of whether students can stack more pennies when they use their dominant hand than when they use their nondominant hand. In her first-hour class, she posed the question:

How many pennies can you stack using your dominant hand?

In her second-hour class, she posed this question:

How many pennies can you stack using your nondominant hand?

In both classes, students were told: "You can touch pennies only with the one hand you are using; you have to place each penny on the stack without touching others; and once you let go of a penny, it cannot be moved. Your score is the number of pennies you had stacked before a penny falls."

Students in each class counted the number of pennies they stacked and prepared a plot of their data. The plot from the first-hour class is shown below. A value on the line between two bars (such as stacking 24 pennies) goes into the bar on the right.

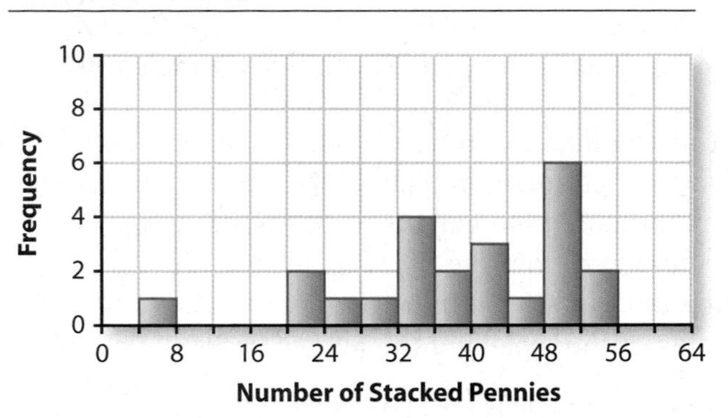

Think About This Situation

Examine the distribution of the number of pennies stacked by students in the first-hour class using their dominant hand.

a How many students were in the first-hour class? What percentage of the students stacked 40 or more pennies using their dominant hand?

b What do you think the plot for the second-hour class might look like?

c Check your conjecture in Part b by having your class stack pennies using your nondominant hands. Make a plot of the numbers stacked by your class using the same scale as that for the dominant hand plot above.

d Compare the shape, center, and spread of the plot from your class with the plot of the first-hour class on the previous page. What conclusions, if any, can you draw?

e Why might comparing the results of first- and second-hour students not give a good answer to this teacher's question? Can you suggest a better design for her study?

In this lesson, you will learn how to make and interpret graphical displays of data so they can help you make decisions involving data.

Investigation 1 Shapes of Distributions

Every day, people are bombarded by data on television, on the Internet, in newspapers, and in magazines. For example, states release report cards for schools and statistics on crime and unemployment, and sports writers report batting averages and shooting percentages. Making sense of data is important in everyday life and in most professions today. Often a first step to understanding data is to analyze a plot of the data. As you work on the problems in this investigation, look for answers to this question:

How can you produce and interpret plots of data and use those plots to compare distributions?

1 As part of an effort to study the wild black bear population in Minnesota, Department of Natural Resources staff anesthetized and then measured the lengths of 143 black bears. (The length of a bear is measured from the tip of its nose to the tip of its tail.) The following **dot plots** (or *number line plots*) show the distributions of the lengths of the male and the female bears.

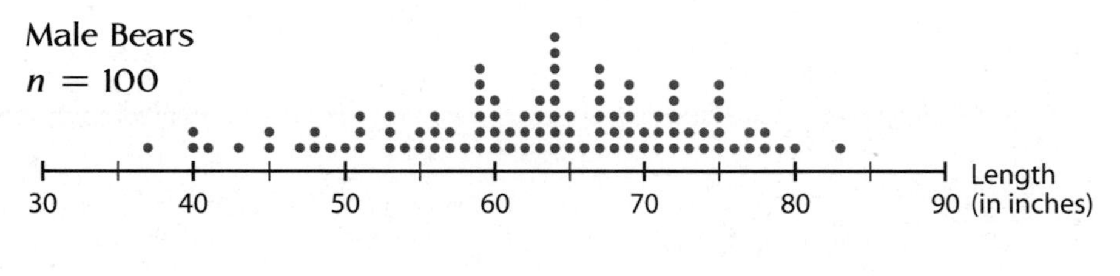

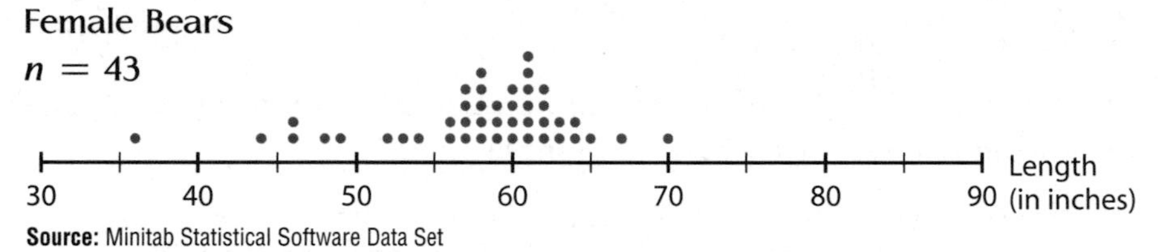

Source: Minitab Statistical Software Data Set

a. Compare the shapes of the two distributions. When asked to *compare*, you should discuss the similarities and differences between the two distributions, not just describe each one separately.

i. Are the shapes of the two distributions fundamentally alike or fundamentally different?

ii. How would you describe the shapes?

b. Are there any lengths that fall outside the overall pattern of either distribution?

c. Compare the centers of the two distributions.

d. Compare the spreads of the two distributions.

2 When describing a distribution, it is important to include information about its *shape*, its *center*, and its *spread*.

a. *Describing shape.* Some distributions are **approximately normal** or *mound-shaped,* where the distribution has one peak and tapers off on both sides. Normal distributions are **symmetric**—the two halves look like mirror images of each other. Some distributions have a **tail** stretching towards the larger values. These distributions are called **skewed to the right** or **skewed toward the larger values**. Distributions that have a tail stretching toward the smaller values are called **skewed to the left** or **skewed toward the smaller values**.

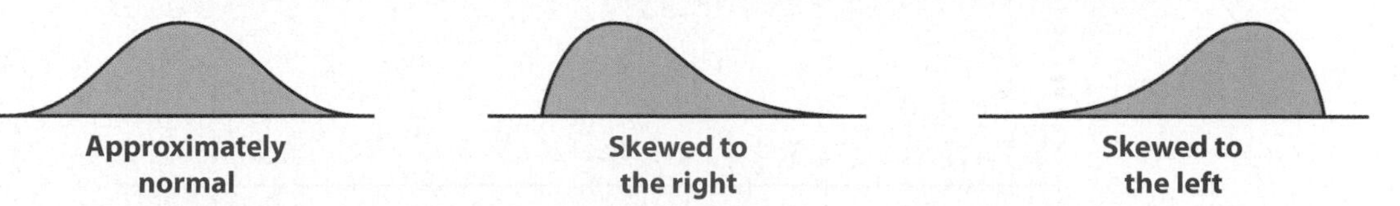

A description of shape should include whether there are two or more *clusters* separated by gaps and whether there are *outliers*. **Outliers** are unusually large or small values that fall outside the overall pattern.

- How would you use the ideas of skewness and outliers to describe the shape of the distribution of lengths of female black bears in Problem 1?

b. *Describing center.* The measure of center that you are most familiar with is the *mean* (or average).

- How could you estimate the mean length of the female black bears?

c. *Describing spread.* You may also already know one measure of spread, the **range**, which is the difference between the maximum value and the minimum value:

$$range = maximum\ value - minimum\ value$$

- What is the range of lengths of the female black bears?

d. Use these ideas of shape, center, and spread to describe the distribution of lengths of the male black bears.

Measures of center (mean and median) and measures of spread (such as the range) are called **summary statistics** because they help to summarize the information in a distribution.

In the late 1940s, scientists discovered how to create rain in times of drought. The technique, dropping chemicals into clouds, is called "cloud seeding." The chemicals cause ice particles to form, which become heavy enough to fall out of the clouds as rain.

To test how well silver nitrate works in causing rain, 25 out of 50 clouds were selected at random to be seeded with silver nitrate. The remaining 25 clouds were not seeded. The amount of rainfall from each cloud was measured and recorded in acre-feet (the amount of water to cover an acre 1 foot deep). The results are given in the following dot plots.

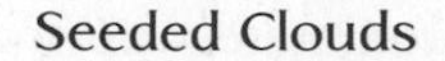

Seeded Clouds

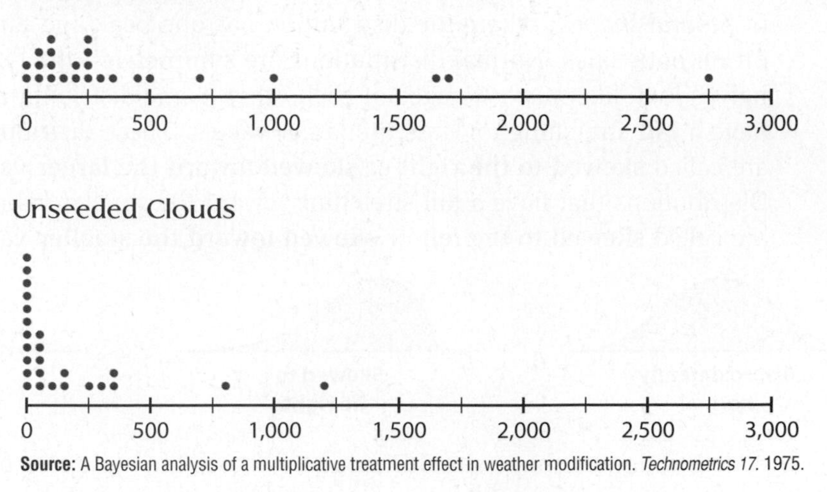

Source: A Bayesian analysis of a multiplicative treatment effect in weather modification. *Technometrics 17.* 1975.

a. Describe the shapes of these two distributions.

b. Which distribution has the larger mean?

c. Which distribution has the larger spread in the values?

d. Does it appear that the silver nitrate was effective in causing more rain? Explain.

Dot plots can be used to get quick visual displays of data. They enable you to see patterns or unusual features in the data. They are most useful when working with small sets of data. **Histograms** can be used with data sets of any size. In a histogram, the horizontal axis is marked with a numerical scale. The height of each bar represents the **frequency** (count of how many values are in that bar). A value on the line between two bars (such as 100 on the following histogram) is counted in the bar on the right.

4 *Pollstar* estimates that revenue from all major North American concerts in 2005 was about $3.1 billion. The histogram below shows the average ticket price for the top 20 North American concert tours.

Concert Tours

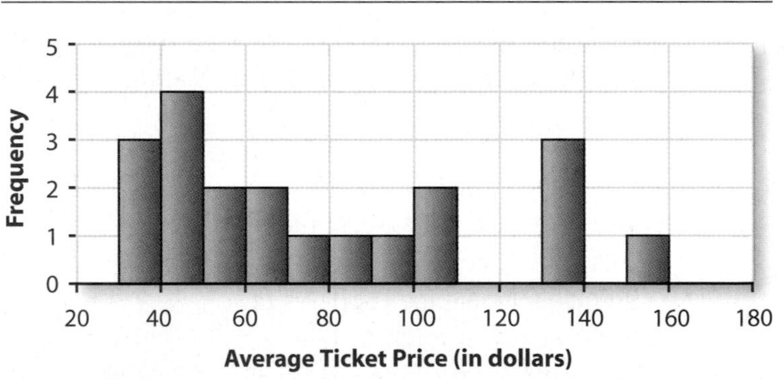

Source: www.pollstaronline.com

a. For how many of the concert tours was the average price $100 or more?

b. Barry Manilow had the highest average ticket price.

  i. In what interval does that price fall?

ii. The 147,470 people who went to Barry Manilow concerts paid an average ticket price of $153.93. What was the total amount paid (*gross*) for all of the tickets?

c. The lowest average ticket price was for Rascal Flatts.

i. In what interval does that price fall?

ii. Their concert tour sold 807,560 tickets and had a gross of $28,199,995. What was the average price of a ticket to one of their concerts?

d. Describe the distribution of these average concert ticket prices.

5 Sometimes it is useful to display data showing the percentage or proportion of the data values that fall into each category. A **relative frequency histogram** has the proportion or percentage that fall into each bar on the vertical axis rather than the frequency or count. Shown below is the start of a relative frequency histogram for the average concert ticket prices in Problem 4.

a. Since prices between $30 and $40 happened 3 out of 20 times, the relative frequency for the first bar is $\frac{3}{20}$ or 0.15. Complete a copy of the table and relative frequency histogram. Just as with the histogram, an average price of $50 goes into the interval 50–60 in the table.

Average Price (in $)	Frequency	Relative Frequency
30–40	3	$\frac{3}{20} = 0.15$
40–50		
50–60		
60–70		
70–80		
80–90		
90–100		
100–110		
110–120		
120–130		
130–140		
140–150		
150–160		
Total		

Concert Tours

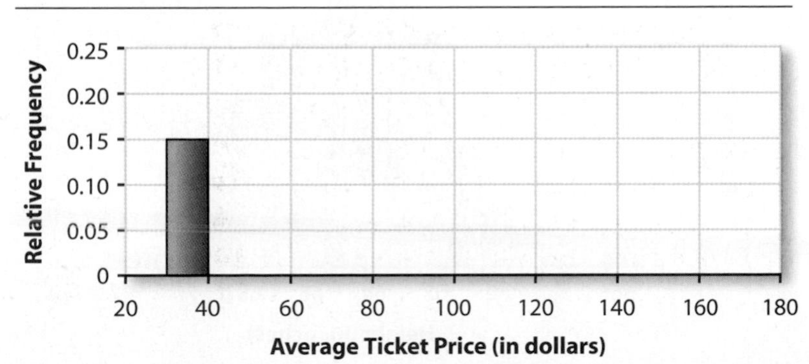

b. When would it be better to use a relative frequency histogram for the average concert ticket prices rather than a histogram?

6 To study connections between a histogram and the corresponding relative frequency histogram, consider the histogram below showing Kyle's 20 homework grades for a semester. Notice that since each bar represents a single whole number (6, 7, 8, 9, or 10), those numbers are best placed in the middle of the bars on the horizontal axis. In this case, Kyle has one grade of 6 and five grades of 7.

a. Make a relative frequency histogram of these grades by copying the histogram but making a scale that shows proportion of all grades on the vertical axis rather than frequency.

b. Compare the shape, center, and spread of the two histograms.

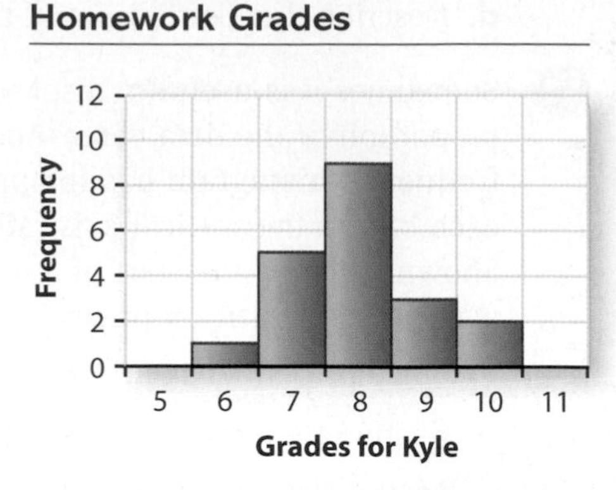

7 The relative frequency histograms below show the heights (rounded to the nearest inch) of large samples of young adult men and women in the United States.

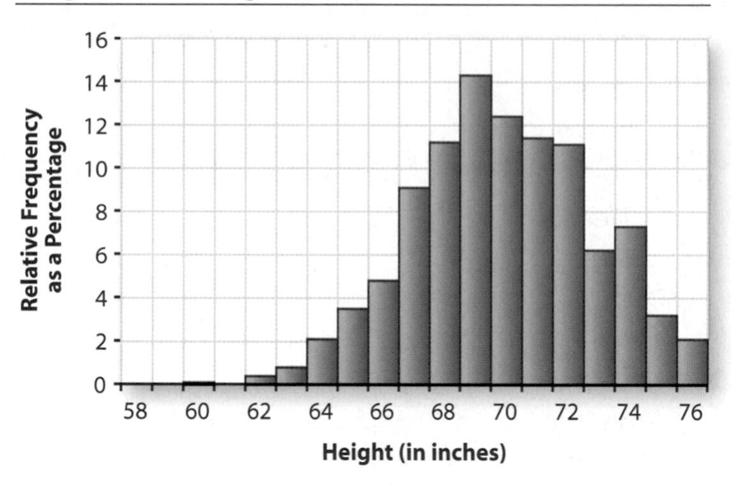

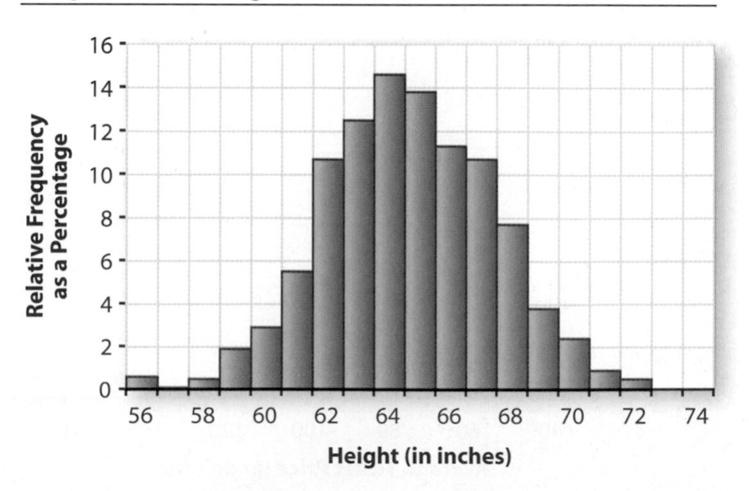

a. About what percentage of these young men are 6 feet tall? About what percentage are at least 6 feet tall?

b. About what percentage of these young women are 6 feet tall? About what percentage are 5 feet tall or less?

c. If there are 5,000 young men in this sample, how many are 5 feet, 9 inches tall? If there are 5,000 young women in this sample, how many are 5 feet, 9 inches tall?

d. Walt Disney World recently advertised for singers to perform in *Beauty and the Beast—Live on Stage*. Two positions were Belle, with height 5'5"–5'8", and Gaston, with height 6'1" or taller. What percentage of these young women would meet the height requirements for Belle? What percentage of these young men would meet the height requirements for Gaston? (Source: corporate.disney.go.com/auditions/disneyworld/roles_dancersinger.html)

Producing a graphical display is the first step toward understanding data. You can use data analysis software or a graphing calculator to produce histograms and other plots of data. This generally requires the following three steps.

- After clearing any unwanted data, enter your data into a list or lists.
- Select the type of plot desired.
- Set a *viewing window* for the plot. This is usually done by specifying the minimum and maximum values and scale on the horizontal (x) axis. Depending on the type of plot, you may also need to specify the minimum and maximum values and scale on the vertical (y) axis. Some calculators and statistical software will do this automatically, or you can use a command such as **ZoomStat**.

Examples of the screens involved are shown here. Your calculator or software may look different.

Producing a Plot

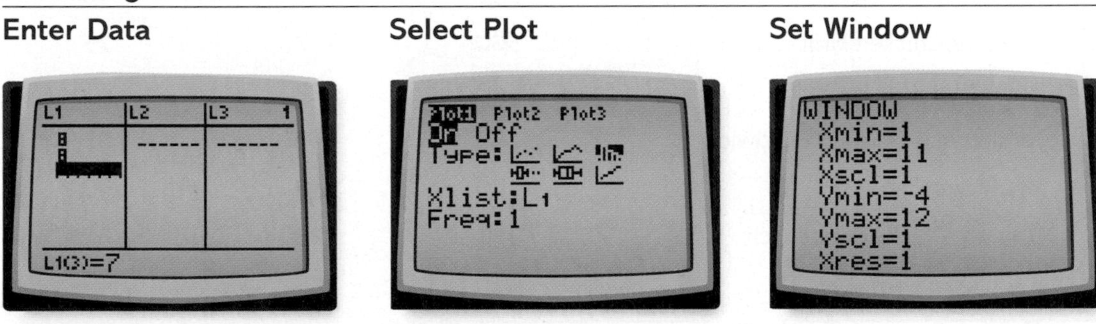

Choosing the width of the bars (Xscl) for a histogram determines the number of bars. In the next problem, you will examine several possible histograms of the same set of data and decide which you think is best.

8 The following table gives nutritional information about some fast-food sandwiches: total calories, amount of fat in grams, and amount of cholesterol in milligrams.

a. Use your calculator or data analysis software to make a histogram of the total calories for the sandwiches listed. Use the values Xmin = 300, Xmax = 1100, Xscl = 100, Ymin = −2, Ymax = 10, and Yscl = 1. Experiment with different choices of Xscl. Which values of Xscl give a good picture of the distribution?

b. Describe the shape, center, and spread of the distribution.

c. If your calculator or software has a "Trace" feature, use it to display values as you move the cursor along the histogram. What information is given for each bar?

d. Investigate if your calculator or data analysis software can create a relative frequency histogram.

How Fast-Food Sandwiches Compare

Company	Sandwich	Total Calories	Fat (in grams)	Cholesterol (in mg)
McDonald's	Cheeseburger	310	12	40
Wendy's	Jr. Cheeseburger	320	13	40
McDonald's	Quarter Pounder	420	18	70
McDonald's	Big Mac	560	30	80
Burger King	Whopper Jr.	390	22	45
Wendy's	Big Bacon Classic	580	29	95
Burger King	Whopper	700	42	85
Hardee's	1/3 lb Cheeseburger	680	39	90
Burger King	Double Whopper w/Cheese	1,060	69	185
Hardee's	Charbroiled Chicken Sandwich	590	26	80
Hardee's	Regular Roast Beef	330	16	40
Wendy's	Ultimate Chicken Grill	360	7	75
Wendy's	Homestyle Chicken Fillet	540	22	55
Burger King	Tendercrisp Chicken Sandwich	780	45	55
McDonald's	McChicken	370	16	50
Burger King	Original Chicken Sandwich	560	28	60
Subway	6" Chicken Parmesan	510	18	40
Subway	6" Oven Roasted Chicken Breast	330	5	45
Arby's	Regular Roast Beef	320	13	45
Arby's	Super Roast beef	440	19	45

Source: *McDonald's Nutrition Facts*, McDonald's Corporation, 2005; *U.S. Nutrition Information*, Wendy's International, Inc., 2005; *Nutrition Data*, Burger King Corp., 2005; *Nutrition*, Hardee's Food Systems, Inc., 2005; *Subway Nutrition Facts-US*, Subway, 2005; *Arby's Nutrition Information*, Arby's, Inc., 2005.

9 Now consider the amounts of cholesterol in the fast-food sandwiches.

a. Make a histogram of the amounts. Experiment with setting a viewing window to get a good picture of the distribution.

b. Describe the distribution of the amount of cholesterol in these sandwiches.

c. What stands out as the most important thing to know for someone who is watching cholesterol intake?

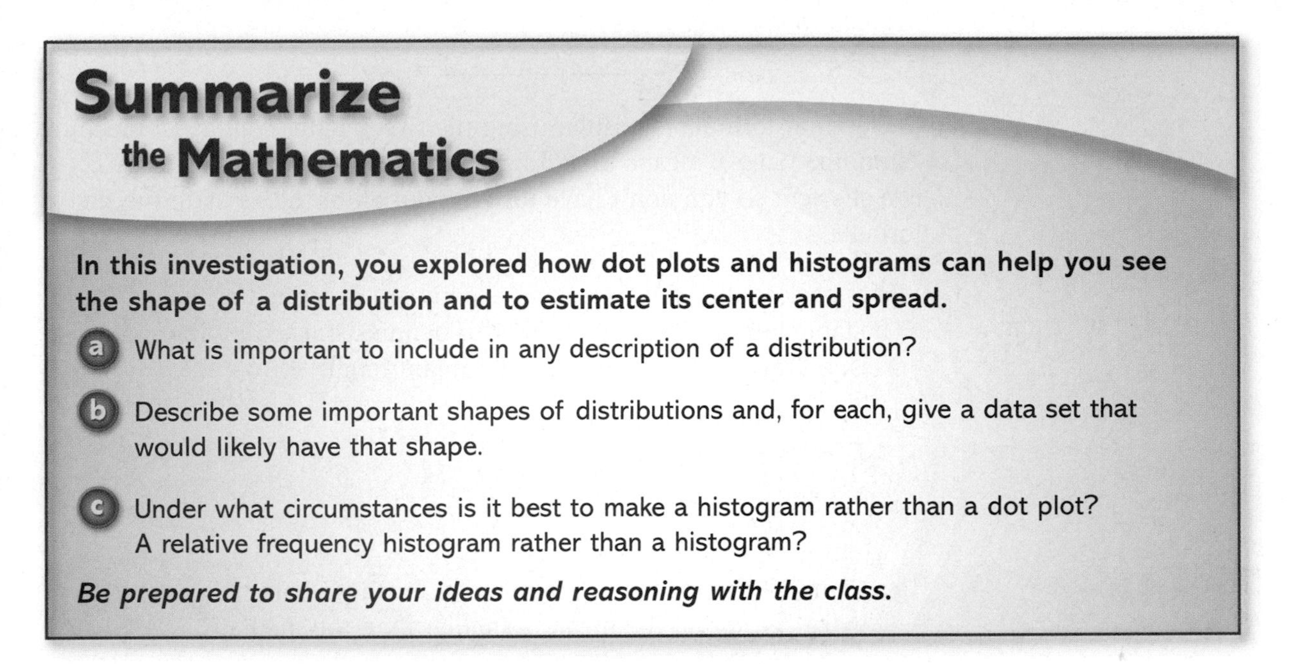

Summarize the Mathematics

In this investigation, you explored how dot plots and histograms can help you see the shape of a distribution and to estimate its center and spread.

a What is important to include in any description of a distribution?

b Describe some important shapes of distributions and, for each, give a data set that would likely have that shape.

c Under what circumstances is it best to make a histogram rather than a dot plot? A relative frequency histogram rather than a histogram?

Be prepared to share your ideas and reasoning with the class.

Check Your Understanding

Consider the amount of fat in the fast-food sandwiches listed in the table on page 82.

a. Make a dot plot of these data.

b. Make a histogram and then a relative frequency histogram of these data.

c. Write a short description of the distribution so that a person who had not seen the distribution could draw an approximately correct sketch of it.

Investigation 2 Measures of Center

In the previous investigation, you learned how to describe the shape of a distribution. In this investigation, you will review how to compute the two most important measures of the center of a distribution—the mean and the median—and explore some of their properties. As you work on this investigation, think about this question:

How do you decide whether to use the mean or median in summarizing a set of data?

Here, for your reference, are the definitions of the median and the mean.

- The **median** is the midpoint of an *ordered* list of data—at least half the values are at or below the median and at least half are at or above it. When there are an odd number of values, the median is the one in the middle. When there are an even number of values, the median is the average of the two in the middle.
- The **mean**, or arithmetic average, is the sum of the values divided by the number of values. When there are n values, $x_1, x_2, \ldots, x_n$, the formula for the mean $\bar{x}$ is

$$\bar{x} = \frac{x_1 + x_2 + \cdots + x_n}{n}, \text{ or } \bar{x} = \frac{\Sigma x}{n}.$$

The second formula is written using the Greek letter *sigma*, Σ, meaning "sum up." So Σx means to add up all of the values of x. Writing Σx is a shortcut so you don't have to write out all of the xs as in the first formula.

1 Refer back to the penny-stacking experiment described on pages 74–75. The table below gives the number of pennies stacked by the first-hour class in Traverse City with their dominant hand.

Dominant Hand

27	35	41	36	34	6	42	20
47	41	51	48	49	32	29	21
50	51	49	35	36	53	54	

a. Compute the median and the mean for these data. Why does it make sense that the mean is smaller than the median?

b. Now enter into a list in your calculator or statistical software the data your class collected on stacking pennies with your nondominant hand. Learn to use your calculator or statistical software to calculate the mean and median.

c. Compare the mean and median of the dominant hand and nondominant hand distributions. When stacking pennies, does it appear that use of the dominant or nondominant hand may make a difference? Explain your reasoning.

d. In what circumstances would you give the mean when asked to summarize the numbers of stacked pennies in the two experiments? The median?

2 Without using your calculator, find the median of these sets of consecutive whole numbers.

a. 1, 2, 3, … , 7, 8, 9

b. 1, 2, 3, … , 8, 9, 10

c. 1, 2, 3, … , 97, 98, 99

d. 1, 2, 3, … , 98, 99, 100

e. Suppose n numbers are listed in order from smallest to largest. Which of these expressions gives the *position* of the median in the list?

$$\frac{n}{2} \qquad \frac{n}{2} + 1 \qquad \frac{n+1}{2}$$

3. Now examine this histogram, which shows a set of 40 integer values.

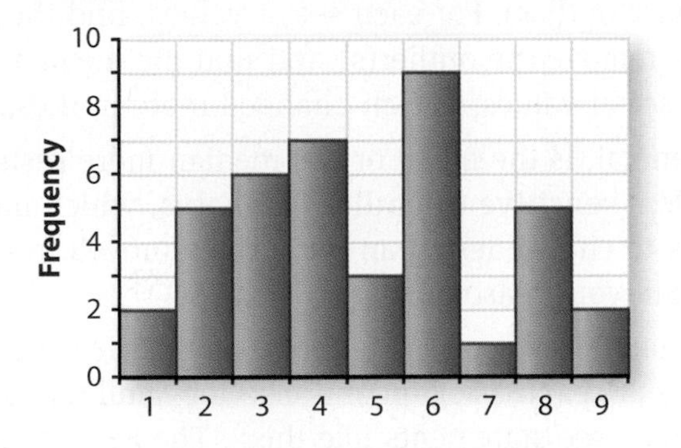

   a. What is the position of the median when there are 40 values? Find the median of this set of values. Locate the median on the horizontal axis of the histogram.

   b. Find the area of the bars to the left of the median. Find the area of the bars to the right of the median. How can you use area to estimate the median from a histogram?

4. The mean lies at the "balance point" of the plot. That is, if the histogram were made of blocks stacked on a lightweight tray, the mean is where you would place one finger to balance the tray. Is the median of the distribution below to the left of the mean, to the right, or at the same place? Explain.

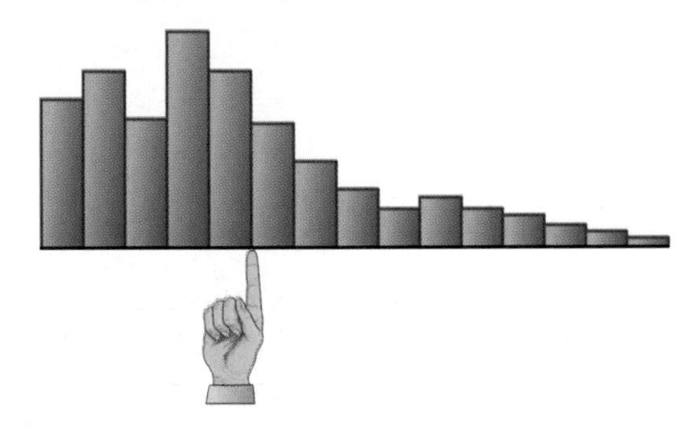

5. The histogram at the right shows the ages of the 78 actresses whose performances won in the Best Leading Role category at the annual Academy Awards (Oscars) 1929–2005. (Ages were calculated by subtracting the birth year of the actress from the year of her award.)

   a. Describe the shape of this distribution.

   b. Estimate the mean age and the median age of the winners. Write a sentence describing what each tells about the ages.

   c. Use the "Estimate Center" custom tool to check your estimate of the mean.

Age of Best Actress

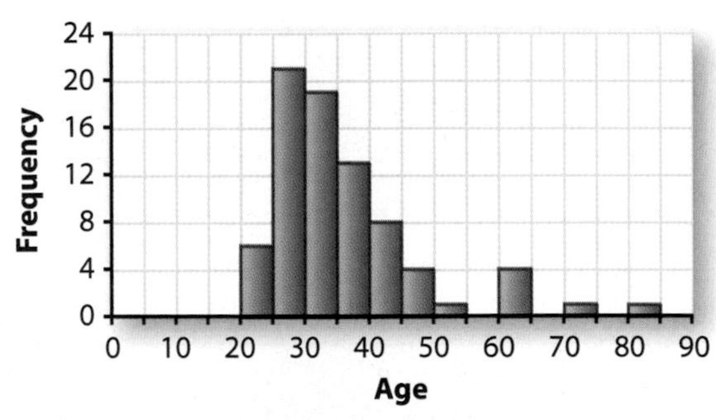

Source: www.oscars.com; www.imdb.com

6. Find the mean and median of the following set of values: 1, 2, 3, 4, 5, 6, 70.

   a. Remove the outlier of 70. Then find the mean and median of the new set of values. Which changed more, the mean or the median?

b. Working with others, create three different sets of values with one or more outliers. For each set of values, find the mean and median. Then remove the outlier(s) and find the mean and median of the new set of values. Which changed more in these cases?

c. In general, is the mean or the median more **resistant to outliers** (or, less **sensitive to outliers**)? That is, which measure of center tends to change less if an outlier is removed from a set of values? Explain your reasoning.

d. The median typically is reported as the measure of center for house prices in a region and also for family incomes. For example, you may see statements like this: "The *Seattle Times* analyzed county assessor's data on 83 neighborhoods in King County and found that last year a household with median income could afford a median-priced home in 49 of them." Why do you think medians are used in this story rather than means? (Source: seattletimes.nwsource.com/homes/html/affo05.html)

7 Make a copy of each of the distributions below. For each distribution, indicate the relationship you would expect between the mean and median by marking and labeling their approximate positions on the distribution.

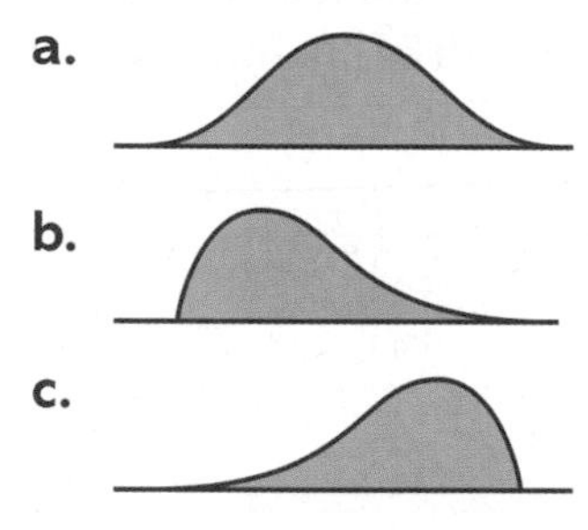

8 In a competitive candy sale, the six students in the Drama Club at Sparta High School sold a mean of 14 bars each; the eight students in the Math Club sold a mean of 11 bars each.

a. The winner of the competition is the club that sells more candy bars. Which club was the winner?

b. Construct an example, giving the number of bars sold by each student, where the median for the six students in the Drama Club is 14 bars, the median for the eight students in the Math Club is 11 bars, and the Drama Club wins the competition.

c. Now construct an example where the median for the six students in the Drama Club is 14 bars, the median for the eight students in the Math Club is 11 bars, but the Math Club is the winner this time.

d. Does knowing only the two medians let you determine which club won? Does knowing only the two means?

e. Which of the following formulas would you use to find the *total* (or *sum*) of a set of numbers if you know the mean $\bar{x}$ and the number of values n?

$total = \frac{\bar{x}}{n}$ $\quad$ $total = n \cdot \bar{x}$ $\quad$ $total = \bar{x} + n$ $\quad$ $total = \bar{x} - n \cdot \bar{x}$

9 When a distribution has many identical values, it is helpful to record them in a **frequency table**, which shows each value and the number of times (*frequency* or *count*) that it occurs. The following frequency table gives the number of goals scored per game during a season of 81 soccer matches. For example, the first line means that there were 5 matches with no goals scored.

Goals per Match

Goals Scored	Number of Matches (frequency)
0	5
1	7
2	28
3	10
4	15
5	8
6	5
7	1
8	1
9	1

a. What is the median number of goals scored per match?

b. What is the total number of goals scored in all matches?

c. What is the mean number of goals scored per match?

d. Think about how you computed the mean number of goals per match in Part c. Which of the following formulas summarizes your method?

$$\frac{\Sigma \text{ goals scored}}{10} = \frac{0 + 1 + 2 + \cdots + 8 + 9}{10}$$

$$\frac{\Sigma \text{ number of matches}}{10} = \frac{5 + 7 + 28 + \cdots + 1 + 1}{10}$$

$$\frac{\Sigma(\text{goals scored})(\text{number of matches})}{\Sigma \text{ number of matches}} = \frac{0 \cdot 5 + 1 \cdot 7 + 2 \cdot 28 + \cdots + 8 \cdot 1 + 9 \cdot 1}{5 + 7 + 28 + \cdots + 1 + 1}$$

$$\frac{\Sigma \text{ goals scored}}{\Sigma \text{ number of matches}} = \frac{0 + 1 + 2 + \cdots + 8 + 9}{5 + 7 + 28 + \cdots + 1 + 1}$$

10 Suppose that, to estimate the mean number of children per household in a community, a survey was taken of 114 randomly selected households. The results are summarized in this frequency table.

Household Size

Number of Children	Number of Households
0	15
1	22
2	36
3	21
4	12
5	6
7	1
10	1

a. How many of the households had exactly 2 children?

b. Make a histogram of the distribution. Estimate the mean number of children per household from the histogram.

c. Calculate the mean number of children per household. You can do this on some calculators and spreadsheet software by entering the number of children in one list and the number of households in another list. The following instructions work with some calculators.

- Enter the number of children in L1 and the number of households in L2.
- Position the cursor on top of L3 and type L1 [×] L2. Then press [ENTER]. What appears in list L3?
- Using the sum of list L3, and the sum of list L2, find the mean number of children per household.

d. How will a frequency table of the number of children in the households of the students in your class be different from the one above? To check your answer, make a frequency table and describe how it differs from the one from the community survey. Would your class be a good sample to use to estimate the mean number of children per household in your community?

Summarize the Mathematics

Whether you use the mean or median depends on the reason that you are computing a measure of center and whether you want the measure to be resistant to outliers.

a. In what situations would you use the mean to summarize a set of data? The median?

b. Describe how to estimate the mean and median from a histogram.

c. Describe how to find the mean and median from a frequency table.

d. What is the relationship between the sum of the values and their mean?

Be prepared to share your examples and ideas with the class.

Check Your Understanding

Leslie, a recent high school graduate seeking a job at United Tool and Die, was told that "the mean salary is over $31,000." Upon further inquiry, she obtained the following information about the number of employees at various salary levels.

Type of Job	Number Employed	Individual Salary
President/Owner	1	$210,000
Business Manager	1	70,000
Supervisor	2	55,000
Foreman	5	36,000
Machine Operator	50	26,000
Clerk	2	24,000
Custodian	1	19,000

a. What percentage of employees earn over $31,000?

b. What is the median salary? Write a sentence interpreting this median.

c. Verify whether the reported mean salary is correct.

d. Suppose that the company decides not to include the owner's salary. How will deleting the owner's salary affect the mean? The median?

e. In a different company of 54 employees, the median salary is $24,000 and the mean is $26,000. Can you determine the total payroll?

On Your Own

Applications

1 The following table gives average hourly compensation costs for production workers from 24 countries. Hourly compensation costs include hourly salary, vacation, holidays, benefits, and other costs to the employer.

Average Hourly Compensation Costs for Production Workers
(in U.S. dollars for selected countries, 2004)

Country	Cost	Country	Cost	Country	Cost	Country	Cost
Australia	23.09	Finland	30.67	Japan	21.90	Spain	17.10
Austria	28.29	France	23.89	Mexico	2.50	Sweden	28.42
Belgium	29.98	Germany	32.53	Netherlands	30.76	Switzerland	30.26
Brazil	3.03	Hong Kong	5.51	New Zealand	12.89	Taiwan	5.97
Canada	21.42	Ireland	21.94	Norway	34.64	United Kingdom	24.71
Denmark	33.75	Italy	20.48	Singapore	7.45	United States	23.17

Source: U.S. Bureau of Labor Statistics, www.bls.gov/news.release/ichcc.t02.htm

a. What is the average yearly compensation cost for a Japanese worker who gets paid for a 40-hour week, 52 weeks a year?

b. Make a dot plot of the costs. Describe how U.S. average hourly compensation costs compare to those of the other countries.

c. Make a histogram of the average hourly compensation costs. Write a summary of the information conveyed by the histogram.

2 In 2004, a family of four was considered to be living in poverty if it had income less than $18,850 per year. The percentage of persons who live below the poverty level varies from state to state. The histogram shows these percentages for the fifty states in 2004.

Percentage of Persons Under Poverty Level, by State

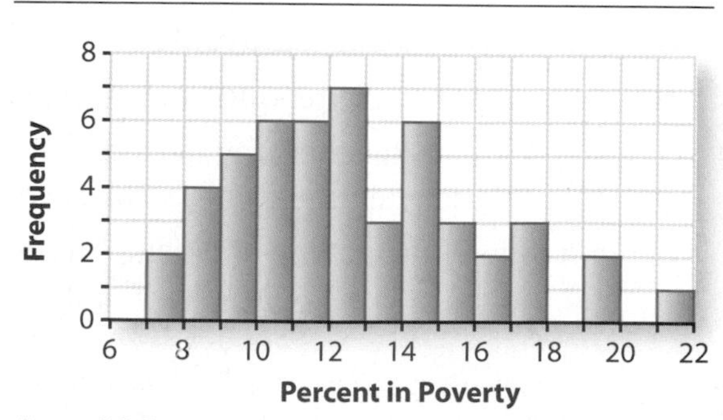

Source: U.S. Census Bureau, *2004 American Community Survey* at www.factfinder.census.gov

a. In how many states do at least 17% of the people live in poverty? In how many states do 15% or more of the people live in poverty?

b. The highest poverty rate is in Mississippi. In what interval does that rate fall? The population of Mississippi is about 2,794,925, and 603,954 live in poverty. Compute the poverty rate for Mississippi. Is this consistent with the interval you selected?

c. The lowest poverty rate is 7.6%, in New Hampshire. About 94,924 people live in poverty in New Hampshire. About how many people live in New Hampshire?

d. About 37,161,510 people in the United States, or 13.1%, are in poverty. Where would this rate fall on the histogram? About how many people live in the United States?

e. Describe the distribution of these percentages.

3 Make a rough sketch of what you think each of the following distributions would look like. Describe the shape, center, and spread you would expect.

a. the last digits of the phone numbers of students in your school

b. the heights of all five-year-olds in the United States (*Hint:* the mean is about 44 inches)

c. the weights of all dogs in the United States

d. the ages of all people who died last week in the United States

4 The two distributions below show the highest and the lowest temperatures on record at 289 major U.S. weather-observing stations in all 50 states, Puerto Rico, and the Pacific Islands.

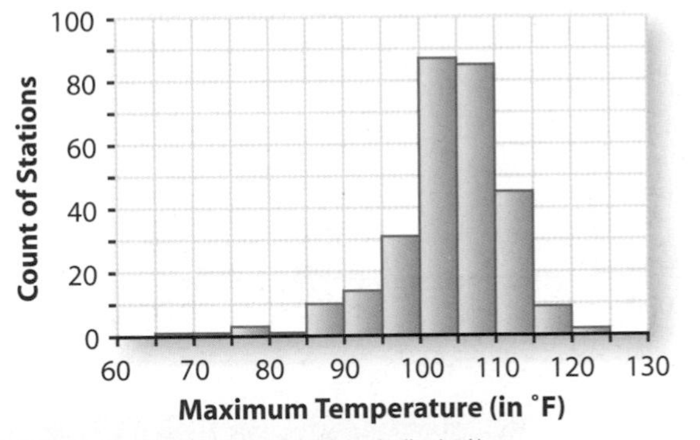

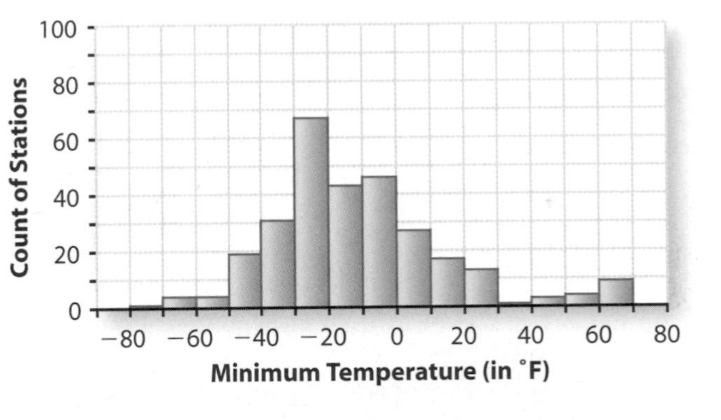

Source: www.ncdc.noaa.gov/oa/climate/online/ccd/

a. Yuma, Arizona, has the highest maximum temperature ever recorded at any of these stations. In what interval does that temperature fall? The coldest ever temperature was recorded at McGrath, Alaska. What can you say about that temperature?

b. About how many stations had a record minimum temperature from –40°F up to –30°F? About how many had a record maximum temperature less than 90°F?

c. Describe the shapes of the two distributions. What might account for the cluster in the tail on the right side of the distribution of minimum temperatures?

d. Without computing, estimate the mean temperature in each distribution.

e. Which distribution has the greater spread of temperatures?

On Your Own

5 For each of the following two distributions:

i. Describe the shape of the distribution.

ii. Estimate the median and write a sentence describing what the median tells about the data.

iii. Estimate the mean and write a sentence describing what the mean tells about the data.

a. This histogram displays the vertical jump, in inches, of 27 basketball players in an NBA draft.

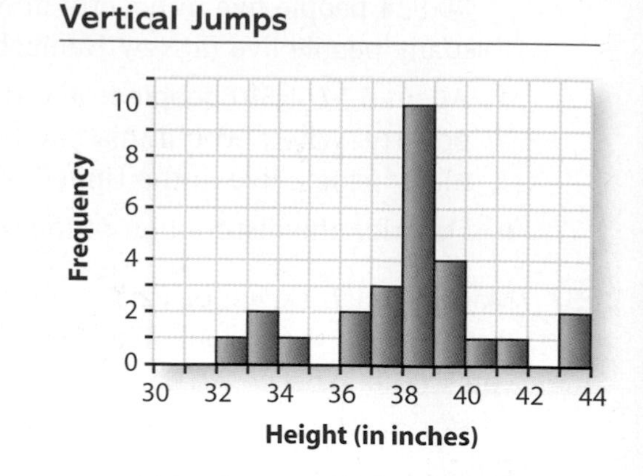

b. This histogram displays the number of video games that are available for each of 43 different platforms (computer operating systems, console systems, and handhelds). The platform with the largest number of games has 3,762. (Source: www.mobygames.com/moby_stats)

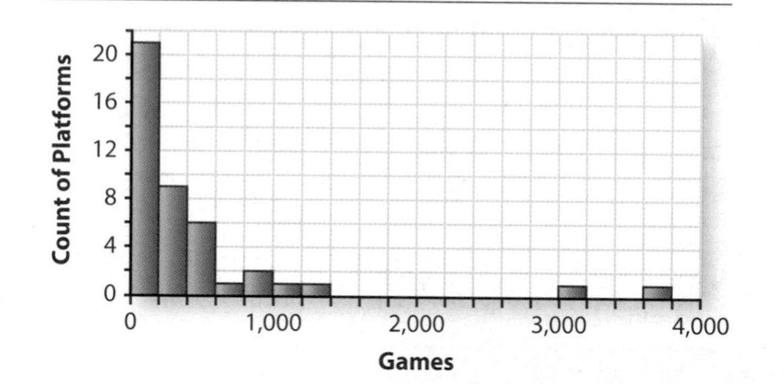

6 As a hobby, a student at the University of Alabama, Huntsville, rated the projection quality of nearby movie theaters. For each showing, a point was deducted for such things as misalignment, misframing, or an audio problem. He visited one theater in Huntsville 92 times in the first $5\frac{1}{2}$ years it was open. A frequency table of the number of points deducted per showing is at the left.

Ratings of Movie Showings

Points Deducted	Frequency
0	38
1	14
2	14
3	9
4	7
5	3
6	3
7	1
8	1
9	0
10	0
11	0
12	2

Source: hsvmovies.com/generated_subpages/ratings_table/ratings_table.html

a. Without sketching it, describe the shape of this distribution.

b. Find the median number and the mean number of points deducted for this theater (which was a relatively good one and given an A rating). Is the mean typical of the experience you would expect to have in this theater? Explain your answer.

7 Suppose your teacher grades homework on a scale of 0 to 4. Your grades for the semester are given in the following table.

Homework Grades

Grade	Frequency
0	5
1	7
2	9
3	10
4	16

a. What is your mean homework grade?

b. When computing your final grade in the course, would you rather have the teacher use your median grade? Explain.

c. Suppose that your teacher forgot to record one of your grades in the table above. After it is added to the table, your new mean is 2.50. What was that missing grade?

Connections

8 The two histograms below display the heights of two groups of tenth-graders.

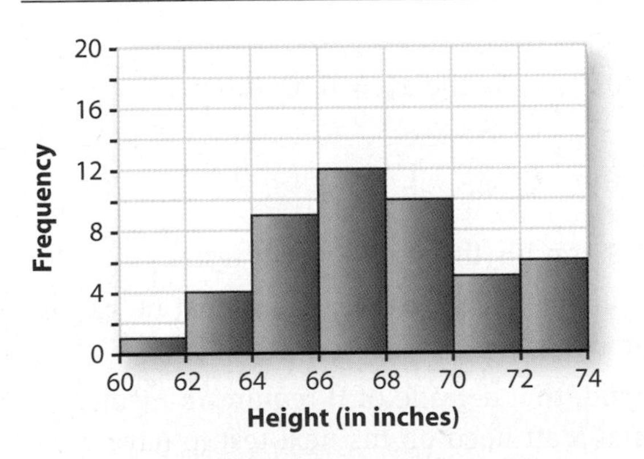

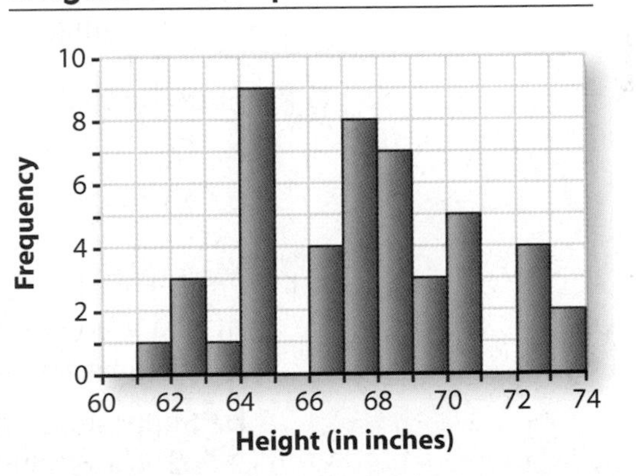

a. Compare their shapes, centers, and spreads.

b. Remake the histogram on the right so that the bars have the same 2-inch width as the one on the left. Now compare the two histograms once again.

On Your Own

9 Another measure of center that you may have previously learned is the **mode**. It is the value or category that occurs the most frequently. The mode is most useful with **categorical data**, data that is grouped into categories. The following table is from a study of the passwords people use. For example, only 2.3% of all passwords that people generate refer to a friend.

Entity Referred to	Percentage of All Passwords
Self	66.5
Relative	7.0
Animal	4.7
Lover	3.4
Friend	2.3
Product	2.2
Location	1.4
Organization	1.2
Activity	0.9
Celebrity	0.1
Not specified	4.3
Random	5.7

Source: "Generating and remembering passwords," *Applied Cognitive Psychology 18.* 2004.

a. What is the *modal* category?

b. Use this category in a sentence describing this distribution of types of passwords.

c. When someone says that a typical family has two children, is he or she probably referring to the mean, median, or mode? Explain your reasoning.

10 Suppose that $x_1 = 2$, $x_2 = 10$, $x_3 = 5$, and $x_4 = 6$. Compute:

a. Σx

b. Σx^2

c. $\Sigma(x - 2)$

d. $\Sigma \frac{1}{x}$

11 Matt received an 81 and an 83 on his first two English tests.

a. If a grade of B requires a mean of at least 80, what must he get on his next test to have a grade of B?

b. Suppose, on the other hand, that a grade of B requires a median of at least 80. What would Matt need on his next test to have a grade of B?

12 The scatterplot below shows the maximum and minimum record temperatures for the 289 stations from Applications Task 4. What information is lost when you see only the histograms? What information is lost when you see only the scatterplot?

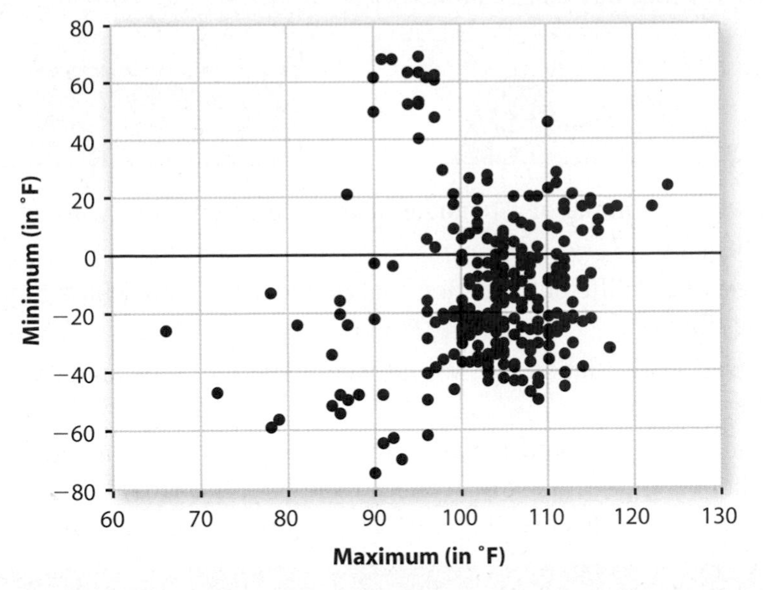

13 The term *median* is also used in geometry. A **median of a triangle** is the line segment joining a vertex to the midpoint of the opposite side. The diagram below shows one median of $\triangle ABC$.

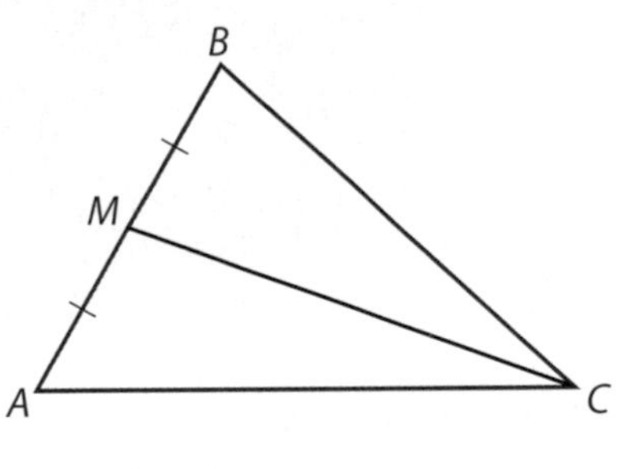

a. On a copy of the diagram, draw the other medians of this triangle.

b. On a sheet of posterboard, draw and cut out a right triangle and a triangle with an obtuse angle. Then draw the three medians of each triangle.

c. What appears to be true about the medians of a triangle?

d. Try balancing each posterboard triangle on the tip of a pencil. What do you notice?

e. Under what condition(s) will the median of a set of data be the balance point for a histogram of that data? Give an example.

Reflections

14 Distributions of real data tend to follow predictable patterns.

a. Most distributions of real data that you will see are skewed right rather than skewed left. Why? Give an example of a distribution not in this lesson that is skewed left.

b. If one distribution of real data has a larger mean than a second distribution of similar data, the first distribution tends also to have the larger spread. Find examples in this lesson where that is the case.

On Your Own

15 Sometimes a distribution has two distinct peaks. Such a distribution is said to be **bimodal**. Bimodal distributions often result from the mixture of two populations, such as adults and children or men and women. Some distributions have no peaks. These distributions are called **rectangular** distributions.

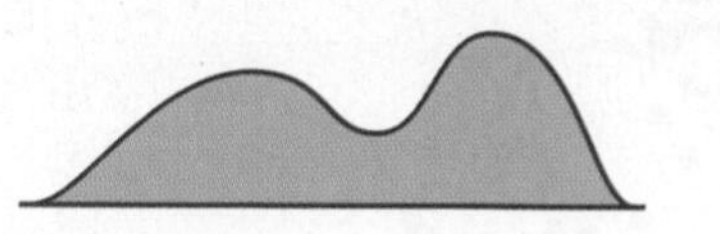

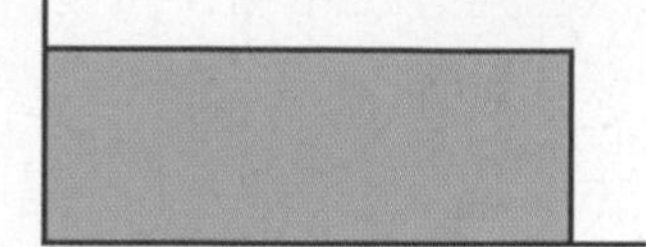

a. Give an example of a bimodal distribution from your work in this lesson.

b. Describe a different situation that would yield data with a bimodal distribution.

c. Describe a situation that would yield data with a rectangular distribution.

d. The following photo shows a "living histogram" of the heights of students in a course on a college campus. How would you describe the shape of this distribution? Why might this be the case?

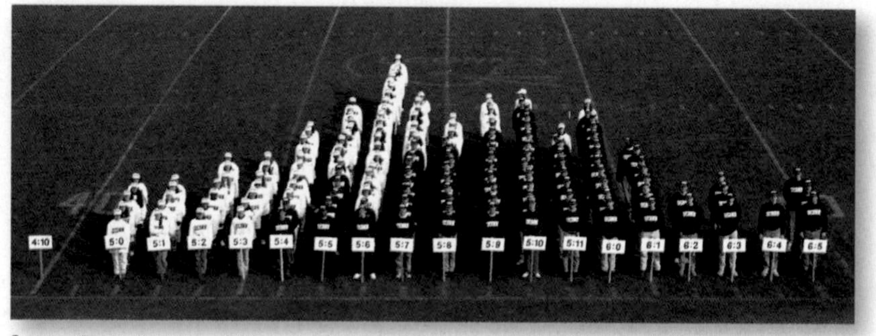

Source: *The Hartford Courant,* "Reaching New Heights," November 23, 1996. Photo by K. Hanley.

16 Suppose that you want to estimate the total weekly allowance received by students in your class.

a. Should you start by estimating the mean or the median of the weekly allowances?

b. Make a reasonable estimate of the measure of center that you selected in Part a and use it to estimate the total weekly allowance received by students in your class.

17 A soccer goalie's statistics for the last three matches are: saved 9 out of 10 shots on goal, saved 8 out of 9 shots on goal, and saved 3 out of 5 shots on goal.

a. Which of the following computations gives the mean percentage saved per match?

$$\frac{20}{24} = 83\% \qquad \frac{\frac{9}{10} + \frac{8}{9} + \frac{3}{5}}{3} \approx 79.6\%$$

b. What does the other computation tell you?

Extensions

18 To test the statement, "The mean of a set of data is the balance point of the distribution," first get a yardstick and a set of equal weights, such as children's cubical blocks or small packets of sugar.

a. Place two weights at 4 inches from one end and two weights at 31 inches. If you try to balance the yardstick with one finger, where should you place your finger?

b. What if you place one weight at 4 inches and two weights at 31 inches?

c. Experiment by placing more than three weights at various positions on the yardstick and finding the balance point.

d. What rule gives you the balance point?

19 In this lesson, you saw that for small data sets, dot plots provide a quick way to get a visual display of the data. **Stemplots** (or *stem-and-leaf plots*) provide another way of seeing patterns or unusual features in small data sets. The following stemplot shows the amount of money in cents that each student in one class of 25 students carried in coins. The stems are the hundreds and tens digits and the leaves are the ones digits.

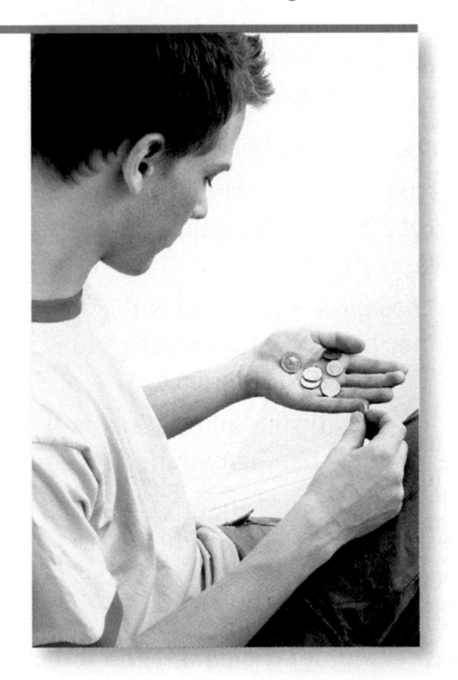

Amount of Money in Coins (in cents)

Stem	Leaves
0	0 0 0 0 0
1	2 7 8 9
2	0 5 5 8
3	4 4 7
4	5 6 6 9
5	
6	7
7	3
8	
9	0
10	
11	4
12	
13	
14	
15	2

11 | 4 represents 114¢

a. How many students had less than 30¢ in coins?

b. Where would you record the amount for another student who had $1.37 in coins? Who had 12¢?

c. Stemplots make it easy to find the median. What was the median amount of change?

20 Sometimes a **back-to-back stemplot** is useful when comparing two distributions. The back-to-back stemplot below shows the ages of the 78 actors and 78 actresses who have won an Academy Award for best performance. The tens digit of the age is given in the middle column and the ones digit is given in the left column for actors and in the right column for actresses. The youngest actor to win an Academy Award was 29, and the youngest actress was 21. This stemplot has **split stems** where, for example, the ages from 20 through 24 are put on the first stem, and the ages from 25 through 29 are put on the second stem. (Source: www.oscars.com; www.imdb.com)

Ages of Academy Award Winners

Age of Actor		Age of Actress
	2	1 2 4 4 4 4
9	2	5 5 6 6 6 6 6 6 7 7 7 7 8 8 8 9 9 9 9 9
4 4 3 2 2 1 1 0 0	3	0 0 0 1 1 1 1 2 3 3 3 3 3 4 4 4 4 4 4
9 8 8 8 8 8 7 7 7 6 6 6 5 5 5 5	3	5 5 5 5 5 7 7 7 8 8 8 8 9
4 4 4 3 3 3 3 3 3 2 2 2 1 1 1 1 0 0 0 0 0 0 0	4	0 1 1 1 1 1 2 2
9 9 8 8 8 7 7 6 6 6 6 5	4	5 7 9 9
4 3 3 2 2 2 1 1	5	4
6 6 6	5	
2 2 1 0 0	6	0 1 1 3
	6	
	7	4
6	7	
	8	0

| 2 | 1 means 21 years of age

a. What would have happened if the stems hadn't been split?

b. An article on salon.com in March 2000 reported a study that was published in the journal *Psychological Reports*. The article discusses only the difference in the mean ages. For example,

> "The study, from the journal *Psychological Reports*, says the average age of a best actress winner in the past 25 years is 40.3. The average age for men is 45.6—a five-year difference.
>
> "While the gap isn't enormous, it is significant, and for actors it grew even larger when nominees, rather than just winners, were analyzed."

Do you think the means are the best ways to compare the ages? If not, explain what measure of center would be better to use and why.

c. Write a paragraph giving your interpretation of the data. (The stemplot includes all winners, not just those from 1975 to 2000, so you will get different values for the means than those reported.)

21 Read the following table about characteristics of public high schools in the United States.

National Public High School Characteristics 2002–2003

Characteristic	Mean	Median
Enrollment size	754	493
Percent minority	31.0	17.9

Source: Pew Hispanic Center analysis of U.S. Department of Education, Common Core of Data (CCD), Public Elementary/Secondary School Universe Survey, 2002–03. *The High Schools Hispanics Attend: Size and Other Key Characteristics*, Pew Hispanic Center Report, November 1, 2005.

a. The mean high school size is larger than the median. What does this tell you about the distribution of the sizes of high schools?

b. There are about 17,505 public high schools in the United States. About how many high school students are there in these schools?

c. A footnote to the table above says, "The mean school characteristics are the simple average over all high schools. These are not enrollment weighted. A small high school receives the same weight as a large high school." Suppose that there are four high schools in a district, with the following enrollments and percent minority.

High School	Enrollment	Percent Minority
Alpha	1,000	14
Beta	1,500	20
Gamma	2,000	15
Delta	3,500	35

i. What is the median percent minority if computed as described above? Interpret this percent in a sentence.

ii. What is the mean percent minority if computed as described above? Interpret this percent in a sentence.

iii. What percentage of students in the district are minority? Interpret this percent in a sentence.

d. From the information in the first table and in Part b, can you determine the percentage of U.S. public high school students who are minority? Explain.

22 Examine the Fastest-Growing Franchise data set in your data analysis software. That data set includes the rank, franchise name, type of service, and minimum startup costs for the 100 fastest growing franchises in the United States.
(Source: www.entrepreneur.com)

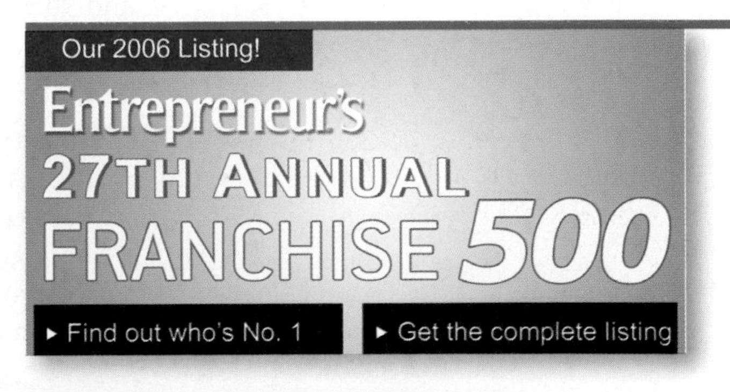

a. What kinds of businesses occur most often in that list? What are some possible reasons for their popularity?

b. Use data analysis software to make an appropriate graph for displaying the distribution of minimum startup costs.

c. Describe the shape, center, and spread of the distribution. Use the "Estimate Center" custom tool.

d. Why might a measure of center of minimum startup costs be somewhat misleading to a person who wanted to start a franchise?

23 The **relative frequency table** below shows (roughly) the distribution of the proportion of U.S. households that own various numbers of televisions.

Household Televisions

Number of Televisions, x	Proportion of Households, p
1	0.2
2	0.3
3	0.3
4	0.1
5	0.1

a. What is the median of this distribution?

b. To compute the mean of this distribution, first imagine that there are only 10 households in the United States. Convert the relative frequency table to a frequency table and compute the mean.

c. Now imagine that there are only 20 households in the United States. Convert the relative frequency table to a frequency table and compute the mean.

d. Use the following formula to compute the mean directly from the relative frequency table.

$$\overline{x} = x_1 \cdot p_1 + x_2 \cdot p_2 + x_3 \cdot p_3 + \cdots + x_k \cdot p_k \text{ or } \overline{x} = \Sigma x_i \cdot p_i$$

e. Explain why this formula works.

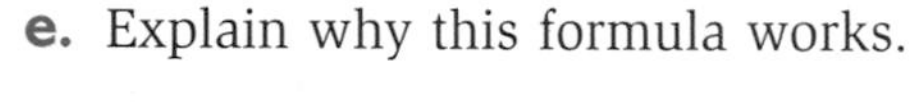

24 Suppose your grade is based 50% on tests, 30% on homework, and 20% on the final exam. So far in the class you have 82% on the tests and 90% on homework.

a. Compute your overall percentage (called a **weighted mean**) if you get 65% on the final exam. If you get 100% on the final exam.

b. Your teacher wants to use a spreadsheet to calculate weighted means for the students in your class in order to assign grades. She uses column A for names, column B for test score, column C for homework percentage, column D for final exam, and column E for the weighted mean. Give the function she would use to calculate the values in column E.

25 Many people who have dropped out of the traditional school setting earn an equivalent to a high school diploma. A GED (General Educational Development Credential) is given to a person who passes a test for a course to complete high school credits.

There were 501,000 people in the United States and its territories who received GEDs in 2000. The following table gives the breakdown by age of those taking the test.

Taking the GED

Age	19 yrs and under	20–24 yrs	25–29 yrs	30–34 yrs	35 yrs and over
% of GED Takers	42%	26%	11%	8%	14%

Source: American Council on Education, General Educational Development Testing Service, *Who took the GED?* Statistical Report, August 2001.

a. Estimate the median age of someone who takes the test and explain how you arrived at your estimate.

b. Estimate the mean age of someone who takes the test and explain how you arrived at your estimate.

Review

26 Given that $4.2 \cdot 5.5 = 23.1$, use mental computation to evaluate each of the following.

a. $-4.2 \cdot 5.5$

b. $-4.2(-5.5)$

c. $\frac{23.1}{-5.5}$

d. $4.2(-55)$

e. $\frac{-23.1}{2.1}$

27 When an object is dropped from some high spot, the distance it falls is related to the time it has been falling by the formula $d = 4.9t^2$, where t is time in seconds and d is distance in meters. Suppose a ball falls 250 meters down a mineshaft. To estimate the time, to the nearest second, it takes for the ball to fall this distance:

a. What possible calculator or computer tools could you use?

b. Could you answer this question without the aid of technology tools? Explain.

c. What solution method *would* you use? Why?

d. What is your estimate of the time it takes for the ball to fall the 250 meters?

e. How could you check your estimate?

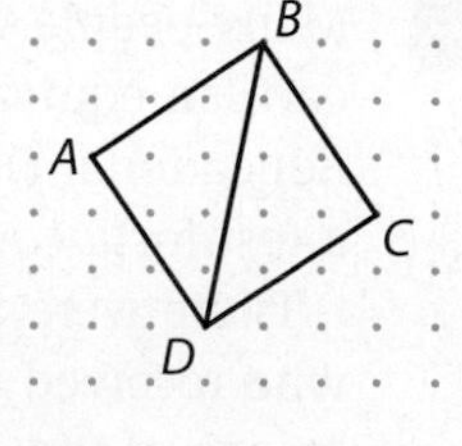

28 Consider the square shown at the right.

a. Find the area of square *ABCD*.

b. Find the length of $\overline{BD}$.

c. Find the area of $\triangle BDC$.

29 Evaluate each expression when $x = 3$.

a. 2^x b. $5 \cdot 2^x$ c. $(5 \cdot 2)^x$

d. $(-x)^2$ e. $(-2)^{x+1}$ f. -2^{x+1}

30 If the price of an item that costs \$90 in 2005 increases to \$108 by 2006, we say that the percent increase is 20%.

a. Assuming that the percent increase is the same from 2006 to 2007, what will be the cost of this same item in 2007?

b. If this percent increase continues, how long will it take for the price to double?

c. Use the words *NOW* and *NEXT* to write a rule that shows how to use the price of the item in one year to find the price of the item in the next year.

31 Trace each diagram onto your paper and then complete each shape so the indicated line is a symmetry line for the shape.

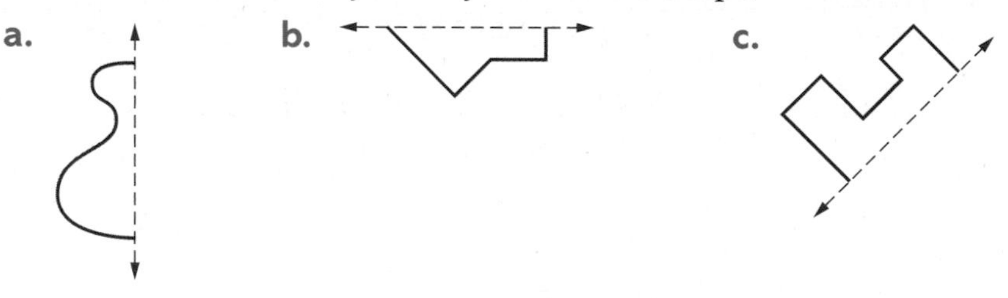

32 The temperature in Phoenix, Arizona, on one October day is shown in the graph below.

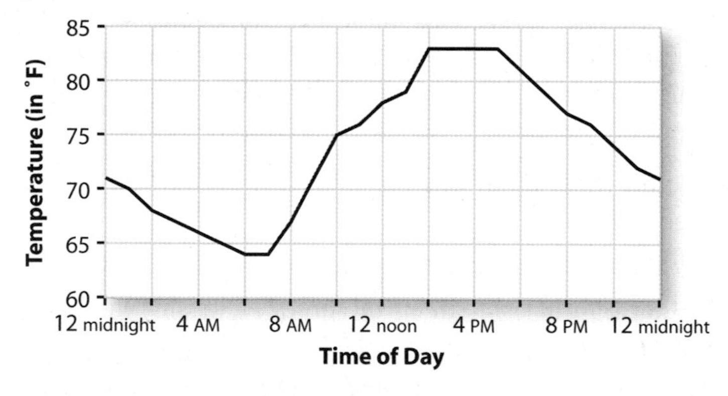

a. What was the high temperature on this day and approximately when did it occur?

b. What was the low temperature on this day and approximately when did it occur?

c. During what part(s) of the day was the temperature less than 75°?

d. During what time period(s) was the temperature increasing? Decreasing? How is this reflected in the graph?

LESSON 2

Measuring Variability

The observation that no two snowflakes are alike is somewhat amazing. But in fact, there is *variability* in nearly everything. When a car part is manufactured, each part will differ slightly from the others. If many people measure the length of a room to the nearest millimeter, there will be many slightly different measurements. If you conduct the same experiment several times, you will get slightly different results. Because variability is everywhere, it is important to understand how variability can be measured and interpreted.

People vary too and height is one of the more obvious variables. The growth charts on page 105 come from a handbook for doctors. The plot *on the left* gives the mean height of boys at ages 0 through 14 and the plot *on the right* gives the mean height for girls at the same ages.

Heights from Birth to 14 Years of Age

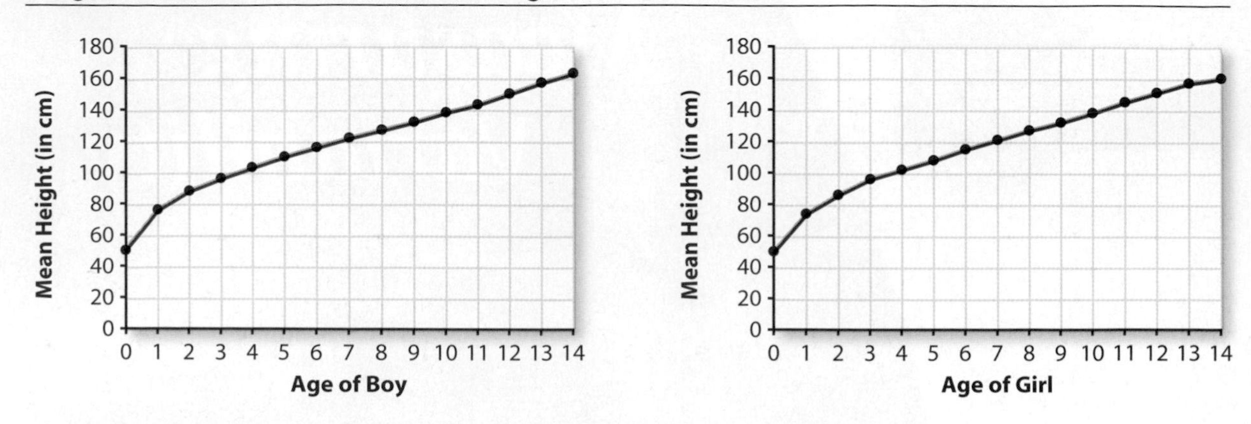

Think About This Situation

Use the plots above to answer the following questions.

a Is it reasonable to call a 14-year-old boy "taller than average" if his height is 170 cm? Is it reasonable to call a 14-year-old boy "tall" if his height is 170 cm? What additional information about 14-year-old boys would you need to know to be able to say that he is "tall"?

b From what you know about people's heights, is there as much variability in the heights of 2-year-old girls as in the heights of 14-year-old girls? Can you use this chart to answer this question?

c During which year do children grow most rapidly in height?

In this lesson, you will learn how to find and interpret measures of position and measures of variability in a distribution.

Investigation 1 Measuring Position

If you are at the 40th **percentile** of height for your age, that means that 40% of people your age are your height or shorter than you are and 60% are taller. Percentiles, like the median, describe the position of a value in a distribution. Your work in this investigation will help you answer this question:

How do you find and interpret percentiles and quartiles?

The physical growth charts on page 105 display two sets of curved lines. The curved lines at the top give height percentiles, while the curved lines at the bottom give weight percentiles. The percentiles are the small numbers 5, 10, 25, 50, 75, 90, and 95 on the right ends of the curved lines.

Boys' Physical Growth Percentiles, (2 to 20 Years)

Girls' Physical Growth Percentiles, (2 to 20 Years)

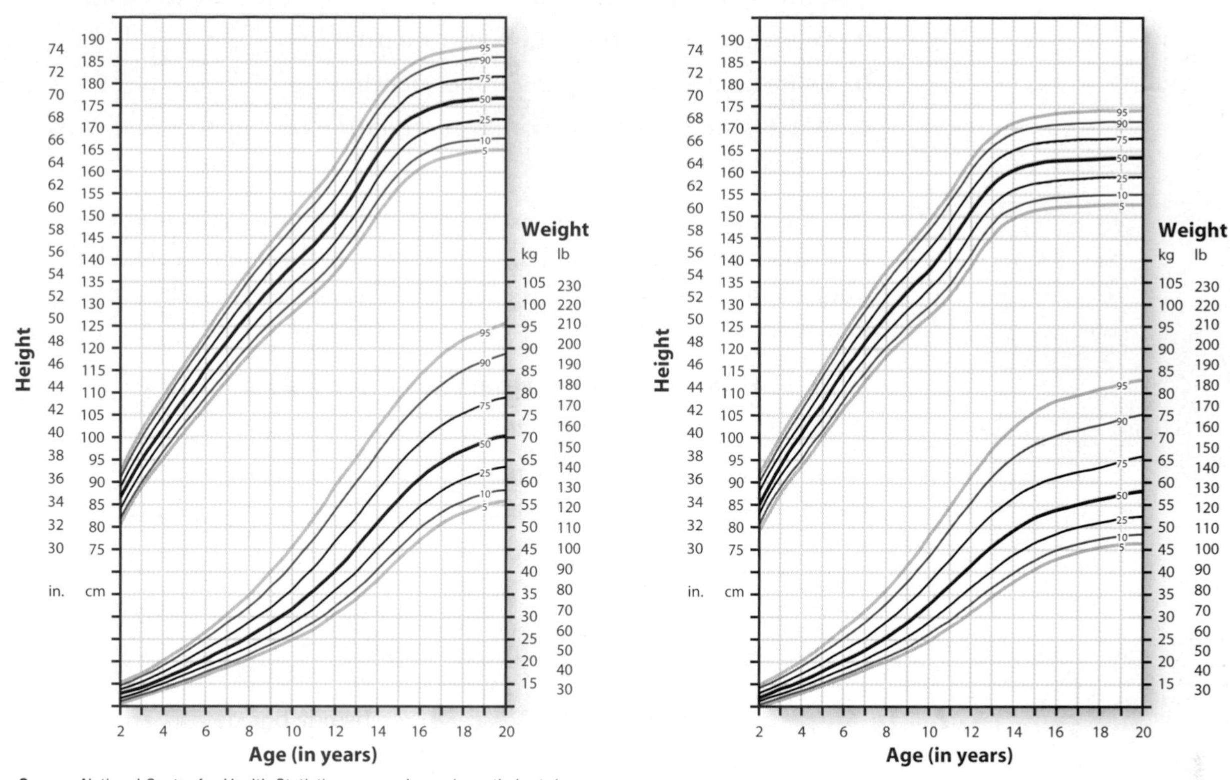

Source: National Center for Health Statistics, www.cdc.gov/growthcharts/

1. Suppose John is a 14-year-old boy who weighs 45 kg (100 pounds). John is at the 25th percentile of weight for his age. Twenty-five percent of 14-year-old boys weigh the same or less than John and 75% weigh more than John. If John's height is 170 cm (almost 5'7"), he is at the 75th percentile of height for his age. Based on the information given about John, how would you describe John's general appearance?

2. Growth charts contain an amazing amount of information. Use the growth charts to help you answer the following questions.
   - **a.** What is the approximate percentile for a 9-year-old girl who is 128 cm tall?
   - **b.** What is the 25th percentile of height for 4-year-old boys? The 50th percentile? The 75th percentile?
   - **c.** About how tall does a 12-year-old girl have to be so that she is as tall or taller than 75% of the girls her age? How tall does a 12-year-old boy have to be?
   - **d.** How tall would a 14-year-old boy have to be so that you would consider him "tall" for his age? How did you make this decision?
   - **e.** According to the chart, is there more variability in the heights of 2-year-old girls or 14-year-old girls?
   - **f.** How can you tell from the height and weight chart when children are growing the fastest? When is the increase in weight the greatest for girls? For boys?

Some percentiles have special names. The 25th percentile is called the **lower** or **first quartile**. The 75th percentile is called the **upper** or **third quartile**. Find the heights of 6-year-old girls on the growth charts.

a. Estimate and interpret the lower quartile.

b. Estimate and interpret the upper quartile.

c. What would the *middle* or *second quartile* be called? What is its percentile?

4 The histogram below displays the results of a survey filled out by 460 varsity athletes in football and women's and men's basketball from schools around Detroit, Michigan. These results were reported in a school newspaper.

Hours Spent on Homework per Day

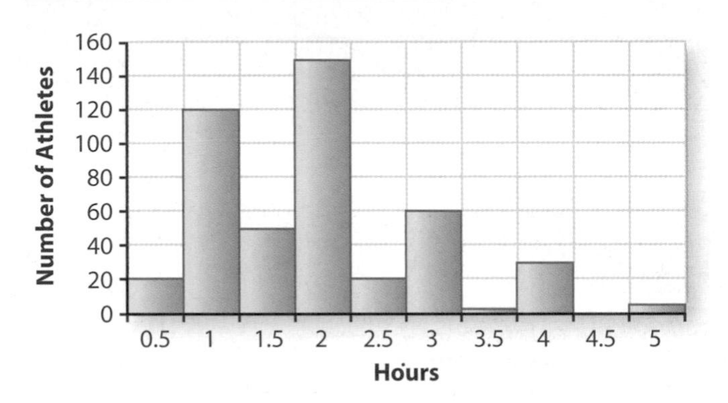

a. What is an unusual feature of this distribution? What do you think is the reason for this?

b. Estimate the median and the quartiles. Use the upper quartile in a sentence that describes this distribution.

c. Estimate the percentile for an athlete who studied 3.5 hours.

5 Suppose you get 40 points out of 50 on your next math test. Can you determine your percentage correct? Your percentile in your class? If so, calculate them. If not, explain why not.

The math homework grades for two ninth-grade students at Lakeview High School are given below.

Susan's Homework Grades

8, 8, 7, 9, 7, 8, 8, 6, 8, 7,
8, 8, 8, 7, 8, 8, 10, 9, 9, 9

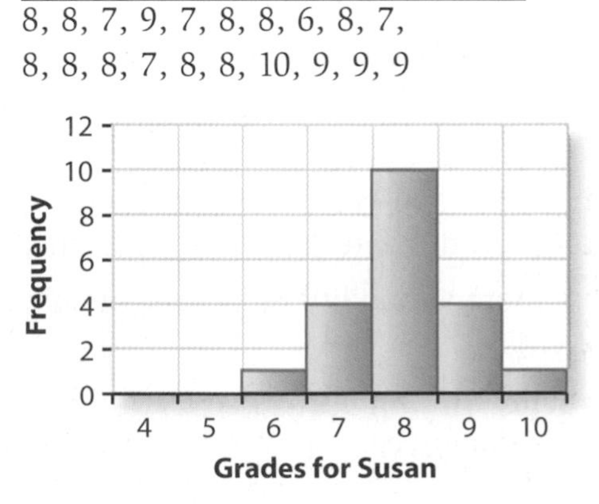

Jack's Homework Grades

10, 7, 7, 9, 5, 8, 7, 4, 7,
5, 8, 8, 8, 4, 5, 6, 5, 8, 7

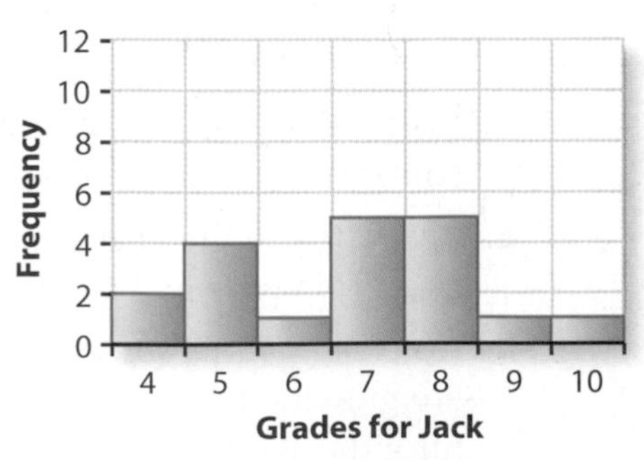

a. Which of the students has greater variability in his or her grades?

b. Put the 20 grades for Susan in an ordered list and find the median.

   i. Find the quartiles by finding the medians of the lower and upper halves.

   ii. Mark the positions of the median and quartiles on your ordered list of grades.

c. Jack has 19 grades. Put them in an ordered list and find the median.

   i. To find the first and third quartiles when there are an odd number of values, one strategy is to leave out the median and then find the median of the lower values and the median of the upper values. Use this strategy to find the quartiles of Jack's grades.

   ii. Mark the positions of the median and quartiles on your ordered list of Jack's grades.

d. For which student are the lower and upper quartiles farther apart? What does this tell you about the variability of the grades of the two students?

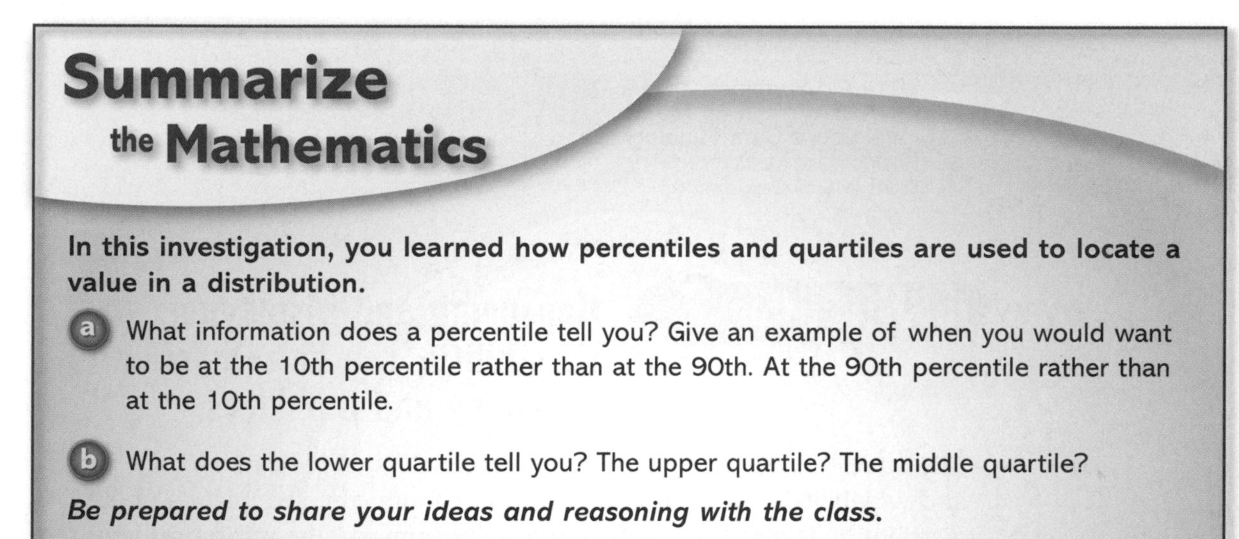

Summarize the Mathematics

In this investigation, you learned how percentiles and quartiles are used to locate a value in a distribution.

(a) What information does a percentile tell you? Give an example of when you would want to be at the 10th percentile rather than at the 90th. At the 90th percentile rather than at the 10th percentile.

(b) What does the lower quartile tell you? The upper quartile? The middle quartile?

Be prepared to share your ideas and reasoning with the class.

Check Your Understanding

The table on page 108 gives the price per ounce of each of the 16 sunscreens rated as giving excellent protection by *Consumer Reports*.

a. Find the median and quartiles of the distribution. Explain what the median and quartiles tell you about the distribution.

b. Which sunscreen is at about the 70th percentile in price per ounce?

Best Sunscreens

Brand	Price Per Ounce
Banana Boat Baby Block Sunblock	$1.13
Banana Boat Kids Sunblock	0.90
Banana Boat Sport Sunblock	0.92
Banana Boat Sport Sunscreen	4.91
Banana Boat Ultra Sunblock	0.91
Coppertone Kids Sunblock With Parsol 1789	1.25
Coppertone Sport Sunblock	4.79
Coppertone Sport Ultra Sweatproof Dry	2.02
Coppertone Water Babies Sunblock	1.17
Hawaiian Tropic 15 Plus Sunblock	0.81
Hawaiian Tropic 30 Plus Sunblock	0.90
Neutrogena UVA/UVB Sunblock	2.17
Olay Complete UV Protective Moisture	1.59
Ombrelle Sunscreen	2.17
Rite Aid Sunblock	0.50
Walgreens Ultra Sunblock	0.68

Source: www.consumerreports.org

Measuring and Displaying Variability: The Five-Number Summary and Box Plots

The quartiles together with the median give a good indication of the center and variability (spread) of a set of data. A more complete picture of the distribution is given by the **five-number summary**, the **minimum value**, the **lower quartile** (**Q_1**), the **median** (**Q_2**), the **upper quartile** (**Q_3**), and the **maximum value**. The distance between the first and third quartiles is called the **interquartile range** ($IQR = Q_3 - Q_1$).

As you work on the following problems, look for answers to these questions:

How can you use the interquartile range to measure variability?

How can you use plots of the five-number summary to compare distributions?

Refer back to the growth charts on page 105.

a. Estimate the five-number summary for 13-year-old girls' heights. For 13-year-old boys' heights.

b. Estimate the interquartile range of the heights of 13-year-old girls. Of 13-year-old boys. What do these IQRs tell you about heights of 13-year-old girls and boys?

c. What happens to the interquartile range of heights as children get older? In general, do boys' heights or girls' heights have the larger interquartile range, or are they about the same?

d. What happens to the interquartile range of weights as children get older? In general, do boys' weights or girls' weights have the larger interquartile range, or are they about the same?

2 Find the range and interquartile range of the following set of values.

1, 2, 3, 4, 5, 6, 70

a. Remove the outlier of 70. Find the range and interquartile range of the new set of values. Which changed more, the range or the interquartile range?

b. In general, is the range or interquartile range more resistant to outliers? In other words, which measure of spread tends to change less if an outlier is removed from a set of values? Explain your reasoning.

c. Why is the interquartile range more informative than the range as a measure of variability for many sets of data?

The five-number summary can be displayed in a **box plot**. To make a box plot, first make a number line. Above this line draw a narrow box from the lower quartile to the upper quartile; then draw line segments connecting the ends of the box to each **extreme value** (the maximum and minimum). Draw a vertical line in the box to indicate the location of the median. The segments at either end are often called **whiskers**, and the plot is sometimes called a **box-and-whiskers plot**.

The following box plot shows the distribution of hot dog prices at Major League Baseball parks.

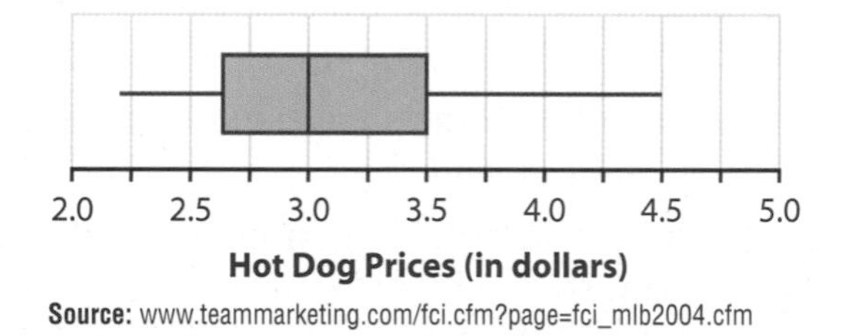

Source: www.teammarketing.com/fci.cfm?page=fci_mlb2004.cfm

a. Is the distribution skewed to the left or to the right, or is it symmetric? Explain your reasoning.

b. Estimate the five-number summary. Explain what each value tells you about hot dog prices.

4 Box plots are most useful when the distribution is skewed or has outliers or if you want to compare two or more distributions. The math homework grades for five ninth-grade students at Lakeview High School—Maria (M), Tran (T), Gia (G), Jack (J), and Susan (S)—are shown with corresponding box plots.

Maria's Grades

8, 9, 6, 7, 9, 8, 8, 6, 9, 9,
8, 7, 8, 7, 9, 9, 7, 7, 8, 9

Tran's Grades

9, 8, 6, 9, 7, 9, 8, 4, 8, 5,
9, 9, 9, 6, 4, 6, 5, 8, 8, 8

Gia's Grades

8, 9, 9, 9, 6, 9, 8, 6, 8, 6,
8, 8, 8, 6, 6, 6, 3, 8, 8, 9

Jack's Grades

10, 7, 7, 9, 5, 8, 7, 4, 7,
5, 8, 8, 8, 4, 5, 6, 5, 8, 7

Susan's Grades

8, 8, 7, 9, 7, 8, 8, 6, 8, 7,
8, 8, 8, 7, 8, 8, 10, 9, 9, 9

Math Homework Grades

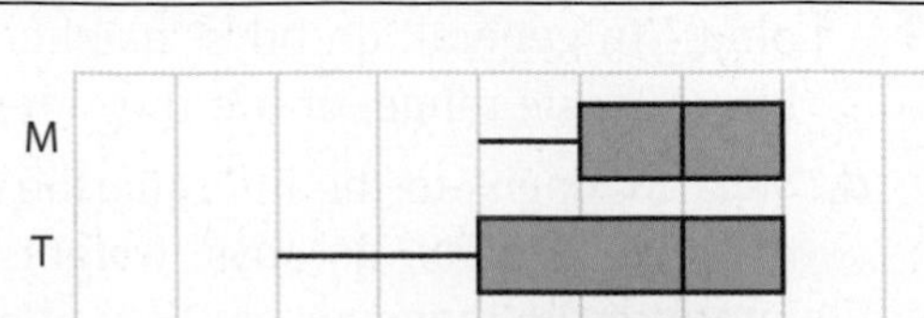
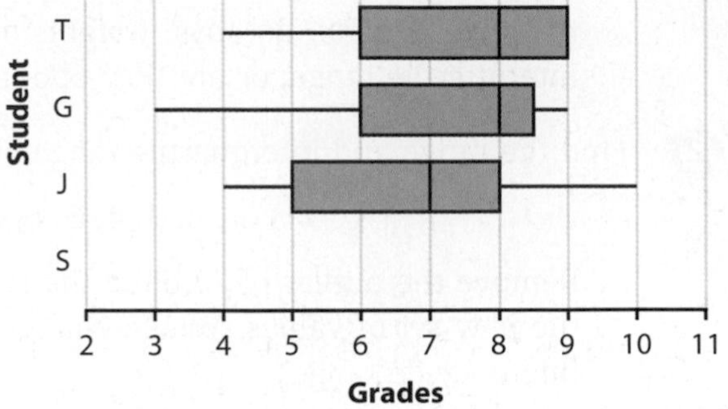

a. On a copy of the plot, make a box plot for Susan's homework grades.

b. Why do the plots for Maria and Tran have no whisker at the upper end?

c. Why is the lower whisker on Gia's box plot so long? Does this mean there are more grades for Gia in that whisker than in the shorter whisker?

d. Which distribution is the most symmetric? Which distributions are skewed to the left?

e. Use the box plots to determine which of the five students has the lowest median grade.

f. Use the box plots to determine which students have the smallest and largest interquartile ranges.

- **i.** Does the student with the smallest interquartile range also have the smallest range?
- **ii.** Does the student with the largest interquartile range also have the largest range?

g. Based on the box plots, which of the five students seems to have the best record?

5 You can produce box plots on your calculator by following a procedure similar to that for making histograms. After entering the data in a list and specifying the viewing window, select the box plot as the type of plot desired.

a. Use your calculator to make a box plot of Susan's grades from Problem 4.

b. Use the Trace feature to find the five-number summary for Susan's grades. Compare the results with your computations in the previous problem.

6 Resting pulse rates have a lot of variability from person to person. In fact, rates between 60 and 100 are considered normal. For a highly conditioned athlete, "normal" can be as low as 40 beats per minute. Pulse rates also can vary quite a bit from time to time for the same person. (Source: www.nlm.nih.gov/medlineplus/ency/article/003399.htm)

a. Take your pulse for 20 seconds, triple it, and record your pulse rate (in number of beats per minute).

b. If you are able, do some mild exercise for 3 or 4 minutes as your teacher times you. Then take your pulse for 20 seconds, triple it, and record this exercising pulse rate (in number of beats per minute). Collect the results from all students in your class, keeping the data paired (*resting*, *exercising*) for each student.

c. Find the five-number summary of resting pulse rates for your class. Repeat this for the exercising pulse rates.

d. Above the same scale, draw box plots of the resting and exercising pulse rates for your class.

e. Compare the shapes, centers, and variability of the two distributions.

f. What information is lost when you make two box plots for the resting and exercising pulse rates for the same people?

g. Make a scatterplot that displays each person's two pulse rates as a single point. Can you see anything interesting that you could not see from the box plots?

h. Make a box plot of the differences in pulse rates, (*exercising* − *resting*). Do you see anything you didn't see before?

Summarize the Mathematics

In this investigation, you learned how to use the five-number summary and box plots to describe and compare distributions.

a What is the five-number summary and what does it tell you?

b Why does the interquartile range tend to be a more useful measure of variability than the range?

c How does a box plot convey the shape of a distribution?

d What does a box plot tell you that a histogram does not? What does a histogram tell you that a box plot does not?

Be prepared to share your ideas and reasoning with the class.

✓ Check Your Understanding

People whose work exposes them to lead might inadvertently bring lead dust home on their clothes and skin. If their child breathes the dust, it can increase the level of lead in the child's blood. Lead poisoning in a child can lead to learning disabilities, decreased growth, hyperactivity, and impaired hearing. A study compared the level of lead in the blood of two groups of children—those who were exposed to lead dust from a parent's workplace and those who were not exposed in this way.

The 33 children of workers at a battery factory were the "exposed" group. For each "exposed" child, a "matching" child was found of the same age and living in the same area, but whose parents did not work around lead. These 33 children were the "control" group. Each child had his or her blood lead level measured (in micrograms per deciliter).

Blood Lead Level (in micrograms per deciliter)

Exposed	Control	Exposed	Control
10	13	34	25
13	16	35	12
14	13	35	19
15	24	36	11
16	16	37	19
17	10	38	16
18	24	39	14
20	16	39	22
21	19	41	18
22	21	43	11
23	10	44	19
23	18	45	9
24	18	48	18
25	11	49	7
27	13	62	15
31	16	73	13
34	18		

Source: "Lead Absorption in children of employees in a lead-related industry," *American Journal of Epidemiology 155.* 1982.

a. On the same scale, produce box plots of the lead levels for each group of children. Describe the shape of each distribution.

b. Find and interpret the median and the interquartile range for each distribution.

c. What conclusion can you draw from this study?

Investigation 3 Identifying Outliers

When describing distributions in Lesson 1, you identified any **outliers**—values that lie far away from the bulk of the values in a distribution. You should pay special attention to outliers when analyzing data.

As you work on this investigation, look for answers to this question:

What should you do when you identify one or more outliers in a set of data?

1 Use the algorithm below to determine if there are any outliers in the resting pulse rates of your class from Problem 6 (page 111) of the previous investigation.

Step 1: Find the quartiles and then subtract them to get the interquartile range, IQR.

Step 2: Multiply the IQR by 1.5.

Step 3: Add the value in Step 2 to the third quartile.

Step 4: Check if any pulse rates are larger than the value in Step 3. If so, these are outliers.

Step 5: Subtract the value in Step 2 from the first quartile.

Step 6: Check if any pulse rates are smaller than the value in Step 5. If so, these are outliers.

2 Reproduced below is the dot plot of lengths of female bears from Lesson 1.

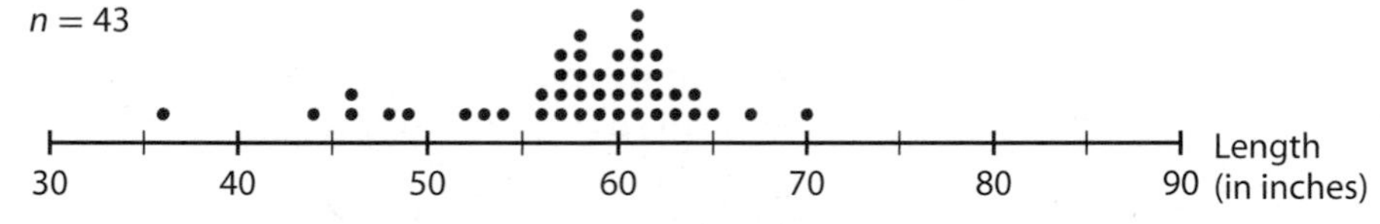

a. Do there appear to be any outliers in the data?

b. The five-number summary for the lengths of female bears is:

minimum = 36, Q_1 = 56.5, median = 59, Q_3 = 61.5, maximum = 70.

i. Use the steps above to identify any outliers on the high end.

ii. Are there any outliers on the low end?

c. The box plot below (often referred to as a **modified box plot**) shows how the outliers in the distribution of the lengths of female bears may be indicated by a dot. The whiskers end at the last length that is not an outlier. What lengths of female bears are outliers?

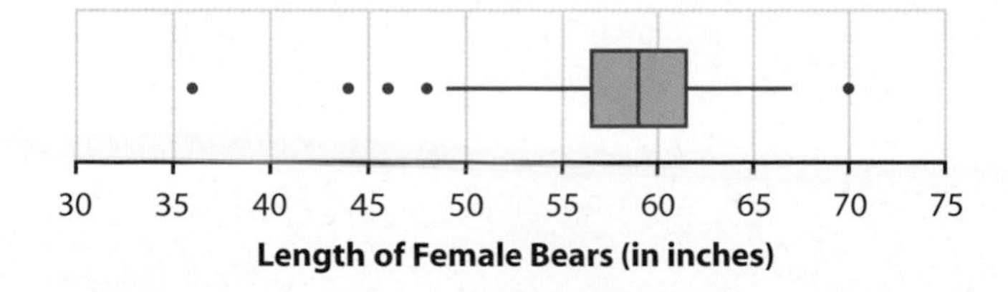

In the Check Your Understanding of Investigation 1 (page 107), you found that the quartiles for the price per ounce of sunscreens with excellent protection were $Q_1 = \$0.90$ and $Q_3 = \$2.095$.

a. Identify any outliers in the distribution of price per ounce of these sunscreens.

b. Make a modified box plot of the data, showing any outliers.

c. Here is the box plot of the prices per ounce for the sunscreens that offered less than excellent protection. Compare this distribution with the distribution from Part b.

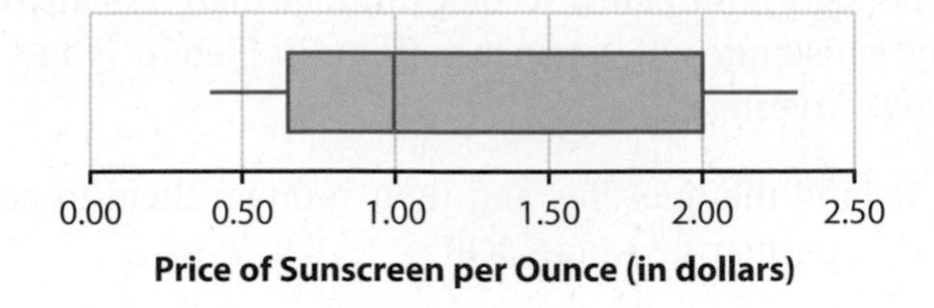

i. Do you tend to get better protection when you pay more?

ii. Do you always get better protection when you pay more?

4 Jolaina found outliers by using a box plot. She measured the length of the box and marked off 1 box length to the right of the original box and 1 box length to the left of the original box. If any of the values extended beyond these new boxes, these points were considered outliers.

a. Jolaina had a good idea but made one mistake. What was it? How can Jolaina correct her mistake?

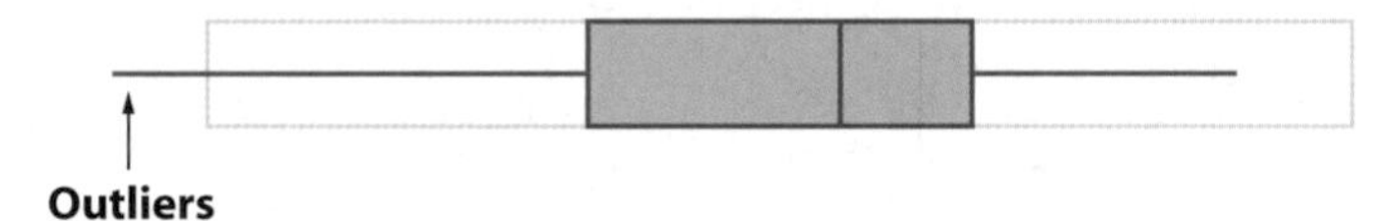

b. Using the corrected version of Jolaina's method, determine if there should be any outliers displayed by these box plots.

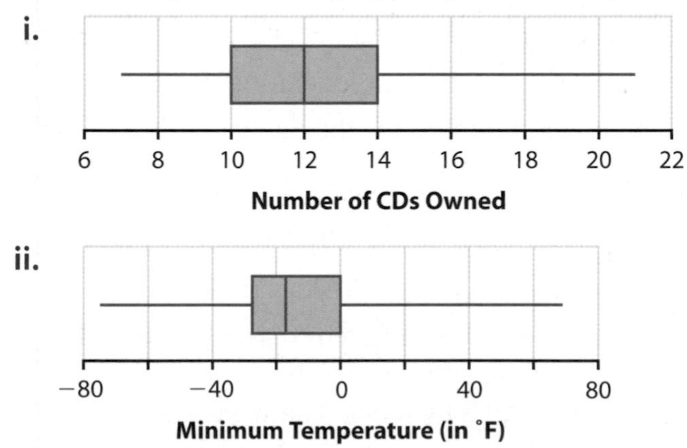

c. Jolaina then made symbolic rules for finding possible outliers in a data set. She says that outliers are values that are

$$\text{larger than} \quad Q_3 + 1.5 \cdot (Q_3 - Q_1) = Q_3 + 1.5 \cdot \text{IQR}$$
$$\text{or smaller than} \quad Q_1 - 1.5 \cdot (Q_3 - Q_1) = Q_1 - 1.5 \cdot \text{IQR}.$$

Are Jolaina's formulas correct? If so, use them to determine if there are any outliers in the data on lengths of female bears in Problem 2. If not, correct the formulas and then use them to find if there are any outliers in these data.

Whether to leave an outlier in the analysis depends on close inspection of the reason it occurred. If it was the result of an error in data collection or if it is fundamentally unlike the other values, it should be removed from the data set. If it is simply an unusually large or small value, you have two choices:

- Report measures of center and measures of variability that are resistant to outliers, such as the median, quartiles, and interquartile range.
- Do the analysis twice, with and without the outlier, and report both.

5 Decide what you would do about possible outliers in each of these situations.

a. The District of Columbia has a far higher number of physicians per 100,000 residents than does any state. That rate, shown on the box plot below, is 683 physicians per 100,000 residents. Why might you not want to include the District of Columbia in this data set of the 50 states?

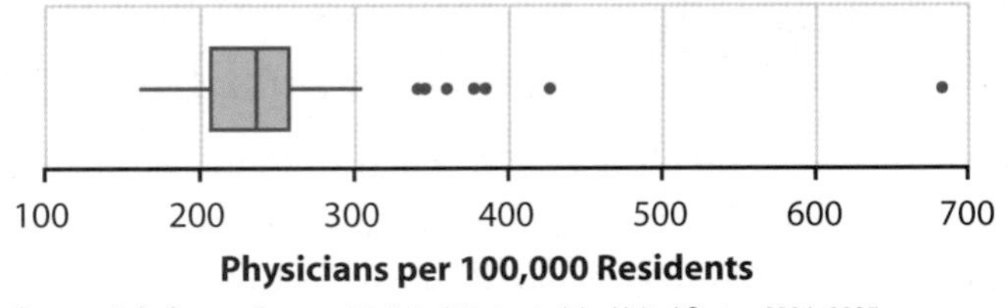

Source: U.S. Census Bureau, *Statistical Abstract of the United States: 2004–2005*

b. The box plots below show the number of grams of fat in chicken (C) and beef (B) sandwiches. Check the table of data on page 82 and identify the sandwich that is the outlier.

i. Do you know of any reason to exclude it from the analysis?

ii. Compute the mean and median of the grams of fat in the beef sandwiches only. Now compute them excluding the outlier. How much does the outlier affect them?

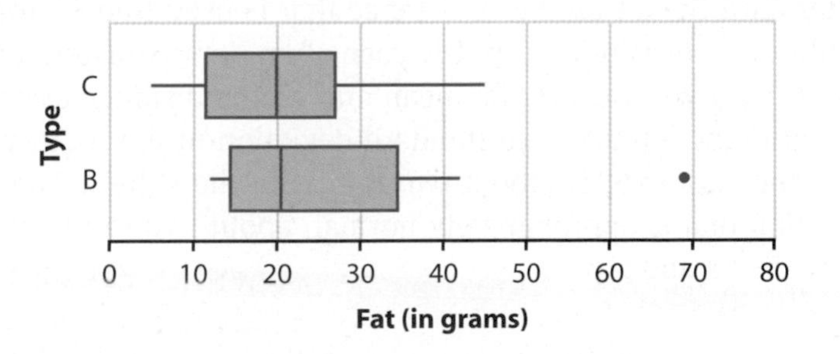

Summarize the Mathematics

Most calculators and statistical software show outliers on modified box plots with a dot.

a Describe in words the rule for identifying outliers. Describe it geometrically. Finally, write the formula.

b How do you decide what to do when you find an outlier in a set of data?

Be prepared to share your ideas and reasoning with the class.

Check Your Understanding

Refer back to the data on lead levels in the two groups of children on page 112. Use the five-number summary you calculated to complete the following tasks.

a. Identify any outliers in these two distributions. What should you do about them?

b. Make a box plot that shows any outliers.

Investigation 4 Measuring Variability: The Standard Deviation

In the previous investigation, you learned how to use the five-number summary and interquartile range (IQR) to describe the variability in a set of data. The IQR is based on the fact that half of the values fall between the upper and lower quartiles. Because it ignores the tails of the distribution, the IQR is very useful if the distribution is skewed or has outliers.

For data that are approximately normal—symmetric, mound-shaped, without outliers—a different measure of spread called the *standard deviation* is typically used. As you work on the problems of this investigation, keep track of answers to this question:

How can you determine and interpret the standard deviation of an approximately normal distribution?

The **standard deviation** s is a distance that is used to describe the variability in a distribution. In the case of an approximately normal distribution, if you start at the mean and go the distance of one standard deviation to the left and one standard deviation to the right, you will enclose the middle 68% (about two-thirds) of the values. That is, in a distribution that is approximately normal, about two-thirds of the values lie between $\bar{x} - s$ and $\bar{x} + s$.

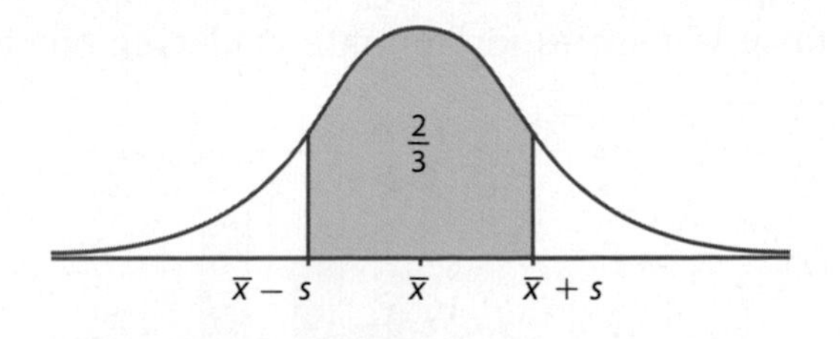

1 On each of the following distributions, the arrows enclose the middle two-thirds of the values. For each distribution:

- **i.** Estimate the mean $\bar{x}$.
- **ii.** Estimate the distance from the mean to one of the two arrows. This distance is (roughly) the standard deviation.

a. Heights of a large sample of young adult women in the United States

Heights of Young Adult Women

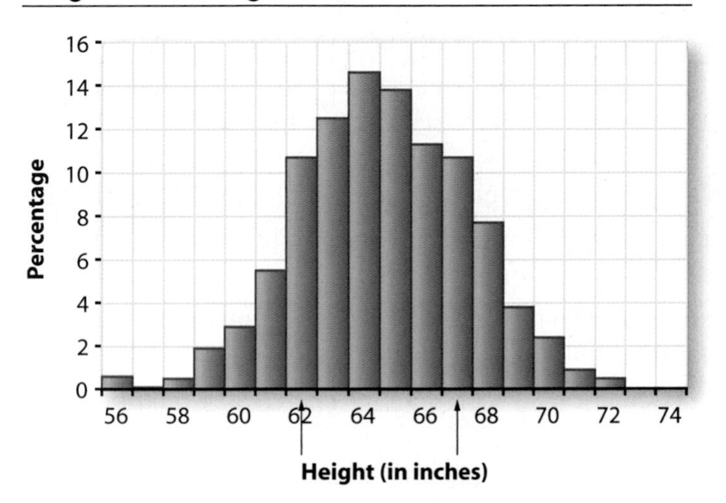

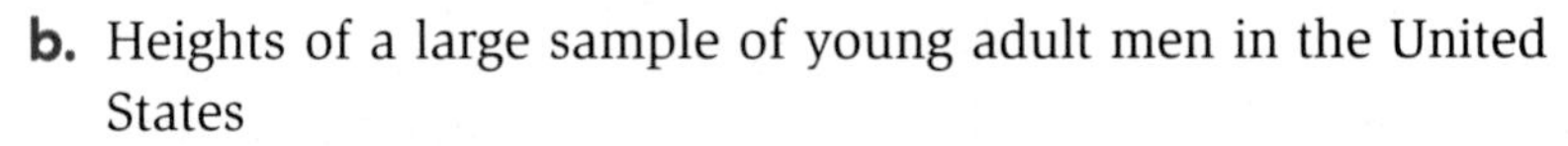

b. Heights of a large sample of young adult men in the United States

Heights of Young Adult Men

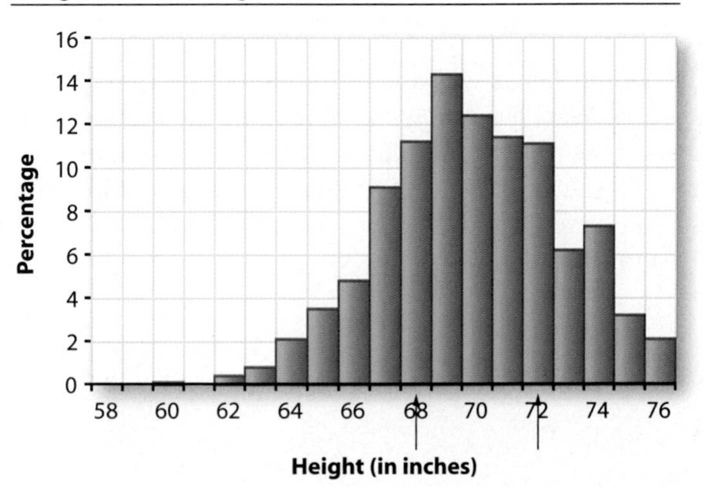

c. Achievement test scores for all ninth graders in one high school

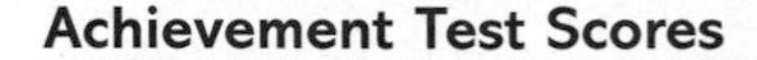

Achievement Test Scores

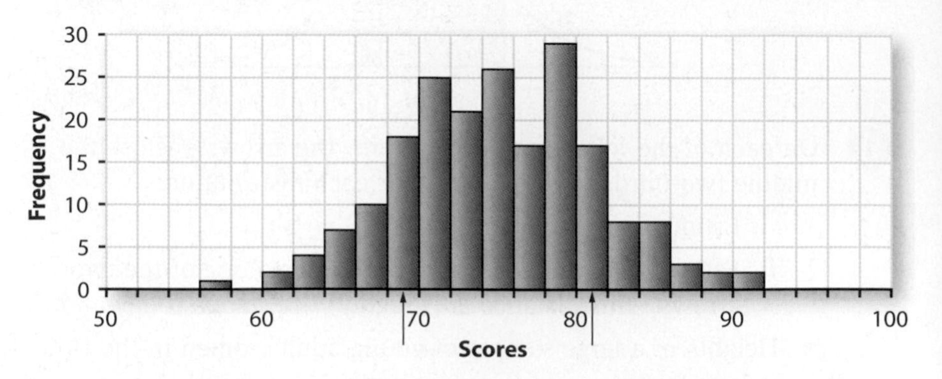

d. Use the "Estimate Center and Spread" custom tool to check your estimates of the mean and standard deviation in Parts a–c.

2 The sophomores who took the PSAT/NMSQT test in 2004 had a mean score of 44.2 on the mathematics section, with a standard deviation of 11.1. The distribution of scores was approximately normal. The highest possible score was 80 and the lowest was 20. (Source: www.collegeboard.com/researchdocs/2004_psat.html)

a. Sketch the shape of the histogram of the distribution of scores, including a scale on the x-axis.

b. A sophomore who scored 44 on this exam would be at about what percentile?

c. A sophomore who scored 33 on this exam would be at about what percentile?

d. A sophomore who scored 55 on this exam would be at about what percentile?

Another measure of where a value x lies in a distribution is its **deviation from the mean**.

$$\text{deviation from mean} = \text{value} - \text{mean} = x - \bar{x}$$

3 In 2003, LeBron James was a first-round draft pick and NBA Rookie of the Year. The following table gives the number of points he scored in the seven games he played in the first month of his freshman season at St. Vincent-St. Mary High School in Akron, Ohio. That season he led his high school team to a perfect 27-0 record and the Division III state title.

Points Scored by LeBron James in His First Month

Date	Opponent	Total Points
Dec. 3	Cuyahoga Falls	15
Dec. 4	Cleveland Central Catholic	21
Dec. 7	Garfield	11
Dec. 17	Benedictine	27
Dec. 18	Detroit Redford	18
Dec. 28	Mansfield Temple Christian	20
Dec. 30	Mapleton	21

Source: www.cleveland.com/hssports/lebron/agate.ssf?/hssports/lebron/lebron_stats.html

a. Find the mean number of points scored per game.

  i. For each game, find the deviation from the mean.

  ii. For which game(s) is James's total points farthest from the mean?

  iii. For which game(s) is James's total points closest to the mean?

b. For which game would you say that he has the most "typical" deviation (not unusually far or unusually close to the mean)?

c. In James's rookie season with the Cleveland Cavaliers, he averaged 20.9 points per game.

  i. The highest number of points he scored in a game that season was 20.1 points above his season average. How many points did he score in that game?

  ii. In his first game in his rookie season for the Cavaliers, he scored 25 points. What was the deviation from his season average for that game?

  iii. In one game that season, James had a deviation from his season average of −12.9 points. How many points did he score in that game?

4 The fact that the mean is the balance point of the distribution is related to a fact about the sum of all of the deviations from the mean.

a. Find the sum of the deviations in Problem 3.

b. Make a set of values with at least five different values. Find the mean and the deviations from the mean. Then find the sum of the deviations from the mean.

c. Check with classmates to see if they found answers similar to yours in Parts a and b. Then make a conjecture about the sum of the deviations from the mean for any set of values.

d. Complete the rule below. (Recall that the symbol Σ means to add up all of the following values. In this case, you are adding up all of the deviations from the mean.)

$$\Sigma(x - \bar{x}) =$$

e. Using the data sets from Parts a and b, do you think there is a rule about the sum of the deviations from the median? Explain your reasoning.

While the standard deviation is most useful when describing distributions that are approximately normal, it also is used for distributions of other shapes. In these cases, the standard deviation is given by a formula. The formula is based on the deviations of the values from their mean.

5 Working in groups of four to six, measure your handspans. Spread your right hand as wide as possible, place it on a ruler, and measure the distance from the end of your thumb across to the end of your little finger. Measure to the nearest tenth of a centimeter.

a. Find the mean of the handspans of the students in your group. Find the deviation from the mean of each student's handspan. Check that the sum of the deviations is 0.

b. Roughly, what is a typical distance from the mean for your group?

c. Compute the standard deviation of your group's handspans by using the steps below. Fill in a copy of the chart as you work, rounding all computations to the nearest tenth of a centimeter.

- In the first column, fill in your group's handspans.
- Write the mean of your group's handspans on each line in the second column.
- Write the deviations from the mean in the third column.
- Write the squares of these deviations in the last column.
- Find the sum of the squared deviations.
- Divide by the number n in your group minus one.
- Take the square root. This final number is the standard deviation.

Span	Mean	Deviation (Span − Mean)	Squared Deviation $(\text{Span} - \text{Mean})^2$
		Add the squared deviations:	
		Divide the sum by $n - 1$:	
		Take the square root:	

d. Have each group write its mean and standard deviation on a piece of paper. Give them to one person who will mix up the papers and write the paired means and standard deviations where everyone in your class can see them. Try to match each pair of statistics with the correct group.

e. Kelsi wrote this sentence: The handspans of our group average 21.2 cm with a handspan typically being about 2.6 cm from average. Write a similar sentence describing your group's handspans.

6 Now consider how standard deviation can be used in the comparison of performance data. Here are Susan's and Jack's homework grades.

Susan's Homework Grades	Jack's Homework Grades
8, 8, 7, 9, 7, 8, 8, 6, 8, 7,	10, 7, 7, 9, 5, 8, 7, 4, 7,
8, 8, 8, 7, 8, 8, 10, 9, 9, 9	5, 8, 8, 8, 4, 5, 6, 5, 8, 7

a. Find the set of deviations from the mean for Susan and the set of deviations for Jack. Who tends to deviate the most from his or her mean?

b. Roughly, what is a typical distance from the mean for Susan? For Jack?

c. Compute the standard deviation of each set of grades. Were these close to your estimates of a typical distance from the mean in Part b?

d. Which student had the larger standard deviation? Explain why that makes sense.

e. Write a sentence about Susan's homework grades that is similar to Kelsi's statement in Problem 5 Part e. Write a similar sentence about Jack's homework grades.

7 Think about the process of computing the standard deviation.

a. What is accomplished by squaring the deviations before adding them?

b. What is accomplished by dividing by the number of deviations (minus 1)?

c. What is accomplished by taking the square root?

d. What unit of measurement should be attached to the standard deviation of a distribution?

8 Look back at your calculations of the standard deviation in Problem 5. Which of the following is the formula for the standard deviation, s?

$$s = \sqrt{\frac{\Sigma(x - \bar{x}^2)}{n - 1}} \qquad s = \sqrt{\frac{\Sigma(x - \bar{x})^2}{n - 1}} \qquad s = \sqrt{\frac{(\Sigma x - \bar{x})^2}{n - 1}} \qquad s = \sqrt{\Sigma\left(\frac{x - \bar{x}}{n - 1}\right)^2}$$

9 Without calculating, match the sets of values below, one from column A and one from column B, that have the same standard deviations.

	Column A
a.	1, 2, 3, 4, 5
b.	2, 4, 6, 8, 10
c.	2, 2, 2, 2, 2
d.	2, 6, 6, 6, 10
e.	2, 2, 6, 10, 10

	Column B
f.	10, 10, 10, 10, 10
g.	4, 6, 8, 10, 12
h.	4, 5, 6, 7, 8
i.	16, 16, 20, 24, 24
j.	4, 8, 8, 8, 12

10 Consider the heights of the people in the following two groups.

- the members of the Chicago Bulls basketball team, and
- the adults living in Chicago.

a. Which group would you expect to have the larger mean height? Explain your reasoning.

b. Which group would you expect to have the larger standard deviation? Explain.

11 Graphing calculators and statistical software will automatically calculate the standard deviation.

a. Enter the handspans for your entire class into a list and use your calculator or software to find the mean $\bar{x}$ and the standard deviation s. Write a sentence using the mean and standard deviation to describe the distribution of handspans.

b. Which handspan is closest to one standard deviation from the mean?

c. If the distribution is approximately normal, determine how many handspans of your class should be in the interval $\bar{x} - s$ to $\bar{x} + s$. How many handspans actually are in this interval?

d. Is the standard deviation of the class larger or smaller than the standard deviation of your group? What characteristic of the class handspans compared to the group handspans could explain the difference?

12 Find the standard deviation of the following set of values.

$$1, 2, 3, 4, 5, 6, 70$$

a. Remove the outlier of 70. Then find the standard deviation of the new set of values. Does the standard deviation appear to be resistant to outliers?

b. Test your conjecture in Part a by working with others to create three different sets of values with one or more outliers. In each case, find the standard deviation. Then remove the outlier(s) and find the standard deviation of the new set of values. Summarize your findings, telling exactly what it is about the formula for the standard deviation that causes the results.

Summarize the Mathematics

In this investigation, you learned how to find and interpret the standard deviation.

a What does the standard deviation tell you about a distribution that is approximately normal? Compare this to what the interquartile range tells you.

b Describe in words how to find the standard deviation.

c Which measures of variation (range, interquartile range, standard deviation) are resistant to outliers? Explain.

d If a deviation from the mean is positive, what do you know about the value? If the deviation is negative? If the deviation is zero? What do you know about the sum of all of the deviations from the mean?

Be prepared to share your thinking and description with the class.

Check Your Understanding

Use the following data on U.S. weather to check your understanding of the standard deviation.

a. The histogram below shows the percentage of time that sunshine reaches the surface of the Earth in January at 174 different major weather-observing stations in all 50 states, Puerto Rico, and the Pacific Islands. The two stations with the highest percentages are Tucson and Yuma, Arizona. The station with the lowest percentage is Quillayute, Washington.

i. Estimate the mean and standard deviation of these percentages, including the units of measurement.

ii. About how many standard deviations from the mean are Tucson and Yuma?

iii. Use the "Estimate Center and Spread" custom tool to check your estimates of the mean and standard deviation.

January Sunshine

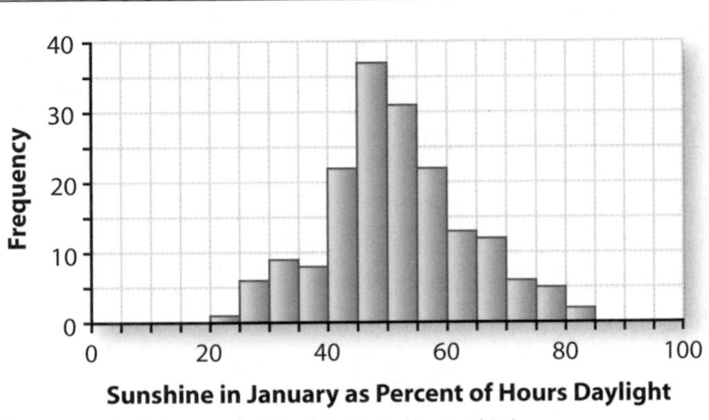

Source: www.ncdc.noaa.gov/oa/climate/online/ccd/avgsun.html

b. The normal monthly precipitation (rain and snow) in inches for Concord, New Hampshire, and for Portland, Oregon, is given in the table below.

i. Using the same scale, make histograms of the precipitation for each of the cities. By examining the plots, how do you think the mean monthly amount of precipitation for the cities will compare? The standard deviation?

ii. Calculate the mean and standard deviation of the normal monthly precipitation for each city. Write a comparison of the rainfall in the two cities, using the mean and the standard deviation.

	Jan	Feb	Mar	Apr	May	June	July	Aug	Sept	Oct	Nov	Dec
Concord	2.97	2.36	3.04	3.07	3.33	3.10	3.37	3.21	3.16	3.46	3.57	2.96
Portland	5.07	4.18	3.71	2.64	2.38	1.59	0.72	0.93	1.65	2.88	5.61	5.71

Source: National Climate Data Center, 2005

Investigation 5 Transforming Measurements

Like all events in life, data do not always come in the most convenient form. For example, sometimes you may want to report measurements in feet rather than meters or percentage correct rather than points scored on a test. Transforming data in this way has predictable effects on the shape, center, and spread of the distribution. As you work on the following problems, look for answers to this question:

What is the effect on a distribution of adding or subtracting a constant to each value and of multiplying or dividing each value by a positive constant?

1 Select 10 members of your class to measure the length of the same desk or table to the nearest tenth of a centimeter. Each student should do the measurement independently and not look at the measurements recorded by other students.

a. As a class, make a dot plot of the measurements.

b. Calculate the mean $\bar{x}$ and standard deviation s of the measurements. Mark the mean on the dot plot. Then mark $\bar{x} + s$ and $\bar{x} - s$.

c. What does the standard deviation tell you about the precision of the students' measurements?

2 Suppose that a group of 10 students would have gotten exactly the same measurements as your class did in Problem 1, except the end of their ruler was damaged. Consequently, their measurements are exactly 0.2 cm longer than yours.

a. What do you think they got for their mean and standard deviation?

b. Using lists on your calculator, transform your list of measurements into theirs. If M stands for the original measurement and D stands for the corresponding measurement made with the damaged ruler, write a rule that describes how you made this transformation.

c. Make a dot plot of the transformed measurements and compare its shape to the plot made in Problem 1.

d. Compute the mean and standard deviation of the transformed measurements.

e. How is the mean of the transformed measurements related to the original mean?

f. How is the standard deviation of the transformed measurements related to the original standard deviation?

3 Now examine the effect of transforming the measurements in Problem 1 from centimeters to inches.

a. Let C stand for a measurement in centimeters and I stand for a measurement in inches. Write a rule that you can use to transform the measurements in Problem 1 from centimeters to inches. (*Note:* There are approximately 2.54 centimeters in an inch.)

b. Make a dot plot of the transformed data and compare its shape to the plot made in Problem 1.

c. Compute the mean and the standard deviation of the transformed measurements.

d. Write a rule that relates the mean of the transformed measurements $\bar{x}_T$ to the original mean $\bar{x}$.

e. Write a rule that relates the standard deviation of the transformed measurements s_T to the original standard deviation s.

f. Suppose that one student mistakenly multiplied by 2.54 when transforming the measurements. What do you think this student got for the mean and standard deviation of the transformed measurements? Check your prediction.

4 Ms. Brenner polled her mathematics classes to find out the hourly wage of students who had baby-sitting jobs. The results are shown in the following table and histogram.

Student	Hourly Wage (in dollars)
Neil	4.00
Bill	4.25
Dimitri	4.30
José	4.50
Keri	4.75
Emerson	4.75
Rashawnda	4.75
Katie	4.85
Clive	5.00
Jan	5.10
Kyle	5.25
Mike	5.25
Toby	5.25
Nafikah	5.30
Robert	5.30

Student	Hourly Wage (in dollars)
Mia	5.50
Tasha	5.50
Sarah	5.50
Vanita	5.60
Silvia	5.60
Olivia	5.75
Katrina	5.80
Deeonna	5.80
Jacob	6.00
Rusty	6.00
Jennifer	6.25
Phuong	6.25
Corinna	6.30
John	6.50

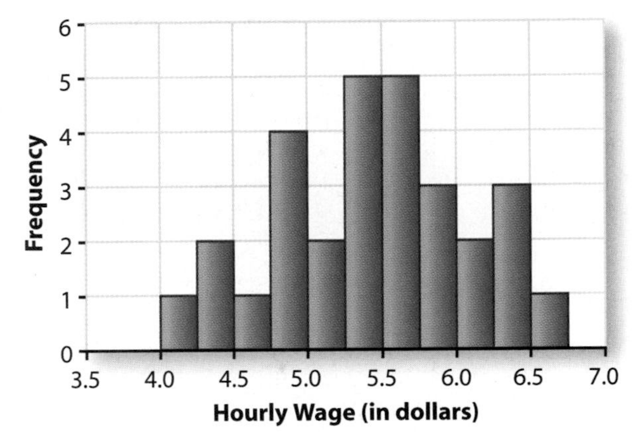

a. Use the histogram to estimate the average hourly wage of these students. Estimate the standard deviation. Using the values in the table, compute the mean and standard deviation. How close were your estimates?

b. Write a sentence or two describing the distribution. Use the mean and standard deviation in your description.

c. Keri decided that it was too much work to enter the decimal point in the wages each time in her calculator list, so she entered each wage without it.

  i. Will the shape of her histogram be different from the given histogram? Explain.

  ii. Predict the mean and standard deviation for Keri's wage data. Check your predictions.

d. Suppose each student gets a 4% raise. How will the shape of the histogram of the new hourly wages be different from the original one? Predict the mean and standard deviation for the new wages of the students. Check your predictions.

e. Suppose that instead of a 4% raise, each student gets a raise of 25¢ per hour. Will the shape of the histogram of the new hourly wages be different from the original one? How will the mean and standard deviation change?

f. Let W_O represent the original hourly wage and W_N represent the new hourly wage. Write a rule that can be used to compute the new wage from the original wage

  i. for the case of a 4% raise, and

  ii. for the case of a 25¢ per hour raise.

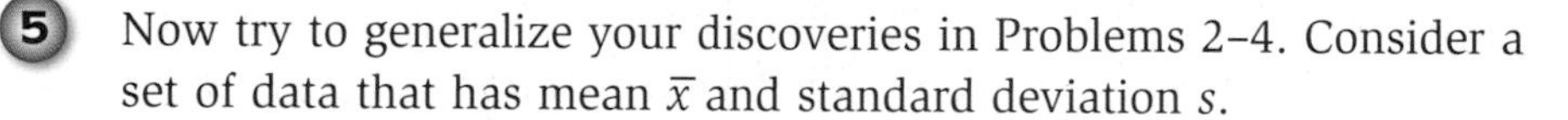

5 Now try to generalize your discoveries in Problems 2–4. Consider a set of data that has mean $\bar{x}$ and standard deviation s.

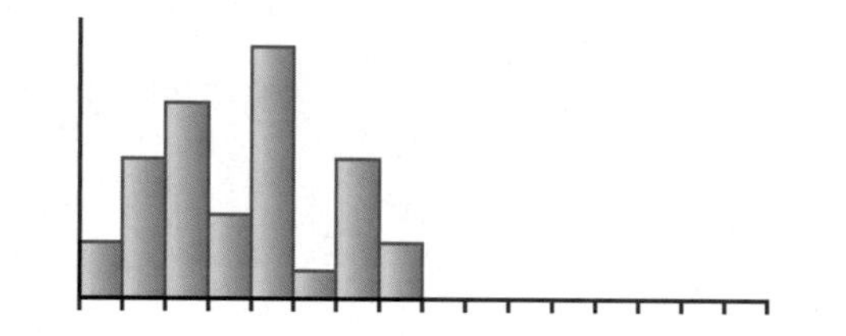

a. Suppose you add the same positive number d to each value. Use the histogram below to explain why the shape of the distribution does not change, the mean of the transformed data will be $\bar{x} + d$, and the standard deviation will remain s.

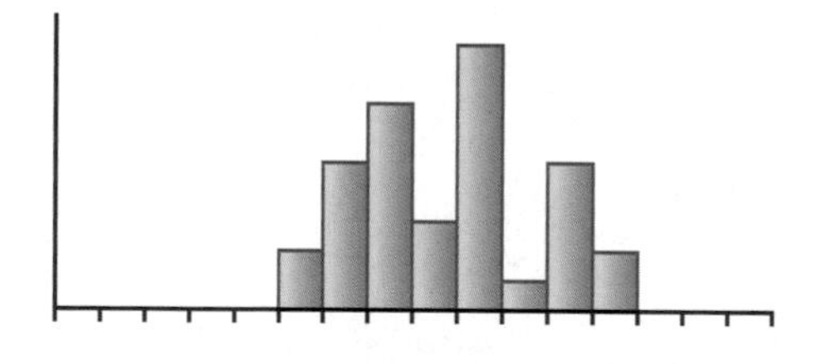

b. Write a summary statement about shape, center, and spread similar to that in Part a for the case of subtracting the same positive number from each value. Illustrate this by showing the effect of such a transformation on a histogram.

c. Write a summary statement similar to that in Part a for the case of multiplying each value by a positive number. Explain how the effect of such a transformation is illustrated by the histogram below.

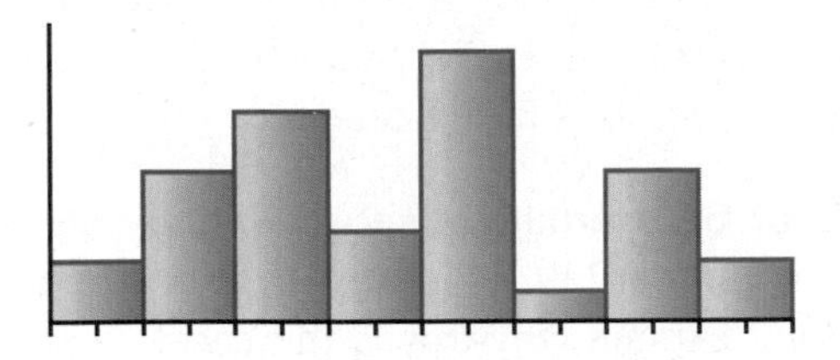

d. Write a similar statement for the case of dividing each value by a positive number. Illustrate this by showing the effect of such a transformation by drawing on a histogram.

e. Does the name "slide" or "stretch" best describe the transformation in Part a? In Part b? In Part c? In Part d?

6 One of the most common transformations is changing points scored to percentages such as on tests. The following display gives the points scored by a class of 32 students on a test with 75 possible points.

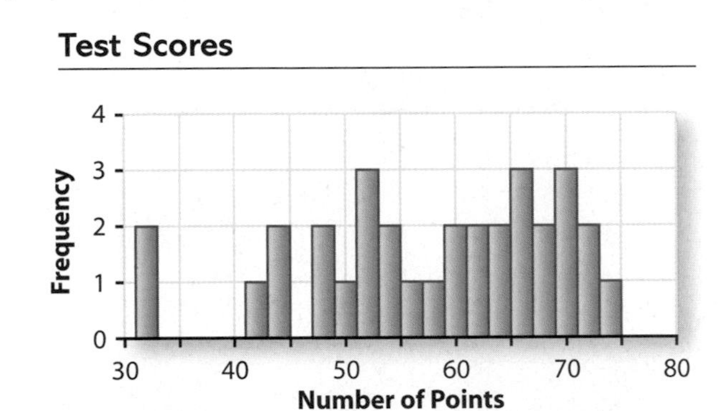

Summary Statistics	
Mean	59.34
Median	62
Stand Dev	11.29
IQR	16

a. Kim earned a score of 54 and Jim earned a score of 65. Change their scores to percentages (to the nearest tenth of a percent) of the possible points.

b. Describe the transformation you used by writing a formula. Be sure to define your variables.

c. Make a new table of summary statistics, using the percentages rather than the number of points scored.

Summarize the Mathematics

In this investigation, you discovered that transforming each value of a set of data affects the shape of the distribution, its center, and its spread in predictable ways.

a. What is the effect on the mean of transforming a set of data by adding or subtracting the same number to each value? What is the effect on the standard deviation? On the shape of the distribution? Explain why this is the case.

b. What is the effect on the mean of transforming a set of data by multiplying or dividing each value by the same positive number? What is the effect on the standard deviation? On the shape of the distribution? Explain why this is the case.

Be prepared to share your ideas and reasoning with the class.

Check Your Understanding

In the Carlyle family, the mean age is 26 with a standard deviation of 22.3 years.

a. What will be their mean age in 5 years? Their standard deviation in 5 years?

b. What is their mean age now in months? Their standard deviation in months?

c. The ages of the people in the Carlyle family are 1, 5, 9, 28, 31, 50, and 58.

- **i.** Compute the mean and standard deviation of their ages in 5 years. Was your prediction in Part a correct?
- **ii.** Compute the mean and standard deviation of their current ages in months. Was your prediction in Part b correct?

On Your Own

Applications

1. The table below gives the percentiles of recent SAT mathematics scores for national college-bound seniors. The highest possible score is 800 and the lowest possible score is 200. Only scores that are multiples of 50 are shown in the table, but all multiples of 10 from 200 to 800 are possible.

College-Bound Seniors

SAT Math Score	Percentile	SAT Math Score	Percentile
750	98	450	28
700	93	400	15
650	85	350	7
600	74	300	3
550	60	250	1
500	43	200	0

Source: *2005 College-Bound Seniors Total Group Profile Report*, The College Board

a. What percentage of seniors get a score of 650 or lower on the mathematics section of the SAT?

b. What is the lowest score a senior could get on the mathematics section of the SAT and still be in the top 40% of those who take the test?

c. Estimate the score a senior would have to get to be in the top half of the students who take this test.

d. Estimate the 25th and 75th percentiles. Use these quartiles in a sentence that describes the distribution.

2. In a physical fitness test, the median time it took a large group of students to run a mile was 10.2 minutes. The distribution had first and third quartiles of 7.1 minutes and 13.7 minutes. Faster runners (shorter times) were assigned higher percentiles.

a. Sheila's time was at the 25th percentile. How long did it take Sheila to run the mile?

b. Mark was told that his time was at the 16th percentile. Write a sentence that tells Mark what this means.

3 The histogram below gives the marriage rate per 1,000 people for 49 U.S. states in 2004. (Nevada, with a rate of 62 marriages per 1,000 people, was left off so the plot would fit on the page.)

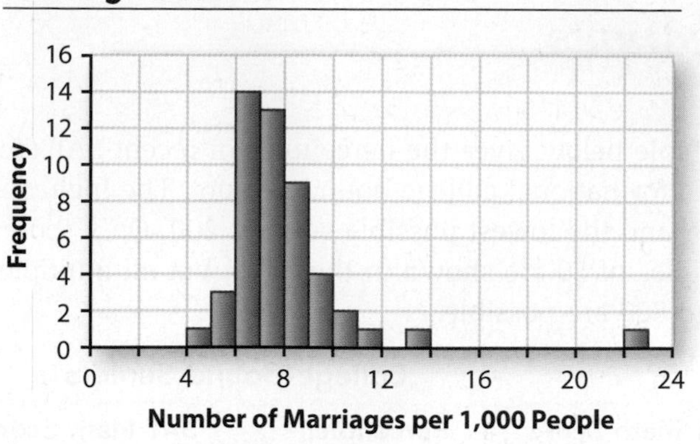

Source: Division of Vital Statistics, National Center for Health Statistics

a. Hawaii had about 1,262,840 residents and 28,793 marriages. What is the marriage rate per 1,000 residents for Hawaii? Where is Hawaii located on the histogram?

b. New York had about 130,744 marriages and 19,227,088 residents. What is the marriage rate per 1,000 residents for New York? Where is it located on the histogram?

c. Why do you think that Nevada's rate of 62 marriages per 1,000 people can't be interpreted as "62 out of every 1,000 residents of Nevada were married in 2004"?

d. The quartiles, including the median, divide a distribution, as closely as possible, into four equal parts. Estimate the median and lower and upper quartiles of the distribution and make a box plot of the distribution. Include Nevada in the distribution.

e. Now estimate the percentile for the following states.

i. Tennessee, with a marriage rate of 11.4

ii. Minnesota, with a marriage rate of 6.0

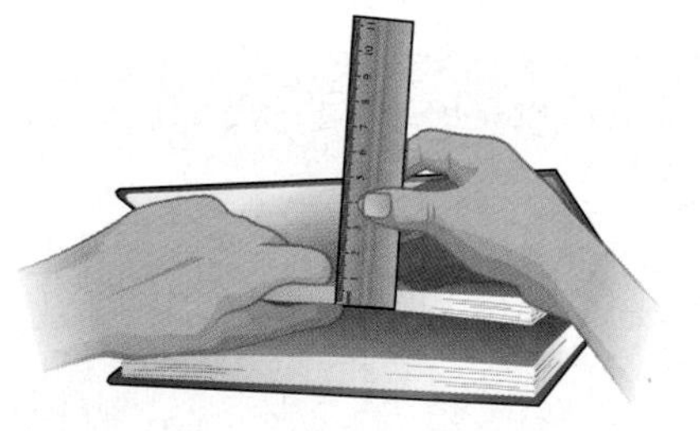

4 Suppose that you want to estimate the thickness of a piece of paper in your textbook. Compress more than a hundred pages from the middle of the book and measure the thickness to the nearest half of a millimeter. Divide by the number of sheets of paper. Round the result to four decimal places.

a. How can you determine the number of sheets of paper by using the page numbers?

b. Make ten more estimates, taking a different number of pages each time. Record your measurements on a dot plot.

c. What is the median of your measurements? What is the interquartile range?

d. Write a sentence or two reporting what you would give as an estimate of the thickness of the piece of paper.

5 The table below gives the price and size of 20 different boxed assortments of chocolate as reported in *Consumer Reports*.

Boxed Assortments of Chocolate

Brand	Price (in $)	Size (in oz)	Cost per oz
John & Kira's Jubilee Wood Gift Box	65	18	3.69
Martine's Gift Box Assorted with Creams	63	16	3.93
Norman Love Confections	37	8	4.62
Candinas	40	16	2.50
La Maison du Chocolat Coffret Maison with assorted chocolates	76	21	3.59
Moonstruck Classic Truffle Collections	70	20	3.50
Jacques Torres Jacques' Assortment	43	16	2.69
Fran's Assorted Truffles Gift Box	58	16	3.63
Godiva Gold Ballotin	35	16	
Leonidas Pralines General Assortment		16	1.75
See's Famous Old Time Assorted	13	16	
Ethel M Rich Deluxe Assortment	26	16	1.62
Lake Champlain Selection Fine Assorted	40	18	2.22
Rocky Mountain Chocolate Factory Gift Assortment Regular	19	14.5	1.31
Hershey's Pot of Gold Premium Assortment	8	14.1	0.57
Russell Stover Assorted	8	16	0.50
Whitman's Sampler Assorted	10	16	0.62
Rocky Mountain Chocolate Factory Sugar-Free Regular Gift Assortment	19	14.5	1.31
Russell Stover Net Carb Assorted	8	8.25	0.97
Ethel M Sugar-Free Truffle Collection	32	15	2.13

Source: *Consumer Reports*, February 2005

a. The cost per ounce is missing for Godiva and for See's. The price is missing for Leonidas Pralines. Compute those values.

b. The histogram to the right shows the cost-per-ounce data. Examine the histogram and make a sketch of what you think the box plot of the same data will look like. Then, make the box plot and check the accuracy of your sketch.

Boxed Assortments of Chocolate

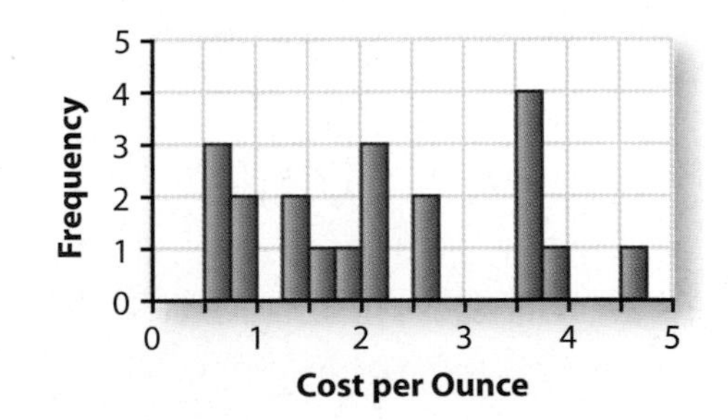

c. Identify any outliers in the cost-per-ounce data.

d. What information about boxed assortments of chocolate can you learn from the histogram that you cannot from the box plot? What information about boxed assortments of chocolate can you learn from the box plot that you cannot from the histogram?

e. Why is it more useful to plot the cost-per-ounce data than the price data?

6. The table below shows the total points scored during the first eight years of the NBA careers of Kareem Abdul-Jabbar and Michael Jordan.

Two Shooting Stars

Kareem Abdul-Jabbar		Michael Jordan	
Year	Points Scored	Year	Points Scored
1970	2,361	1985	2,313
1971	2,596	1986	408
1972	2,822	1987	3,041
1973	2,292	1988	2,868
1974	2,191	1989	2,633
1975	1,949	1990	2,753
1976	2,275	1991	2,580
1977	2,152	1992	2,404

a. Which player had the higher mean number of points per year?

b. What summary statistics could you use to measure consistency in a player? Which player was more consistent according to each of your statistics?

c. Use the rule to determine if there are any outliers for either player.

d. Jordan had an injury in 1986. If you ignore his performance for that year, how would you change your answers to Parts a and b?

7. The histogram below gives the scores of the ninth-graders at Lakeside High School on their high school's exit exam.

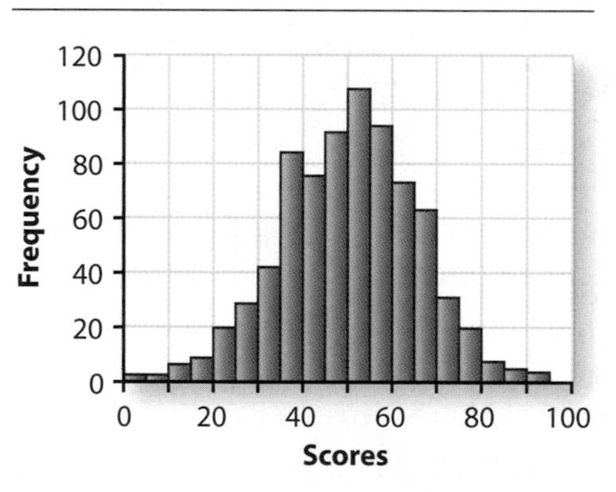

a. Estimate the mean and the standard deviation of the scores.

b. Estimate the percentile of a student whose score is one standard deviation below the mean. Then estimate the percentile corresponding to a score one standard deviation above the mean.

8 The numbers below are the play times (using the battery) in hours of 19 models of MP3 players. (Source: *Consumer Reports*, December 2005)

63, 45, 32, 30, 26, 18, 17, 17, 16,
16, 14, 14, 13, 10, 10, 10, 10, 9, 7

a. Compute the median and interquartile range and the mean and standard deviation of the play times.

b. One MP3 player has a play time of 10 hours. What is the deviation from the mean for that MP3 player?

c. Remove the 63 from the list and recompute the summary statistics in Part a.

d. How do you think the play times should be summarized? Explain.

9 In an experiment to compare 2 fertilizers, 12 trees were treated with Fertilizer A, and a different 12 trees were treated with Fertilizer B. The table below gives the number of kilograms of oranges produced per tree.

Kilograms of Oranges per Tree

Fertilizer A	Fertilizer B
3	14
14	116
19	33
0	40
96	10
92	72
11	8
24	10
5	2
31	13
84	15
15	44

a. Make a back-to-back stemplot of the number of kilograms of oranges produced by trees with Fertilizer A and with Fertilizer B. (See pages 97 and 98 for examples of stemplots.)

b. Use the stemplot to estimate the mean of each group.

c. Which group appears to have the larger standard deviation? How can you tell?

d. Compute the mean and standard deviation of each group. How close were your estimates in Part b?

e. What are the shape, mean, and standard deviation of the distribution of the number of *pounds* of oranges for Fertilizer A? For Fertilizer B? (There are about 2.2 pounds in a kilogram.)

f. Is a scatterplot an appropriate plot for these data? Why or why not?

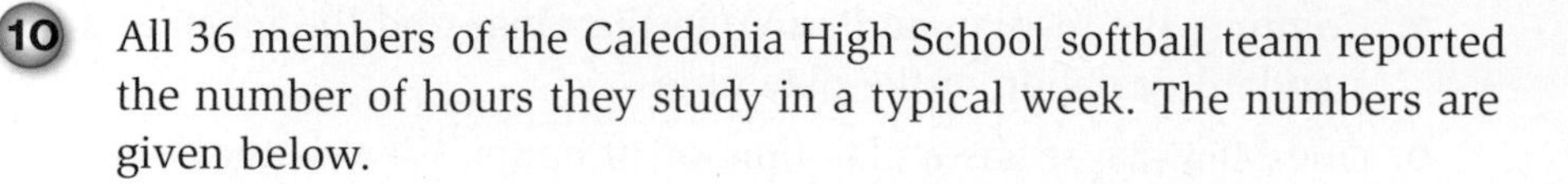

10 All 36 members of the Caledonia High School softball team reported the number of hours they study in a typical week. The numbers are given below.

5	5	5	6	10	11
12	12	12	13	14	15
15	16	16	16	17	17
17	17	18	19	19	20
20	20	20	20	20	23
25	25	25	27	28	40

Study Time of Softball Team Members

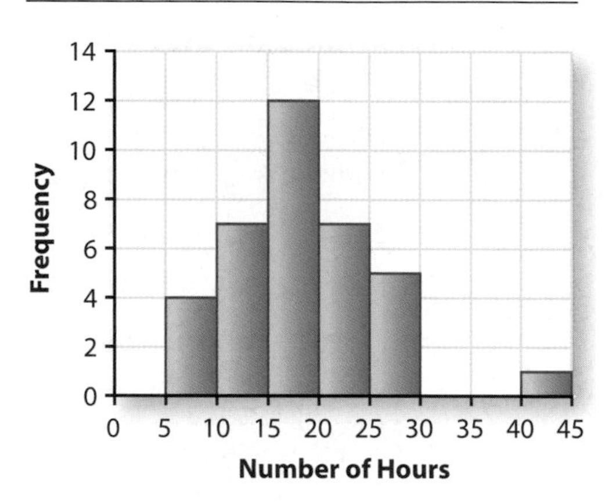

a. Estimate the mean and standard deviation of the distribution from the histogram.

b. Compute the mean and standard deviation of the distribution. How close were your estimates in Part a?

c. Akemi is the student who studies 40 hours a week. She is thinking of quitting the softball team. How will the mean and standard deviation change if Akemi quits and her number of hours is removed from the set of data?

d. Describe two ways to find the mean and standard deviation of the number of hours studied *per semester* (20 weeks) by these students. Find the mean and standard deviation using your choice of method.

e. The softball coach expects team members to practice a total of 10 hours. If practice hours are added to the weekly study hours for each student, how will the mean and standard deviation change?

Connections

11 Consider the box plot below.

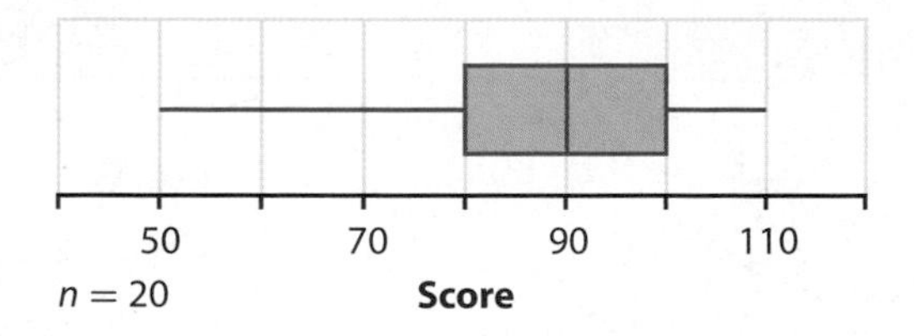

a. What does the "$n = 20$" below the plot mean?

b. About how many scores are between 50 and 80? Between 80 and 100? Greater than 80?

c. Is it possible for the box plot to be displaying the scores below? Explain your reasoning.

50, 60, 60, 75, 80, 80, 82, 83, 85, 90, 90,
91, 91, 94, 95, 95, 98, 100, 106, 110

d. Create a set of scores that could be the ones displayed by this box plot.

12 The box plots below represent the amounts of money (in dollars) carried by the people surveyed in four different places at a mall.

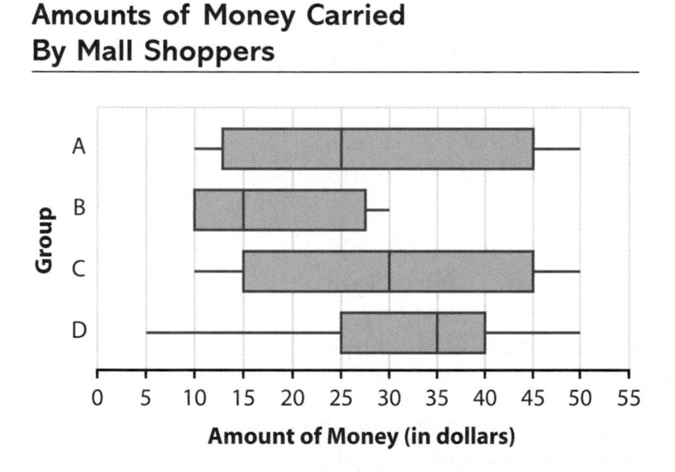

a. Which group of people has the smallest range? The largest?

b. Which group of people has the smallest interquartile range? The largest?

c. Which group of people has the largest median amount of money?

d. Which distribution is most symmetric?

e. Which group of people do you think might be high school students standing in line for tickets at a movie theater on Saturday night? Explain your reasoning.

f. Match each of the groups A, B, C, and D with its histogram below.

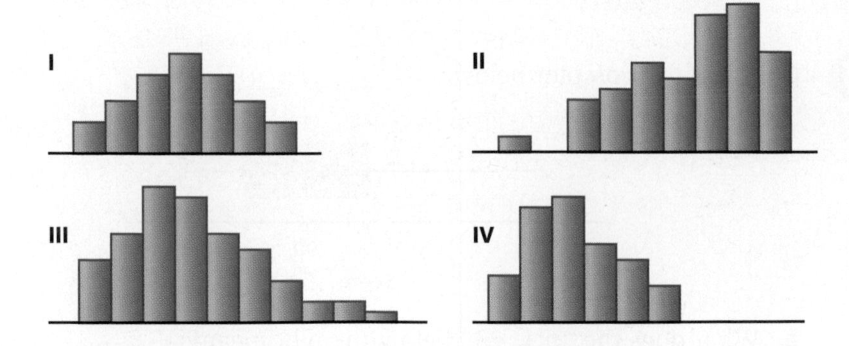

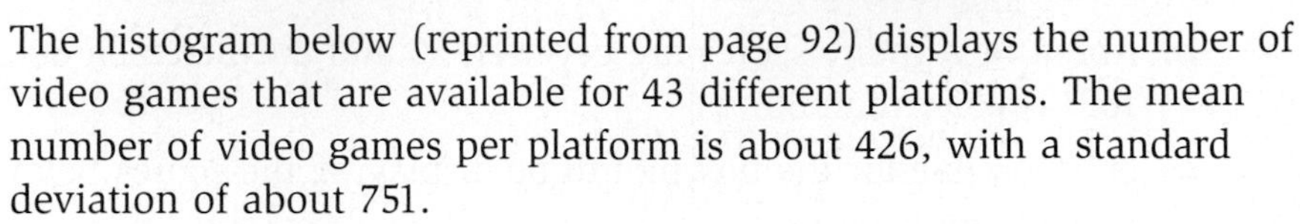

13 The histogram below (reprinted from page 92) displays the number of video games that are available for 43 different platforms. The mean number of video games per platform is about 426, with a standard deviation of about 751.

Number of Video Games

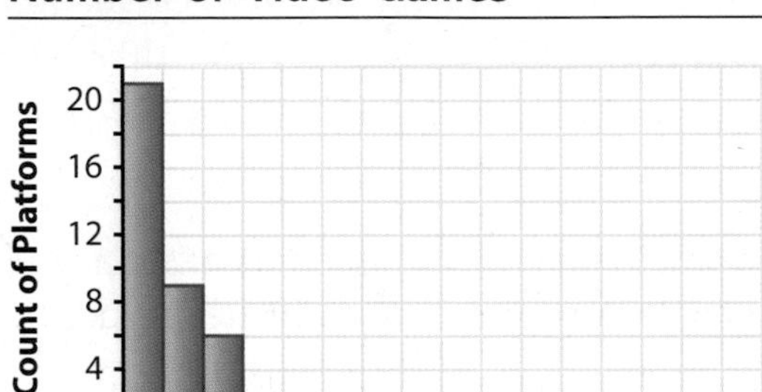

a. Do about 68% of the platforms fall within one standard deviation of the mean?

b. Why aren't the mean and standard deviation very informative summary statistics for these data?

14 Give counterexamples that show the statements below are not true in general.

a. If two sets of numbers have the same range, you should consider them to have the same variability.

b. If two sets of numbers have the same mean and the same standard deviation, they have the same distribution.

15 Refer to Problem 1 of Investigation 5 (page 124) for the 10 measurements of a desk or table.

a. Find the median and interquartile range of the original measurements.

b. Find the median and interquartile range after each measurement is transformed to inches.

i. How do the median and interquartile range of the transformed data compare to those of the original data?

ii. In general, what is the effect on the median and interquartile range if you divide each value in a data set by the same number?

c. Find the median and interquartile range after adding 0.2 cm to each original measurement.

   i. How do the median and interquartile range of the transformed data compare to those of the original data?

   ii. In general, what is the effect on the median and interquartile range of adding the same number to each value of a data set? Explain your reasoning.

Reflections

16 On page 112, you read about a study of lead in the blood of children. Each child who had been exposed to lead on the clothing of a parent was paired with a child who had not been exposed. A complete analysis should take this pairing into account. One way of doing that is to subtract the lead level of the control child from the lead level of the exposed child.

a. Find these differences and make a box plot of the differences.

b. If the exposure to lead makes no difference in the level of lead in the blood, where would the box plot be centered?

c. What conclusion can you draw from examining the box plot of the differences?

d. What additional information does this analysis take into account that the analysis in the Check Your Understanding did not?

e. Another way to look at these data is to make a scatterplot. What can you learn from the scatterplot below that you could not see from the other plots?

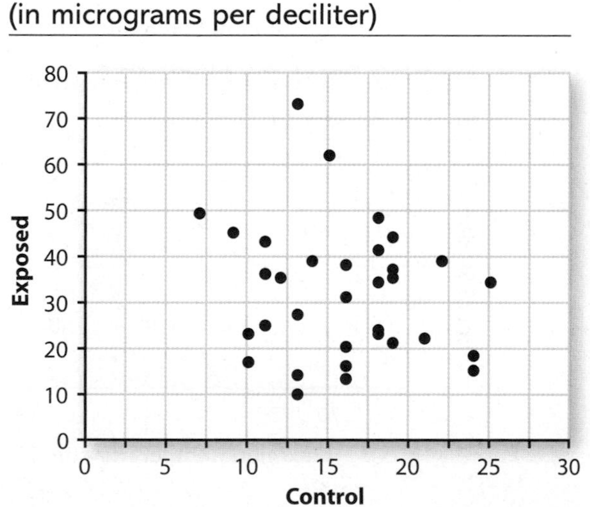

On Your Own

17 These box plots represent the scores of 80 seniors and 80 juniors on a fitness test. List the characteristics you know will be true about a box plot for the combined scores of the seniors and the juniors. For example, what will the minimum be?

Fitness Test Scores

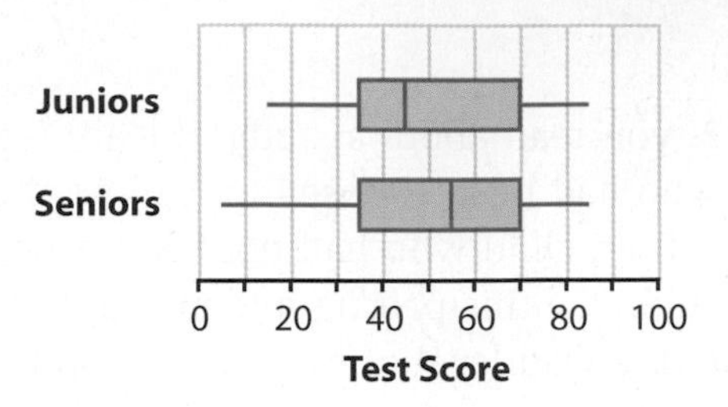

18 John Tukey, the same statistician who invented stemplots and box plots, established the standard rule for identifying possible outliers. When asked why he used 1.5 rather than some other number, he replied that 1 was too small and 2 was too big. Explain what he meant.

19 There are 15 outliers on the low end plus 2 outliers on the high end in the box plot below that shows the maximum temperatures ever recorded at the 289 U.S. weather stations in the 50 states, Puerto Rico, and the Pacific Islands.

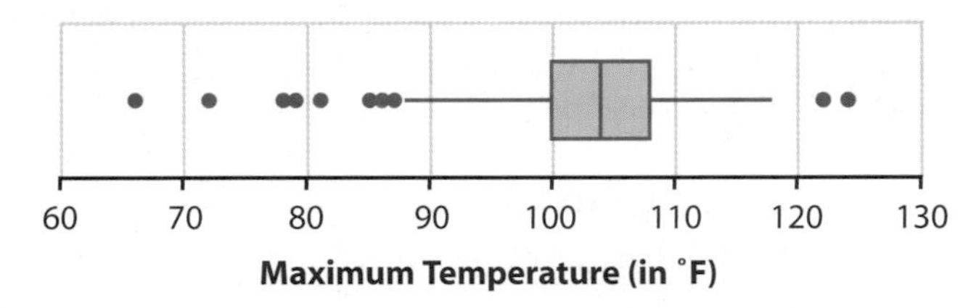

Maximum Recorded Temperature at U.S. Weather Stations

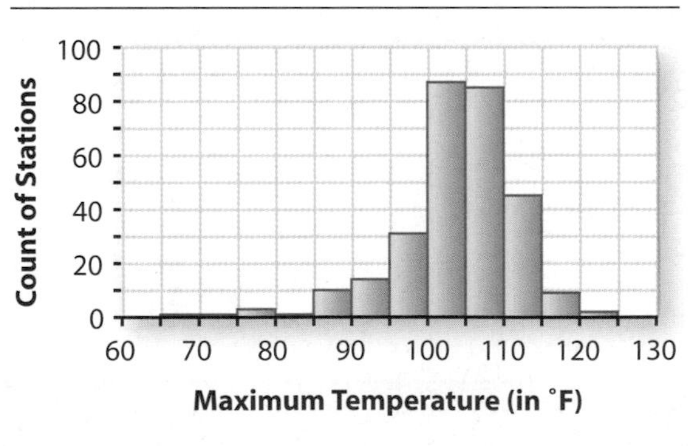

a. Study the histogram of the maximum temperatures. How can you tell from this histogram that there are outliers?

b. What are some geographical explanations for why there are so many outliers?

20 Is a minimum or a maximum value always an outlier? Is an outlier always a maximum or minimum value? Explain your answers.

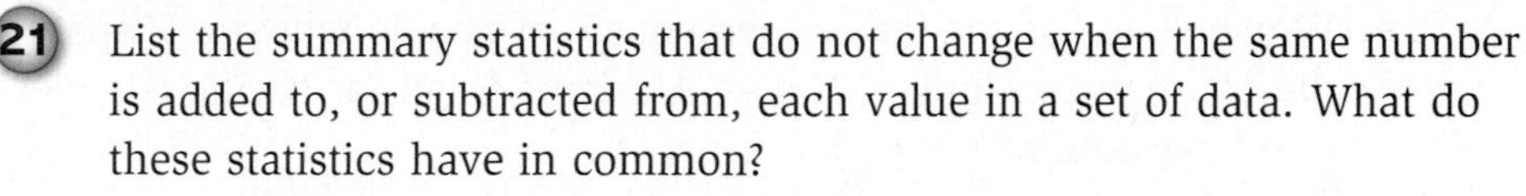

21 List the summary statistics that do not change when the same number is added to, or subtracted from, each value in a set of data. What do these statistics have in common?

22 When Nikki looked at the summary statistics for the 32 student tests in Problem 6 of Investigation 5 on page 127, she said, "These statistics can't be right. One standard deviation either side of the mean captures $\frac{2}{3}$ of the data and the IQR captures 50% of the data. So, the standard deviation should be larger than the IQR." Do you agree or disagree? Explain your reasoning.

Extensions

23 If your family has records of your growth, plot your own height over the years on a copy of the appropriate National Center for Health Statistics growth chart. How much has the percentile for your age varied over your lifetime?

24 Consider the position of the lower quartile for data sets with n values.

a. When n is odd and the values are placed in order from smallest to largest, explain why the position of the lower quartile is

$$\frac{n+1}{4}.$$

b. What is the position of the lower quartile when n is even?

25 Examine the 1999 U.S. Population by Race data set in your data analysis software. That data set includes the percentage of the population in each state and the District of Columbia who are Hispanic, Black, American Indian, Native Alaskan, Asian, or Pacific Islander. It also indicates which presidential candidate got the majority of votes cast in the 2000 presidential election in each state.

United States of America

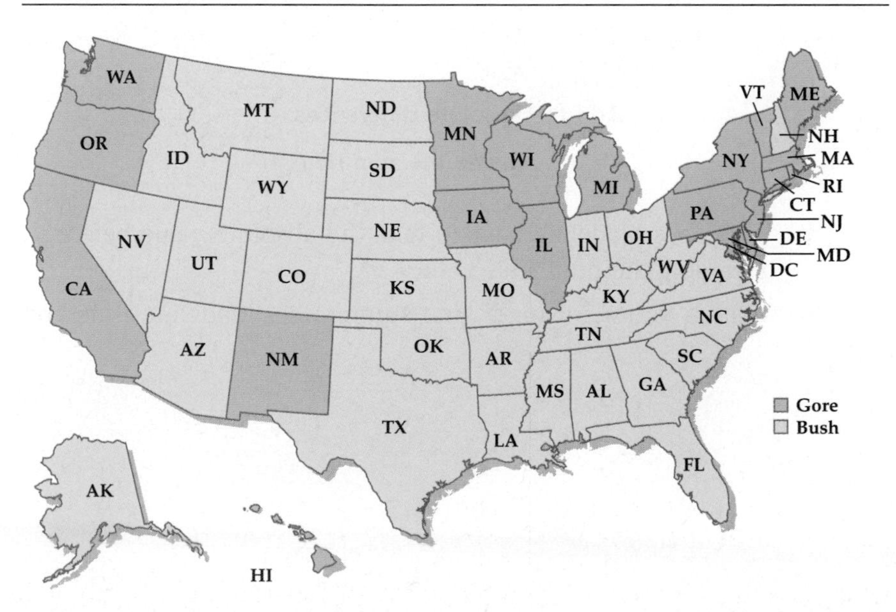

a. Which state has the largest percentage of people who are Hispanic, Black, American Indian, Native Alaskan, Asian, or Pacific Islander? The largest number?

b. If you find the mean of these 51 percentages, will that necessarily give you the percentage for the United States as a whole? Give a small example to illustrate your answer.

c. Make box plots of the percentages for states that favored Bush in the 2000 election and for the states (and Washington, D.C.) that favored Gore.

d. Describe the differences between the box plots. Why do you think there are these differences?

e. What other plot could be used to compare the two distributions? Make this plot that shows the two distributions. Can you see anything interesting that you could not see from the box plots?

f. Write a brief report that compares the two distributions. Explain your choice of summary statistics.

26 Madeline thought that a good measure of spread would be simply to

- find the deviations from the mean,
- take the absolute value of each one (Recall that if $a \geq 0$, then $|a| = a$ and if $a < 0$, then $|a| = -a$.)
- average them.

Madeline calls her method the MAD: **Mean Absolute Deviation.**

a. Use Madeline's method and the table below to find the MAD for these numbers:

1, 4, 6, 8, 9, 14

Number	Mean	(Number − Mean)	|Number − Mean|
		Add the absolute differences:	
		Divide the sum by n:	

b. Why does Madeline have to take the absolute value before averaging the numbers?

c. Write a formula using Σ that summarizes Madeline's method.

d. Compute the standard deviation of the number above. How does the standard deviation compare to the MAD?

e. Madeline has indeed invented an appealing measure for describing spread. However, the MAD does not turn out to be as useful a summary statistic as the standard deviation, so it does not have a central place in the theory or practice of statistics. Does your calculator or statistical software have a function for the MAD?

27 Suppose each of the 32 students at Price Lab School tried to cut a square out of cardboard that was 24.6 mm on each side. A histogram of the actual perimeters of their squares is displayed below. The mean perimeter was 98.42 mm.

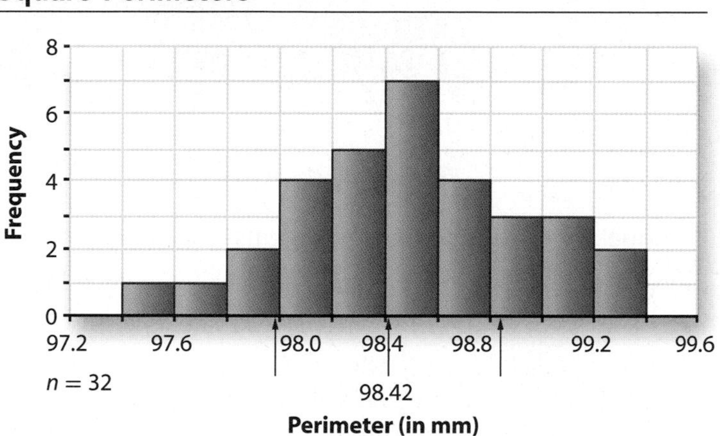

a. What might explain the variability in perimeters?

b. An interval one standard deviation above and below the mean is marked by arrows on the histogram. Use it to estimate the standard deviation.

c. Estimate the percentage of the perimeters that are within one standard deviation of the mean.

d. Estimate the percentage of the perimeters that are within two standard deviations of the mean. (For a normal distribution, this is about 95%.)

On Your Own

28 The following are the resting pulse rates of a group of parents of ninth-graders at Beaverton High School.

Parent Pulse Rates

60	62	67	68	70	72	74	75
76	76	77	78	80	81	81	82
83	84	84	84	86	88	88	88
89	90	91	94	107			

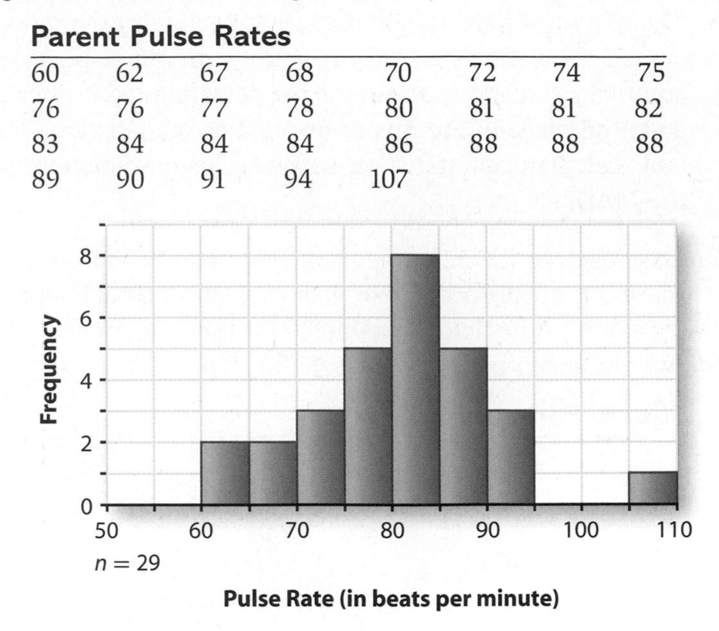

a. Describe the distribution of the parents' pulse rates.

b. Dion's mother thinks she may have lost count since her pulse rate was the lowest, 60. Does her reported pulse rate look unusually low to you? Explain.

c. The teacher said that anyone with a resting rate of more than two standard deviations from the mean should repeat the test to check the results. How many students had their parents repeat the test? Did Dion's mother need to repeat the test?

d. The parents did two minutes of mild exercise, then counted their rates again. This time the mean was 90, and the standard deviation was 12.4. Cleone's mother had a pulse rate of 95 after exercising.

   i. Is her rate unusually high? How can the standard deviation help explain your answer?

   ii. What would your conclusion be about Cleone's mother's pulse rate if the standard deviation was 1.24?

Review

29 The number 20,000 can be written as $2(10{,}000) = 2 \cdot 10^4$ and 2,000,000 can be written as $2(1{,}000{,}000) = 2 \cdot 10^6$.

a. On your calculator when you multiply 20,000 by 2,000,000 you get "4E10." What does this mean?

b. Predict what you will get when you use your calculator to multiply 2,000,000 by 4,000,000.

c. Predict what you will get when you use your calculator to multiply 2,400,000 by 20,000. What rule does the calculator appear to be using to format the answer?

30 If 10% of a number is 20, use mental computation to find the following.

a. 30% of the number
b. 150% of the number
c. One half of the number
d. 35% of the number

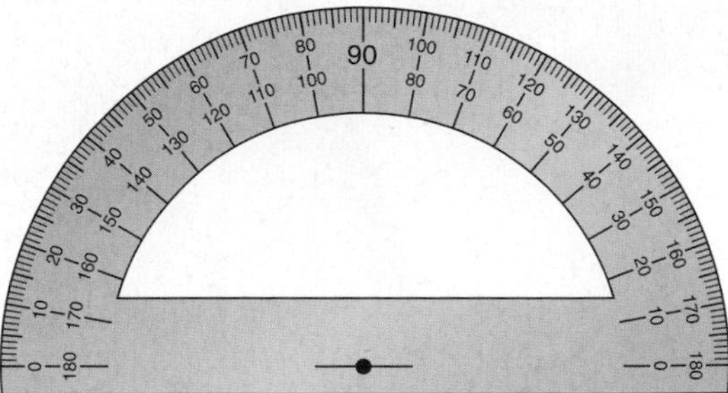

31 Using a protractor, draw and label each angle. If you do not have a protractor, place a sheet of paper over the protractor to the right.

a. $m\angle BAC = 90°$
b. $m\angle FDE = 30°$
c. $m\angle PQR = 120°$
d. $m\angle XZY = 65°$
e. $m\angle STV = 180°$

32 Find results for each of these calculations.

a. 12 – (–8)
b. –3 – 7
c. –3 – (–7)
d. 8 – 12
e. –8 + (–12)
f. 2.5 – (–1.3)

33 An amusement park reports an increase of 21 bungee customers from Saturday to Sunday. If this represents an increase of 7% in the number of customers:

a. What would a 1% increase be?
b. What would a 10% increase be?
c. What would a 5.1% increase be?
d. What was the original number of customers?
e. What is 5.1% of your answer for Part d?
f. What is 7% of your answer for Part d?

34 Use a protractor (or place a sheet of paper over the protractor in Task 31) to help you draw two lines $\overleftrightarrow{AB}$ and $\overleftrightarrow{CD}$ that intersect at point O so that $m\angle AOC = 52°$. Label your diagram. What are the measures of $\angle COB$, $\angle BOD$, and $\angle DOA$?

35 Express each of these fractions in equivalent simplest form.

a. $\frac{-12}{-30}$
b. $\frac{20}{-12}$
c. $\frac{5-8}{9-5}$
d. $\frac{-5-8}{-5-(-8)}$
e. $\frac{78-6}{9-(-18)}$
f. $\frac{5-7}{10+14}$

36 Mike has the following coins in his pocket: a penny, a nickel, a dime, and a quarter. Two of these coins fall out of his pocket. What is the probability that their total value is less than fifteen cents?

37 Suppose that you have twelve 1-inch square tiles.

a. Sketch diagrams of all possible ways that you can arrange the tiles so that they form a rectangle. Each rectangle must be completely filled in with tiles.
b. Find the perimeter of each rectangle in Part a.

38 Without computing, determine if each expression is greater than 0, equal to 0, or less than 0.

a. –5.75(–0.35)
b. $(-1.56)^4 - 123$
c. –5,768 + 10,235
d. $\frac{783(-52.6)}{-12.85}$

Looking Back

In this unit, you learned how to display data using dot plots, histograms, and box plots. Examination of these plots gave you information about the shape, the center, and the variability (spread) of the distributions.

You also learned how to compute and interpret common measures of center (mean and median) and common measures of variability (interquartile range and standard deviation).

Finally, you explored the effects on a distribution of transforming by adding a constant and by multiplying by a positive constant. While exploring the following data set you will review these key ideas.

A California psychologist, Robert V. Levine, noticed that the *pace of life* varies from one U.S. city to another and decided to quantify that impression.

For each city, he measured

- how long on average it took bank clerks to make change,
- the average walking speed of pedestrians on an uncrowded downtown street during the summer, and
- the speaking rate of postal clerks asked to explain the difference between regular mail, certified mail, and insured mail.

These three measurements were combined into one total score for each city, given in the table at the top of the next page. A higher total score means a faster pace of life.

Pace of Life in U.S. Cities

Total Score	City	Region	Total Score	City	Region
83	Boston	NE	75	Houston	SO
76	Buffalo	NE	79	Atlanta	SO
71	New York	NE	67	Louisville	SO
75	Worcester	NE	58	Knoxville	SO
80	Providence	NE	70	Chattanooga	SO
79	Springfield, MA	NE	54	Shreveport	SO
78	Paterson, NJ	NE	67	Dallas	SO
62	Philadelphia	NE	70	Nashville	SO
73	Rochester	NE	66	Memphis	SO
79	Columbus	MW	60	San Jose	WE
66	Canton	MW	79	Salt Lake City	WE
60	Detroit	MW	72	Bakersfield	WE
74	Youngstown	MW	61	San Diego	WE
72	Indianapolis	MW	59	San Francisco	WE
72	Chicago	MW	61	Oxnard	WE
77	Kansas City	MW	61	Fresno	WE
68	East Lansing	MW	50	Sacramento	WE
68	St. Louis	MW	45	Los Angeles	WE

Source: The Pace of Life, *American Scientist, 78.* September–October 1990.

1 The histogram below shows the distribution of the total scores for the pace of life in the 36 cities.

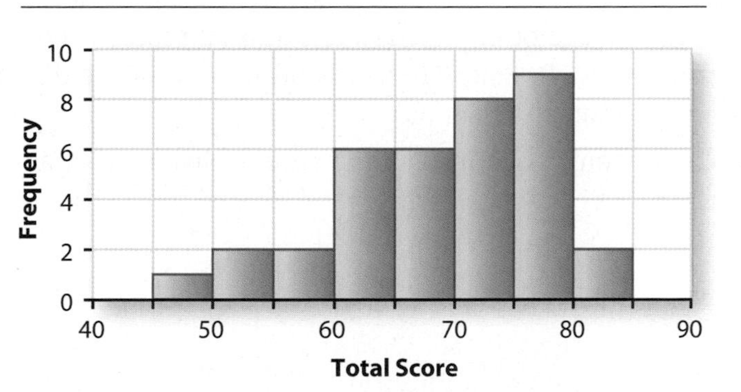

a. Describe the shape of the distribution.

b. Estimate the five-number summary from the histogram.

c. The mean of the distribution is 68.5.

  i. What is the deviation from the mean for Philadelphia? For New York?

  ii. Which of the 36 cities has a total score that is the largest deviation from the mean?

d. Without computing, is the standard deviation closer to 5, or 10, or 20? Explain.

2 The box plots below show the nine cities in each of three regions: the Midwest, the South, and the West.

Pace of Life by Geographic Region

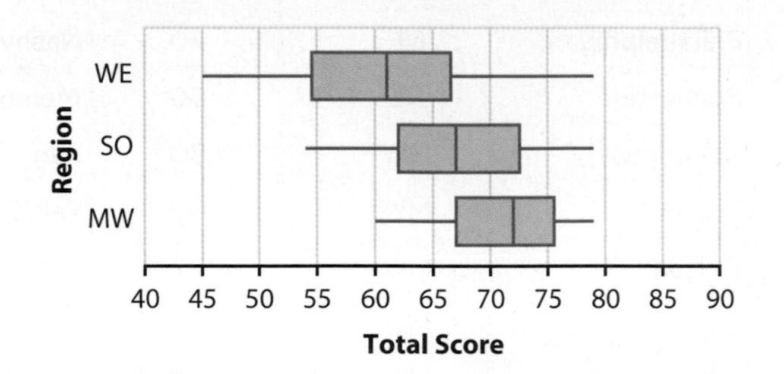

a. The box plot for the Northeast is missing. Find the five-number summary for the Northeast and determine if there are any outliers. Then make the box plot, showing any outliers.

b. If the cities selected are typical, in which region of the country is the pace of life fastest? Explain your reasoning.

c. Without computing, how can you tell which region has the largest standard deviation? Compute and interpret the standard deviation for that region.

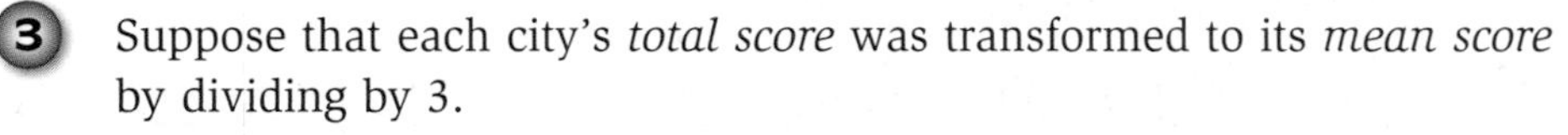

3 Suppose that each city's *total score* was transformed to its *mean score* by dividing by 3.

a. The average of the distribution of total scores is 68.5 and the median is 70. Find the mean and median of the distribution of mean scores.

b. How would each measure of spread change, if at all?

Summarize the Mathematics

Patterns in data can be seen in graphical displays of the distribution and can be summarized using measures of center and spread.

a Describe the kinds of information you can get by examining:

i. a dot plot,

ii. a histogram and a relative frequency histogram, and

iii. a box plot.

b Describe the most common measures of center, how to find them, and what each one tells you.

c Describe the most common measures of variability, how to find them, and what each one tells you.

d What measures can you use to tell someone the position of a value in a distribution?

e How do you identify outliers and what should you do once you identify them? Which summary statistics are resistant to outliers?

f What is the effect on measures of center of transforming a set of data by adding a constant to each value or multiplying each value by a positive constant? On measures of variation?

Be prepared to share your ideas and reasoning with the class.

✓ *Check Your Understanding*

Write, in outline form, a summary of the important mathematical concepts and methods developed in this unit. Organize your summary so that it can be used as a quick reference in future units and courses.

LINEAR FUNCTIONS

In the *Patterns of Change* unit, you explored a variety of situations in which variables change over time or in response to changes in other variables. Recognizing patterns of change enabled you to make predictions and to understand situations better.

In this unit of *Core-Plus Mathematics*, you will focus on patterns in tables, graphs, and rules of the simplest and one of the most important relationships among variables, linear functions. The understanding and skill needed to analyze and use linear functions will develop from your work on problems in three lessons.

Lessons

1 *Modeling Linear Relationships*

Identify problem conditions, numeric patterns, and symbolic rules of functions with graphs that are straight lines. Write rules for linear functions given a problem situation or data in a table or a graph. Fit lines and function rules to data patterns that are approximately linear.

2 *Linear Equations and Inequalities*

Express questions about linear functions as equations or inequalities. Use function tables, graphs, and symbolic reasoning to answer those questions.

3 *Equivalent Expressions*

Use context clues and algebraic properties of numbers and operations to recognize and write equivalent forms of symbolic representations of linear functions.

LESSON 1

Modeling Linear Relationships

In the *Patterns of Change* unit, you studied a variety of relationships between quantitative variables. Among the most common were **linear functions**—those with straight-line graphs, data patterns showing a constant rate of change in the dependent variable, and rules like $y = a + bx$.

For example, Barry represents a credit card company on college campuses. He entices students with free gifts—hats, water bottles, and T-shirts—to complete a credit card application. The graph on the next page shows the relationship between Barry's daily pay and the number of credit card applications he collects. The graph pattern suggests that *daily pay* is a linear function of *number of applications*.

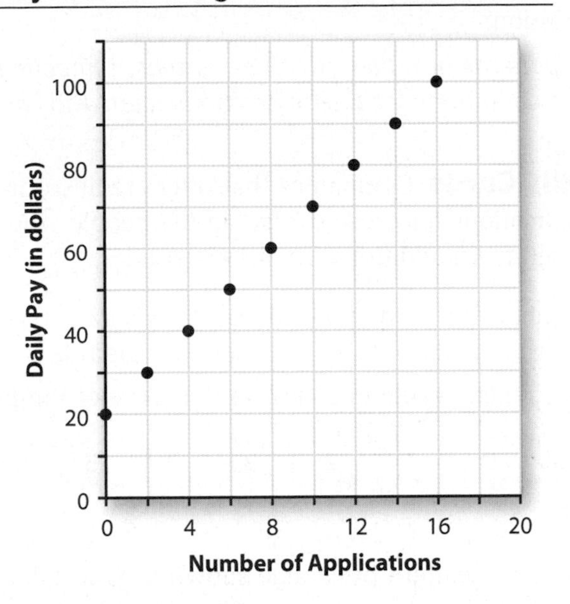

Think About This Situation

Think about the connections among graphs, data patterns, function rules, and problem conditions for linear relationships.

a How does Barry's daily pay change as the number of applications he collects increases? How is that pattern of change shown in the graph?

b If the linear pattern shown by the graph holds for other (*number of applications, daily pay*) pairs, how much would you expect Barry to earn for a day during which he collects just 1 application? For a day he collects 13 applications? For a day he collects 25 applications?

c What information from the graph might you use to write a rule showing how to calculate daily pay for any number of applications?

Working on the problems of this lesson will develop your ability to recognize linear relationships between variables and to represent those relationships with graphs, tables of values, and function rules.

Investigation 1 Getting Credit

Information about a linear function may be given in the form of a table or graph, a symbolic rule, or a verbal description that explains how the dependent and independent variables are related. To be proficient in answering questions about linear functions, it helps to be skillful in translating given information from one form into another.

As you work on problems of this investigation, look for clues to help you answer this question:

How are patterns in tables of values, graphs, symbolic rules, and problem conditions for linear functions related to each other?

Selling Credit Cards Companies that offer credit cards pay the people who collect applications for those cards and the people who contact current cardholders to sell them additional financial services.

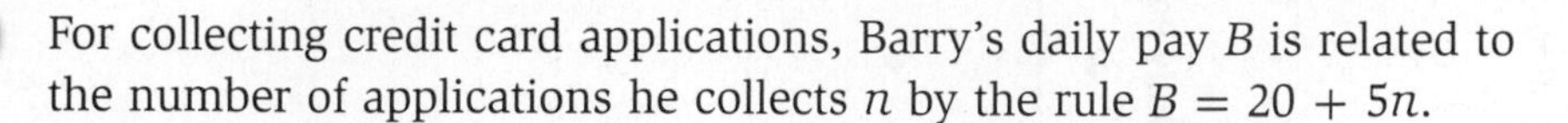

1 For collecting credit card applications, Barry's daily pay B is related to the number of applications he collects n by the rule $B = 20 + 5n$.

a. Use the function rule to complete this table of sample (n, B) values:

Number of Applications	0	1	2	3	4	5	10	20	50
Daily Pay (in dollars)									

b. Compare the pattern of change shown in your table with that shown in the graph on the preceding page.

c. How much will Barry earn on a day when he does not collect any credit card applications? How can this information be seen in the rule $B = 20 + 5n$? In the table of sample (n, B) values? In the graph on the preceding page?

d. How much additional money does Barry earn for each application he collects? How can this information be seen in the rule $B = 20 + 5n$? In the table? In the graph?

e. Use the words *NOW* and *NEXT* to write a rule showing how Barry's daily pay changes with each new credit card application he collects.

2 Cheri also works for the credit card company. She calls existing customers to sell them additional services for their account. The next table shows how much Cheri earns for selling selected numbers of additional services.

Number of Services Sold	10	20	30	40	50
Daily Pay (in dollars)	60	80	100	120	140

a. Does Cheri's daily pay appear to be a linear function of the number of services sold? Explain.

b. Assume that Cheri's daily pay is a linear function of the number of services she sells, and calculate the missing entries in the next table.

Number of Services Sold	0	10	15	20	25	30	40	50	100	101
Daily Earnings (in dollars)		60		80		100	120	140		

A key feature of any function is the way the value of the dependent variable changes as the value of the independent variable changes. Notice that as the number of services Cheri sells increases from 30 to 40, her pay increases from \$100 to \$120. This is an increase of \$20 in pay for an increase of 10 in the number of services sold, or an average of \$2 per sale. Her pay increases at a *rate* of \$2 per service sold.

c. Using your table from Part b, study the *rate of change* in Cheri's daily pay as the number of services she sells increases by completing entries in a table like the one below.

Change in Sales	Change in Pay (in \$)	Rate of Change (in \$ per sale)
10 to 20		
20 to 25		
25 to 40		
50 to 100		

What do you notice about the rate of change in Cheri's daily pay as the number of services she sells increases?

d. Use the words *NOW* and *NEXT* to write a rule showing how Cheri's pay changes with each new additional service she sells.

e. Consider the following function rules.

$C = 2 + 40n$ $\qquad C = n + 2$ $\qquad C = 40 + 2n$

$C = 50 + \frac{n}{2}$ $\qquad C = 2n + 50$

i. Which of the rules show how to calculate Cheri's daily pay C for any number of services n she sells? How do you know?

ii. What do the numbers in the rule(s) you selected in part i tell you about Cheri's daily pay?

3 The diagram below shows graphs of pay plans offered by three different banks to employees who collect credit card applications.

Atlantic Bank: $A = 20 + 2n$

Boston Bank: $B = 20 + 5n$

Consumers Bank: $C = 40 + 2n$

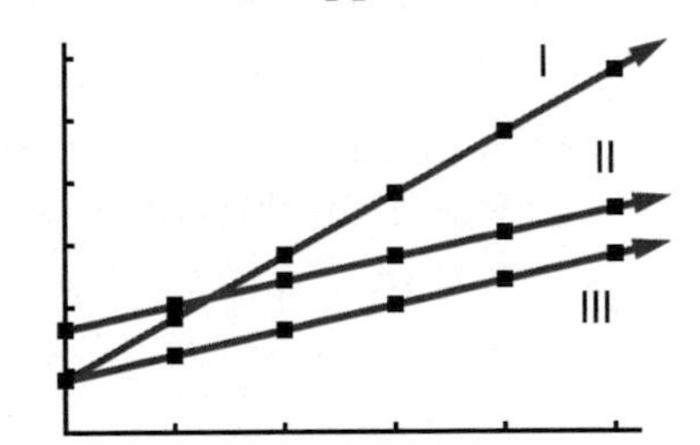

a. Match each function rule with its graph. Explain how you can make the matches without calculations or graphing tool help.

b. What do the numbers in the rule for the pay plan at Atlantic Bank tell you about the relationship between daily pay and number of credit card applications collected?

Buying on Credit Electric Avenue sells audio/video, computer, and entertainment products. The store offers 0% interest for 12 months on purchases made using an Electric Avenue store credit card.

Emily purchased a television for $480 using an Electric Avenue store credit card. Suppose she pays the minimum monthly payment of $20 each month for the first 12 months.

a. Complete a table of (*number of monthly payments, account balance*) values for the first 6 months after the purchase, then plot those values on a graph.

Number of Monthly Payments	0	1	2	3	4	5	6
Account Balance (in dollars)							

b. Will Emily pay off the balance within 12 months? How do you know?

c. If you know Emily's account balance *NOW*, how can you calculate the *NEXT* account balance, after a monthly payment?

d. Which of the following function rules gives Emily's account balance E after m monthly payments have been made?

$E = 20m - 480$ $\qquad E = m - 20$ $\qquad E = -20m + 480$

$E = 480 + 20m$ $\qquad E = 480 - 20m$

e. Determine the rate of change, including units, in the account balance as the number of monthly payments increases from:

0 to 2;

2 to 3;

3 to 6.

i. How does the rate of change reflect the fact that the account balance *decreases* as the number of monthly payments increases?

ii. How can the rate of change be seen in the graph from Part a? In the function rule(s) you selected in Part c?

f. How can the starting account balance be seen in the table in Part a? In the graph? In the function rule(s) you selected in Part d?

5 The diagram below shows graphs of account balance functions for three Electric Avenue customers.

Emily: $E = 480 - 20m$
Darryl: $D = 480 - 40m$
Felicia: $F = 360 - 40m$

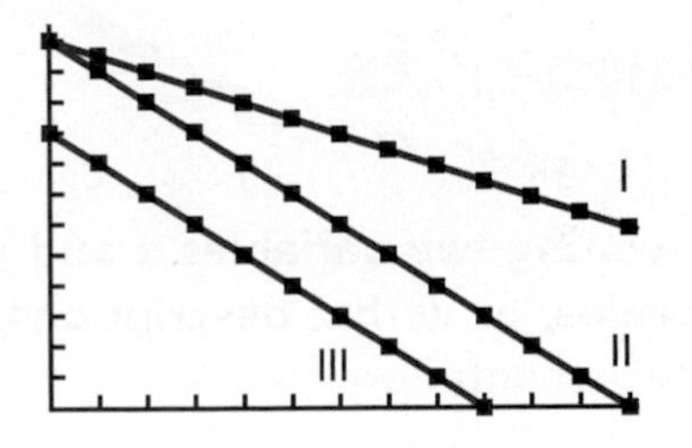

a. Match each function rule with its graph. Explain how you could make the matches without calculations or graphing tool help.

b. What do the numbers in the rules for Darryl's and Felicia's account balances tell you about the values of their purchases and their monthly payments?

Linear Functions Without Contexts When studying linear functions, it helps to think about real contexts. However, the connections among graphs, tables, and symbolic rules are the same for linear functions relating *any* two variables.

You've probably noticed by now that the rate of change of a linear function is constant and that the rate of change corresponds to the direction and steepness of the graph, or the *slope* of the graph.

You can determine the **rate of change** of y as x increases, or the **slope** of the graph between two points, using the ratio:

$$\frac{\text{change in } y}{\text{change in } x} \text{ or } \frac{\Delta y}{\Delta x}.$$

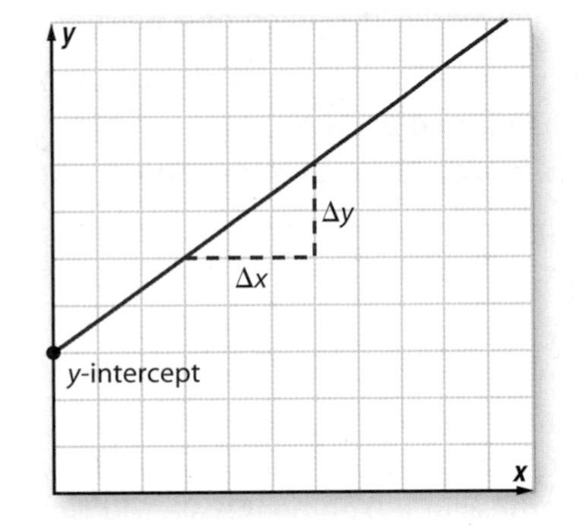

(Δ is the Greek letter "delta," which is used to represent "difference" or "change.")

Another key feature of a linear function is the **y-intercept** of its graph, the point where the graph intersects the y-axis.

6 Draw a graph for each function on a separate set of coordinate axes.

a. $y = 1 + \frac{2}{3}x$

b. $y = 2x$

c. $y = 2x - 3$

d. $y = 2 - \frac{1}{2}x$

Then analyze each function rule and its graph as described below.

i. Label the coordinates of three points A, B, and C on each graph. Calculate the slopes of the segments between points A and B, between points B and C, and between points A and C.

ii. Label the coordinates of the y-intercept on each graph.

iii. Explain how the numbers in the symbolic rule relate to the graph.

Summarize the Mathematics

Linear functions relating two variables *x* and *y* can be represented using tables, graphs, symbolic rules, or verbal descriptions. Key features of a linear function can be seen in each representation.

a How can you determine whether a function is linear by inspecting a:

i. table of (x, y) values?

ii. graph of the function?

iii. symbolic rule relating y to x?

iv. *NOW-NEXT* rule?

b How can the rate of change and the slope of the graph for a linear function be found from a:

i. table of (x, y) values?

ii. graph of the function?

iii. symbolic rule relating y to x?

iv. *NOW-NEXT* rule?

c How can the y-intercept of the graph of a function be seen in a:

i. table of (x, y) values?

ii. graph of the function?

iii. symbolic rule relating y to x?

Be prepared to share your ideas and reasoning with the class.

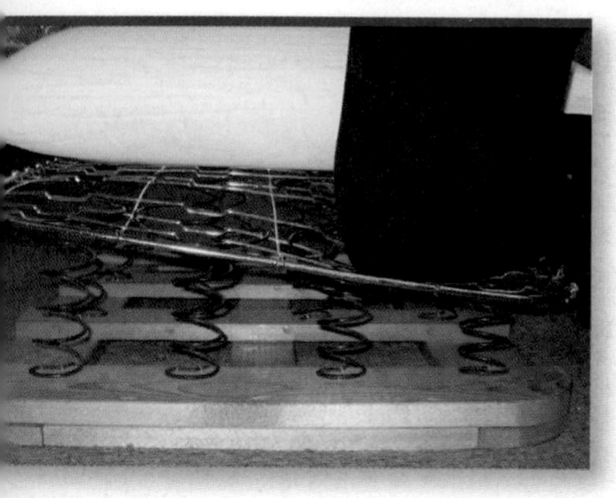

Check Your Understanding

Linear functions can be used to describe the action of springs that stretch, like those in telephone cords, and springs that compress, like those in a mattress or a bathroom scale. Hooke's Law in science says that, for an ideal coil spring, the relationship between weight and length is perfectly linear, within the elastic range of the spring.

The table below shows data from an experiment to test Hooke's Law on different coil springs.

Spring 1	
Weight (ounces)	Length (inches)
0	12
4	14
8	16
12	18
16	20

Spring 2	
Weight (ounces)	Length (inches)
0	5
2	7
4	9
6	11
8	13

Spring 3	
Weight (ounces)	Length (inches)
0	18
3	15
6	12
9	9
12	6

Spring 4	
Weight (ounces)	Length (inches)
0	12
4	10
8	8
12	6
16	4

For each spring:

a. Identify the length of the spring with no weight applied.

b. Describe the rate of change of the length of the spring as weight is increased. Indicate units.

c. Decide whether the experiment was designed to measure spring stretch or spring compression.

d. Write a rule using *NOW* and *NEXT* to show how the spring length changes with each addition of one ounce of weight.

e. Match the spring to the rule that gives its length ℓ in inches when a weight of w ounces is applied.

$\ell = 12 - \frac{1}{2}w$ $\quad \ell = 12 + \frac{1}{2}w$ $\quad \ell = 5 + w$ $\quad \ell = 18 - w$

f. Match the spring to the graph in the diagram below that shows ℓ as a function of w.

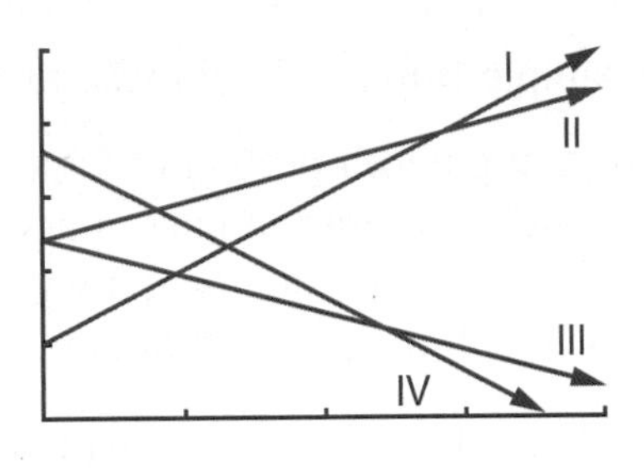

Investigation 2 Symbolize It

A symbolic rule showing how values of one variable can be calculated from values of another is a concise and simple way to represent a function. Mathematicians typically write the rules for linear functions in the form $y = mx + b$. Statisticians prefer the general form $y = a + bx$. In a linear function rule like $y = 3x + 7$, or equivalently $y = 7 + 3x$, the number 3 is called the **coefficient of x** and the number 7 is called the **constant term**.

You probably have several strategies for finding values of the constant term and the coefficient of x in rules for particular linear functions. As you complete the problems in this investigation, look for clues that will help you answer this basic question:

How do you use information in a table, a graph, or the conditions of a problem to write a symbolic rule for a linear function?

1 **Dunking Booth Profits** The student council at Eastern High School decided to rent a dunking booth for a fund-raiser. They were quite sure that students would pay for chances to hit a target with a ball to dunk a teacher or administrator in a tub of cold water.

The dunking booth costs $150 to rent for the event, and the student council decided to charge students $0.50 per throw.

a. How do you know from the problem description that *profit* is a linear function of the *number of throws*?

b. Use the words *NOW* and *NEXT* to write a rule showing how fund-raiser profit changes with each additional customer.

c. Write a rule that shows how to calculate the profit P in dollars if t throws are purchased. Explain the thinking you used to write the rule.

d. What do the coefficient of t and the constant term in your rule from Part c tell about:

- **i.** the graph of profit as a function of number of throws?
- **ii.** a table of sample (*number of throws, profit*) values?

The description of the dunking booth problem included enough information about the relationship between number of customers and profit to write the profit function. However, in many problems, you will have to reason from patterns in a data table or graph to write a function rule.

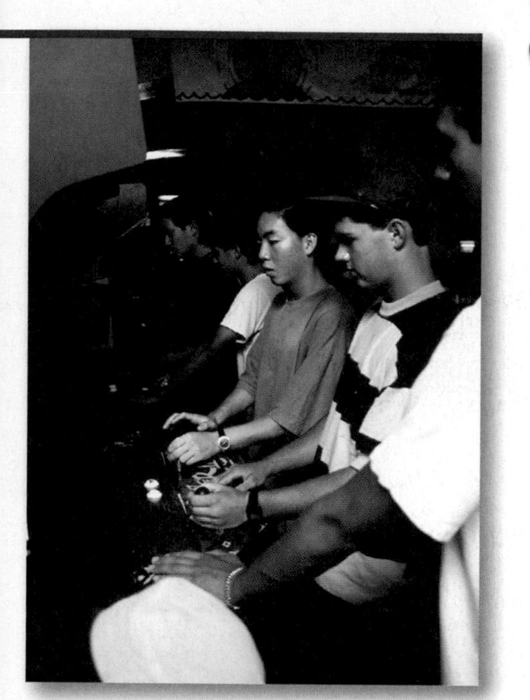

2 **Arcade Prices** Every business has to deal with two important patterns of change, called *depreciation* and *inflation*. When new equipment is purchased, the resale value of that equipment declines or depreciates as time passes. The cost of buying new replacement equipment usually increases due to inflation as time passes.

The owners of Game Time, Inc. operate a chain of video game arcades. They keep a close eye on prices for new arcade games and the resale value of their existing games. One set of predictions is shown in the graph below.

a. Which of the two linear functions predicts the future price of new arcade games? Which predicts the future resale value of arcade games that are purchased now?

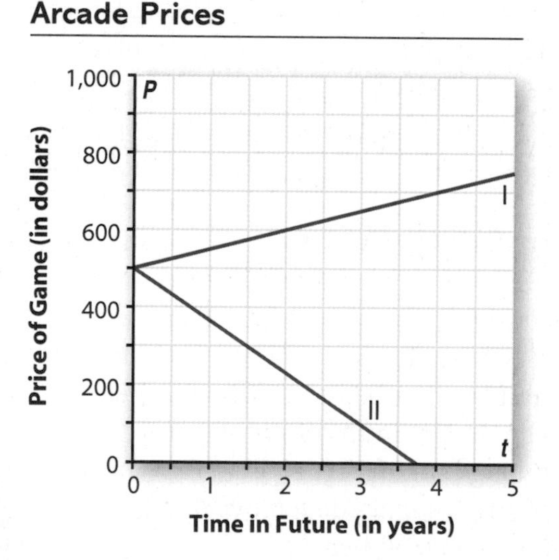

b. For each graph:

  i. Find the slope and y-intercept. Explain what these values tell about arcade game prices.

  ii. Write a rule for calculating game price P in dollars at any future time t in years.

3 **Turtles** The Terrapin Candy Company sells its specialty—turtles made from pecans, caramel, and chocolate—through orders placed online. The company web page shows a table of prices for sample orders. Each price includes a fixed shipping-and-handling cost plus a cost per box of candy.

Number of Boxes	1	2	3	4	5	10
Price (in dollars)	20	35	50	65	80	155

a. Explain why that price seems to be a linear function of the number of boxes ordered.

b. What is the rate of change in order price as the number of boxes increases?

c. Write a rule for calculating the price P in dollars for n boxes of turtle candies.

d. Use your rule to find the price for 6 boxes and the price for 9 boxes of turtle candies.

4 **Drink Sales** The Washington High School store sells bottled drinks before and after school and during lunch.

During the first few weeks of school, the store manager set a price of $1.25 per bottle, and daily sales averaged 85 bottles per day. She then increased the price to $1.75 per bottle, and sales decreased to an average of 65 bottles per day.

a. What is the rate of change in average daily sales as the price per bottle increases from $1.25 to $1.75? What units would you use to describe this rate of change?

b. Assume that sales are a linear function of price. Use the rate of change you found in Part a to reason about expected daily "sales" for a price of $0. Then explain why you would or would not have much confidence in that prediction.

c. Use your answers to Parts a and b to write a rule for calculating expected sales y for any price x in dollars. Check that your rule matches the given information.

d. Use your rule to estimate the expected daily sales if the price is set at $0.90 per bottle.

5 **Alternate Forms** It is natural to write rules for many linear functions in **slope-intercept form** like $y = a + bx$ or $y = mx + b$. In some problems, the natural way to write the rule for a linear function leads to somewhat different symbolic forms. It helps to be able to recognize those alternate forms of linear functions. Several rules are given in Parts a–e. For each rule:

i. Decide if it represents a linear function. Explain your reasoning.

ii. If the rule defines a linear function, identify the slope and the y-intercept of the function's graph. Write the rule in slope-intercept form.

a. $y = 10 + 2(x - 4)$

b. $m = n(n - 5)$

c. $y = 2x^2 - 3$

d. $p = (2s + 4) + (3s - 1)$

e. $y = \frac{2}{x + 1}$

6 **Given Two Points** Each pair of points listed below determines the graph of a linear function. For each pair, give the following.

i. the slope of the graph

ii. the y-intercept of the graph

iii. a rule for the function

a. (0, 5) and (2, 13)

b. (−3, 12) and (0, 10)

c. (−1, 6) and (1, 7)

d. (3, 9) and (5, 5)

Summarize the Mathematics

There are several different methods of writing rules for linear functions.

a To write a rule in the form $y = a + bx$ or $y = mx + b$, how can you use information about:

i. slope and y-intercept of the graph of that function?

ii. rate of change and other information in a table of (x, y) values?

b How can you determine the rate of change or slope if it's not given directly?

c How can you determine the y-intercept if it's not given directly?

d What is the *NOW-NEXT* rule for a linear function with rule $y = mx + b$? For a function with rule $y = a + bx$?

Be prepared to share your ideas and reasoning with the class.

Check Your Understanding

Write rules in the *NOW-NEXT* and $y = mx + b$ forms for the linear functions that give the following tables and graphs. For the graphs, assume a scale of 1 on each axis.

a.

x	y
5	20
15	40
25	60
35	80

b.

x	y
−1	8
0	5
1	2
2	−1

c.

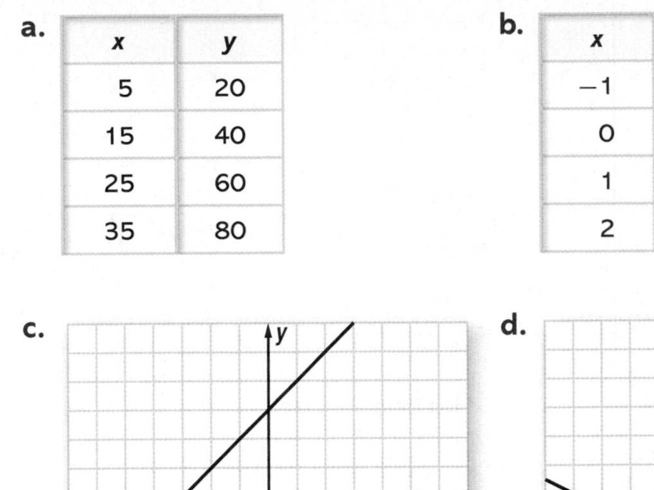

d.

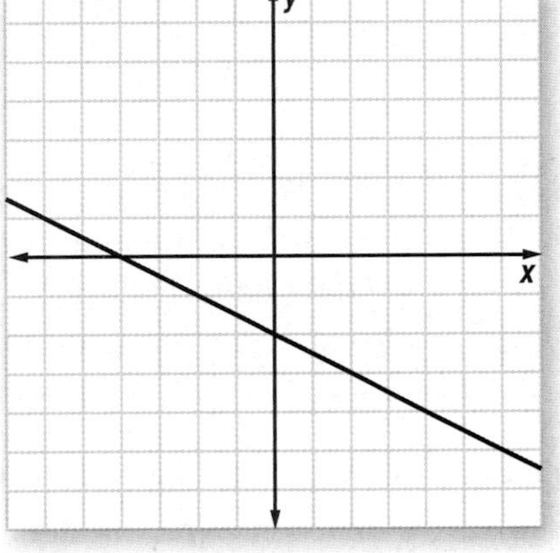

Investigation 3 Fitting Lines

Linear functions provide useful representations for relationships between variables in many situations, including cases in which data patterns are only approximately linear. As you work on this investigation, look for clues that will help you answer this question:

How can you produce and use function rules to represent data patterns that are not perfectly linear?

Shadows On sunny days, every vertical object casts a shadow that is related to its height. The following graph shows data from measurements of flag height and shadow location, taken as a flag was raised up its pole. As the flag was raised higher, the location of its shadow moved farther from the base of the pole.

Although the points do not all lie on a straight line, the data pattern can be closely approximated by a line.

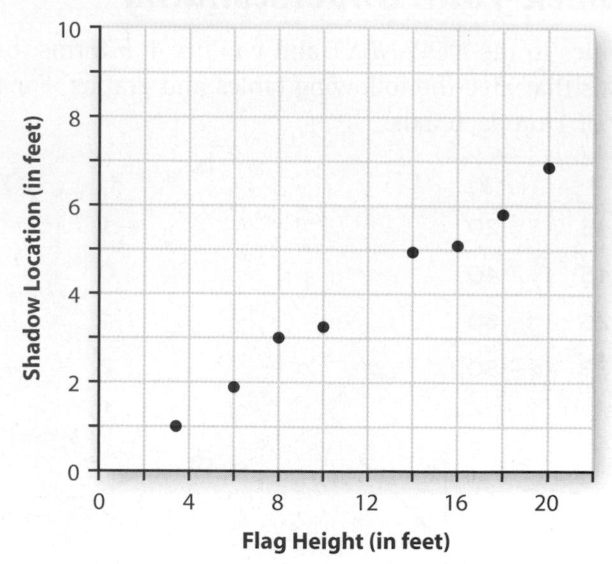

1 Consider the (*flag height, shadow location*) data plotted above.

a. On a copy of the plot, use a straight edge to find a line that fits the data pattern closely. Compare your line with those of your classmates. Discuss reasons for any differences.

b. Write the rule for a function that has your line as its graph.

The line and the rule that match the (*flag height, shadow location*) data pattern are **mathematical models** of the relationship between the two variables. Both the graph and the rule can be used to explore the data pattern and to answer questions about the relationship between flag height and shadow location.

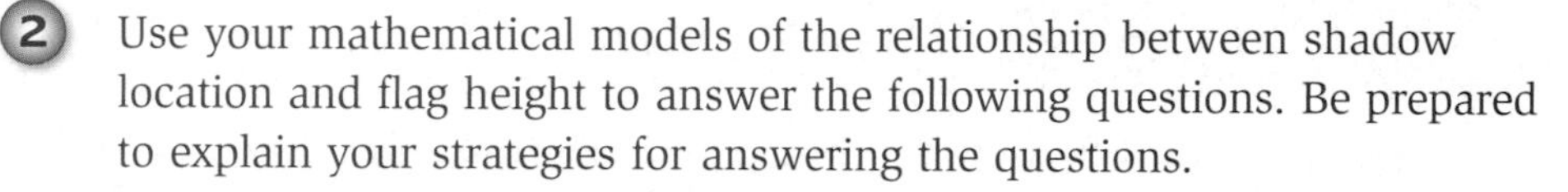

2 Use your mathematical models of the relationship between shadow location and flag height to answer the following questions. Be prepared to explain your strategies for answering the questions.

a. What shadow location would you predict when the flag height is 12 feet?

b. What shadow location would you predict when the flag height is 25 feet?

c. What flag height would locate the flag shadow 6.5 feet from the base of the pole?

d. What flag height would locate the flag shadow 10 feet from the base of the pole?

Time Flies Airline passengers are always interested in the time a trip will take. Airline companies need to know how flight time is related to flight distance. The following table shows published distance and time data for a sample of United Airlines nonstop flights to and from Chicago, Illinois.

Nonstop Flights to and from Chicago

Travel Between Chicago and:	Distance (in miles)	Flight Time (in minutes)	
		Westbound	Eastbound
Boise, ID	1,435	220	190
Boston, MA	865	160	140
Cedar Rapids, IA	195	55	55
Frankfurt, Germany	4,335	550	490
Hong Kong, China	7,790	950	850
Las Vegas, NV	1,510	230	210
Paris, France	4,145	570	500
Pittsburgh, PA	410	95	85
San Francisco, CA	1,845	275	245
Tokyo, Japan	6,265	790	685

Source: www.uatimetable.com

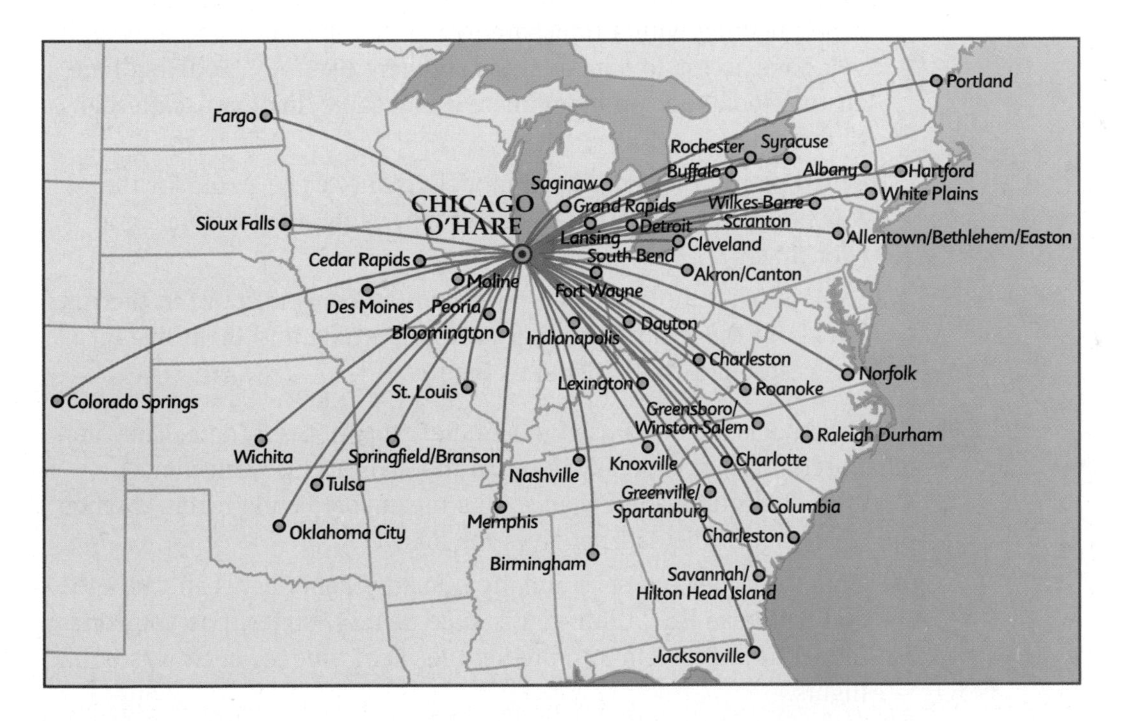

Scheduled flight time for a given distance depends on many factors, but the factor that has the greatest effect is the speed of prevailing winds. As you can see in the table, westbound flights generally take longer, since the prevailing wind patterns blow from west to east. Therefore, it makes sense to consider westbound flights and eastbound flights separately.

 To analyze the relationship between westbound flight time and flight distance, study the following scatterplot of the data on westbound flight distance and flight time.

Westbound Flight Distance and Time

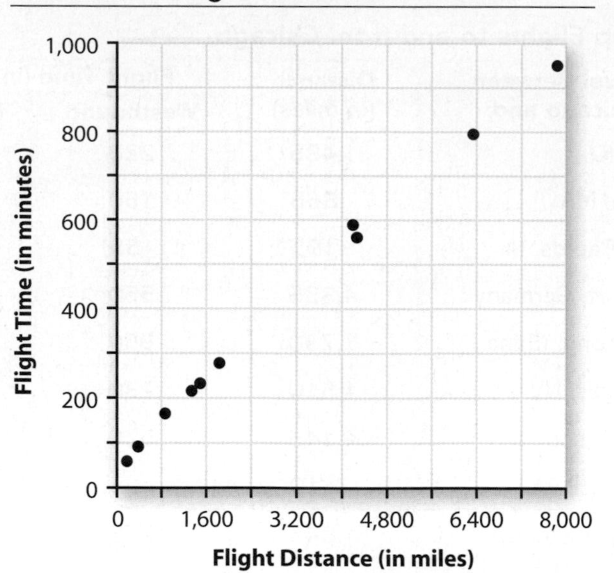

a. On a copy of the plot, locate a line that you believe is a good model for the trend in the data. You might find a good modeling line by experimenting with a transparent ruler and pencil. Alternatively, if you have access to data analysis software like the "Modeling Line" custom tool, you can manipulate a moveable line on a scatterplot of the data.

When you've located a good modeling line, write a rule for the function that has that line as its graph, using d for distance and t for time.

b. Explain what the coefficient of d and the constant term in the rule tell about the relationship between flight time and flight distance for westbound United Airlines flights.

4 Linear models are often used to summarize patterns in data. They are also useful in making predictions. In the analysis of connections between flight time and distance, this means predicting t from d when no (d, t) pair in the data set gives the answer.

a. United Airlines has several daily nonstop flights between Chicago and Salt Lake City, Utah—a distance of 1,247 miles. Use your linear model from Problem 3 to predict the flight time for such westbound flights.

b. The scheduled flight times for Chicago to Salt Lake City flights range from 3 hours and 17 minutes to 3 hours and 33 minutes. Compare these times to the prediction of your linear model. Explain why there might be differences between the predicted and scheduled times.

How's the Weather Up There? Linear functions are also useful for modeling patterns in climate data. You may have noticed that mountain tops can remain snow-covered long after the snow has melted in the areas below. This is because, in general, the higher you go above sea level, the colder it gets.

Extreme adventurers, such as those who attempt to climb Mt. Everest or those who jump from planes at high altitudes, must protect themselves from harsh temperatures as well as from the lack of oxygen at high altitudes. As skydiver Michael Wright explains about skydiving from 30,000 feet, "Cool? Yes it is. Cold? You bet. Typically 25 below zero (don't be concerned if that is °F or °C, it's still cold)."

An airplane descending to Los Angeles International Airport might record data showing a pattern like that in the next table.

Airplane Altitude and Temperature Data Above Los Angeles

Altitude (in 1,000s of feet)	Temperature (in °F)
34.6	−58
27.3	−35
20.5	−14
13.0	13
9.5	27
6.6	39
4.2	49
2.1	57
0.6	63
0.1	65

5 When working with paired data, it is helpful to use the list operations provided by calculators and statistical software. To get started, you need to enter the altitude data in one list and the temperature data in another list. Select an appropriate viewing window and produce a plot of the *(altitude, temperature)* data.

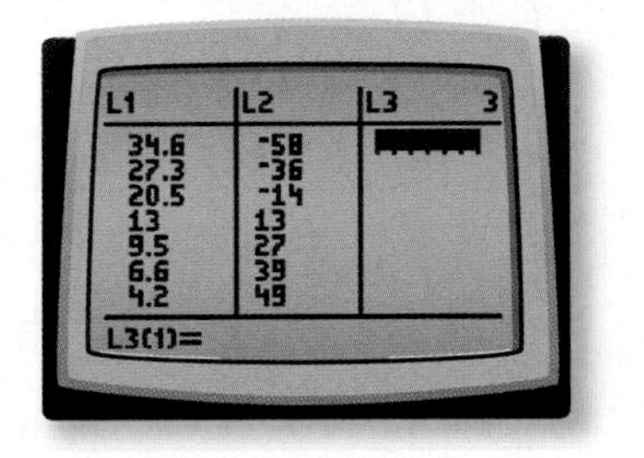

a. Describe the overall pattern of change in temperature as altitude increases.

b. Use two data points to estimate the rate of change in temperature as altitude (in thousands of feet) increases.

c. Use the data to make a reasonable estimate of the temperature at an altitude of 0 feet. Then use this value, together with the estimated rate of change from Part b, to write a rule for calculating temperature T as a function of altitude x (in thousands of feet).

d. Graph the function from Part c on a scatterplot of the data. Adjust the constant term and the coefficient of x in your rule until you believe the graph of your function closely matches the pattern in the data. Explain how you decided when the line was a good fit.

e. The highest elevation in Los Angeles is 5,080 feet at Elsie Peak. Use your linear model from Part d to predict the temperature at Elsie Peak on the day that the other data were collected.

Linear regression is a branch of statistics that helps in studying relationships between variables. It uses a mathematical algorithm to fit linear models to scatterplot patterns. You will learn more about the algorithm in the Course 2 unit on *Regression and Correlation*. But the algorithm is programmed in most graphing calculators and statistical software for computers, so you can put it to use in mathematical modeling right now.

6. Use the linear regression algorithm available on your calculator or computer to find a linear function that models the pattern in the (*altitude, temperature*) data, rounding the coefficient of x and the constant term to the nearest tenth.

   a. Display the graph of this function on your data plot and compare its fit to that of the function you obtained in Problem 5, Part d.

   b. What do the coefficient of x and the constant term in the linear regression rule tell you about the relationship between altitude and temperature?

7. Skydivers typically jump from altitudes of 10,000 to 15,000 feet. However, high altitude jumping, from 24,000 to 30,000 feet, is becoming popular. Use your linear regression model from Problem 6 to study the temperatures experienced by skydivers at different altitudes on the day the data were collected.

   a. Estimate the temperature at altitudes of 10,000, 15,000, 24,000, and 30,000 feet.

   b. What change in temperature can be expected as altitude decreases from 30,000 to 24,000 feet? From 24,000 to 15,000 feet? From 15,000 to 10,000 feet? What is the rate of change in temperature as altitude changes in each situation?

   c. Frostbite does not occur at temperatures above 28° Fahrenheit. Estimate the altitudes at which temperature is predicted to be at least 28°F.

   d. The current world record for skydiving altitude is 102,800 feet, set by Joe Kittinger Jr. in 1960. What temperature does your model predict for an altitude of 102,800 feet? What does this prediction suggest about limits on the linear model for predicting temperature from altitude?

8. Look back at the scatterplot of United Airlines westbound flight distances and times on page 164. The linear regression model for westbound flight time as a function of flight distance to and from Chicago is approximately $t = 0.12d + 52$.

   a. What do the coefficient of d and the constant term in this rule tell you about the relationship between westbound flight time and flight distance?

b. Display the graph of the linear regression model on the scatterplot of (*flight distance, flight time*) data. Compare its fit to that of the modeling function you developed in Problem 3 Part a.

c. Describing the rate of change in flight time as flight distance increases in terms of hundredths of a minute per mile is not very informative.

  i. Rewrite the rate of change 0.12 as a fraction and explain what the numerator and the denominator of this fraction tell about the relationship between flight time and flight distance.

  ii. Express the rate of change in flight time as flight distance increases in terms of minutes per 100 miles.

  iii. Write the fraction from part i in an equivalent form that shows the rate of change in minutes per 500 miles. Explain what this fraction suggests about the average speed of westbound planes.

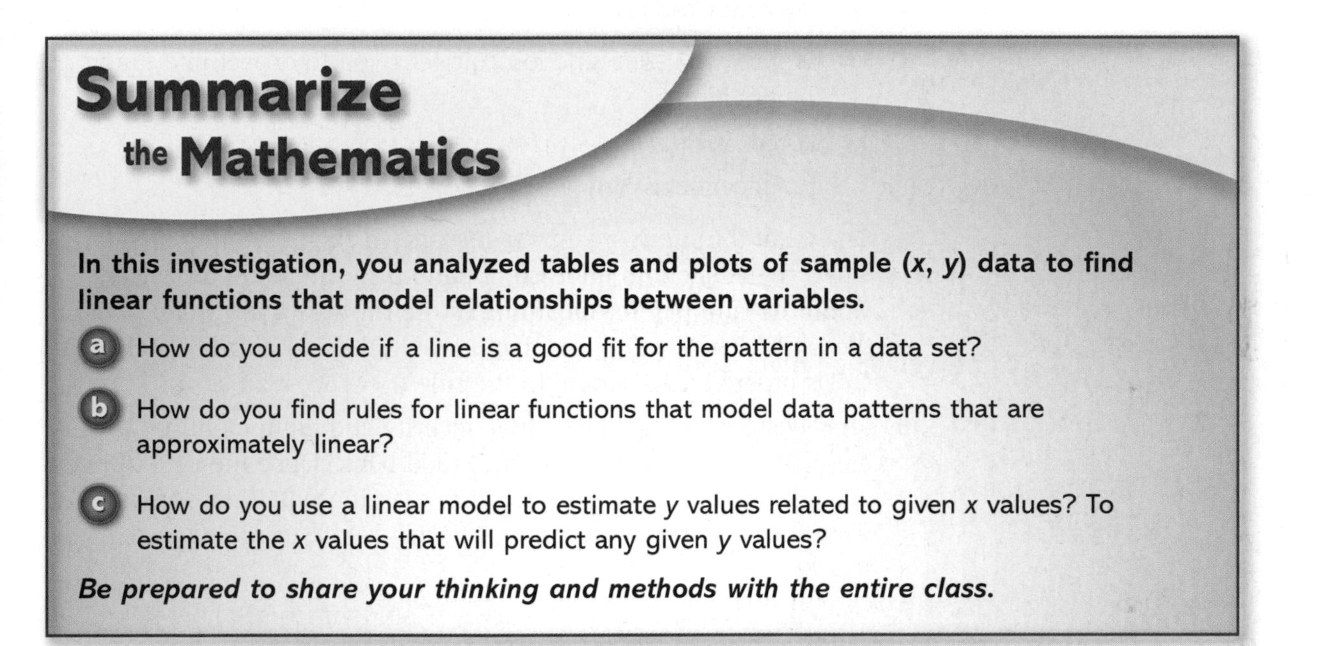

Summarize the Mathematics

In this investigation, you analyzed tables and plots of sample (x, y) data to find linear functions that model relationships between variables.

a How do you decide if a line is a good fit for the pattern in a data set?

b How do you find rules for linear functions that model data patterns that are approximately linear?

c How do you use a linear model to estimate y values related to given x values? To estimate the x values that will predict any given y values?

Be prepared to share your thinking and methods with the entire class.

Check Your Understanding

Look back at the United Airlines eastbound flight data on page 15. Since these flights have a tailwind as opposed to a headwind, they take less time.

a. On a plot of the eastbound (*flight distance, flight time*) data, locate a line that you believe is a good fit for the pattern in the data.

b. What linear function has the line you located in Part a as its graph? What do the coefficient of the independent variable and the constant term in the rule for that function tell about flight distance and flight time for eastbound flights?

c. United Airlines has nonstop flights between Chicago and Portland, Maine—a distance of 898 miles. Use your linear model to predict the time for an eastbound flight of 898 miles.

d. The scheduled flight times for the Chicago to Portland flights range from 2 hours and 10 minutes to 2 hours and 31 minutes. If your prediction was not in that range, what factors might explain the error of prediction?

On Your Own

Applications

1 Lake Aid is an annual benefit talent show produced by the students of Wilde Lake High School to raise money for the local food bank. Several functions that relate to Lake Aid finances are described in Parts a–c. For each function:

- **i.** Explain what the numbers in the function rule tell about the situation.
- **ii.** Explain what the function rule tells you to expect in tables of values for the function.
- **iii.** Explain what the function rule tells you to expect in a graph of the function.
- **iv.** Write a *NOW-NEXT* rule to describe the pattern of change in the dependent variable.

a. Several of the show organizers researched the possibility of selling DVDs of the show to increase donations to the food bank. They would have to pay for recording of the show and for production of the DVDs. The cost C (in dollars) would depend on the number of DVDs ordered n according to the rule $C = 150 + 2n$.

b. Proceeds from ticket sales, after security and equipment rental fees are paid, are donated to the local food bank. Once the ticket price was set, organizers determined that the proceeds P (in dollars) would depend on the number of tickets sold t according to the rule $P = 6t - 400$.

c. The organizers of the event surveyed students to see how ticket price would affect the number of tickets sold. The results of the survey showed that the number of tickets sold T could be predicted from the ticket price p (in dollars) using the rule $T = 950 - 75p$.

2 Given below are five functions and at the right five graphs. Without doing any calculating or graphing yourself, match each function with the graph that most likely represents it. In each case, explain the clues that helped you make the match.

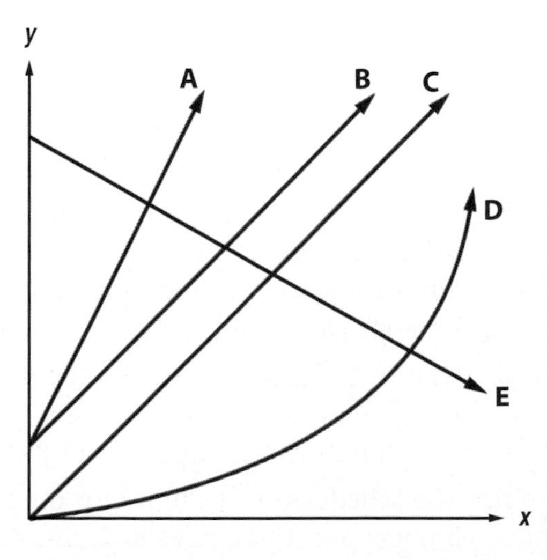

a. $y = x$

b. $y = 2x + 2$

c. $y = 0.1x^2$

d. $y = x + 2$

e. $y = 9 - 0.5x$

3 The graph below shows the relationship between weekly profit and the number of customers per week for Skate World Roller Rink.

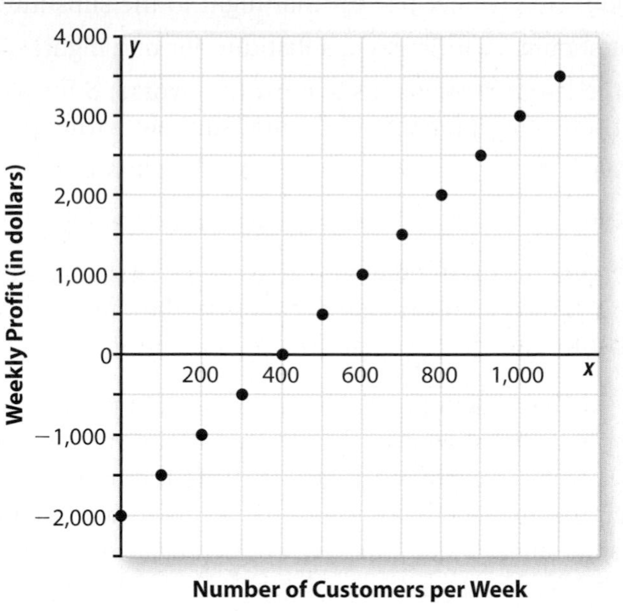

a. Determine the slope and y-intercept of the line that fits this data pattern.

b. Explain what the slope and y-intercept of the line tell you about the relationship between Skate World profit and number of customers per week.

c. If Skate World reached maximum capacity during each skating session for a week, admissions for that week would total 2,400 customers. Estimate the rink's profit in this situation. Explain your reasoning.

4 The table below gives the amount of money spent on national health care for every ten years from 1960 to 2000.

U.S. Health-Care Expenditures, 1960–2000 (in billions of dollars)

1960	1970	1980	1990	2000
26.7	73.1	245.8	696.0	1,299.5

Source: *The World Almanac and Book of Facts 2003.* New York, NY: World Almanac Education Group, Inc. 2003.

a. Was the amount of money spent on national health care a linear function over time from 1960 to 2000? Explain how you could tell without plotting the data.

b. What is the rate of change in health-care expenditures from 1960 to 1970? From 1970 to 1980? From 1980 to 1990? From 1990 to 2000? From 1960 to 2000? What does this suggest about the probable shape of a plot of the data?

5 Victoria got a job at her school as scorekeeper for a summer basketball league. The job pays $450 for the summer and the league plays on 25 nights. Some nights Victoria will have to get a substitute for her job and give her pay for that night to the substitute.

a. What should Victoria pay a substitute for one night?

b. Use the letters n for nights a substitute works, S for pay to the substitute, and E for Victoria's total summer earnings.

i. Write a rule for calculating S as a function of n.

ii. Write a rule for calculating E as a function of n.

c. Sketch graphs of the functions that relate total substitute pay and Victoria's total summer earnings to the number of nights a substitute works. Compare the patterns in the two graphs.

6 Some of the best vacuum cleaners are only sold door-to-door. The salespeople demonstrate the cleaning ability of the appliance in people's homes to encourage them to make the purchase. Michael sells vacuum cleaners door-to-door. He earns a base salary plus a commission on each sale. His weekly earnings depend on the number of vacuum cleaners he sells as shown in the table below.

Michael's Earnings

Number of Vacuum Cleaners Sold in a Week	2	4	6	8
Weekly Earnings (in dollars)	600	960	1,320	1,680

a. Verify that weekly earnings are a linear function of the number of vacuum cleaners sold.

b. Determine the rate of change in earnings as sales increase. What part of Michael's pay does this figure represent?

c. What would Michael's earnings be for a week in which he sold 0 vacuum cleaners?

d. Use your answers to Parts b and c to write a rule that shows how Michael's weekly earnings E can be calculated from the number of vacuum cleaners sold in a week S.

e. Company recruiters claim that salespeople sell as many as 15 vacuum cleaners in a week. What are the weekly earnings for selling 15 vacuum cleaners?

7 The table below shows the pattern of growth for one bean plant grown under special lighting.

Day	**Height** (in cm)
3	4.2
4	4.7
5	5.1
7	6.0

a. Plot the (*day, height*) data and draw a line that is a good fit for the trend in the data.

b. Write a function rule for your linear model. What do the numbers in the rule tell about days of growth and height of the bean plant?

c. Predict the height of the plant on day 6 and check to see if that prediction seems to fit the pattern in the data table.

8 For each of these function rules, explain what the constant term and the coefficient of the independent variable tell about the tables and graphs of the function.

a. $y = -4 + 2x$

b. $p = 7.3n + 12.5$

c. $y = 200 - 25x$

d. $d = -9.8t + 32$

9 Write rules for linear functions with graphs containing the following pairs of points.

a. (0, 3) and (6, 6)

b. (0, −4) and (5, 6)

c. (−4, −3) and (2, 3)

d. (−6, 4) and (3, −8)

10 The Riverdale Adventure Club is planning a spring skydiving lesson and first jump. Through the club newsletter, club members were asked to take a poll as to whether or not they would purchase a video of their jump for various prices.

The results of the poll are shown in the table below.

Cost (in dollars)	25	30	35	40	50	60	75
Number of Buyers	93	89	77	71	64	55	38

a. Create a linear model for the (*cost, number of buyers*) data. Represent your linear model as a graph and as a function rule.

b. Use your linear model from Part a to predict the number of members who would purchase a video of their jump for \$45. For \$70. For \$90. For \$10. Which estimates would you most trust? Why?

c. Should you use your model to predict the number of buyers if videos cost \$125? Why or why not?

d. For what cost of a video would you predict 50 buyers? 75 buyers? 100 buyers?

11 The snowy tree cricket is known as the "thermometer cricket" because it is possible to count its chirping rate and then estimate the temperature.

The table below shows the rate of cricket chirps at various temperatures.

Temperature (in °F)	50	54	58	61	66	70	75	78	83	86
Chirps per Minute	41	57	78	90	104	120	144	160	178	192

a. Use a calculator or computer software to determine the linear regression model for chirp rate C as a function of temperature T.

b. At what rate do you predict crickets will chirp if the temperature is 70°F? 90°F?

c. Now find the linear regression model for temperature T as a function of chirp rate C.

d. Use the linear regression model from Part c to predict the temperature when crickets are chirping at a rate of 150 chirps per minute. At a rate of 10 chirps per minute. Which prediction would you expect to be more accurate? Why?

e. Caution must be exercised in using linear regression models to make predictions that go well beyond the data on which the models are based.

i. For what range of temperatures would you expect your linear model in Part a to give accurate chirping rate predictions?

ii. For what range of chirping rates would you expect your linear model in Part c to give accurate temperature predictions?

12 Many Americans love to eat fast food, but also worry about weight. Many fast-food restaurants offer "lite" items in addition to their regular menu items. Examine these data about the fat and calorie content of some fast foods.

Item	Grams of Fat	Calories
McDonald's		
Grilled Chicken Bacon Ranch Salad	9	260
Grilled Chicken Caesar Salad	6	220
McChicken	16	370
Hardee's		
Charbroiled Chicken Sandwich	26	590
Regular Roast Beef	16	330
Wendy's		
Mandarin Chicken Salad	2	170
Jr. Cheeseburger	13	320
Ultimate Chicken Grill Sandwich	7	360
Taco Bell		
Ranchero Chicken Soft Taco, "Fresco Style"	4	170
Grilled Steak Taco, "Fresco Style"	5	170

Source: www.mcdonalds.com; www.hardees.com; www.wendys.com; www.tacobell.com

a. Make a scatterplot of the data relating calories to grams of fat in the menu items shown.

b. Draw a modeling line using the points (6, 220) and (16, 370). Write a rule for the corresponding linear function. Explain what the constant term and the coefficient of x in that rule tell about the graph and about the relation between calories and grams of fat.

c. Use your calculator or computer software to find the linear regression model for the (*grams of fat, calories*) data in the table. Compare this result to what you found in Part b.

13 Over the past 40 years, more and more women have taken full-time jobs outside the home. There has been controversy about whether they are being paid fairly. The table below shows the median incomes for men and women employed full-time outside the home from 1970 to 2003. These data do not show pay for comparable jobs, but median pay for all jobs.

On Your Own

Median Income (in dollars)

Year	Men	Women	Year	Men	Women
1970	6,670	2,237	1990	20,293	10,070
1975	8,853	3,385	1995	22,562	12,130
1980	12,530	4,920	2000	28,343	16,063
1985	16,311	7,217	2003	29,931	17,259

Source: www.census.gov/hhes/income/histinc/p02.html

a. What do you believe are the most interesting and important patterns in these data?

b. Did women's incomes improve in relation to men's incomes between 1970 and 2003?

c. The diagram below shows a plot of the (*years since 1970, median income*) data for women, using 0 for the year 1970. A linear model for the pattern in those data is drawn on the coordinate grid. Write a function rule for this linear model.

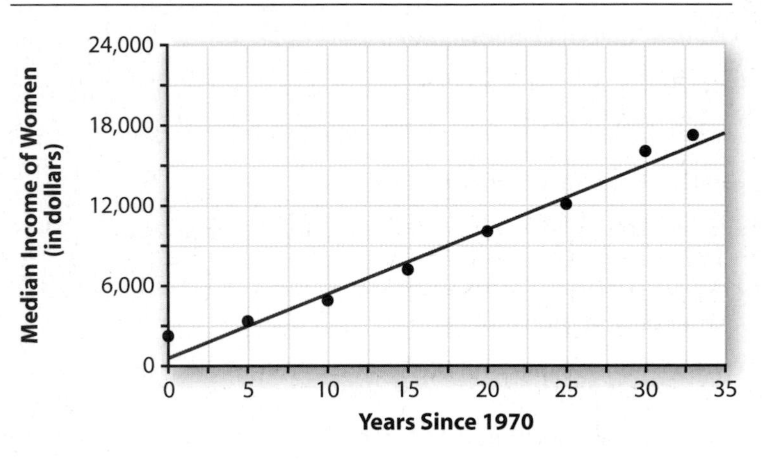

d. Using the linear model, estimate the median income of women in 1983 and 2007.

e. Is it reasonable to use the model to estimate the median income of women in 1963? Explain.

f. Now find the linear regression model for the (*years since 1970, median income*) data. (The data set is in *CPMP-Tools*.) Compare the predictions of that model with your results from Part d.

g. What do the coefficient and the constant term in the linear model of Part f tell about the pattern of change in median income as time passed between 1970 and 2000?

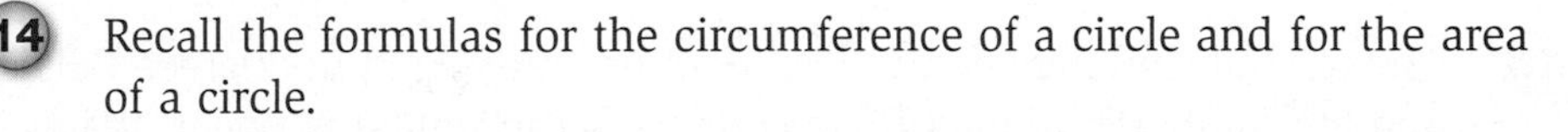

Connections

14 Recall the formulas for the circumference of a circle and for the area of a circle.

Circumference: $C = 2\pi r$ Area: $A = \pi r^2$

a. Is circumference a linear function of the radius of a circle? Explain how you know.

b. What does the formula for circumference tell about how circumference of a circle changes as the radius increases?

c. Is area a linear function of the radius of a circle? Explain how you know.

d. What does the formula for area tell about how area of a circle changes as the radius of a circle increases?

15 On hilly roads, you sometimes see signs warning of steep grades ahead. What do you think a sign like the one at the right tells you about the slope of the road ahead?

16 The diagram at the right shows four linear graphs. For each graph I–IV, do the following.

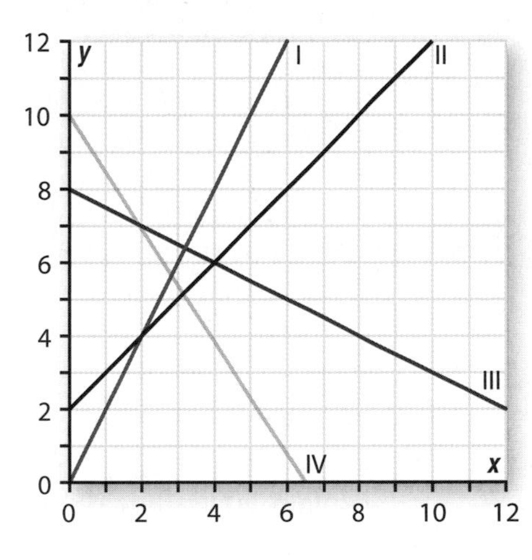

a. Find the rate at which y changes as x increases.

b. Write a *NOW-NEXT* rule that describes the pattern of change shown by the graph.

c. Write a rule for calculating y as a function of x.

d. Explain how your answers relate to each other.

17 For each table of values (in Parts a and b) use a spreadsheet to reproduce the table.

- Enter the first value for x from the table in cell A1 and the corresponding value for y in cell B1.
- Plan the spreadsheet so that: (1) the values of x appear in column A; (2) the corresponding values of y are calculated using *NOW-NEXT* reasoning in column B; and (3) the corresponding values of y are calculated with an appropriate "$y = ...$" formula in column C.
- Enter formulas in cells A2, B2, and C1 from which the rest of the cell formulas can be generated by application of "fill down" commands.

a.

x	0	1	2	3
y	7	10	13	16

b.

x	5	10	15	20
y	30	20	10	0

18 The relationship between the temperature measured in degrees Celsius and the temperature measured in degrees Fahrenheit is linear. Water boils at 100°C, or 212°F. Water freezes at 0°C, or 32°F.

a. Use this information to write a rule for calculating the temperature in degrees Fahrenheit as a function of the temperature in degrees Celsius.

b. Write a rule for calculating the temperature in degrees Celsius as a function of the temperature in degrees Fahrenheit.

c. Recall the quote from skydiver Michael Wright on page 16 about skydiving from 30,000 feet, "Cool? Yes it is. Cold? You bet. Typically 25 below zero (don't be concerned if that is °F or °C, it's still cold)."

i. Use your rule from Part a to calculate the equivalent of −25°C in degrees Fahrenheit.

ii. Use your rule from Part b to calculate the equivalent of −25°F in degrees Celsius.

iii. Use a table or graph to determine when it really doesn't matter whether one is talking about °F or °C—when the temperature is the same in both scales.

19 The table below shows data from the "Taking Chances" investigation (page 9) in Unit 1.

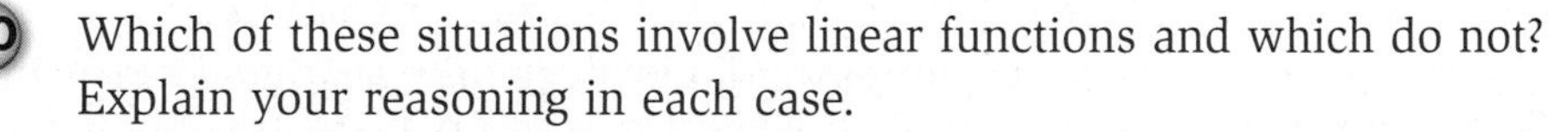

Number of Trials	20	40	60	80	100	120
Cumulative Profit (in $)	3	7	11	13	15	19

a. Explain why a linear model is reasonable for these data.

b. Is cumulative profit an exact linear function of the number of trials? Explain why or why not.

c. Use a graphing calculator or computer software to find a linear model for the (*number of trials, cumulative profit*) data.

d. What is the coefficient of the independent variable in your model of Part c? What does it tell you about the relationship between cumulative profit and number of trials?

20 Which of these situations involve linear functions and which do not? Explain your reasoning in each case.

a. If a race car averages 150 miles per hour, the distance d covered is a function of driving time t.

b. If the length of a race is 150 miles, time t to complete the race is a function of average speed s.

c. If the length of a race is 150 miles, average speed s for the race is a function of race time t.

21 When Robin and Mike had to find a linear function with graph passing through the two points $A(-3, 12)$ and $Q(4, -2)$, they produced the following work.

The rule will be in the form $y = mx + b$.	
The slope of the line is -2.	(1)
So, $y = -2x + b$.	(2)
Since $A(-3, 12)$ is on the line, $12 = -2(-3) + b$	(3)
So, $6 = b$	(4)
So, the rule is $y = -2x + 6$	(5)

a. Did Robin and Mike find the correct function rule? If so, what do you think their reasoning was at each step? If not, where did they make an error?

b. Use reasoning similar to that of Robin and Mike to find a function rule for the line through the points $(-2, 2)$ and $(6, 10)$.

c. Use similar reasoning to find a function rule for the line through the points $(3, 5)$ and $(8, -15)$.

22 The diagram at the right shows four *parallel lines*.

a. For each line I–IV, find its slope and a rule for the function with that graph.

b. Write rules for two different linear functions with graphs parallel to the given graphs. Explain how you know that your lines are parallel to the given lines.

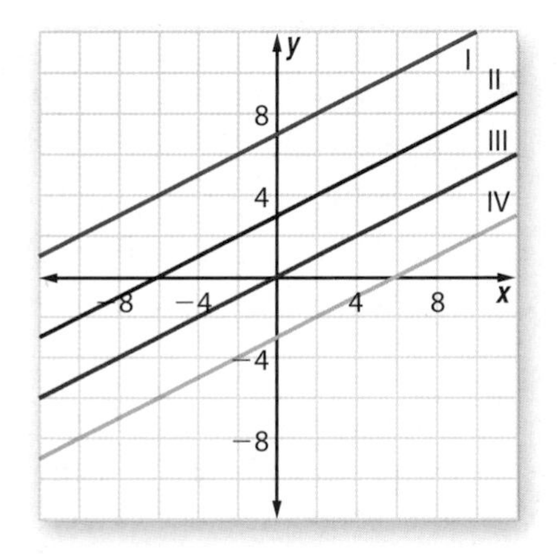

Reflections

23 To use linear functions wisely it helps to be in the habit of asking, "What sorts of numbers would make sense in this situation?" For example, in the function relating profit P to the number of customers n at the Starlight Cinema, it would not make much sense to substitute negative values for n in the formula $P = 6.5n - 2{,}500$. In each of the following situations, decide what range of values for the variables would make sense. It may be helpful to examine tables of values or graphs for some of the functions.

a. Suppose a ball is tossed into the air with an upward velocity of 40 feet per second. Its upward velocity is a function of time in flight, according to the formula $V = 40 - 32T$. Velocity V is in feet per second and time T is in seconds. What range of values for T and V make sense in this context?

b. The resale value R in dollars of an arcade game is given by $R = 500 - 133T$, where T is time in years after the purchase of the new equipment. What range of values for R and T make sense?

c. In one apartment building, new renters are offered \$150 off their first month's rent, then they pay a normal rate of \$450 per month. The total rent R paid for an apartment in that building is given by $R = 450m - 150$, where m is the number of months. What range of values for R and m make sense?

24 Think about how the values of the constant term and the coefficient are related to the graphs of linear functions. Suppose you enter the rule $y = 2 + 1.5x$ in your graphing calculator and produce a graph in the standard viewing window.

a. How will the graph that you see be different from that of $y = 2 + 1.5x$ if you:

i. increase or decrease the coefficient of x?

ii. increase or decrease the constant term?

b. Draw sketches that show possible graphs of functions $y = a + bx$ for each of these cases.

i. $a < 0$ and $b > 0$

ii. $a > 0$ and $b < 0$

iii. $a < 0$ and $b < 0$

iv. $a > 0$ and $b > 0$

25 Answer each of the following questions. In each case, explain how the answers can be determined without actually graphing any functions.

a. Is the point $(-3, -4)$ on the graph of the line $y = \frac{4}{3}x$?

b. Will the graphs of $y = 3x + 7$ and $y = 2 + 3x$ intersect?

c. Will the graphs of $y = 3x + 7$ and $y = 2 - 3x$ intersect?

26 In finding a linear function that models a data pattern, sometimes students simply draw a line connecting two points that are at the left and right ends of the scatterplot. Sketch a scatterplot showing how this simple strategy can produce quite poor models of data patterns.

27 Investigate the linear regression procedure for finding a linear model to fit data patterns.

a. For each of the following data sets, use your calculator or computer software to make a data plot. Then use linear regression to find a linear model and compare the graph produced by the linear regression model to the actual data pattern.

i.

x	0	1	2	3	4	5	6	7	8
y	3	5	7	9	11	13	15	17	19

ii.

x	0	1	2	3	4	5	6	7	8
y	1	5	7	9	11	13	15	17	29

iii.

x	0	1	2	3	4	5	6	7	8
y	4	6	8	10	12	14	16	18	20

iv.

x	0	1	2	3	4	5	6	7	8
y	1	2	5	10	17	26	37	50	65

v.

x	0	1	2	3	4	5	6	7	8
y	3	8	11	12	12	11	8	3	−4

b. What limitations of the linear regression procedure are suggested by the results of your work in Part a?

28 Consider the information needed to draw a line or find the equation for a line. How many different lines are determined by each of these conditions?

a. pass through the points (−4, 1) and (2, 4)

b. pass through the point (2, 1) and have slope −2

c. pass through the point (2, 4)

d. have slope −2

e. pass through the points (0, 0), (1, 1), and (2, 3)

f. pass through the points (0, 1), (1, 2), (2, 3), and (3, 4)

Extensions

29 In Connections Task 17, you wrote spreadsheet programs to produce tables of x and y values for two linear functions. You can extend those ideas to produce tables of values for any linear function when given only the starting x and y values and the pattern of change in the x and y values, Δx and Δy.

Complete the spreadsheet program begun below in a way that will allow you to enter specific start and change values and see the corresponding table of x and y values automatically.

- Assume that the starting x value will be entered in cell E1, the starting y value in cell E2, Δx in cell E3, and $\Delta y/\Delta x$ in cell E4.
- Plan the spreadsheet so that: (1) the values of x appear in column A; (2) the corresponding values of y are calculated using *NOW-NEXT* reasoning in column B; and (3) the corresponding values of y are calculated with an appropriate "$y = \ldots$" formula in column C.
- Enter formulas in cells A1, A2, B1, B2, and C1 from which the rest of the cell formulas can be generated by application of "fill down" commands.

Table of *x* and *y* Values.xls

◇	A	B	C	D	E
1				Start x =	
2				Start y =	
3				Δx =	
4				$\Delta y/\Delta x$ =	
5					

On Your Own

30 Carefully graph the function $y = \frac{2}{3}x + 1$ on grid paper.

a. What is the slope of this line?

b. Using a protractor or mira, carefully draw a line perpendicular to the graph of $y = \frac{2}{3}x + 1$ through the point (0, 1).

c. What is the slope of the perpendicular line? How does the slope of this line compare to the slope you determined in Part a?

d. Write an equation for the perpendicular line.

e. Will all lines having the slope you determined in Part c be perpendicular to all lines having the slope you determined in Part a? Explain why or why not.

f. Carefully draw another pair of perpendicular lines on grid paper and compare their slopes. Explain your conclusions.

31 The graph below illustrates the relationship between time in flight and height of a soccer ball kicked straight up in the air. The relation is given by $H = -4.9t^2 + 20t$, where t is in seconds and H is in meters.

a. What could it mean to talk about the slope of this curved graph? How could you estimate the slope of the graph at any particular point?

b. How would you measure the rate of change in the height of the ball at any point in its flight?

c. What would rate of change in height or slope of the graph tell about the motion of the ball at any point in its flight?

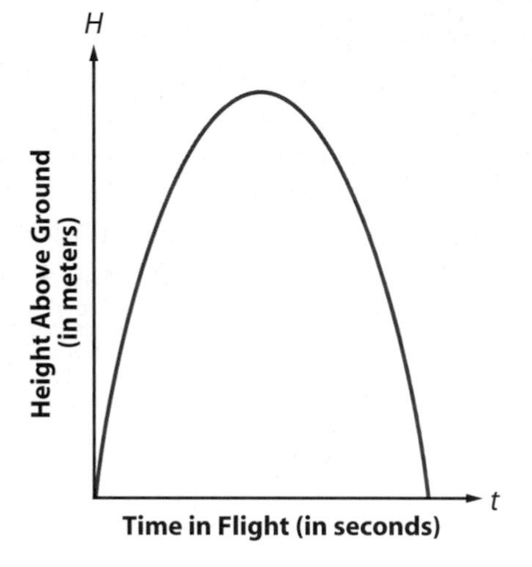

32 The following scatterplot shows grade point averages of some Wisconsin students in their eighth- and ninth-grade school years. The graph of $y = x$ is drawn on the plot.

a. What is true about the students represented by points that lie on the line $y = x$? That lie above the line $y = x$? That lie below the line $y = x$?

b. The linear regression model for these data is approximately $y = 0.6x + 1.24$. What do the numbers 0.6 and 1.24 tell you about the relationship between eighth- and ninth-grade averages for the sample of students in this study?

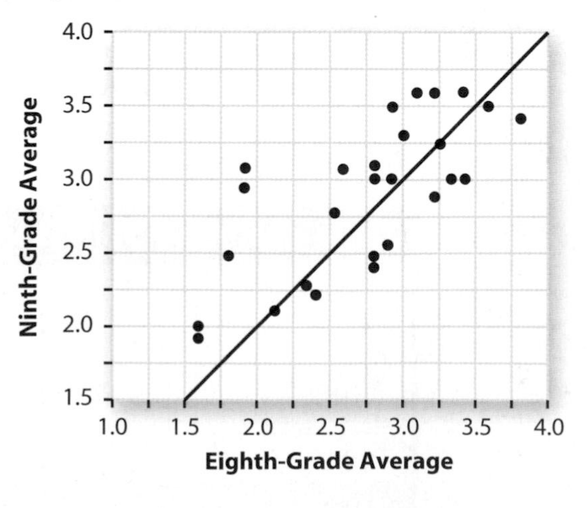

33 In this lesson, you fitted linear function models to data patterns. You can also use lines to analyze data that do not have a functional relationship. For example, consider the next scatterplot that shows average maximum temperatures in January and July for selected cities around the world.

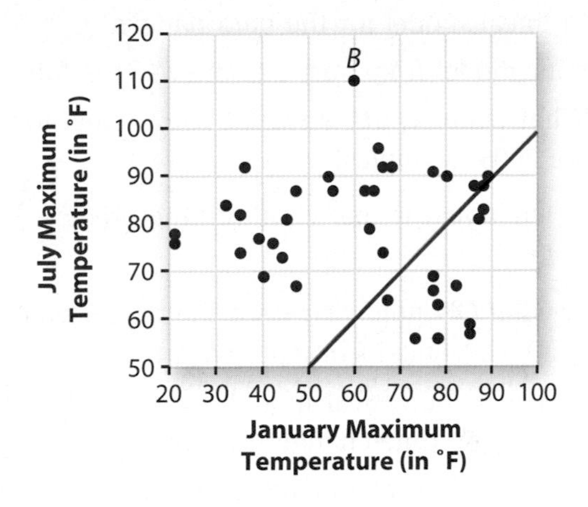

a. What rule describes the line drawn on the scatterplot? If a city is represented by a point on the line, what is true about that city?

b. What is true about the cities represented by points located below the line? Where do you think these cities are located geographically?

c. What is the difference in average temperature in July versus January for the city that is represented by point *B*?

d. Would it be useful to use linear regression on this data set? Explain your reasoning.

34 The 100-meter run for men has been run in the Olympics since 1896. The winning times for each of the years through 2004 are given in the following table.

Winning Times for Men: Olympic 100 Meters

Year	Time (sec)	Year	Time (sec)	Year	Time (sec)
1896	12.0	1936	10.3	1980	10.25
1900	10.8	1948	10.3	1984	9.99
1904	11.0	1952	10.4	1988	9.92
1908	10.8	1956	10.5	1992	9.96
1912	10.8	1960	10.2	1996	9.84
1920	10.8	1964	10.0	2000	9.87
1924	10.6	1968	9.95	2004	9.85
1928	10.8	1972	10.14		
1932	10.3	1976	10.06		

Source: *The World Almanac and Book of Facts 2001*. Mahwah, NJ: World Almanac Education Group, Inc. 2001; www.olympics.com

a. There are no 100-meter race times for 1916, 1940, and 1944. Why are these data missing?

b. Make a plot of the (*year*, *time*) data using 1890 as year 0. Then decide whether you think a linear model is reasonable for the pattern in your plot. Explain your reasoning.

c. Find a linear model for the data pattern.

d. Use your model from Part c to answer the following questions.

   i. What winning times would you predict for the 1940 and for the 2008 Olympics?

   ii. In what year is the winning time predicted to be 9.80 seconds or less?

   iii. In what Olympic year does the model predict a winning time of 10.4 seconds? Compare your prediction to the actual data.

e. According to your linear model, by about how much does the men's winning time change from one Olympic year to the next?

f. What reasons can you imagine for having doubts about the accuracy of predictions from the linear model for change in winning time as years pass?

35 Women began running 100-meter Olympic races in 1928. The winning times for women are shown in the table below.

Winning Times for Women: Olympic 100 Meters

Year	Time (sec)	Year	Time (sec)
1928	12.2	1972	11.07
1932	11.9	1976	11.08
1936	11.5	1980	11.60
1948	11.9	1984	10.97
1952	11.5	1988	10.54
1956	11.5	1992	10.82
1960	11.0	1996	10.94
1964	11.4	2000	10.75
1968	11.0	2004	10.93

Source: *The World Almanac and Book of Facts 2001*. Mahwah, NJ: World Almanac Education Group, Inc. 2001; www.olympics.com

a. Study the data and describe patterns you see in change of winning race time as years pass.

b. Make a plot and then find a linear model for the data pattern. Use 1900 as year 0.

c. Use your linear model to answer each of the following questions. For questions ii–iv, compare your predictions to the actual data.

- **i.** What winning time would you predict for 1944?
- **ii.** What winning time does the model predict for 1996?
- **iii.** In what Olympic year does the model suggest there will be a winning time of 10.7 seconds?
- **iv.** According to the model, when should a winning time of 11.2 seconds have occurred?

d. According to the model, by about how much does the women's winning time change from one Olympic year to the next? Compare this to your answer for Part e of Extensions Task 34.

Review

36 Identify all pairs of similar triangles. Then for each pair of similar triangles identify the scale factor.

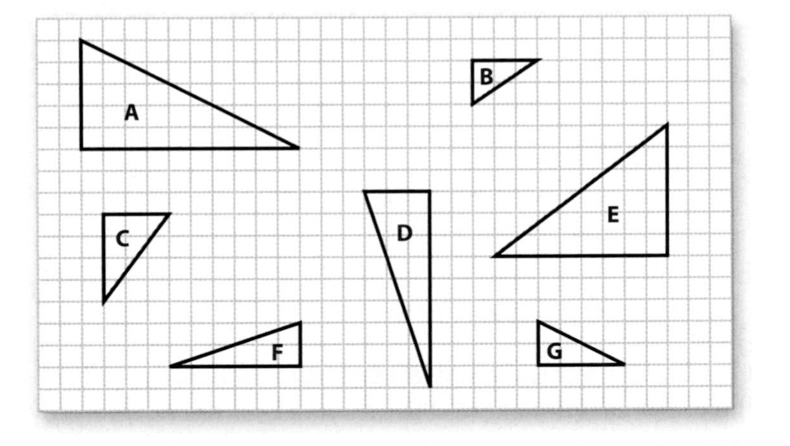

37 Solve each of the following equations for x.

a. $3x = 1$ **b.** $\frac{4}{3}x = 1$ **c.** $4 \div \frac{4}{3} = x$

d. $\frac{4}{5} \div \frac{1}{3} = x$ **e.** $x \div \frac{3}{13} = 1$ **f.** $\frac{1}{5} \div \frac{3}{13} = x$

38 Translating problem conditions into mathematical statements is an important skill.

a. Which of these mathematical statements uses the letters S for number of students and T for number of teachers to express correctly the condition, "At Hickman High School there are 4 student parking places for every teacher parking place"?

$4S = T$ $S = 4T$ $S = T + 4$ $T = S + 4$

b. Which of these mathematical statements correctly uses T for tax and P for price to express the fact that, "In California stores, an 8% sales tax is charged on the price of every purchase"?

$0.8T = P$ $\quad T = 0.8P$ $\quad P + 0.08 = T$ $\quad T = 0.08P$

39 Consider the pentagon shown below.

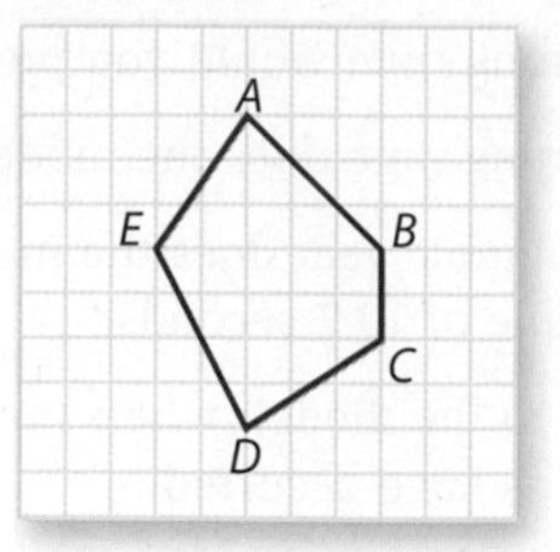

a. How many diagonals does this shape have? Name them.

b. Do any of the diagonals seem to bisect each other? Explain your reasoning.

c. Is $\overline{AD} \perp \overline{BE}$? Explain your reasoning.

40 Smart supermarket shoppers use unit prices to compare values for products.

a. For each of the following comparisons, decide which item is the better buy by finding the unit prices of each.

i. an 18-ounce box of cereal for \$3.50, or a 24-ounce box for \$4.50?

ii. a 32-ounce jar of spaghetti sauce for \$4.25, or a 20-ounce jar for \$2.50?

iii. a 6-pack of 20-ounce soft drink bottles for \$3.25, or a 12-pack of 12-ounce cans for \$4?

b. What is the connection between unit prices and slopes or rates of change for linear functions?

41 Carlos surveyed 120 ninth graders at his school and asked what is their least favorite chore. The results of his survey are provided in the table below.

Chore	Number of People
Cleaning bathroom	45
Mowing lawn	30
Walking the dog	15
Raking leaves	10
Doing the dishes	20

a. What percentage of the people surveyed said that cleaning the bathroom is their least favorite chore?

b. Make a bar graph of the survey results.

c. Make a circle graph that displays the results of Carlos's survey.

42 In the figures below, tell whether the gold shape appears to be the reflected image of the green shape across the given line. If it is not, explain how you know.

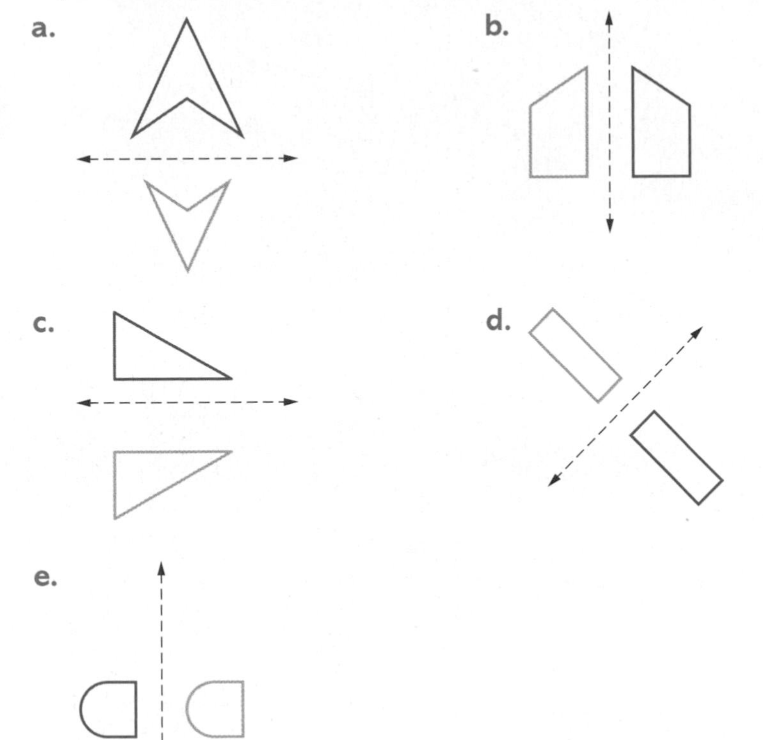

43 On the first three tests in a marking period, D'Qwell has scores of 85, 90, and 75. To earn a B in the course, he needs a mean test score of at least 85, while an A requires a minimum mean score of 93.

a. What score must D'Qwell get on the final test to earn a B?

b. What score, if any, would earn him an A?

c. If D'Qwell has a quiz average of 7 on the first 9 quizzes of a marking period, how will his quiz average change if he gets a 10 on the next quiz?

LESSON 2

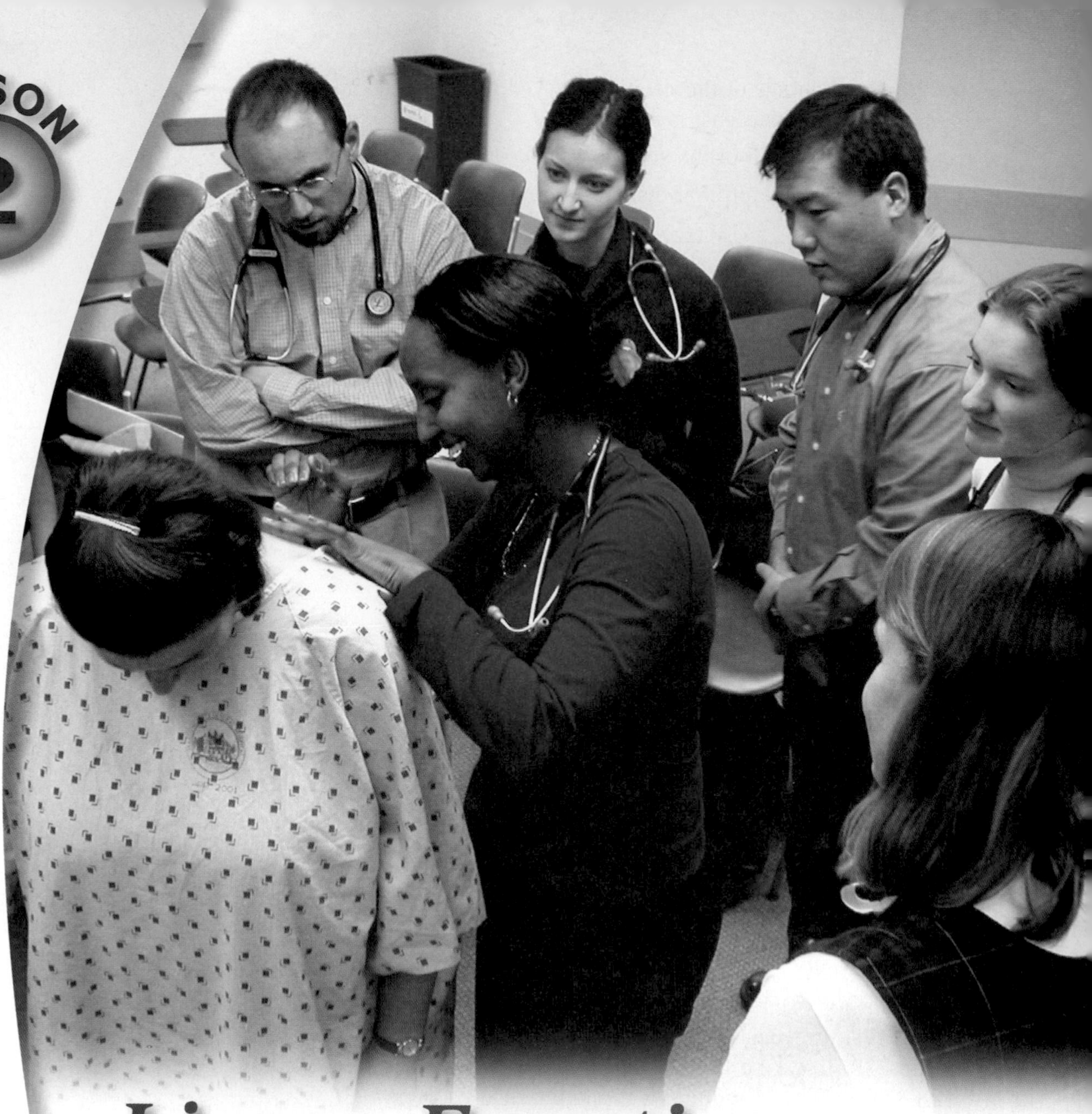

Linear Equations and Inequalities

For most of the twentieth century, the vast majority of American medical doctors were men. However, during the past 40 years there has been a significant increase in the number of women graduating from medical schools. As a result, the percent of doctors who are women has grown steadily to nearly 25% in 2000. The graph on the next page shows this trend.

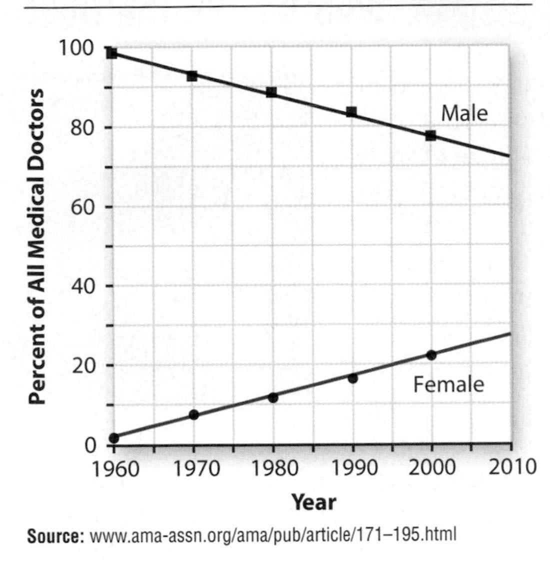

Source: www.ama-assn.org/ama/pub/article/171–195.html

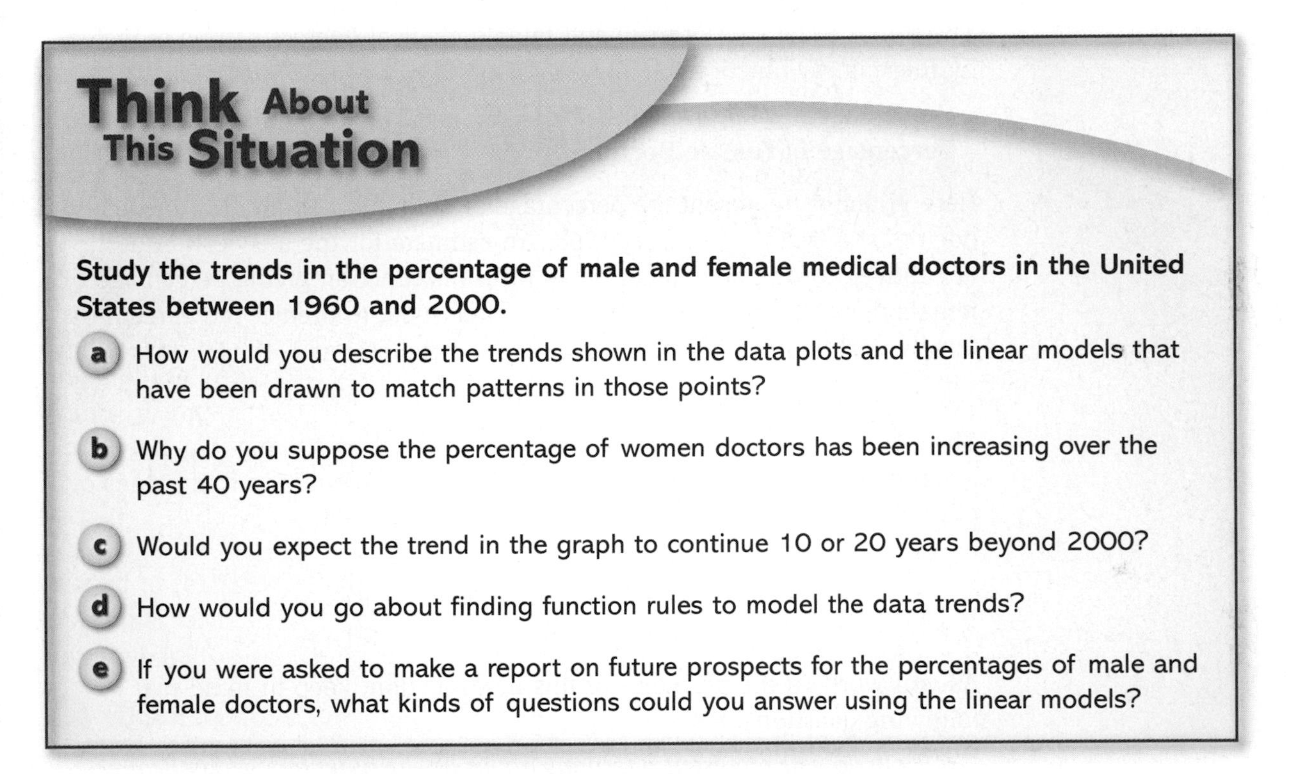

Think About This Situation

Study the trends in the percentage of male and female medical doctors in the United States between 1960 and 2000.

a How would you describe the trends shown in the data plots and the linear models that have been drawn to match patterns in those points?

b Why do you suppose the percentage of women doctors has been increasing over the past 40 years?

c Would you expect the trend in the graph to continue 10 or 20 years beyond 2000?

d How would you go about finding function rules to model the data trends?

e If you were asked to make a report on future prospects for the percentages of male and female doctors, what kinds of questions could you answer using the linear models?

In this lesson, you will explore ways to express questions about linear functions as equations or inequalities. You will use tables, graphs, and symbolic reasoning to solve those equations and inequalities and to interpret your solutions in problem contexts.

Investigation 1 Who Will Be the Doctor?

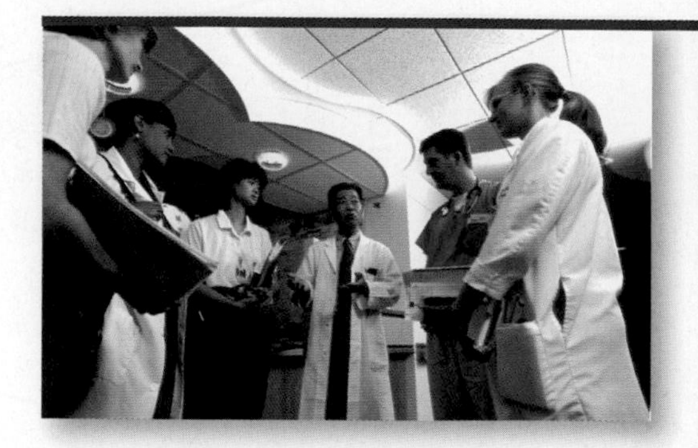

Several kinds of questions occur naturally in thinking about trends in the percentage of male and female medical doctors. To plan for future educational programs and medical services, medical schools, hospitals, and clinics might wonder:

(1) In 2020, what percent of U.S. medical doctors will be female?
(2) When will the percent of female doctors reach 40%?
(3) When will the percent of male and female doctors be equal?
(4) How long will the percent of male doctors remain above 70%?

The trends in percent of male and female medical doctors can be modeled by these linear functions.

Percentage of Male Doctors: $y_1 = 98 - 0.54t$
Percentage of Female Doctors: $y_2 = 2 + 0.54t$

Here y_1 and y_2 represent the percentage of male and female U.S. medical doctors at a time t years after 1960. An estimate for the answer to question (1) above can be calculated directly from the function giving percentage of female doctors. Since 2020 is 60 years after 1960, to predict the percent of female doctors in that year we evaluate the expression $2 + 0.54t$ for $t = 60$.

$$y_2 = 2 + 0.54(60), \text{ or } y_2 = 34.4$$

The other three questions above can be answered by solving two algebraic equations and an inequality. In each case, the problem is to find values of t (years since 1960) when the various conditions hold.

$$2 + 0.54t = 40 \quad (2)$$
$$98 - 0.54t = 2 + 0.54t \quad (3)$$
$$98 - 0.54t > 70 \quad (4)$$

As you work on the problems of this investigation, keep in mind the following questions:

How do you represent questions about linear functions symbolically?

How can you use tables and graphs to estimate solutions of equations and inequalities?

1 Write equations or inequalities that can be used to estimate answers for each of these questions about the percentage of male and female medical doctors in the United States.

a. In 1985, what percent of U.S. medical doctors were male?

b. When will the percent of male doctors fall to 40%?

c. How long will the percent of female doctors remain below 60%?

d. When will the percent of male doctors decline to only double the percent of female doctors?

2 Write questions about trends in percent of male and female medical doctors that can be answered by solving these equations and inequalities.

a. $98 - 0.54t = 65$

b. $y_2 = 2 + 0.54(50)$

c. $2 + 0.54t < 30$

d. $98 - 0.54t > 2 + 0.54t$

e. $98 - 0.54t = 4(2 + 0.54t)$

Writing equations and inequalities to match important questions is only the first task in solving the problems they represent. The essential next step is to **solve the equations** or to **solve the inequalities**. That is, find values of the variables that satisfy the conditions.

One way to estimate solutions for equations and inequalities that match questions about percentages of male and female medical doctors is to make and study tables and graphs of the linear models.

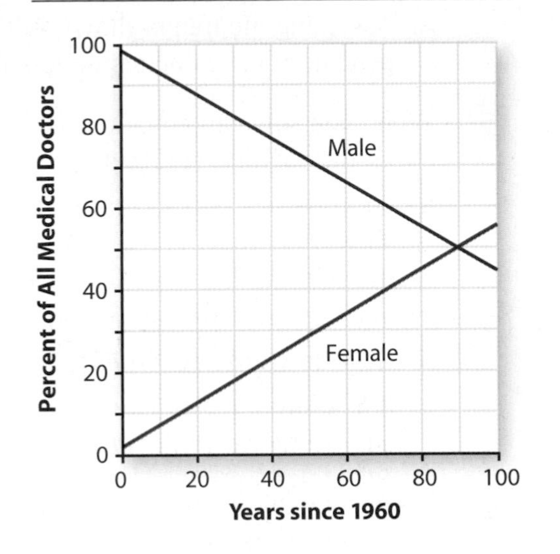

$$y_1 = 98 - 0.54t \quad \text{and} \quad y_2 = 2 + 0.54t$$

t	y_1	y_2
0	98.0	2.0
10	92.6	7.4
20	87.2	12.8
30	81.8	18.2
40	76.4	23.6

t	y_1	y_2
50	71.0	29.0
60	65.6	34.4
70	60.2	39.8
80	54.8	45.2
90	49.4	50.6

3 For the next equations and inequalities:

- Use the tables and graphs above to estimate the value or range of values that satisfy the given condition.
- Explain what each solution tells about the percentages of male and female medical doctors in the United States.
- Be prepared to explain or show how you used a table or graph to estimate the solution.

a. $y_2 = 2 + 0.54(40)$

b. $98 - 0.54t = 90$

c. $98 - 0.54t = 2 + 0.54t$

d. $98 - 0.54t > 80$

e. $y_1 = 98 - 0.54(65)$

f. $2 + 0.54t < 29$

g. $98 - 0.54t = 4(2 + 0.54t)$

h. $70 = 2 + 0.54t$

4 Write equations and inequalities to represent the following questions. Then use tables or graphs to estimate the solutions for the equations and inequalities and explain how those solutions answer the original questions. Be prepared to explain or show how you used a table or graph to estimate the solutions.

a. When will the percent of male doctors decline to 55%?

b. When will the percent of female doctors reach 35%?

c. How long will the percent of male doctors be above 40%?

d. What percent of U.S. medical doctors will be female when you are 20 years old?

e. Assuming the trends shown in the graph on page 187, when will the percent of male doctors be less than the percent of female doctors?

5 When you solve an equation or inequality, it is always a good idea to check the solution you find.

a. Suppose one person told you that the solution to $45 = 98 - 0.54t$ is $t = 100$, and another person told you that the solution is $t = 98$. How could you check to see if either one is correct without using a table or a graph?

b. How do you know whether a solution is *approximate* or *exact*?

c. If a solution for an equation is exact, does that mean that the answer to the prediction question is certain to be true? Explain.

6 If someone told you that the solution to $2 + 0.54t \leq 45$ is $t \leq 80$, how could you check the proposed solution:

a. Using a table?

b. From a graph?

c. Using a computer algebra system?

d. Without using a table, a graph, or a computer algebra system?

e. If you wanted to see if a solution is *exact*, which method of checking would you use?

Summarize the Mathematics

Many questions about linear relationships require solution of linear equations or inequalities, such as $50 = 23 + 5.2x$ or $45 - 3.5x < 25$.

a What does it mean to solve an equation or inequality?

b How could you use tables and graphs of linear functions to solve the following equation and inequality?

i. $50 = 23 + 5.2x$

ii. $45 - 3.5x < 25$

c How can you check a solution to an equation or inequality?

Be prepared to share your ideas with the class.

Check Your Understanding

Bronco Electronics is a regional distributor of electronic products specializing in graphing calculators. When an order is received, the shipping department packs the calculators in a box. The shipping cost C is a function of the number n of calculators in the box. It can be calculated using the function $C = 4.95 + 1.25n$.

Use your graphing calculator or computer software to make a table and a graph showing the relation between the number of calculators in a box and shipping costs for that box. Include information for 0 to 20 calculators. Use the table, graph, or cost function rule to answer the following questions.

a. How much would it cost to ship an empty box? How is that information shown in the table, the graph, and the cost function rule?

b. How much does the addition of a single calculator add to the cost of shipping a box? How is that information shown in the table, the graph, and the cost function rule?

c. Write and solve equations or inequalities to answer the following questions about Bronco Electronics shipping costs.

i. What is the cost of shipping 8 calculators?

ii. If the shipping cost is $17.45, how many calculators are in the box?

iii. How many calculators can be shipped if the cost is to be held to at most $25?

d. What questions about shipping costs could be answered by solving:

i. $27.45 = 4.95 + 1.25n$?

ii. $4.95 + 1.25n \leq 10$?

Investigation 2 Using Your Head

It is often possible to solve problems that involve linear equations without the use of tables, graphs, or computer algebra systems. Solving equations by symbolic reasoning is called solving *algebraically*. For example, to solve $3x + 12 = 45$ algebraically you might reason like one of these students.

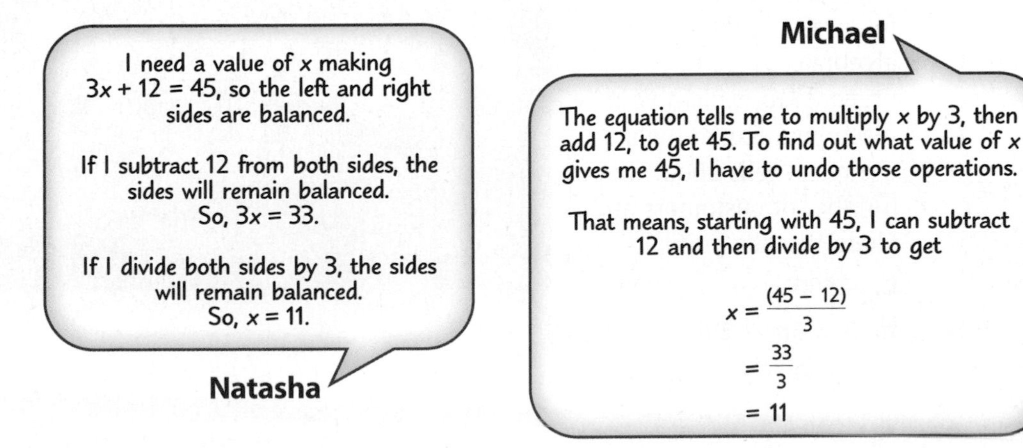

As you work on the problems in this investigation, think about these questions:

Why does solving linear equations by reasoning like that of Natasha and Michael make sense?

How can reasoning like that of Natasha and Michael be used to solve other linear equations algebraically?

1 Analyze the reasoning strategies used by Natasha and Michael by answering the following questions.

a. Why did Natasha subtract 12 from both sides? Why didn't she add 12 to both sides? What if she subtracted 10 from both sides?

b. Why did Natasha divide both sides by 3?

c. What did Michael mean by "undoing" the operations?

d. Why did Michael subtract 12 and then divide by 3? Why not divide by 3 and then subtract 12?

e. Both students found that $x = 11$. How can you be sure the answer is correct?

2 Solve the equation $8x + 20 = 116$ algebraically in a way that makes sense to you. Check your answer.

3 A calculator can help with the arithmetic involved in solving equations.

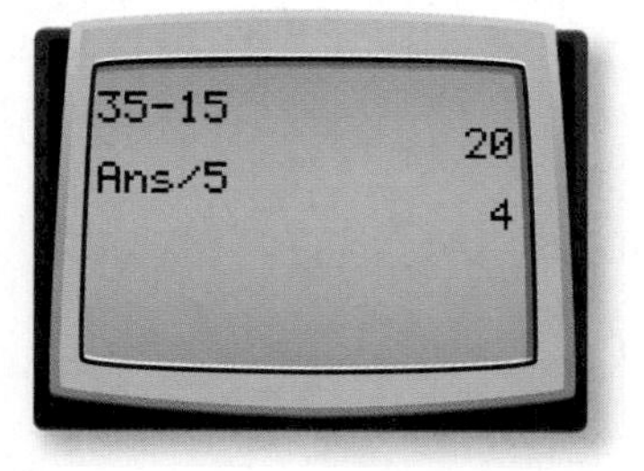

a. When one student used her calculator to solve an equation by undoing the operations, her screen looked like that at the left. What equation could she have been solving?

b. What would appear on your screen if you used a calculator to solve the equation $30x + 50 = 120$ by the "undoing" method?

c. What would appear on your screen if you solved $30x + 50 = 120$ in just one step?

4 Profit P (in dollars) at Skate World is given by $P = 5n - 2{,}000$, where n is the number of customers in a week. Solve each of the following equations algebraically and be prepared to explain your reasoning. Explain what the result tells about the number of customers and Skate World's profit.

a. $-500 = 5n - 2{,}000$

b. $0 = 5n - 2{,}000$

c. $1{,}250 = 5n - 2{,}000$

5 Martin and Anne experimented with the strength of different springs. They found that the length of one spring was a function of the weight upon it according to the function $L = 9.8 - 1.2w$. The length was measured in inches and the weight in pounds. To determine the weight needed to compress the spring to a length of 5 inches, they reasoned as follows.

We need to solve $9.8 - 1.2w = 5$.

Martin:	**Anne:**
If $9.8 - 1.2w = 5$, then	If $9.8 - 1.2w = 5$, then
$9.8 = 5 + 1.2w$.	$9.8 - 5 = 1.2w$.
Then $4.8 = 1.2w$.	This means that $4.8 = 1.2w$.
So, $w = \frac{4.8}{1.2}$,	So, $w = \frac{4.8}{1.2}$,
or $w = 4$.	or $w = 4$.

a. Is each step of their reasoning correct? If so, how would you justify each step? If not, which step(s) contains errors and what are those errors?

b. What does the answer tell about the spring?

6 Bronco Electronics received bids from two shipping companies. For shipping n calculators, Speedy Package Express would charge $3 + 2.25n$ dollars. The Fly-By-Night Express would charge $4 + 2n$ dollars. Solve the equation $3 + 2.25n = 4 + 2n$. Explain what the solution tells about the shipping bids.

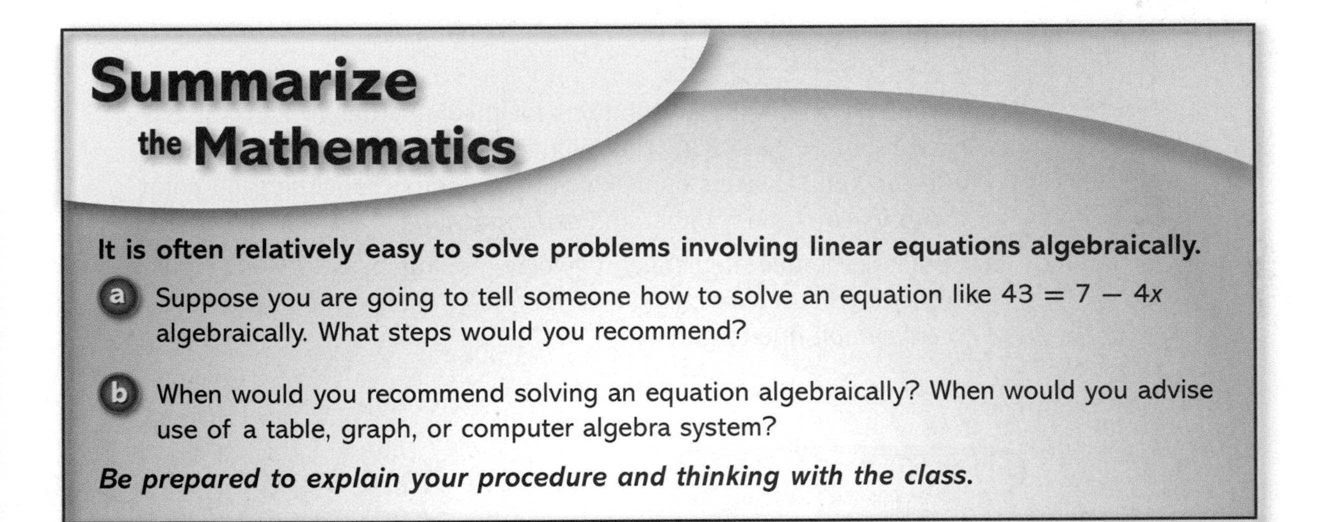

Summarize the Mathematics

It is often relatively easy to solve problems involving linear equations algebraically.

a Suppose you are going to tell someone how to solve an equation like $43 = 7 - 4x$ algebraically. What steps would you recommend?

b When would you recommend solving an equation algebraically? When would you advise use of a table, graph, or computer algebra system?

Be prepared to explain your procedure and thinking with the class.

Check Your Understanding

When a soccer ball, volleyball, or tennis ball is hit into the air, its upward velocity changes as time passes. **Velocity** is a measure of both speed and direction. The ball slows down as it reaches its maximum height and then speeds up in its return flight toward the ground. On its way up, the ball has a positive velocity, and on its way down it has a negative velocity. Suppose the upward velocity of a high volleyball serve is given by the function:

$$v = 64 - 32t$$

where t is time in seconds and v is velocity in feet per second.

a. Solve each of the following equations algebraically. Show your reasoning and explain what each solution tells about the flight of the ball.

- **i.** $16 = 64 - 32t$
- **ii.** $64 - 32t = -24$
- **iii.** $64 - 32t = 0$
- **iv.** $96 = 64 - 32t$

b. If you were to estimate solutions for the equations in Part a using a table of (t, v) values for the function $v = 64 - 32t$, what table entries would provide each solution? Record your answers in a table like this.

Equation	t	v
i		
ii		
iii		
iv		

c. If you were to estimate solutions for the equations in Part a using a graph of $v = 64 - 32t$, what points would provide each solution? Record your answers on a sketch of the graph, labeling each point with its equation number and coordinates.

d. What is the rate of change in velocity as time passes? What units describe this rate of change? (This rate of change represents the *acceleration* due to gravity.)

Investigation 3 Using Your Head ... More or Less

The reasoning you used in Investigation 2 to solve linear equations can be applied to solve linear inequalities algebraically. However, unlike equations, the direction of an inequality matters. If $x = 2$, then $2 = x$. On the other hand, $3 < 5$ is true but $5 < 3$ is not. As you work through the problems of this investigation, make notes of answers to the following question:

How can you solve a linear inequality algebraically?

 Begin by exploring the effect of multiplying both sides of an inequality by a negative number.

a. Consider the following true statements.

$$3 < 7 \qquad -2 < 1 \qquad -8 < -4$$

For each statement, multiply the number on each side by -1. Then indicate the relationship between the resulting numbers using $<$ or $>$.

b. Based on your observations from Part a, complete the statement below.

If $a < b$*, then* $(-1)a$ __?__ $(-1)b$.

c. Next, consider relations of the form $c > d$ and multiplication by -1. Test several examples and then make a conjecture to complete the statement below.

If $c > d$*, then* $(-1)c$ __?__ $(-1)d$.

2 Pairs of numbers are listed below. For each pair, describe how it can be obtained *from the pair above it*. Then indicate whether the direction of the inequality stays the same or reverses. The first two examples have been done for you.

$9 > 4$	Inequality operation	Inequality direction
12 __>__ 7	add 3 to both sides	stays the same
24 __>__ 14	multiply both sides by 2	stays the same
a. 20 __?__ 10		
b. -4 __?__ -2		
c. -2 __?__ -1		
d. 8 __?__ 4		
e. 6 __?__ 2		
f. -18 __?__ -6		
g. 3 __?__ 1		
h. 21 __?__ 7		

3 Look back at your answers to Problem 2 and identify cases where operations reversed the direction of inequality.

a. What operations seem to cause this reversal of inequality relationships?

b. See if you can explain why it makes sense for those operations to reverse inequality relationships. Compare your ideas with those of your classmates and resolve any differences.

4 In Investigation 1, you saw that the percentages of male and female doctors can be estimated from the number of years since 1960 using the following functions.

Percentage of Male Doctors:

$y_1 = 98 - 0.54t$

Percentage of Female Doctors:

$y_2 = 2 + 0.54t$

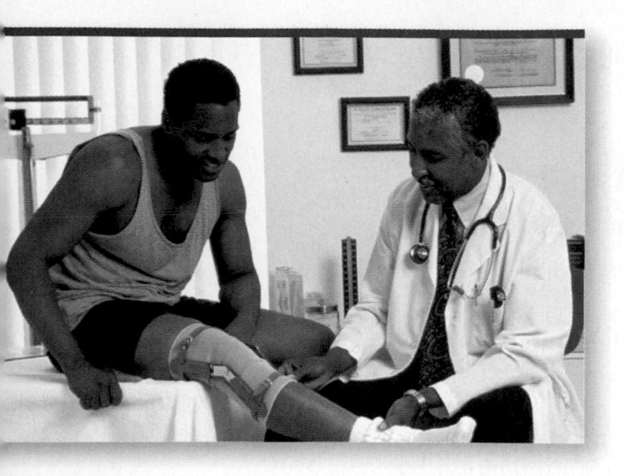

When their class was asked, "For how long will the majority of U.S. medical doctors be male?",

Taylor wrote this inequality: $98 - 0.54t > 50$.
Jamie wrote this inequality: $2 + 0.54t < 50$.

a. Explain the reasoning that Jamie may have used to create her inequality. Do you think that the solution to either inequality will answer the question? Why or why not?

b. Taylor's and Jamie's solutions are given below. Based on what you know about the percentages of male and female doctors in the United States, which answer makes more sense? Why?

Taylor's solution:

I need to solve $98 - 0.54t > 50$.

Subtract 98 from both sides:
$-0.54t > -48$

Then divide both sides by -0.54:
$t > 88.9$

So the majority of U.S. medical doctors will be males beginning approximately 89 years from 1960, or after 2049.

Jamie's solution:

I need to solve $2 + 0.54t < 50$.

Subtract 2 from both sides:
$0.54t < 48$

Then divide both sides by 0.54:
$t < 88.9$

So the majority of U.S. medical doctors will be males for approximately 89 years from 1960, or until 2049.

c. What is the error in the incorrect solution?

5 Solve the following linear inequalities using reasoning similar to that used in solving simple linear equations algebraically. Pay careful attention to the direction of the inequality. Be sure to check your solutions.

a. $1.5t - 150 > 450$ **b.** $4.95 + 1.25n \leq 10$ **c.** $45 - 3.5x < 25$
d. $32 \leq 6p - 10$ **e.** $100 > 250 - 7.5d$

Summarize the Mathematics

Just as with linear equations, it is often relatively easy to solve linear inequalities algebraically.

a Suppose you are going to tell someone how to solve an inequality like $7 - 4x > 43$ algebraically. What steps would you recommend? Why?

b How would you check your solution to an inequality like $7 - 4x > 43$? To an inequality like $7 - 4x \geq 43$?

c How is solving a linear inequality algebraically similar to, and different from, solving a linear equation algebraically?

d When would you recommend solving an inequality like the ones you've seen so far algebraically? When would you advise use of a table, graph, or computer algebra system?

Be prepared to explain your procedures and reasoning.

Check Your Understanding

In Lesson 1, you examined the effects of inflation and depreciation. You developed the following functions to model the change over time in the price of a new video arcade game and the change over time in the resale value of a game purchased new in 2002. Here, x represents years since 2002.

Price of New Game: $y_1 = 500 + 50x$

Resale Value of Used Game: $y_2 = 500 - 133x$

Solve each of the following inequalities algebraically.

- Show your reasoning in finding the solutions.
- Check your solutions.
- Explain what each solution tells about game prices.
- Make a table and sketch a graph of the price functions for $0 \leq x \leq 5$. Highlight the table entries and graph points that indicate the solutions for each inequality.

a. $500 + 50x > 600$

b. $500 - 133x < 100$

c. $700 \geq 500 + 50x$

d. $300 \leq 500 - 133x$

Investigation 4 Making Comparisons

In many problems involving linear functions, the key question asks for comparison of two different functions. For example, in Investigation 1 of this lesson you used linear models to compare the patterns of change in percentage of female and male doctors in the United States. In this investigation, you will examine methods for making sense of situations modeled by other *systems of linear equations*.

Increasing numbers of businesses, including hotels and cafés, are offering access to computers with high-speed Internet. Suppose that while on vacation Jordan would like to read and send e-mail, and two nearby businesses, Surf City Business Center and Byte to Eat Café, advertise their Internet services as shown at the top of the next page.

Surf City Business Center

Surf the Internet ~
just \$3.95 per day
plus \$0.05 per minute.

Byte to Eat Café

Stop in for a byte—
high-speed Internet access
for \$2 per day
plus \$0.10 per minute.

As you explore the question of which business offers a more economical deal, keep in mind this question:

How can you represent and solve problems involving comparisons of two linear functions?

1 For both businesses, the daily charge is a function of the number of minutes of Internet use.

a. For each business, write a rule for calculating the daily charge for any number of minutes.

b. What are the daily charges by each business for customers using 30 minutes?

c. How many minutes could Jordan spend on the Internet in a day for \$10 using the pricing plans for each of the two businesses?

d. For what number of minutes of Internet use in a day is Surf City Business Center more economical? For what number of minutes of Internet use in a day is Byte to Eat Café more economical?

e. Do you or someone you know use the Internet? For what purposes? Which pricing plan would cost less for this kind of use of the Internet?

2 To compare the price of Internet access from the two businesses, the key problem is to find the number of minutes for which these two plans give the same daily charge. That means finding a value of x (number of minutes) for which each function gives the same value of y (daily cost).

a. Use tables and graphs to find the number of minutes for which the two businesses have the same daily charge. Indicate both the number of minutes and the daily charge.

b. When one class discussed their methods for comparing the price of Internet access from the two businesses, they concluded, "The key step is to solve the equation $3.95 + 0.05x = 2 + 0.10x$." Is this correct? Explain your reasoning.

c. Solve the equation in Part b algebraically. Show your reasoning.

The questions about daily Internet access charges from two different businesses involve comparisons of two linear functions. The functions can be expressed with rules.

Surf City Business Center: $y_1 = 3.95 + 0.05x$

Byte to Eat Café: $y_2 = 2 + 0.10x$

If you think about what values of x and y will make $y = 3.95 + 0.05x$ and $y = 2 + 0.10x$ true, then you are thinking about $y = 3.95 + 0.05x$ and $y = 2 + 0.10x$ as equations. The pair of linear equations is sometimes called a **system of linear equations**.

3. Finding the pairs of numbers x and y that satisfy both equations is called **solving the system**.

   a. Does the pair (1, 7) satisfy either equation in the system above? If so, what does this solution say about Internet access charges? What about the pairs (10, 3) and (20, 8)?

   b. Find a pair of numbers x and y that satisfies both equations. What does this solution say about Internet access charges?

   c. Is there another solution of the system, that is, another pair of numbers (x, y) that satisfies both equations? How do you know?

Thinking about comparing costs is helpful when developing strategies for solving systems of linear equations. You can use the same strategies for solving when you do not know what x and y represent.

4. Use tables, graphs, or algebraic reasoning to solve each system. Use each method of solution at least once. Check each solution by substituting the values of x and y into both original equations. If a system does not have a solution, explain why.

   a. $\begin{cases} y = 2x + 5 \\ y = 3x + 1 \end{cases}$

   b. $\begin{cases} y = 10 - 1.6x \\ y = 2 + 0.4x \end{cases}$

   c. $\begin{cases} y = 1.5x + 2 \\ y = 5 + 1.5x \end{cases}$

   d. $\begin{cases} y = 2(3 + 0.8x) \\ y = 1.6x + 6 \end{cases}$

5. Describe a situation (like the Internet access situation) that involves comparing two linear functions. Set up and solve a system of linear equations that might model the situation and explain what the solution tells you about the situation.

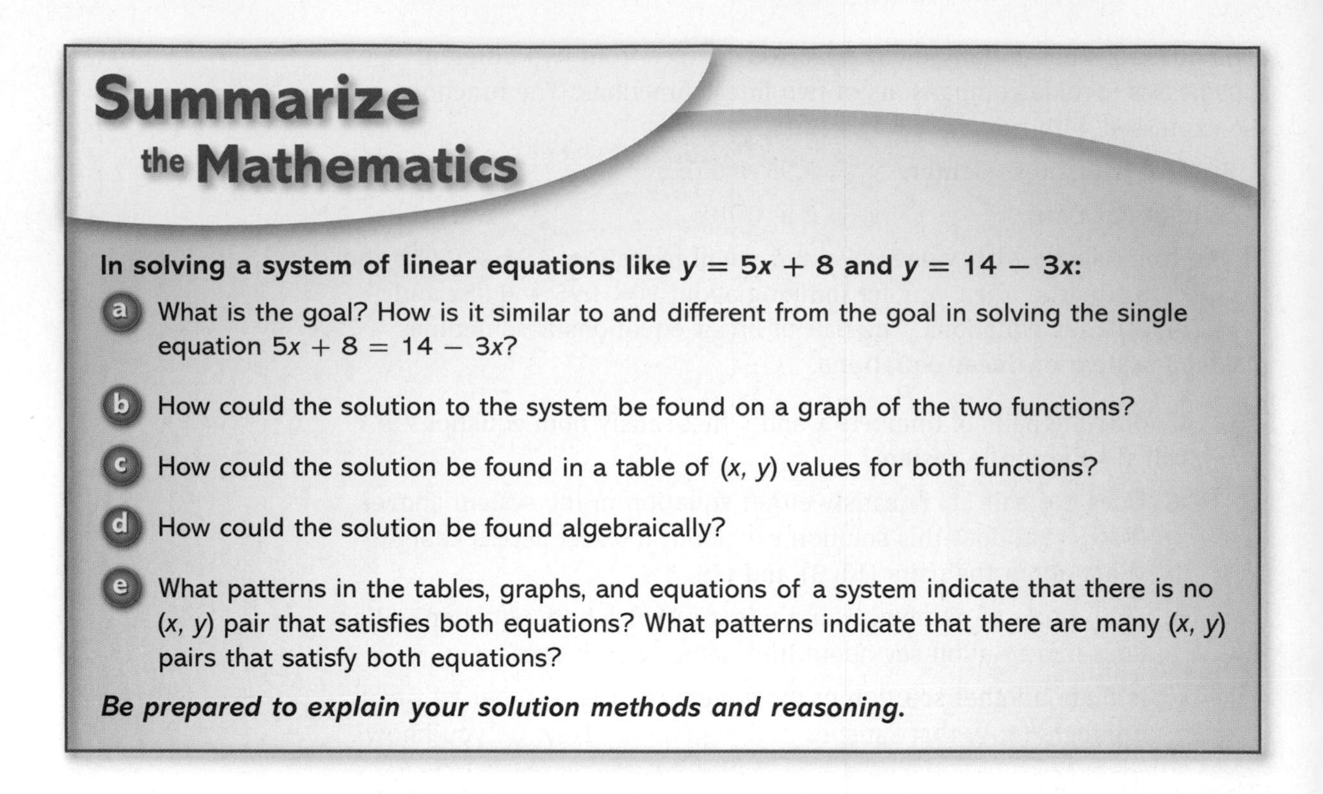

Summarize the Mathematics

In solving a system of linear equations like $y = 5x + 8$ and $y = 14 - 3x$:

a. What is the goal? How is it similar to and different from the goal in solving the single equation $5x + 8 = 14 - 3x$?

b. How could the solution to the system be found on a graph of the two functions?

c. How could the solution be found in a table of (x, y) values for both functions?

d. How could the solution be found algebraically?

e. What patterns in the tables, graphs, and equations of a system indicate that there is no (x, y) pair that satisfies both equations? What patterns indicate that there are many (x, y) pairs that satisfy both equations?

Be prepared to explain your solution methods and reasoning.

Check Your Understanding

Charter-boat fishing for walleyes is popular on Lake Erie. The charges for an eight-hour charter trip are:

Charter Company	Boat Rental	Charge per Person
Wally's	$200	$29
Pike's	$50	$60

Each boat can carry a maximum of ten people in addition to the crew.

a. Write rules for calculating the cost for charter service by Wally's and by Pike's.

b. Determine which service is more economical for a party of 4 and for a party of 8.

c. Assuming you want to minimize your costs, under what circumstances would you choose Wally's charter service? How would you represent your answer symbolically?

On Your Own

Applications

1 Parents often weigh their child at regular intervals during the first several months after birth. The data usually can be modeled well with a linear function. For example, the rule $y = 96 + 2.1x$ gives Rachel's weight in ounces as a function of her age in days.

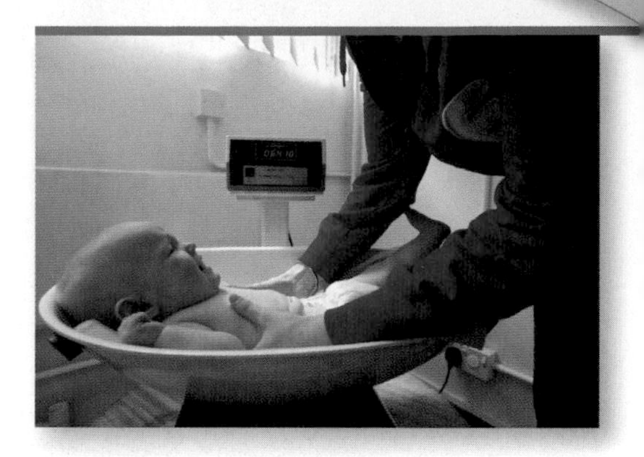

a. How much did Rachel weigh at birth?

b. Make a table and a graph of this function for $0 \leq x \leq 90$ with $\Delta x = 15$.

c. For each equation or inequality below, use the table or graph to estimate the solution of the equation or inequality. Then explain what the solution tells you about Rachel's weight and age.

i. $y = 96 + 2.1(10)$ ii. $159 = 96 + 2.1x$

iii. $264 = 96 + 2.1x$ iv. $96 + 2.1x \leq 201$

2 Mary and Jeff both have jobs at their local baseball park selling programs. They get paid \$10 per game plus \$0.25 for each program they sell.

a. Write a rule for pay earned as a function of number of programs sold.

b. Write equations, inequalities, or calculations that can be used to answer each of the following questions.

i. How many programs does Jeff need to sell to earn \$25 per game?

ii. How much will Mary earn if she sells 75 programs at a game?

iii. How many programs does Jeff need to sell to earn at least \$35 per game?

c. Produce a table and a graph of the relation between program sales and pay from which the questions in Part b can be answered. Use the graph and the table to estimate the answers.

On Your Own

3. Ella works as a server at Pietro's Restaurant. The restaurant owners have a policy of automatically adding a 15% tip on all customers' bills as a courtesy to their servers. Ella works the 4 P.M. to 10 P.M. shift. She is paid $15 per shift plus tips.

   a. Write a rule for Ella's evening wage based on the total of her customers' bills. Use your graphing calculator or computer software to produce a table and a graph of this function.

   b. If the customers' bills total $310, how much will Ella earn?

   c. Write and solve an equation to answer the question, "If Ella's wage last night was $57, what was the total for her customers' bills?"

4. The Yogurt Shop makes several different flavors of frozen yogurt. Each new batch is 650 ounces, and a typical cone uses 8 ounces. As sales take place, the amount A of each flavor remaining from a fresh batch is a function of the number n of cones of that flavor that have been sold. The function relating amount of yogurt left to number of cones sold is $A = 650 - 8n$.

   a. Solve each equation related to sales of chocolate-vanilla swirl yogurt. Show your work. Explain what your solution tells about sales of chocolate-vanilla yogurt and the amount left.

   i. $570 = 650 - 8n$

   ii. $250 = 650 - 8n$

   iii. $A = 650 - 8(42)$

   b. Use the function $A = 650 - 8n$ to write and solve equations to answer the following questions.

   i. How many cones have been sold if 390 ounces remain?

   ii. How much yogurt will be left when 75 cones have been sold?

   iii. If the machine shows 370 ounces left, how many cones have been sold?

5. Victoria can earn as much as $450 as a scorekeeper for a summer basketball league. She learned that she must pay $18 per game for substitutes when she misses a game. So, her summer earnings E will depend on the number of games she misses g according to the function $E = 450 - 18g$. Solve each of the following equations and explain what the solutions tell you about Victoria's summer earnings.

   a. $306 = 450 - 18g$

   b. $360 = 450 - 18g$

   c. $0 = 450 - 18g$

   d. $E = 450 - 18(2)$

   e. $315 = 450 - 18g$

   f. $486 = 450 - 18g$

6 When people shop for cars or trucks, they usually look closely at data on fuel economy. The data are given as miles per gallon in city and in highway driving. Let c stand for miles per gallon in city driving and h stand for miles per gallon in highway driving. Data from tests of 20 of the most popular American cars and trucks show that the function $h = 1.4 + 1.25c$ is a good model for the relation between the variables. Solve the following equations algebraically. Explain what the results tell you about the relation between city and highway mileage. Be prepared to explain the reasoning you used to find each solution.

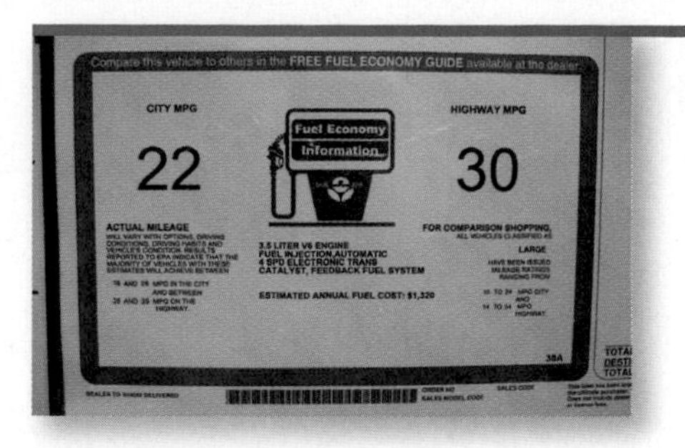

a. $35 = 1.4 + 1.25c$

b. $10 = 1.4 + 1.25c$

c. $h = 1.4 + (1.25)(20)$

7 Describe a problem situation which could be modeled by the function $y = 10 + 4.35x$.

a. What would solving $109 \geq 10 + 4.35x$ mean in your situation?

b. Solve $109 \geq 10 + 4.35x$ algebraically. Then show how the solution could be estimated from a table or graph.

8 Solve each of the following equations and inequalities algebraically and check your answers. Show the steps in your solutions and in your checks.

a. $25 = 13 + 3x$

b. $74 = 8.5x - 62$

c. $34 + 12x < 76$

d. $76 \geq 34 - 12x$

e. $3{,}141 = 2{,}718 + 42x$

9 Refer back to the Internet pricing plans for Surf City Business Center and Byte to Eat Café given on page 198. Suppose Surf City Business Center wants to become more competitive for customers looking for high-speed Internet access. The owner decides to change the daily base charge from \$3.95 to \$2.95, but maintain the \$0.05 per minute charge.

a. Write a rule for the daily charges under the new program.

b. How are the graphs of the new and old Internet access charges related?

c. What would be the daily charge for 30 minutes of Internet use using the new program?

d. How many minutes would one need to spend on the Internet in order for Surf City Business Center's new program to be more economical than Byte to Eat Café?

e. If a customer is charged \$5.20 for one day's Internet use under the new Surf City Business Center pricing plan, how many minutes did he or she spend online?

f. What would Byte to Eat Café have charged for the same number of minutes?

10 Surf City Business Center did not notice any large increase in customers when they changed their base daily charge from \$3.95 to \$2.95. They decided to change it back to \$3.95 and reduce the per-minute charge from \$0.05 to \$0.03.

a. Write a rule that models their new Internet access charge.

b. How are the graphs of this new and the original Internet access charges related?

c. What is the cost of 30 minutes of Internet use under this new plan?

d. For how many minutes of Internet use is this new Surf City Business Center pricing plan competitive with Byte to Eat Café?

e. Compare the cost of Internet access under this plan with that proposed in Applications Task 9. Which plan do you think will attract more customers? Explain your reasoning.

11 Recall Bronco Electronics, a regional distributor of graphing calculators, from the Check Your Understanding on page 191. Their shipping cost C can be calculated from the number n of calculators in a box using the rule $C = 4.95 + 1.25n$. Bronco Electronics got an offer from a different shipping company. The new company would charge based on the rule $C = 7.45 + 1.00n$. Write and solve equations or inequalities to answer the following questions:

a. For what number of calculators in a box will the two shippers have the same charge?

b. For what number of calculators in a box will the new shipping company's offer be more economical for Bronco Electronics?

12 From the situations described below, choose two situations that most interest you. Identify the variables involved and write rules describing one of those variables as a function of the other. In each case, determine conditions for which each business is more economical than the other. Show how you compared the costs.

a. A school club decides to have customized T-shirts made. The Clothing Shack will charge \$30 for setup costs and \$12 for each shirt. The cost of having them made at Clever Creations is a \$50 initial fee for the setup and \$8 for each T-shirt.

b. Speedy Messenger Service charges a \$30 base fee and \$0.75 per ounce for urgent small package deliveries between office buildings. Quick Delivery charges a \$25 base and \$0.90 per ounce.

c. Cheezy's Pies charges \$5 for a 12-inch sausage pizza and \$5 for delivery. The Pizza Palace delivers for free, but they charge \$7 for a 12-inch sausage pizza.

d. The *Evening News* has a minimum charge of \$4 for up to 3 lines and \$1.75 for each additional line of a listing placed in the classified section. The *Morning Journal* charges \$8 for the first 5 lines and \$1.25 for each additional line.

13 Solve each of the following systems of linear equations and check each solution. Among the four problems, use at least two different solution methods (tables, graphs, or algebraic reasoning).

a. $\begin{cases} y = x + 4 \\ y = 2x - 9 \end{cases}$

b. $\begin{cases} y = -2x + 18 \\ y = -x + 10 \end{cases}$

c. $\begin{cases} y = 3x - 12 \\ y = 1.5x + 3 \end{cases}$

d. $\begin{cases} y = x \\ y = -0.4x + 7 \end{cases}$

Connections

14 Recall the formula for the circumference of a circle:

$$C = \pi d.$$

Write equations or inequalities that can be used to answer the following questions. Then find answers to the questions.

a. A 16-inch pizza has a diameter of 16 inches. What is the circumference of a 16-inch pizza?

b. The average arm span of a group of 10 first graders is 47 inches. If they hold hands and stretch to form a circle, what will be the approximate diameter of the circle?

c. If you have 50 inches of wire, what are the diameters of the circles you could make with all or part of the wire?

15 There are two especially useful properties of arithmetic operations. The first relates addition and subtraction. The second relates multiplication and division.

- *For any numbers a, b, and c, $a + b = c$ is true if and only if $a = c - b$.*
- *For any numbers a, b, and c, $a \times b = c$ is true if and only if $a = c \div b$ and $b \neq 0$.*

Erik solved the equation $3x + 12 = 45$ given at the beginning of Investigation 2 as follows:

If $3x + 12 = 45$, then $3x = 45 - 12$, or $3x = 33$.

If $3x = 33$, then $x = 33 \div 3$, or $x = 11$.

a. Explain how the above properties of operations can be used to support each step in Erik's reasoning.

b. Solve the equation $130 + 30x = 250$ using reasoning like Erik's that relies on the connection between addition and subtraction and the connection between multiplication and division. Check your answer.

16 When you know the algebraic operations that you want to use in solving an equation, you can get help with the details from a computer algebra system (CAS). For example, to solve $3x + 11 = 5x + 7$, you can proceed as in the screen below.

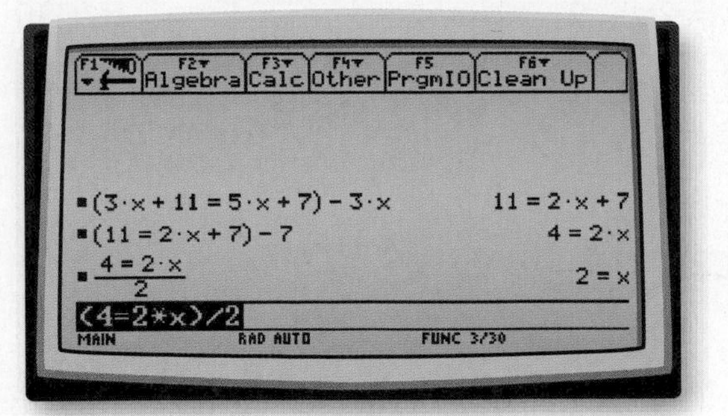

Study this work to figure out what each entered instruction is asking the CAS to do. Then apply your understanding to solve the following equations in a similar way with the CAS available to you. Record the steps you enter at each step in the solution process and the results of those steps.

a. $-3x + 72 = 4x - 5$

b. $\frac{2}{5}x - \frac{9}{5} = \frac{7}{10}$

c. $2.5t - 5.1 = 9.3 - 0.7t$

17 Here are the first three shapes in a geometric pattern of Xs made from identical squares.

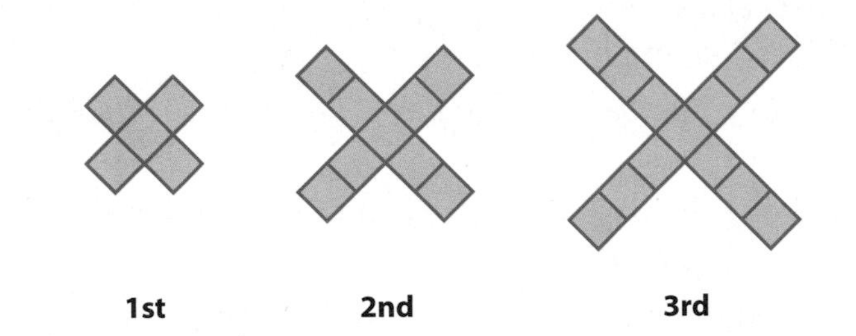

a. Write a rule showing how to calculate the number of squares in the *NEXT* shape from the number of squares *NOW*.

b. Write a rule for the number S of squares in the nth shape.

c. Solve these equations and inequalities, and explain what the solutions tell about the pattern.

i. $4n + 1 = 49$	**ii.** $4n + 1 = 81$
iii. $S = 4(8) + 1$	**iv.** $4n + 1 < 100$

18 The table on page 207 shows winning times for women and men in the Olympic 100-meter freestyle swim for games since 1912.

a. Make plots of the (*year, winning time*) data for men and for women. Use 0 for the year 1900.

b. Find the linear regression model for each data pattern.

c. Which group of athletes has shown a greater improvement in time, men or women? Explain.

d. What are the approximate coordinates of the point where the linear regression lines intersect?

e. What is the significance of the point of intersection of the two lines in Part d? How much confidence do you have that the lines accurately predict the future? Explain.

Olympic 100-meter Freestyle Swim Times

Year	Women's Time (in seconds)	Men's Time (in seconds)
1912	82.2	63.4
1920	73.6	61.4
1924	72.4	59.0
1928	71.0	58.6
1932	66.8	58.2
1936	65.9	57.6
1948	66.3	57.3
1952	66.8	57.4
1956	62.0	55.4
1960	61.2	55.2
1964	59.5	53.4

Year	Women's Time (in seconds)	Men's Time (in seconds)
1968	60.0	52.2
1972	58.59	51.22
1976	55.65	49.99
1980	54.79	50.40
1984	55.92	49.80
1988	54.93	48.63
1992	54.64	49.02
1996	54.50	48.74
2000	53.83	48.30
2004	53.84	48.17

Source: *The World Almanac and Book of Facts 2003.* Mahwah, NJ: World Almanac Education Group, Inc. 2003; www.olympics.com

Reflections

19 The function $y = 43 + 5x$ and the equation $78 = 43 + 5x$ are closely related to each other.

a. How can you use $y = 43 + 5x$ to solve $78 = 43 + 5x$?

b. What does solving $78 = 43 + 5x$ tell you about $y = 43 + 5x$?

20 When solving equations or inequalities that model real situations, why should you check not only the solution, but also whether the solution makes sense? Illustrate your thinking with an example.

On Your Own

21 Consider Mary and Jeff's pay possibilities for selling programs at the ballpark. The rule $P = 10 + 0.25s$ gives their pay P in dollars for selling s programs in a night.

a. What operations are needed to calculate the pay for selling 36 programs in a single night?

b. What operations are needed to solve the equation $10 + 0.25s = 19$? What will the solution tell you?

c. In what sense do the operations in Part b "undo" the operations in Part a?

d. How does the order in which you do the operations in Part a compare with the order in Part b? Why does this make sense?

e. How, if at all, does the procedure for solving change if you are asked to solve the inequality $10 + 0.25s \geq 19$? How does the meaning of the solution change?

22 How could you use a graph of the function $y = 7 - 4x$ to decide, without calculation or algebraic solution, whether the solutions to $7 - 4x > 43$ will be an inequality like $x < a$ or like $x > a$ and whether a will be positive or negative?

23 Consider the inequality $2x + 8 > 5x - 4$.

a. What are two different, but reasonable, first steps in solving $2x + 8 > 5x - 4$?

b. What does the solution to the inequality tell you about the graphs of $y = 2x + 8$ and $y = 5x - 4$?

c. What does the solution tell you about tables of values for $y = 2x + 8$ and $y = 5x - 4$?

24 When asked to solve the system of linear equations

$$\begin{cases} y = 2x + 9 \\ y = 5x - 18 \end{cases}$$

Sabrina reasoned as follows:

I want x so that $2x + 9 = 5x - 18$.

Adding 18 to each side of that equation gives me $2x + 27 = 5x$, and the sides remain balanced.

Subtracting $2x$ from each side of the new equation gives $27 = 3x$, and the sides remain balanced.

Dividing each side of that equation by 3 gives $x = 9$, and the sides remain balanced.

If $x = 9$, then one equation is $y = 2(9) + 9$ and the other equation is $y = 5(9) - 18$. Both equations give $y = 27$.

The solution of the system must be $x = 9$ and $y = 27$.

a. Do you agree with each step of her reasoning? Why or why not?

b. Use reasoning like Sabrina's to solve the following system of linear equations.

$$\begin{cases} y = 8x + 3 \\ y = 2x - 9 \end{cases}$$

Extensions

25 Refer back to Applications Task 3. Suppose a new policy at the restaurant applies an automatic service charge of 15% to each bill. Servers will receive \$20 per shift plus 10% of the bills for customers they serve. Busers will receive \$25 per shift plus 5% of the bills for customers at tables they serve.

a. Write rules for the functions that show how:

i. a server's daily earnings depend on the total of bills for customers at tables he or she serves.

ii. busers' daily earnings depend on the total of bills for customers at tables he or she serves.

Graph these two functions on the same coordinate axes.

b. Write three questions about the wages for wait staff and busers. Write equations or inequalities corresponding to your questions. Then solve the equations or inequalities and answer the questions. Show how you arrived at your solution for each equation or inequality.

26 The diagram at the right shows graphs of two functions:

$y = x + 3$ and $y = x^2 - 3$.

Reproduce the graphs on your graphing calculator or computer and use the graphs to solve each equation or inequality.

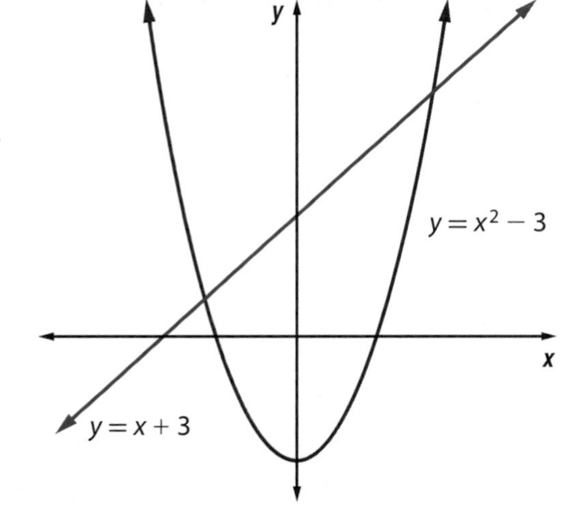

a. $x + 3 = x^2 - 3$

b. $x + 3 \geq x^2 - 3$

c. $x + 3 < x^2 - 3$

27 One linear function relating grams of fat F and calories C in popular "lite" menu items of fast-food restaurants is given by $C = 300 + 16(F - 10)$. Solve each equation or inequality below, and explain what your solution tells you about grams of fat and calories in fast-food. Use each of the following strategies at least once.

- Use a graph of the function.
- Use a table for the function.
- Use algebraic reasoning, as in the examples of this lesson.

a. $430 = 300 + 16(F - 10)$

b. $685 = 300 + 16(F - 10)$

c. $140 = 300 + 16(F - 10)$

d. $685 \geq 300 + 16(F - 10)$

On Your Own

28. Any linear function can be described by a rule of the form $y = a + bx$. Explain, with sketches, how to solve each of the following using graphs of linear functions. Describe the possible number of solutions. Assume $b \neq 0$ and $d \neq 0$.

   a. equations of the form $c = a + bx$

   b. inequalities of the form $c \leq a + bx$

   c. equations of the form $a + bx = c + dx$

   d. inequalities of the form $a + bx \leq c + dx$

29. The student government association at the Baltimore Freedom Academy wanted to order Fall Festival T-shirts for all students. Thrifty Designs charges a one-time art fee of \$20 and then \$6.25 per shirt.

   a. What rule shows how to calculate the cost c_1 of purchasing n shirts from Thrifty Designs?

   b. For each of the following questions:

   - Write an equation or inequality with solutions that will answer the question.
   - Explain how the solution for the equation or inequality is shown in a table or graph of (n, c_1) values.
   - Write an answer to the given question about T-shirt purchase.

   i. How much would it cost to buy T-shirts for 250 people?

   ii. How many T-shirts could be purchased for \$1,000?

   c. Suppose that Tees and More quotes a cost of \$1,540 for 250 T-shirts and \$1,000 for 160 T-shirts.

   i. If the cost of T-shirts from Tees and More is a linear function of the number of shirts purchased, what rule shows the relationship between number of shirts n and cost c_2?

   ii. What one-time art fee and cost per shirt are implied by the Tees and More price quotation?

   iii. Write and solve an inequality that answers the question, "For what numbers of T-shirts will Tees and More be less expensive than Thrifty Designs?" Be sure to explain how the solution is shown in a table or graph of (n, c_1) and (n, c_2) values.

30. Refer back to the Internet access pricing plans for Surf City Business Center and Byte to Eat Café given on page 198. Suppose Surf City Business Center decides to lower its base daily charge to \$2.95 but is unsure what to charge per call. They want to advertise daily charges that are lower than Byte to Eat Café if one spends more than 20 minutes online per day.

   a. To meet their goal, at what point will the Surf City Business Center graph need to cross the Byte to Eat Café graph?

   b. What charge per minute by Surf City Business Center will meet that condition?

c. Suppose a customer spends 60 minutes online per day. By how much is the new Surf City Business Center plan lower than the Byte to Eat Café plan for this many minutes?

d. If a person spends only 10 minutes online per day, how much less will he or she spend by using Byte to Eat Café rather than the new Surf City Business Center plan?

31 Refer back to the Check Your Understanding on page 200. Suppose it was noticed that most fishing parties coming to the dock were 4 or fewer persons.

a. How should Wally revise his boat rental fee so that his rates are lower than the competition's (Pike's) for parties of 3 or more? Write a rule for the new rate system.

b. How much less would a party of four pay by hiring Wally's charter service instead of Pike's?

c. Which service should you hire for a party of 2? How much will you save?

d. Suppose Pike's charter service lowers the per-person rate from $60 to $40. For what size parties would Pike's be less expensive?

e. If Wally wants to change his per-person rate so that both services charge the same for parties of 4, what per-person rate should Wally charge? Write a rule that models the new rate structure.

32 Create a linear system relating cost to number of uses of a service for which Company A's rate per service is 1.5 times that of Company B's, but Company B's service is not more economical until 15 services have been performed.

Review

33 Match each triangle description with the sketch(es) to which it applies.

a. An acute triangle

b. A scalene triangle

c. An obtuse triangle

d. An isosceles triangle

e. An equilateral triangle

f. A right scalene triangle

g. An obtuse isosceles triangle

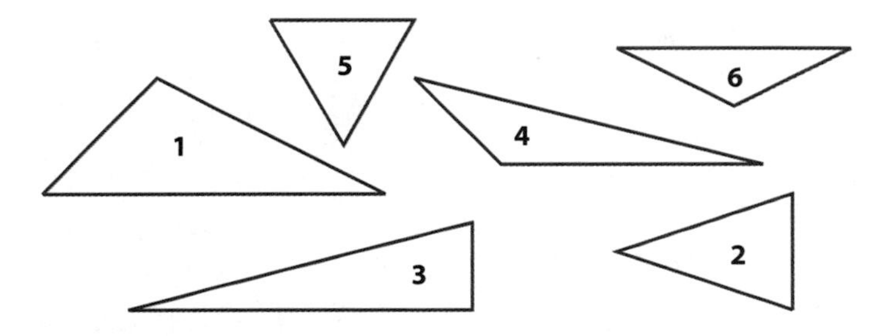

34 Use the fact that 5% of $50,000 = $2,500 to calculate the following percentages.

a. 0.5% of $50,000

b. 5.5% of $50,000

c. 105% of $50,000

d. 95% of $50,000

On Your Own

35 Use the fact that $13 \times 14 = 182$ to mentally calculate the value for each of the following expressions.

a. $(-13)(-14)$

b. $(14)(-13)$

c. $(13)(14) + (-13)(14)$

d. $\dfrac{(13)(-14)}{(-14)(-13)}$

36 Write an equation for the line passing through the points with coordinates $(-1, -2)$ and $(2, 0)$.

37 Find the area and perimeter of each figure. Assume that all angles that look like right angles are right angles and all segments that look parallel are parallel.

a. **b.**

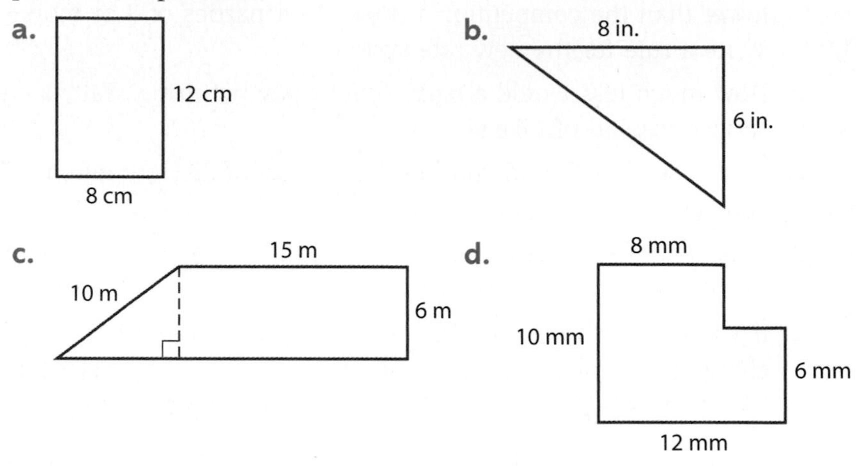

c.

15 m
10 m
6 m

d.

8 mm
10 mm
6 mm
12 mm

38 List all of the 2-digit numbers that can be made from the digits 3, 4, and 5. Digits may be repeated.

a. Suppose that you randomly choose one of the numbers you listed. What is the probability that it is divisible by 5?

b. Suppose that you randomly choose one of the numbers you listed. What is the probability that it is divisible by 3?

c. Suppose that you randomly choose one of the numbers you listed. What is the probability that it is divisible by 5 or 3?

d. Suppose that you randomly choose one of the numbers you listed. What is the probability that it is divisible by 5 and 3?

39 The following four expressions look very similar.

$$2 - x - 5 \qquad 2 - (x - 5) \qquad 2 - (x + 5) \qquad 2 - 5 - x$$

a. Substitute $x = 0$ into each expression to find the value for each expression.

b. Substitute $x = -1$ into each expression to find the value for each expression.

c. Substitute $x = 1$ into each expression to find the value for each expression.

d. Substitute $x = 2$ into each expression to find the value for each expression.

e. Which of the above expressions will always have equal value when you substitute the same number for x in the expressions? Explain.

40 Pentagon *ABCDE* is congruent to pentagon *PQRST*. Find each indicated angle measure or side length.

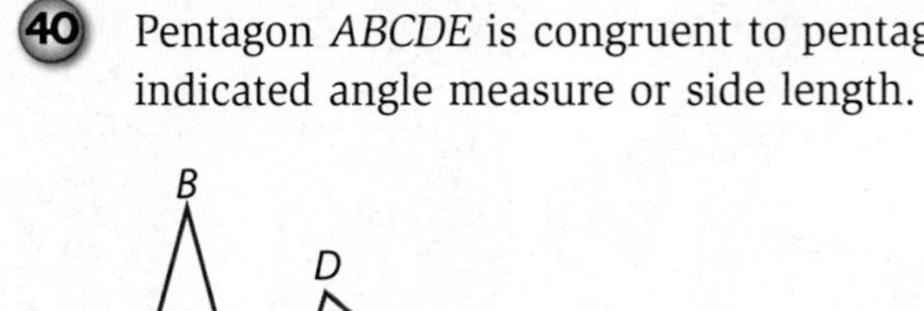

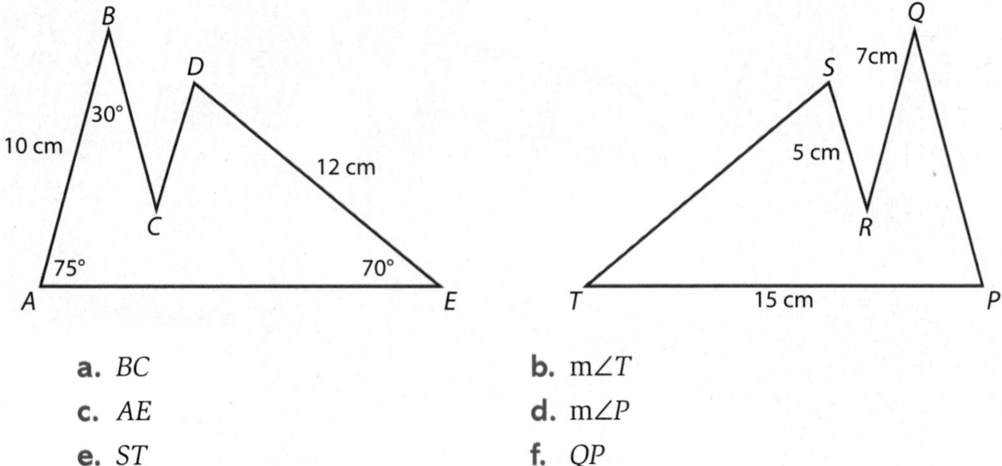

a. *BC*

b. m∠*T*

c. *AE*

d. m∠*P*

e. *ST*

f. *QP*

41 Tom and Jenny each drew a line to fit the same linear pattern of data. They wrote rules to make predictions for the value of y for different values of x. Complete the table below to compare their predictions.

x	0	1	2	3
(Tom's Rule) $y = 0.5x + 7$	7	7.5		
(Jenny's Rule) $y = 0.6x + 7$	7	7.6		
Difference in Predictions	0	0.1		

Why does the difference in predictions change? When is the difference greater than 0.5?

LESSON 3

Equivalent Expressions

Entertainment is a big business in the United States—from network television, movies, and concert tours to local school plays and musical shows. Each live or recorded performance is prepared with weeks, months, or even years of creative work and business planning.

For example, a recording label might have the following costs to produce a CD by a popular music artist.

- \$100,000 to record the tracks;
- \$1.50 per copy for materials and reproduction;
- \$2.25 per copy for royalties to the writers, producers, and performers.

The record label might receive income of about \$5 per copy from the stores that sell the CD.

Think About This Situation

The record label's profit on a CD is a function of the number of copies that are made and sold. A function rule for profit gives an *expression* for calculating profit. This lesson will focus on expressions for calculating various quantities.

a Using the given numbers, how would you calculate the label's net profit for 100,000 copies made and sold? For 1 million copies made and sold? For n copies made and sold?

b One group of students wrote the expression $5n - (3.75n + 100{,}000)$ to calculate the label's net profit on sales of n copies. Is this correct? How do you know? Why might they have expressed profit this way?

c Another group of students wrote the expression $-100{,}000 + 1.25n$ to calculate the label's net profit on sales of n copies. Is this correct? How do you know? Why might they have expressed profit this way?

d Could you represent the profit for n copies in other different ways?

e How could you convince another student that two different expressions represent the same quantity? Consider the expressions for profit you produced or those given in Parts b and c.

Two expressions are **equivalent** if they produce identical outputs when given identical inputs. In this lesson, you will develop your ability to recognize, write, and reason with equivalent linear expressions.

Investigation 1 Different, Yet the Same

Your thinking about possible profit rules for a new CD release showed an important fact about linear expressions: Several different expressions can each represent the same quantity.

Tables and graphs are one way to explore whether two expressions are equivalent, but it is helpful to be able to tell by looking when two expressions are equivalent. As you work on the problems of this investigation, keep in mind the following question:

What operations on linear expressions lead to different, but equivalent, expressions?

Movie Production Studios that make motion pictures deal with many of the same cost and income questions as music producers. Contracts sometimes designate parts of the income from a movie to the writers, directors, and actors. Suppose that for one film those payments are:

4% to the writer of the screenplay;
6% to the director;
15% to the leading actors.

1. What payments will go to the writer, the director, the leading actors, and to all these people combined in the following situations?
   a. The studio receives income of $25 million from the film.
   b. The studio receives income of $50 million from the film.

2. Suppose the studio receives income of I million dollars from the film.
   a. Write an expression for the total payment to the writer, the director, and the leading actors in a form that shows the breakdown to each person or group.
   b. Write another expression for the total payment that shows the combined percent of the film income that is paid out to the writer, the director, and the leading actors.

3. A movie studio will have other costs too. For example, there will be costs for shooting and editing the film. Suppose those costs are $20 million.
   a. Assume that the $20 million for shooting and editing the film and the payments to the writer, the director, and the leading actors are the only costs for the film. What will the studio's profit be if the income from the film is $50 million?
   b. Consider the studio's profit (in millions of dollars) when the income from the film is I million dollars.
      i. Write an expression for calculating the studio's profit that shows the separate payments to the writer, the director, and the leading actors.
      ii. Write another expression for calculating the studio's profit that combines the payments to the writer, the director, and the leading actors.
      iii. Is the following expression for calculating the studio's profit correct? How do you know?

$$I - (20 + 0.25I)$$

      iv. Write another expression for calculating the studio's profit and explain what that form shows.

Movie Theaters For theaters, there are two main sources of income. Money is collected from ticket sales and from concession stand sales.

Suppose that a theater charges $8 for each admission ticket, and concession stand income averages $3 per person.

4 **Income** Consider the theater income during a month when they have n customers.

a. Write an expression for calculating the theater's income that shows separately the income from ticket sales and the income from concession stand sales.

b. Write another expression for calculating income that shows the total income received per person.

5 **Expenses** Suppose that the theater has to send 35% of its income from ticket sales to the movie studio releasing the film. The theater's costs for maintaining the concession stand stock average about 15% of concession stand sales. Suppose also that the theater has to pay rent, electricity, and staff salaries of about $15,000 per month.

a. Consider the theater's expenses when the theater has n customers during a month.

i. How much will the theater have to send to movie studios?

ii. How much will the theater have to spend to restock the concession stand?

iii. How much will the theater have to spend for rent, electricity, and staff salaries?

b. Write two expressions for calculating the theater's total expenses, one that shows the breakdown of expenses and another that is as short as possible.

6 **Profit** Consider next the theater's profit for a month in which the theater has n customers.

a. Write an expression for calculating the theater's profit that shows each component of the income and each component of the expenses.

b. Write another expression for calculating the theater's profit that shows the total income minus the total expenses.

c. Write another expression for calculating the theater's profit that is as short as possible.

7 **Taxes** The movie theater charges $8 per admission ticket sold and receives an average of $3 per person from the concession stand. The theater has to pay taxes on its receipts. Suppose the theater has to pay taxes equal to 6% of its receipts.

a. Consider the tax due if the theater has 1,000 customers.

i. Calculate the tax due for ticket sales and the tax due for concession stand sales, then calculate the total tax due.

ii. Calculate the total receipts from ticket sales and concession stand sales combined, then calculate the tax due.

b. Write two expressions for calculating the tax due if the theater has n customers, one for each way of calculating the tax due described in Part a.

8 In Problem 6, you wrote expressions for the monthly theater profit after all operating expenses. A new proposal will tax profits only, but at 8%. Here is one expression for the tax due under this new proposal.

$$0.08(7.75n - 15{,}000)$$

a. Is the expression correct? How can you be sure?

b. Write an expression for calculating the tax due under the new proposal that is as short as possible. Show how you obtained your expression. Explain how you could check that your expression is equivalent to the one given above.

Summarize the Mathematics

In many situations, two people can suggest expressions for linear functions that look quite different but are equivalent. For example, these two symbolic expressions for linear functions are equivalent.

$$15x - (12 + 7x) \qquad \text{and} \qquad 8x - 12$$

a What does it mean for these two expressions to be equivalent?

b How could you test the equivalence of these two expressions using tables and graphs?

c Explain how you might reason from the first expression to produce the second expression.

Be prepared to explain your responses to the entire class.

Check Your Understanding

Many college basketball teams play in winter tournaments sponsored by businesses that want the advertising opportunity. For one such tournament, the projected income and expenses are as follows.

- Income is \$60 per ticket sold, \$75,000 from television and radio broadcast rights, and \$5 per person from concession stand sales.
- Expenses are \$200,000 for the colleges, \$50,000 for rent of the arena and its staff, and a tax of \$2.50 per ticket sold.

a. Find the projected income, expenses, and profit if 15,000 tickets are sold for the tournament.

b. Write two equivalent expressions for tournament income if n tickets are sold. In one expression, show each source of income. In the other, rearrange and combine the income components to give the shortest possible expression.

c. Write two equivalent expressions for tournament expenses if n tickets are sold. In one expression, show each source of expense. In the other, rearrange and combine the expense components to give the shortest possible expression.

d. Write two equivalent expressions for tournament profit if n tickets are sold. In one expression, show income separate from expenses. In the other, rearrange and combine components to give the shortest possible expression.

Investigation 2 The Same, Yet Different

In Investigation 1, you translated information about variables into expressions and then into different, but equivalent, expressions. You used facts about the numbers and variables involved to guide and check the writing of new equivalent symbolic expressions.

The examples in Investigation 1 suggest some ways to rewrite symbolic expressions that will produce equivalent forms, regardless of the situation being modeled. Think about how you can rewrite expressions involving variables in equivalent forms even if you do not know what the variables represent. As you work on the problems of this investigation, look for answers to this question:

How can algebraic properties of numbers and operations be used to verify the equivalence of expressions and to write equivalent expressions?

1 These expressions might represent the profit for a given number of sales. Using your thinking from Investigation 1 as a guide, write at least two different but equivalent expressions for each.

a. $8x - 3x - 2x - 50$ **b.** $6a - (20 + 4a)$

c. $0.8(10n - 30)$ **d.** $t - 20 - 0.3t$

2 Think about how you might convince someone else that the expressions you wrote in Problem 1 are, in fact, equivalent.

a. How might you use tables and graphs to support your claim?

b. How might you argue that two expressions are equivalent without the use of tables or graphs? What kind of evidence do you find more convincing? Why?

3 Determine which of the following pairs of expressions are equivalent. If a pair of expressions is equivalent, explain how you might justify the equivalence. If a pair is not equivalent, show that the pair is not equivalent.

a. $3.2x + 5.4x$ and $8.6x$ **b.** $3(x - 2)$ and $6 - 3x$

c. $4y + 7y - 12$ and $-12 + 11y$ **d.** $7x + 14$ and $7(x + 2)$

e. $8x - 2(x - 3)$ and $6x - 6$ **f.** $3x + 7y - 21$ and $10xy + (-21)$

g. $\frac{8y + 12}{4}$ and $2y + 3$ **h.** $x + 4$ and $3(2x + 1) - 5x + 1$

There are five properties of numbers and operations that are especially helpful in transforming algebraic expressions to useful equivalent forms.

- **Distributive Property of Multiplication over Addition**—For any numbers a, b, and c:

$$a(b + c) = ab + ac \text{ and } ac + bc = (a + b)c.$$

This property can be applied to write expressions with or without the use of parentheses. For example,

$$5(2x + 3) = 10x + 15.$$

The distributive property can also be applied to the sums of products with common factors. For example,

$$3x + 7x = (3 + 7)x = 10x \quad \text{and} \quad 6\pi + 8\pi = (6 + 8)\pi = 14\pi.$$

- **Commutative Property of Addition**—For any numbers a and b:

$$a + b = b + a.$$

- **Associative Property of Addition**—For any numbers a, b, and c:

$$a + (b + c) = (a + b) + c.$$

The commutative and associative properties are often used together to rearrange the addends in an expression. For example,

$$\begin{aligned} 3x + 5 + 4x + 7 &= (3x + 4x) + (5 + 7) \\ &= 7x + 12. \end{aligned}$$

- **Connecting Addition and Subtraction**—For any numbers a and b:

$$a - b = a + (-b).$$

This property can be applied to rewrite an expression that involves subtraction so that the terms involved can be rearranged using the commutative and associative properties of addition. For example,

$$\begin{aligned} 4x - 3 - 5x + 7 &= 4x + (-3) + (-5x) + 7 \\ &= 4x + (-5x) + (-3) + 7 \\ &= -x + 4. \end{aligned}$$

The property can also be used with the distributive property to expand a product that involves subtraction. For example,

$$\begin{aligned} 5(2x - 3) &= 5(2x + (-3)) \\ &= 10x + (-15) \\ &= 10x - 15. \end{aligned}$$

- **Connecting Multiplication and Division**—For any numbers a and b with $b \neq 0$:

$$\frac{a}{b} = a \cdot \frac{1}{b}.$$

This property can be combined with the distributive property to rewrite an expression that involves division. For example,

$$\begin{aligned} \frac{20 - 15x}{5} &= (20 - 15x)\frac{1}{5} \\ &= 4 - 3x. \end{aligned}$$

4 When you are given an expression and asked to write an equivalent expression that does not contain parentheses, this is called **expanding** the expression. Use the distributive property to rewrite the following expressions in *expanded* form.

a. $4(y + 2)$ **b.** $(5 - x)(3y)$ **c.** $-2(y - 3)$

d. $-7(4 + x)$ **e.** $\frac{(16x - 8)}{4}$ **f.** $\frac{1}{3}(2x + 3)$

5 When you are given an expression and asked to write an equivalent expression that gives a product, this is called **factoring** the expression. For example, in Problem 4 Part a you wrote $4(y + 2) = 4y + 8$. Writing $4y + 8$ as $4(y + 2)$ is said to be writing $4y + 8$ in *factored* form. Use the distributive property to rewrite the following expressions in factored form.

a. $2x + 6$ **b.** $20 - 5y$ **c.** $6y - 9$

d. $8 + 12x$ **e.** $3x + 15y$ **f.** $xy - 7x$

6 Using the distributive property to add or subtract products with common factors is called **combining like terms**. Use the distributive property to rewrite the following expressions in equivalent shorter form by combining like terms.

a. $7x + 11x$ **b.** $7x - 11x$

c. $5 + 3y + 12 + 7y$ **d.** $2 + 3x - 5 - 7x$

e. $\frac{3x}{4} - 2x + \frac{x}{4}$ **f.** $10x - 5y + 3y - 2 - 4x + 6$

7 Write each of the following expressions in its *simplest* equivalent form by expanding and then combining like terms.

a. $7(3y - 2) + 6y$ **b.** $5 + 3(x + 4) + 7x$

c. $2 + 3x - 5(1 - 7x)$ **d.** $10 - (5y + 3)$

e. $10 - \frac{15x - 9}{3}$ **f.** $5(x + 3) - \frac{4x + 2}{2}$

g. $7y + 4(3y - 11)$ **h.** $8(x + 5) - 3(x - 2)$

8 Write each of the following expressions in equivalent form by combining like terms and then factoring.

a. $7 + 15x + 5 - 6x$ **b.** $x - 10 + x + 2$

c. $20x + 10 - 5x$ **d.** $24 - 5x - 4 + 6x - 8 + 2x$

9 When *simplifying* an expression, it is easy to make mistakes. Some of the pairs of expressions below are equivalent and some are not. If a pair of expressions is equivalent, describe the properties of numbers and operations that justify the equivalence. If a pair is not equivalent, correct the mistake.

a. $2(x - 1)$ and $2x - 1$ **b.** $4(3 + 2x)$ and $12 + 8x$

c. $9 - (x + 7)$ and $16 - x$ **d.** $\frac{6x + 12}{6}$ and $x + 12$

e. $5x - 2 + 3x$ and $8x - 2$ **f.** $4x - x + 2$ and 6

Computer algebra systems (CAS) have been programmed to use properties of numbers and operations like those described early in this investigation to expand, factor, and simplify algebraic expressions. To use a CAS for this purpose you need only enter the expression accurately and then apply the "expand" or "factor" commands from the CAS algebra menu.

You can also check equivalence of two given expressions by entering each as part of an equation and pressing the ENTER key. If the expressions are equivalent, the CAS will respond with "true."

10 Compare the CAS output shown below with your answers to the following problems. Discuss and reconcile any differences.

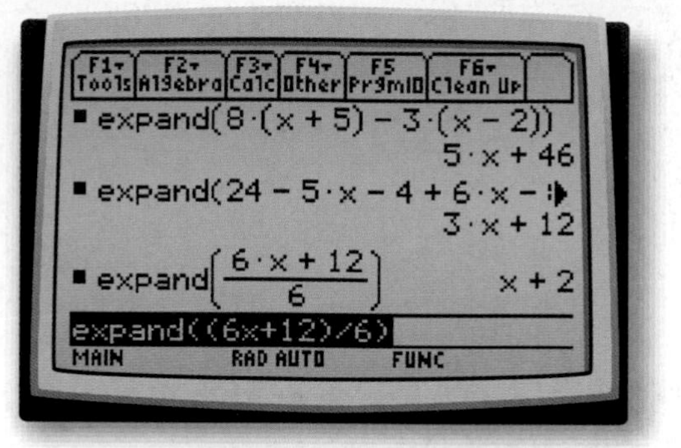

a. Problem 7 Part h

b. Problem 8 Part d

c. Problem 9 Part d

11 Use the CAS that is available to you to perform the following algebraic procedures. Compare the CAS output with what you expect from your knowledge of algebraic manipulations and reconcile any differences.

a. What is the shortest expression equivalent to $25x - 4(3x + 7) + 106$?

b. What expression equivalent to that in Part a is in simplest factored form?

c. What expression is equivalent in expanded form to $5(7 - 3x) - (8x + 12)23$?

d. What results from asking your CAS to factor $a \cdot x + a \cdot c$?

e. What results from asking your CAS to factor $a \cdot (x + b) - c \cdot (x + b)$?

f. What results from asking your CAS to expand $a \cdot (x + b) - c \cdot (x + b)$?

Summarize the Mathematics

In this investigation, you applied key properties of numbers and operations to evaluate the equivalence of expressions and to create equivalent expressions.

a Summarize these algebraic properties in your own words and give one example of how each property is used in writing equivalent expressions.

b How can you tell if expressions such as those in this investigation are in simplest form?

c What are some easy errors to make that may require careful attention when writing expressions in equivalent forms? How can you avoid making those errors?

Be prepared to explain your thinking and examples to the class.

Check Your Understanding

For each of the following expressions:

- Write two expressions equivalent to the original—one that is as short as possible.
- Describe the algebraic reasoning used to obtain each expression.
- Test the equivalence of each expression to the original by comparing tables and graphs.

a. $9x + 2 - x$

b. $3(x + 7) - 6$

c. $\frac{20 - 4(x - 1)}{2}$

d. $2(1 + 3x) - (5 - 6x)$

On Your Own

Applications

1. To advertise a concert tour, the concert promoter paid an artist \$2,500 to design a special collector's poster. The posters cost \$2.50 apiece to print and package in a cardboard cylinder. They are to be sold for \$7.95 apiece.

   a. Write expressions that show how to calculate cost, income, and profit if n posters are printed and sold.

   b. Write two expressions that are different from, but equivalent to, the profit expression you wrote in Part a. Explain why you are sure they are equivalent.

2. The video game industry is a big business around the world. Development of a new game might cost millions of dollars. Then to make and package each game disc will cost several more dollars per copy. Suppose the development cost for one game is \$5,000,000; each disc costs \$4.75 to make and package; and the wholesale price is set at \$35.50 per disc.

   a. Write expressions that show how to calculate the cost of designing and making n discs and the income earned from selling those n discs.

   b. Write two different but equivalent expressions for profit from selling n discs.

   c. Use evidence in tables, graphs, or properties of numbers and operations to justify equivalence of the two expressions from Part b.

3. The historic Palace Theater offers students and seniors a \$2 discount off the regular movie ticket price of \$7.50. The theater has 900 seats and regularly sells out on the weekends. Marcia and Sam wrote the following expressions for income from ticket sales based on the number x of discounted tickets sold for a sold-out show.

   Marcia's expression: $5.5x + 7.5(900 - x)$
   Sam's expression: $900(7.5) - 2x$

   a. Explain how Marcia and Sam may have reasoned in writing their expressions.

b. Use tables and graphs to check whether Marcia's and Sam's expressions are equivalent.

c. Write a simpler expression for income based on the number x of discounted tickets sold for a sold-out show. Show how you could reason from Marcia's and Sam's expressions to this simpler expression.

4 In art class, students are framing square mirrors with hand-painted ceramic tiles as shown at the right. Each tile is one inch by one inch.

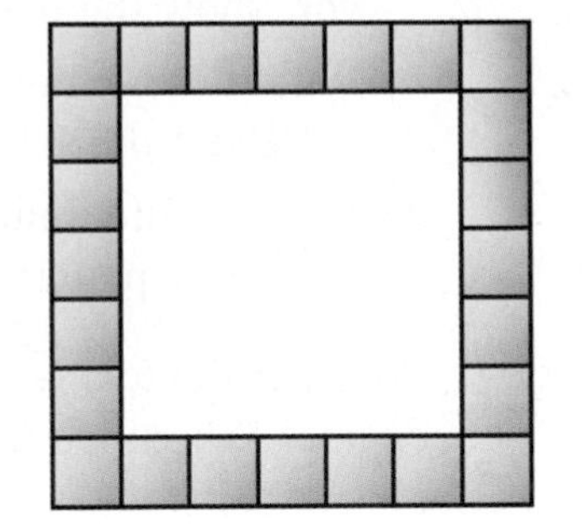

a. How many tiles are needed to frame a square mirror with side length 5 inches? 3 inches? 10 inches?

b. Write an expression for the number of tiles needed to frame a square mirror with side length x inches.

c. One group of students came up with the following expressions for the number of tiles needed to frame a square mirror with side length x inches. Explain the thinking that might have led to each of these expressions. Use tables, graphs, or algebraic reasoning to demonstrate the equivalence of the expressions.

i. $(x + 1) + (x + 1) + (x + 1) + (x + 1)$

ii. $(x + 2) + x + (x + 2) + x$

iii. $4x + 4$

iv. $4(x + 1)$

v. $2(x + 2) + 2x$

5 Are the following pairs of expressions equivalent? Explain your reasoning in each case.

a. $7 - 5(x + 4 - 3x)$ and $7 - 5x + 20 - 15x$

b. $7x - 12 + 3x - 8 + 9x - 5$ and $7x - 4 + 2x + 8 + 10x - 13$

6 For each of the following expressions, write an equivalent expression that is as short as possible.

a. $3x + 5 + 8x$

b. $7 + 3x + 12 + 9x$

c. $8(5 + 2x) - 36$

d. $2(5x + 6) + 3 + 4x$

e. $\frac{10x - 40}{5}$

f. $5x + 7 - 3x + 12$

g. $3x + 7 - 4(3x - 6)$

h. $-7x + 13 + \frac{12x - 4}{4}$

7 For each of the following expressions, combine like terms and then write in factored form.

a. $6x + 5 + 9x$

b. $20 + 6x + 4 + 10x$

c. $32 + 20x$

d. $13x + 6 - (2 - 3x)$

Connections

8 One expression for predicting the median salary in dollars for working women since 1970 is $4{,}000 + 750(y - 1970)$, where y is the year.

a. Write an equivalent expression in the form $a + by$. Explain how you know the new expression is equivalent to the original.

b. What do the numbers 4,000, 750, and the expression $(y - 1970)$ tell about the salary pattern?

c. What do the numbers a and b that you found in Part a tell about the salary pattern?

9 Consider the following formula for transforming temperature in degrees Fahrenheit F to temperature in degrees Celsius C.

$$C = \frac{5}{9}(F - 32)$$

a. Use the distributive property to rewrite the expression for temperature in degrees Celsius in the form $aF + b$.

b. Write a question that is more easily answered using the original expression for calculating the temperature in degrees Celsius. Write another question that is more easily answered using the expression from Part a. Explain why you think one expression is better for each question.

10 Consider the following set of instructions.

Pick any number.
Multiply it by 2.
Subtract 10 from the result.
Multiply the result by 3.
Add 30 to the result.
Finally, divide by your original number.

```
                100
Ans-10
                 90
Ans*3
                270
Ans+30
                300
Ans/50
```

a. Repeat the process several times with different starting numbers. What are your answers in each case?

b. Let x represent the starting number. Write an expression showing the calculations for any value of x.

c. Write the expression from Part b in simplest equivalent form. Explain how it makes the results in Part a reasonable.

11 The length of a rectangle is ℓ and the width is w.

a. Write at least three different expressions that show how to calculate the perimeter of the rectangle. Explain how you might reason from a drawing of a rectangle to help you write each expression.

b. Use the properties of numbers and operations, discussed in this lesson, to reason about the equivalence of the expressions you wrote in Part a.

12 Recall the formula $A = \frac{1}{2}bh$ for the area of a triangle where b is the length of the base and h is the height of the triangle. A trapezoid with bases of lengths b_1 and b_2 and height h is shown below.

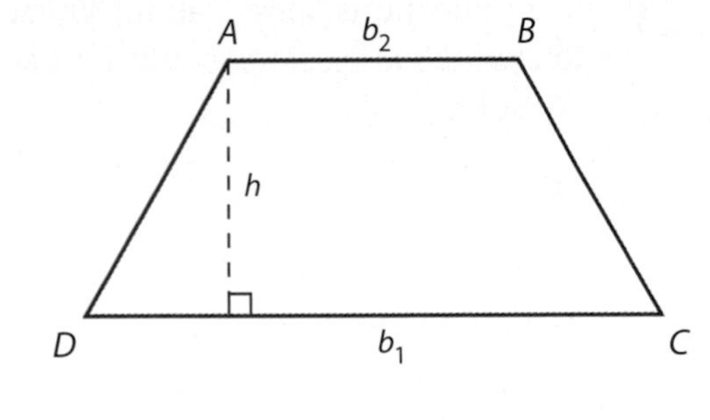

a. Make a copy of the diagram.

b. Draw $\triangle ACD$ and write an expression showing how to calculate its area.

c. Write an expression showing how to calculate the area of $\triangle ABC$.

d. Write an expression showing how to calculate the area of trapezoid *ABCD*.

e. Write a different but equivalent expression for the area of the trapezoid.

13 Refer back to the Check Your Understanding problem of Investigation 1 (page 218). To study the profit prospects of different options in organizing a college basketball tournament, it might be helpful to construct a spreadsheet to calculate income, expenses, and profit that will result from various decisions.

a. Copy the expressions you developed for income, expenses, and profit in terms of the number of tickets sold n based on the following information.

- Income is $60 per ticket sold, $75,000 from television and radio broadcast rights, and $5 per person from concession stand sales.
- Expenses are $200,000 for the colleges, $50,000 for rent of the arena and its staff, and a tax of $2.50 per ticket sold.

b. Use those expressions to complete the following spreadsheet so that you could explore effects of different ticket sale numbers. Enter the required numbers and formulas in column B of the spreadsheet.

Profit Prospects.xls

◇	A	B
1	Tickets Sold =	
2	Income =	
3	Expenses =	
4	Profit =	

c. Expand and modify the spreadsheet so that you could also adjust ticket price, average concession stand income, ticket tax, and television/radio broadcast rights to see immediately how profit changes. (*Hint:* Put the labels for those factors in cells of column C and then the values in adjacent cells of column D.)

Reflections

14 In Applications Task 2 about video game discs, any one of the following expressions could be used to calculate profit when n discs are sold.

$$35.50n - 4.75n - 5{,}000{,}000$$
$$30.75n - 5{,}000{,}000$$
$$35.50n - (5{,}000{,}000 + 4.75n)$$

a. Explain how you can be sure that all three expressions are equivalent.

b. Which expression do you believe shows the business conditions in the best way? Explain the reasons for your choice.

15 Think about a real situation involving a changing quantity. Write at least two different but equivalent expressions that show how to calculate the quantity.

16 How do you prefer to check whether two expressions are equivalent: using tables of values, graphs, or algebraic reasoning? Why? Which gives the strongest evidence of equivalence?

17 Each of the expressions in Applications Tasks 5–7 defines a linear function. However, none of those expressions looks exactly like $a + bx$, the familiar form of an expression for a linear function.

a. What features of expressions like those in the Applications tasks suggest that the graph of the function defined by that expression will be a line?

b. What might appear in an expression that would suggest that the graph of the corresponding function would not be a line? Give some examples and sketch the graphs of those examples.

18 In transforming algebraic expressions to equivalent forms, it's easy to make some mistakes and use "illegal" moves. Given below are six pairs of algebraic expressions. Some are equivalent and some are not.

- Use tables, graphs, and/or algebraic reasoning to decide which pairs are actually equivalent and which involve errors in reasoning.
- In each case of equivalent expressions, describe algebraic reasoning that could be used to show the equivalence.
- In each case of an algebra mistake, spot the error in reasoning. Write an explanation that would help clear up the problem for a student who made the error.

a. Is $3(2x + 8)$ equivalent to $6x + 8$?

b. Is $4x - 6x$ equivalent to $2x$?

c. Is $8(2 - 6x)$ equivalent to $16 - 48x$?

d. Is $10 + 3x - 12$ equivalent to $3x + 2$?

e. Is $\frac{5x + 10}{5}$ equivalent to $x + 10$?

f. Is $-4(x - 3)$ equivalent to $-4x - 12$?

Extensions

19 Solve these equations. Show your work and check your solutions.

a. $3(x - 4) = 12$

b. $2(a + 1) = 8 + a$

c. $13 - (5 + n) = 6$

20 Look back at the statements of the Commutative and Associative Properties of Addition on page 220.

a. There are corresponding properties of multiplication. Write statements for the corresponding properties.

b. Give examples showing how the Commutative and Associative Properties of Multiplication can be used in writing equivalent algebraic expressions.

21 The properties of numbers and operations discussed in this lesson can be applied to write equivalent expressions for any expression, not just linear expressions.

a. Consider the rule suggested for the income I from the Five Star bungee jump as a function of ticket price p from "Physics and Business at Five Star Amusement Park" in Unit 1.

$$I = p(50 - p)$$

Expand the expression $p(50 - p)$.

b. Consider the rule some students wrote for estimating the total population of Brazil *NEXT* year given the population *NOW*.

$$NEXT = 0.01 \cdot NOW + NOW - 0.009 \cdot NOW$$

Combine like terms to write a shorter expression on the right-hand side of the rule.

22 Use the distributive property to write the expression $(x + 2)(x + 3)$ in expanded form. Is this a linear expression? Explain how you know.

23 Write the following expressions in factored form.

a. $8x^2 + 12x$

b. $10x - 15x^2$

c. $3x^3 - 27x^2 + 18x$

24 You have learned how a CAS can be used to expand or factor expressions, test the equivalence of two expressions, and solve equations. How could you solve a system of linear equations using a CAS? Test your ideas using the system below from Lesson 2 (page 199). Compare the solution with the solution you previously found.

Surf City Business Center: $y_1 = 3.95 + 0.05x$

Byte to Eat Café: $y_2 = 2 + 0.10x$

Review

25 Write equations for the lines satisfying these conditions.

a. passing through the point (0, 4) and having slope 3

b. passing through the points (10, 7) and (20, 12)

c. passing through the points (−3, 5) and (1, −3)

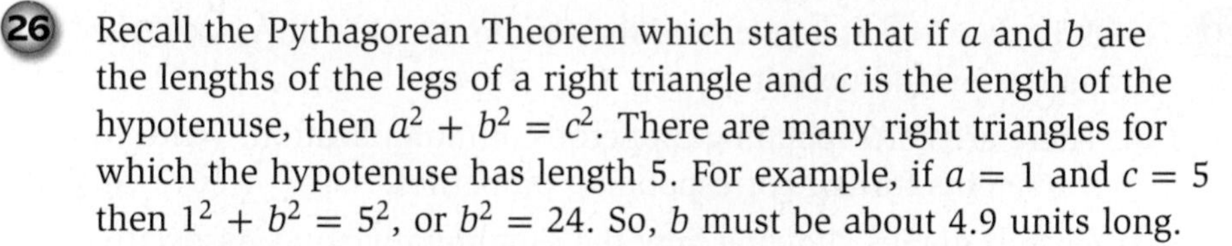

26 Recall the Pythagorean Theorem which states that if a and b are the lengths of the legs of a right triangle and c is the length of the hypotenuse, then $a^2 + b^2 = c^2$. There are many right triangles for which the hypotenuse has length 5. For example, if $a = 1$ and $c = 5$ then $1^2 + b^2 = 5^2$, or $b^2 = 24$. So, b must be about 4.9 units long.

a. Complete the table below by assigning different values to the leg length a and calculating the length b of the other leg.

a	1	2	3	4
b	4.9			
c	5	5	5	5

b. Is the pattern of change in b as a increases a linear pattern? Explain.

27 Consider the two functions $y = 2 + 0.25x$ and $y = -8 + 1.5x$. Use tables or graphs to estimate solutions for these equations and inequalities.

a. $2 + 0.25x = 0$

b. $2 + 0.25x = -8 + 1.5x$

c. $-8 + 1.5x > 0$

d. $2 + 0.25x \leq -8 + 1.5x$

28 Consider the quadrilaterals shown below. Assume that segments that look parallel are parallel, and that segments or angles that look congruent are congruent.

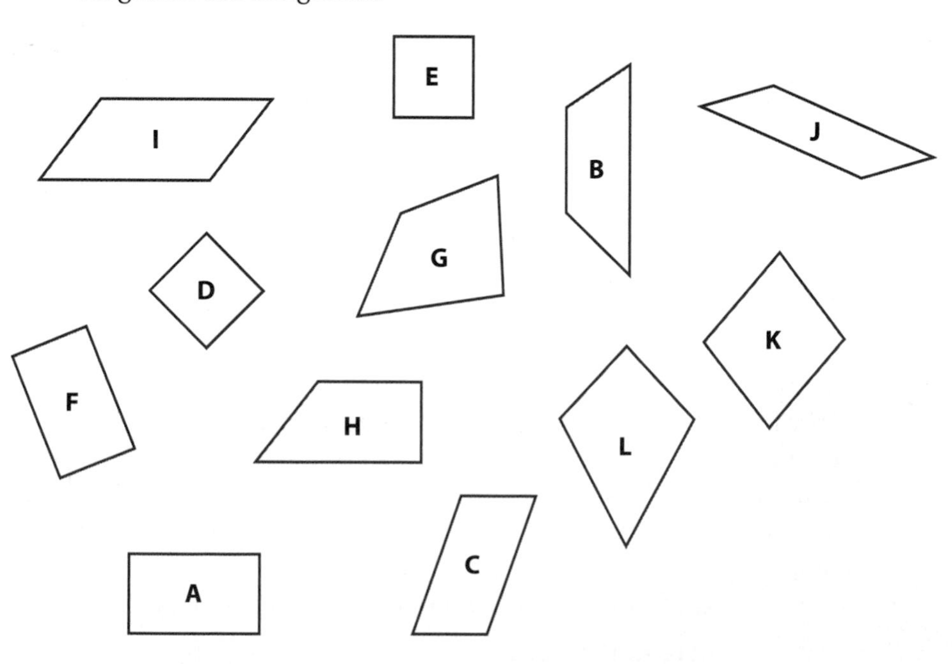

a. Identify all figures that are squares.

b. Identify all figures that are rectangles.

c. Identify all figures that are parallelograms.

d. Identify all figures that are trapezoids.

29 The histogram below displays the number of books read over the summer by all of the ninth grade students at Treadwell High School.

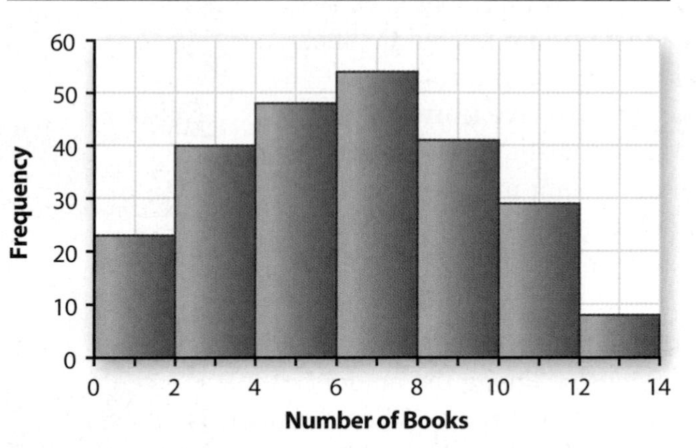

a. Emily estimated that the median of these data is about 8 books. Do you think she is correct? Explain your reasoning.

b. Will estimated that the mean of these data is about 6 books and the standard deviation is about 4 books. Do his estimates seem reasonable? Explain your reasoning.

30 Using the diagram below, determine if each statement is true or false. In each case, explain your reasoning.

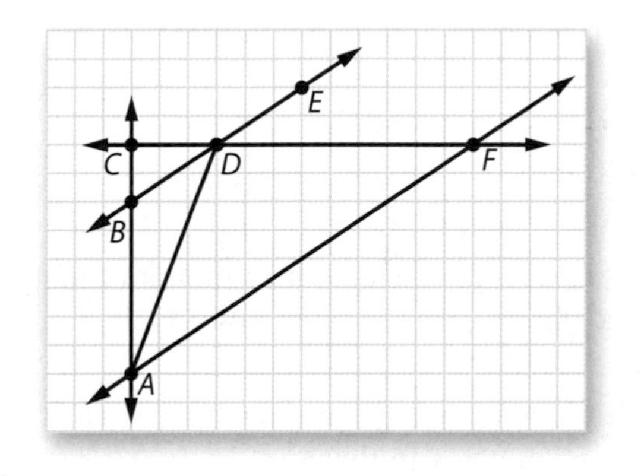

a. $\overleftrightarrow{AC} \perp \overleftrightarrow{DF}$

b. $m\angle FDA < m\angle CDA$

c. $\overleftrightarrow{BE} \parallel \overleftrightarrow{AF}$

d. $AD = DF$

Looking Back

Through the lessons of this unit, you have seen that variables in many situations are related by linear functions. You have learned how to find rules that describe those functions. You also have learned how to use equations to answer questions about the variables and relations. This final lesson of the unit provides problems that will help you review, pull together, and apply your new knowledge.

1. **Fuel Consumption** When private pilots make flight plans for their trips, they must estimate the amount of fuel required to reach their destination airport. When the plane is in flight, pilots watch to see how much fuel they have left in their tanks. The table below shows fuel remaining in the tanks at various times during a flight under constant speed for one type of small plane.

Fuel Consumption

Time in Flight (in minutes)	40	60	80	100	120
Fuel Remaining (in gallons)	50	45	40	35	30

a. Is fuel remaining a linear function of the time in flight? How do you know?

b. Determine the rate of change in fuel remaining as time in flight increases. What units describe this rate of change?

c. How much fuel was in the tanks at the start of the flight?

d. Write a rule that shows how to calculate the amount of fuel remaining in the *NEXT* hour given the amount of fuel remaining in gallons *NOW*.

e. Write a rule that shows how to calculate the amount of fuel remaining F after t minutes in flight.

f. How much fuel remained in the tanks after 1.5 hours? After 3 hours?

g. At one point in the flight, the pilot observes that 5 gallons of fuel remain in the tanks. How much flying time is left?

2. **Health and Nutrition** Even if we do not always eat what is best for us, most Americans can afford nutritious and varied diets. In many countries of the world, life is a constant struggle to find enough food. This struggle causes health problems such as reduced life expectancy and infant mortality.

a. The data in the table below show how average daily food supply (in calories) is related to life expectancy (in years) and infant mortality rates (in deaths per 1,000 births) in a sample of countries. Make scatterplots of the (*daily calories, life expectancy*) and (*daily calories, infant mortality*) data.

Health and Nutrition

Country	Daily Calories	Life Expectancy	Infant Mortality
Argentina	3,136	74	20
Bolivia	2,170	64	56
Dominican Republic	2,316	67	36
Haiti	1,855	53	61
Mexico	3,137	73	28
New Zealand	3,405	78	6
Paraguay	2,485	71	37
United States	3,642	78	7

Source: *The New York Times 2003 Almanac.* New York, NY: Penguin Company, 2002.

Study the patterns in the table and the scatterplots. Then answer these questions.

i. What seems to be the general relation between daily calories and life expectancy in the sample countries?

ii. What seems to be the general relation between daily calories and infant mortality in the sample countries?

iii. What factors other than daily calorie supply might affect life expectancy and infant mortality?

b. Economists might use a linear model to predict the increase of life expectancy or decrease of infant mortality for various increases in food supply.

i. Determine a linear regression model for calculating life expectancy from calories using the (*daily calories, life expectancy*) data pattern.

ii. Determine a linear regression model for calculating infant mortality from calories using the (*daily calories, infant mortality*) data pattern.

iii. What do the slopes of the graphs of your linear models say about the pattern relating life expectancy to daily calories in the sample countries? How about the relationship between infant mortality and daily calories?

c. Average daily calorie supply in Chile is 2,810. What life expectancy and infant mortality would you predict from the calorie data?

d. Brazil has a life expectancy of 68 years.

i. For what daily calorie supply would you predict this life expectancy?

ii. The actual daily calorie supply for Brazil is 2,938 calories. What does the difference between the value suggested by the model and the actual value tell about the usefulness of the model you have found?

e. What life expectancy does your model predict for a daily calorie supply of 5,000? How close to that prediction would you expect the actual life expectancy to be in a country with a daily calorie supply of 5,000?

3 **Popcorn Sales** Many people who go to movies like to have popcorn to munch on during the show. But movie theater popcorn is often expensive. The manager of a local theater wondered how much more she might sell if the price were lower. She also wondered whether such a reduced price would actually bring in more popcorn income.

One week she set the price for popcorn at $1.00 per cup and sold an average of 120 cups per night. The next week she set the price at $1.50 per cup and sold an average of 90 cups per night. She used that information to graph a linear model to predict number of cups sold at other possible prices.

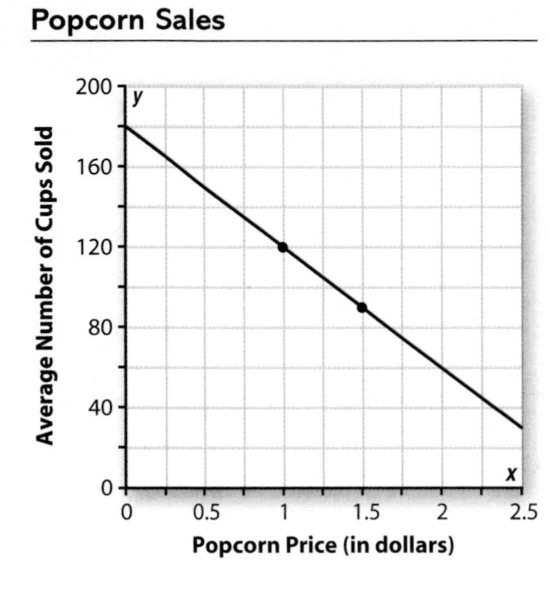

a. Write a rule for the linear model. Explain what the slope and y-intercept of the model tell about the prospective number of popcorn cups sold at various prices.

b. Write and solve equations or inequalities to answer the following questions.

i. At what price does your model predict average daily sales of about 150 cups of popcorn?

ii. At what price does your model predict average daily sales of fewer than 60 cups of popcorn?

iii. How many cups of popcorn does your model predict will be sold per day at a price of $1.80 per cup?

c. Use the rule relating average daily number of cups sold to price to make a table relating price to income from popcorn sales. Explain what the pattern in the table tells about the relation between price and income.

4 Solve the following equations and inequalities. Use each of the following methods of solving—table, graph, and algebraic reasoning—at least twice. Check your answers.

a. $9 + 6x = 24$

b. $286 = 7p + 69$

c. $6 - 4x \leq 34$

d. $8 + 1.1x = -25$

e. $20 = 3 + 5(x - 1)$

f. $17y - 34 = 8y - 16$

g. $1.5x + 8 \leq 3 + 2x$

h. $14 + 3k > 27 - 10k$

5 **Party Planning** The ninth grade class at Freedom High School traditionally has an end-of-year dance party. The class officers researched costs for the dance and came up with these items to consider.

Party Planning

Item	Cost
DJ for the dance	\$350
Food	\$3.75 per student
Drinks	\$1.50 per student
Custodians, Security	\$225

The question is whether the class treasury has enough money to pay for the dance or whether they will have to sell tickets.

a. Which of the following function rules correctly express dance cost C as a function of the number of students N who plan to come to the dance? Explain your reasoning.

$C = 350 + 3.75N + 1.50N + 225$
$C = 5.25N + 575$
$C = 575 + 5.25N$
$C = 580.25N$

b. Write and solve an equation or inequality to determine how many students could come to the dance without a ticket charge if the class treasury has \$950.

c. Write and solve an equation or inequality to determine how many students could come to the dance with a ticket charge of only \$2 if the class treasury has \$950.

6 Using algebraic expressions to help make sense out of problem situations is an important part of mathematics. Writing expressions and function rules is often a first step. Being able to recognize and generate equivalent algebraic expressions is another important skill.

a. Write rules for the linear functions with graphs passing through the indicated points.

i. $(0, -3)$ and $(4, 1)$

ii. $(0, 3)$ and $(6, 0)$

iii. $(2, -6)$ and $(8, 12)$

b. Write rules for the linear functions with graphs having the given slopes and passing through the given points.

i. slope $= 3$; passes through $(4, 12)$

ii. slope $= \frac{2}{3}$; passes through $(-6, -1)$

iii. slope $= -4$; passes through $(17, 82)$

c. Compare the following pairs of linear expressions to see if they are equivalent. Explain your reasoning in each case.

i. $4.2x + 6$ and $(1 - 0.7x)6$

ii. $4C - 3(C + 2)$ and $-6 + C$

iii. $0.3S - 0.4S + 2$ and $\frac{20 - S}{10}$

7 Solve the following system of equations using calculator- or computer-based methods and by algebraic reasoning.

$$\begin{cases} y = 35 + 0.2x \\ y = 85 + 0.7x \end{cases}$$

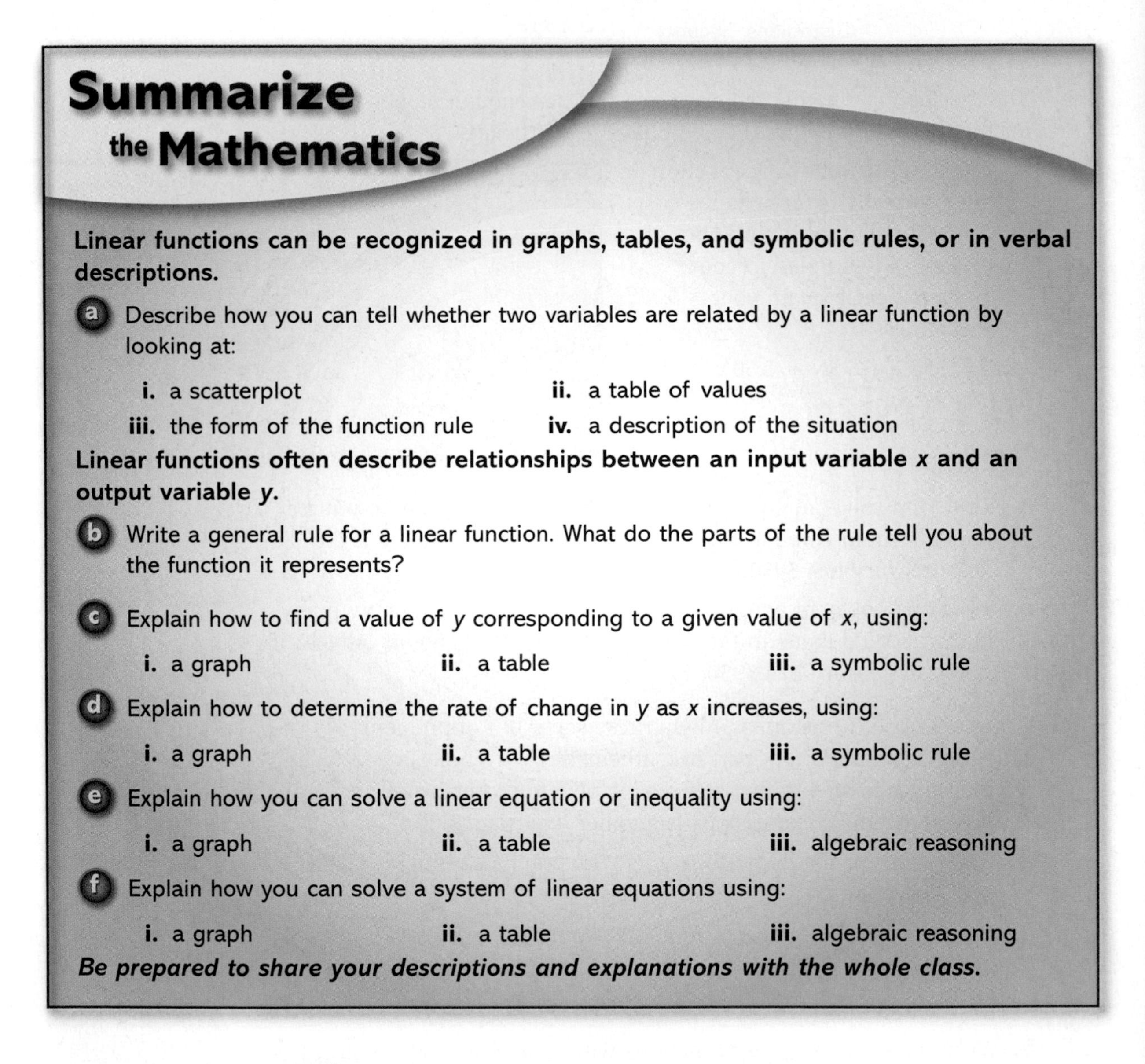

Summarize the Mathematics

Linear functions can be recognized in graphs, tables, and symbolic rules, or in verbal descriptions.

a Describe how you can tell whether two variables are related by a linear function by looking at:

i. a scatterplot **ii.** a table of values

iii. the form of the function rule **iv.** a description of the situation

Linear functions often describe relationships between an input variable *x* and an output variable *y*.

b Write a general rule for a linear function. What do the parts of the rule tell you about the function it represents?

c Explain how to find a value of y corresponding to a given value of x, using:

i. a graph **ii.** a table **iii.** a symbolic rule

d Explain how to determine the rate of change in y as x increases, using:

i. a graph **ii.** a table **iii.** a symbolic rule

e Explain how you can solve a linear equation or inequality using:

i. a graph **ii.** a table **iii.** algebraic reasoning

f Explain how you can solve a system of linear equations using:

i. a graph **ii.** a table **iii.** algebraic reasoning

Be prepared to share your descriptions and explanations with the whole class.

Check Your Understanding

Write, in outline form, a summary of the important mathematical concepts and methods developed in this unit. Organize your summary so that it can be used as a quick reference in future units and courses.

Vertex-Edge Graphs

Many situations involve paths and networks, like bus routes and computer networks. *Vertex-edge graphs* can be used as mathematical models to help analyze such situations. A vertex-edge graph is a diagram consisting of points (vertices) and arcs or line segments (edges) connecting some of the points. Such graphs are part of geometry, as well as part of an important contemporary field called *discrete mathematics*.

In this unit, you will use vertex-edge graphs and Euler circuits to help find optimum paths, such as the best route to collect money from parking meters, deliver newspapers, or plow snow from city streets. You will also use vertex coloring of graphs to avoid conflict among objects, such as scheduling conflicts among meetings or broadcast interference among nearby radio stations.

You will develop the understanding and skill needed to solve problems about optimum paths and conflict through your work in two lessons.

Lessons

1 *Euler Circuits: Finding the Best Path*

Use Euler circuits and their properties to solve problems about optimum circuits.

2 *Vertex Coloring: Avoiding Conflict*

Use vertex coloring to solve problems related to avoiding conflict in a variety of settings.

LESSON 1

Euler Circuits: Finding the Best Path

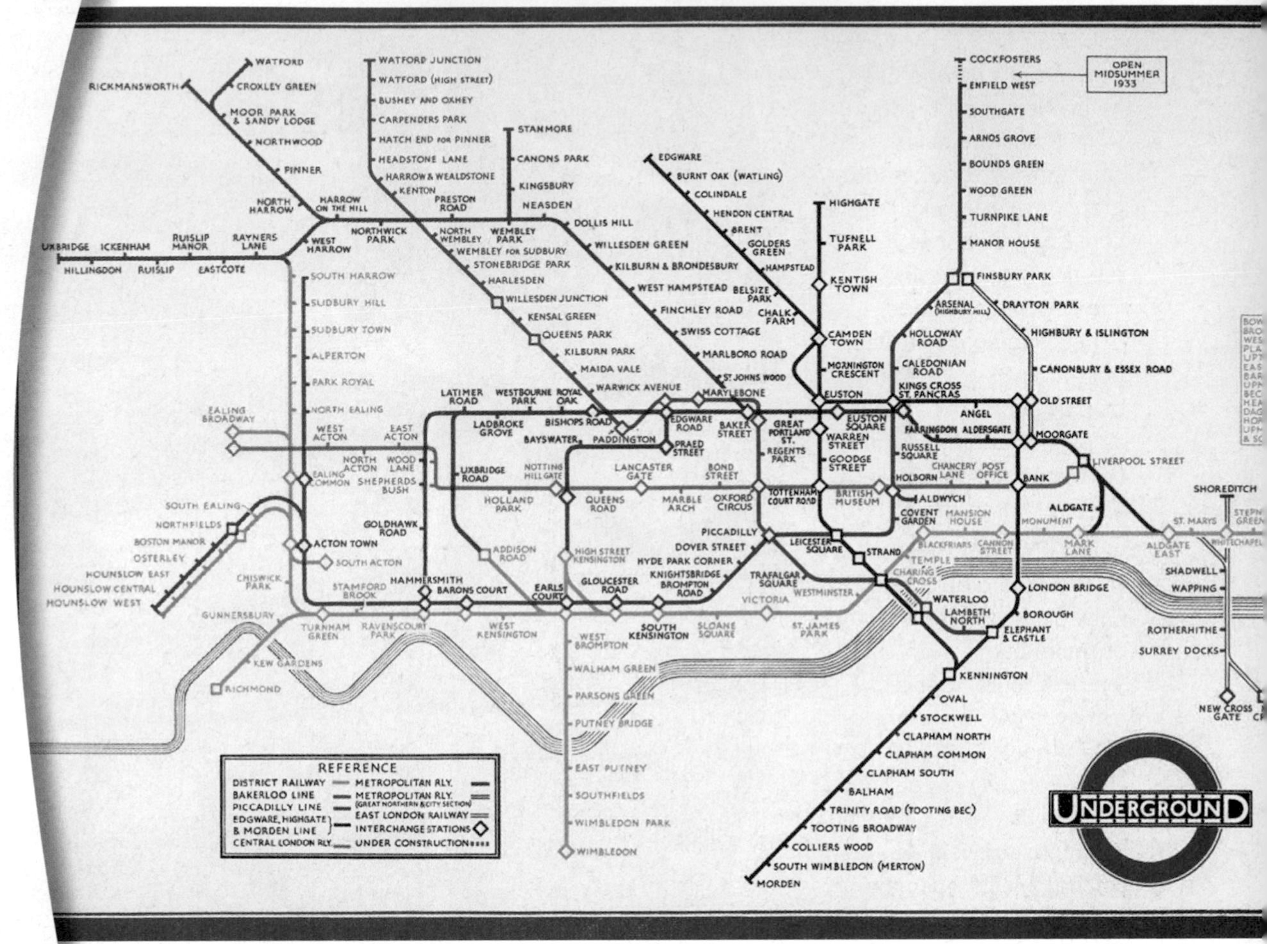

Source: London Transport Museum Guide

Often when solving problems using geometric figures or diagrams, you are concerned with their size, shape, or position. However, sometimes a geometric diagram is used to represent a situation in which size, shape, and position are not important. Instead, connections are what really matter. In this unit, you will study geometric diagrams made up of vertices and edges in which size and shape are not essential characteristics, but how the vertices are connected is very important.

A subway map is one common example of a geometric diagram with vertices and edges for which precise size and shape are not crucial. Perhaps the first such map was the 1933 London Underground map shown above.

Think About This Situation

Examine the Underground map and think about how a visitor to London might have used the map.

a What information is conveyed by the map? What information about the city is not conveyed by the map?

b Why are the size and shape of the map layout not essential?

c What are important features of subway maps like the one above?

d Describe other geometric diagrams with vertices and edges for which *connections* are important, but exact size and shape are not essential.

Geometric diagrams made up of vertices and edges, in which connections are important but exact size, shape, and position are not essential, are sometimes called *vertex-edge graphs*, or simply *graphs*. In this lesson, you will learn how to use vertex-edge graphs to find optimum routes.

Investigation 1 Planning Efficient Routes

You can save time, energy, and expense by studying a complex project before you begin your work. There may be many ways to carry out the project. However, one way may be judged to be the "best" or *optimum*, in some sense. As you work on this investigation, think about this question:

How can you create and use a mathematical model to find an optimum solution to problems such as the following locker-painting problem?

Locker Painting Suppose you are hired to paint all the lockers around eight classrooms on the first floor of a high school. The lockers are located along the walls of the halls as shown in the diagram to the right. Letters are placed at points where you would stop painting one row of lockers and start painting another. Five-gallon buckets of paint, a spray paint compressor, and other equipment are located in the first-floor equipment room *E*. You must move this bulky equipment with you as you paint the lockers. You also must return it to the equipment room when you are finished painting. (The lockers in the center hall must be painted one side at a time.)

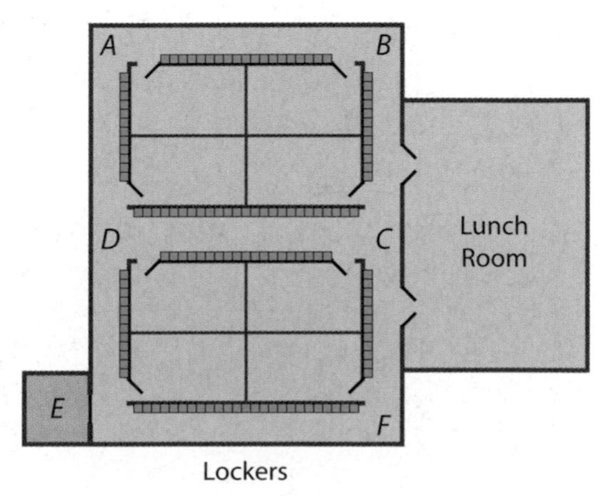

Lockers

1 Since you are being paid for the job, not by the hour, you would like to paint the lockers as quickly and efficiently as possible.

a. Which row would you paint first? Is there more than one choice for the first row to paint?

b. Which row would you paint last? Why?

2 Here are three plans that have been suggested for painting the lockers.

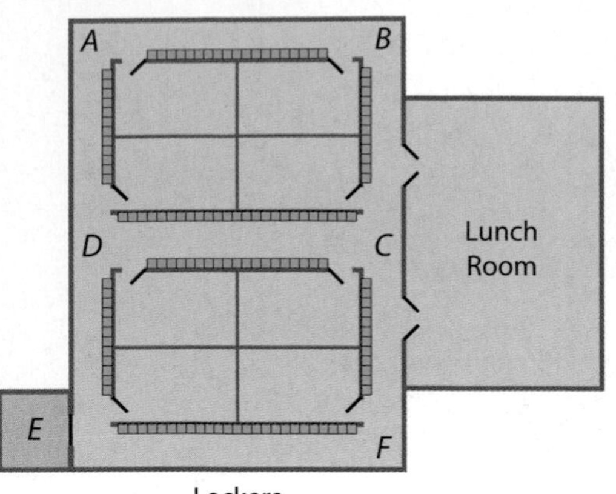

Lockers

Plan I: Paint from *E* to *F*, *F* to *C*, *C* to *D* (one side), *D* to *E*, *D* to *A*, *A* to *B*, *B* to *C*, *C* to *D* (the other side).

Plan II: Paint from *A* to *B*, *B* to *C*, *C* to *D* (one side), *D* to *A*, *D* to *C* (other side), *C* to *F*, *F* to *E*, *E* to *D*.

Plan III: Paint from *E* to *D*, *D* to *A*, *A* to *B*, *B* to *C*, *C* to *F*, *F* to *E*, *D* to *C* (one side), *C* to *D* (other side).

a. Which, if any, of these plans do you think is optimum; that is, the "best" way to do the painting? If a plan is not optimum, explain why not.

b. Without help from your classmates, prepare a plan you think is optimum for painting the lockers.

c. Compare your plan with those of others.

i. How are they alike? How are they different?

ii. Agree on a list of criteria that can be used to decide whether a plan is optimum.

3 A **mathematical model** is a symbolic or pictorial representation including only the essential features of a problem situation. The floor-plan map of the first floor of the school shows the rows of lockers, classrooms, lunch room, equipment room, hallways, and outer walls. There are some features of this map that you do not need in order to solve the locker-painting problem.

a. Which of the features of the map did you use as you tried to solve the locker-painting problem? Which features were not needed?

b. Refer to the first-floor map of the school above. Think about a simplified diagram (a mathematical model) that includes only the essential features of the locker-painting problem. For example, the lettered points on the map are important because *E* is the beginning and ending point, and the other letters mark where one row of lockers ends and another begins. Complete a copy of the diagram below so that it is a mathematical model for the locker-painting problem.

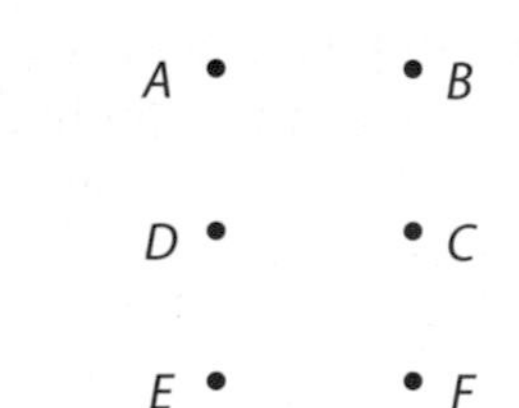

A diagram consisting of a set of points along with segments or arcs joining some of the points is called a **vertex-edge graph**, or simply a **graph**. The points are called **vertices**, and each point is called a **vertex**. The segments or arcs joining the vertices are called **edges**. The word "graph" is used to mean different things at different times in mathematics. In this unit, the word "graph" typically refers to a diagram consisting of vertices and edges.

4 Now examine the vertex-edge graph models drawn by some other students.

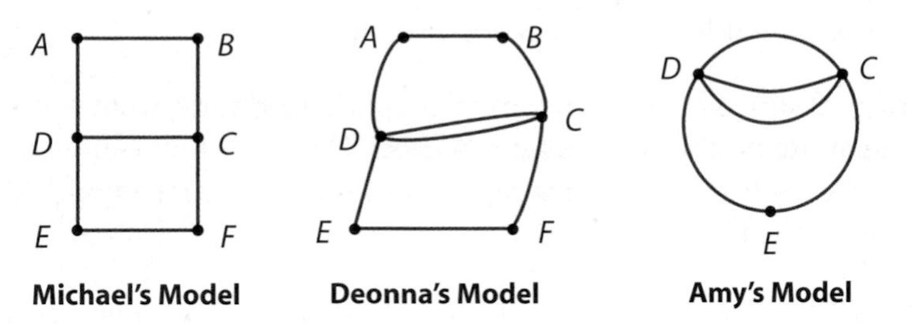

a. Does Michael's vertex-edge graph show all the essential features of the locker-painting problem? If so, explain. If not, describe what is needed.

b. Is Deonna's vertex-edge graph an appropriate model for the locker-painting problem? Why do you think Deonna joined vertices C and D with 2 edges?

c. Is Amy's graph an appropriate model for the locker-painting problem? Explain.

5 In Problem 2, you were asked to find an optimum plan for painting the first-floor lockers.

a. Use that plan to trace an optimum painting route on the vertex-edge graph you drew in Problem 3. If you cannot trace your optimum route on your graph, carefully check both your optimum plan and your graph.

b. Trace the same painting route on Deonna's graph. Does it matter if the vertices are connected by straight line segments or curved arcs? Does it matter how long the edges are?

6 Below is a vertex-edge graph that models a different arrangement of lockers.

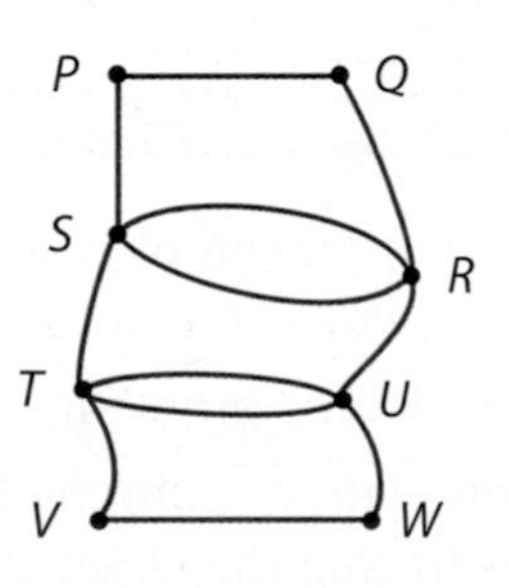

a. Draw a school floor-plan map that corresponds to this graph. Assume that the equipment room is at V.

b. Find, if possible, an optimum route for painting these lockers.

Summarize the Mathematics

In this investigation, you explored how vertex-edge graphs can be used to model situations in which an efficient route is to be found.

a. What is the difference between a floor-plan map of a school showing the lockers to be painted and a mathematical model of the locker-painting problem?

b. A key step in modeling a problem situation with a graph is to decide what the vertices and edges will represent. Refer back to Problem 4. Both Deonna's and Amy's models are appropriate. What do the vertices and edges represent in each of these graphs in terms of the locker-painting problem?

c. Can two vertex-edge graphs that have different shapes represent the same problem situation? Can two graphs that have different numbers of vertices and edges represent the same problem situation? Explain.

d. In Problem 2, you wrote a list of criteria for an optimum locker-painting plan. Restate those criteria in terms of tracing around a vertex-edge graph that models the situation.

Be prepared to share your ideas with the entire class.

Check Your Understanding

Suppose the lockers and an equipment room on the west wing of a high school are located as shown below.

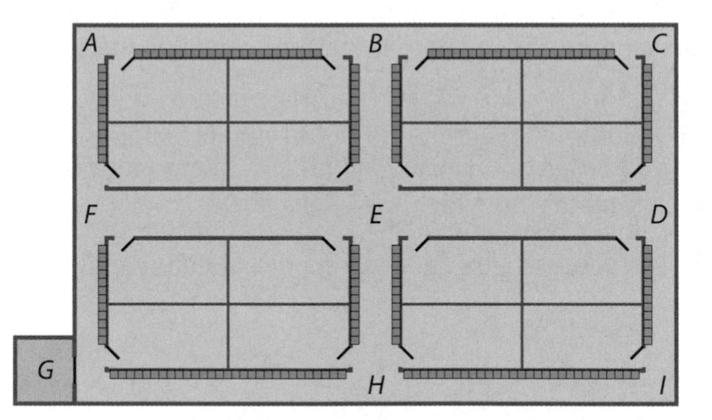

a. If you were to model the problem of painting these lockers with a vertex-edge graph, what would the vertices represent? The edges?

b. Draw a graph that models this problem.

c. Determine an optimum plan for painting the lockers. Check your plan against the criteria for tracing the edges and vertices of a graph that you prepared in Part d of the Summarize the Mathematics.

Investigation 2 Making the Circuit

Your criteria for the optimum sequence for painting the lockers in the previous investigation are the defining characteristics of an important property of a graph. An **Euler** (pronounced *oy' lur*) **circuit** is a route through a connected graph such that (1) each edge of the graph is traced exactly once, and (2) the route starts and ends at the same vertex.

You only consider Euler circuits in connected graphs. A **connected graph** is a graph that is all in one piece. That is, from each vertex there is at least one path to every other vertex. Given a connected graph, it often is helpful to know if it has an Euler circuit. (The name "Euler" is in recognition of the eighteenth-century Swiss mathematician Leonhard Euler. He was the first to study and write about these circuits.) As you work through this investigation, look for clues that help you answer these questions:

How can you tell if a graph has an Euler circuit?
If a graph has an Euler circuit, how can you systematically find it?

1 Graph models of the sidewalks in two sections of a town are shown below. Parking meters are placed along these sidewalks.

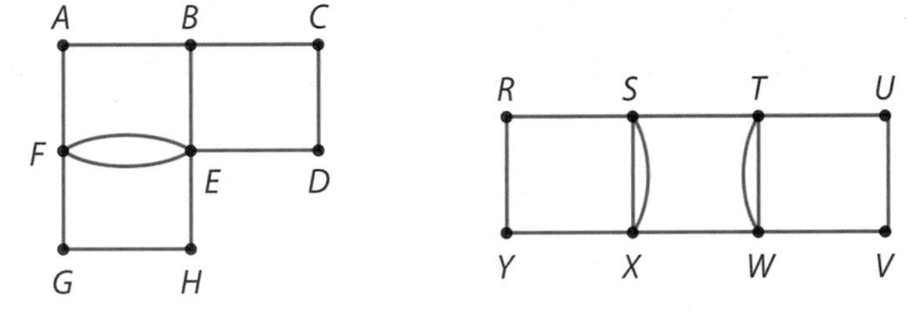

a. Why would it be helpful for a parking-control officer to know if these graphs have Euler circuits?

b. Does the graph that models the east section of town have an Euler circuit? Explain your reasoning.

c. Does the graph that models the west section of town have an Euler circuit? Does it have more than one Euler circuit? Explain.

2 The three graphs at the top of the next page are similar to puzzles enjoyed by people all over the world. In each case, the challenge is to trace the figure. You must trace every edge exactly once without lifting your pencil and return to where you started. That is, the challenge is to trace an Euler circuit through the figure or graph. Place a sheet of paper over each graph and try to trace an Euler circuit. If the graph has an Euler circuit, write down the vertices in order as you trace the circuit. (Note that in the graph in Part c, only the points with letter labels are vertices. The other edge crossings are not vertices; you can think of them like overpasses in a road system.)

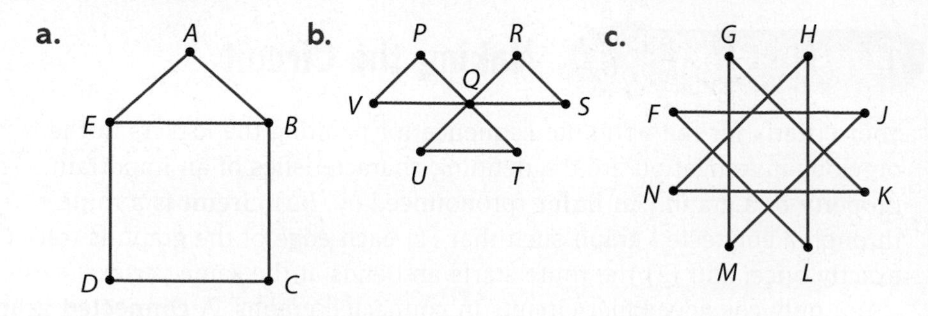

3 By looking at the form of a function rule, you often can predict the shape of the graph of the function without plotting any points. Similarly, it would be helpful to be able to examine a vertex-edge graph and predict if it has an Euler circuit without trying to trace it.

a. Have each member of your group draw a graph with five or more edges that has an Euler circuit. On a separate sheet of paper, have each group member draw a connected graph with five or more edges that does *not* have an Euler circuit. Alternatively, use vertex-edge graph software to generate several graphs that have an Euler circuit and several graphs that do not.

b. Sort your group's graphs into two collections, those that have an Euler circuit and those that do not.

c. Examine the graphs in the two collections. Describe key ways in which graphs that have Euler circuits differ from those that do not.

d. Try to figure out a way to predict if a graph has an Euler circuit simply by examining its vertices.

i. Test your method of prediction using the graphs in Problem 2.

ii. If you have access to vertex-edge graph software, generate several additional general graphs to test your method.

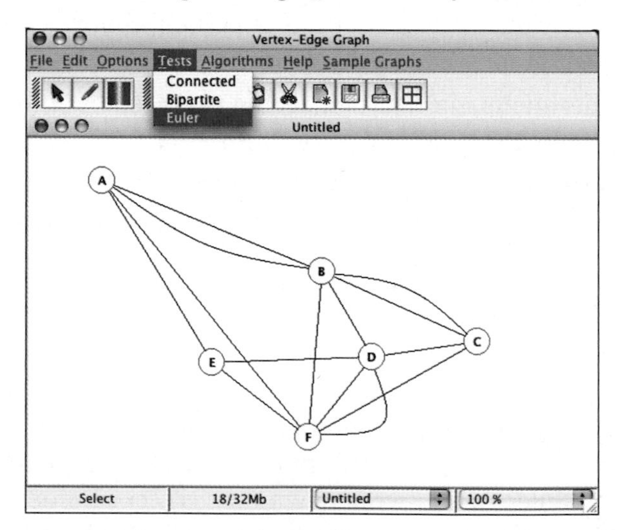

e. Make a conjecture about the properties of a graph that has an Euler circuit. Explain why you think your conjecture is true for *any* graph that has an Euler circuit. Compare and test your conjecture with other students' conjectures and graphs. Modify your conjecture and explanation as needed.

4 In your conjecture from Problem 3 Part e about which graphs have an Euler circuit, you probably counted the number of edges at each vertex of the graph. The number of edges touching a vertex is called the **degree of the vertex**. Restate your conjecture in terms of the degrees of the vertices. (If an edge loops back to the same vertex, that counts as two edge touchings. For an example see Extensions Task 29 on page 262.)

5 Once you can predict whether a graph has an Euler circuit, it is often still necessary to find the circuit. Consider the graphs below.

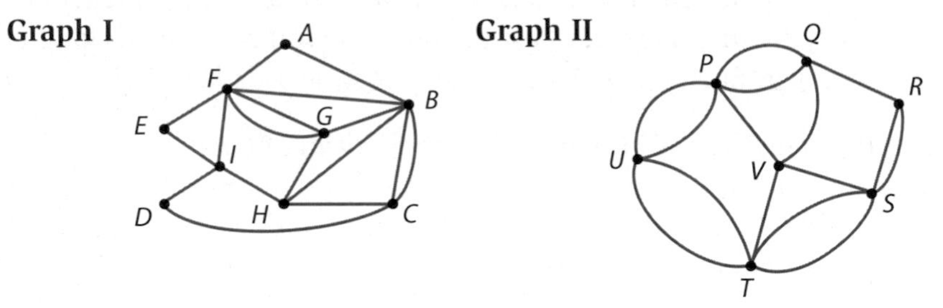

a. For each graph, predict whether it has an Euler circuit.

b. If the graph has an Euler circuit, find it.

c. Describe the method you used to find your Euler circuit. Describe other possible methods for finding Euler circuits.

6 One systematic method for finding an Euler circuit is to trace the circuit in stages. For example, suppose you and your classmates want to find an Euler circuit that begins and ends at *A* in the graph below. You can trace the circuit in several stages.

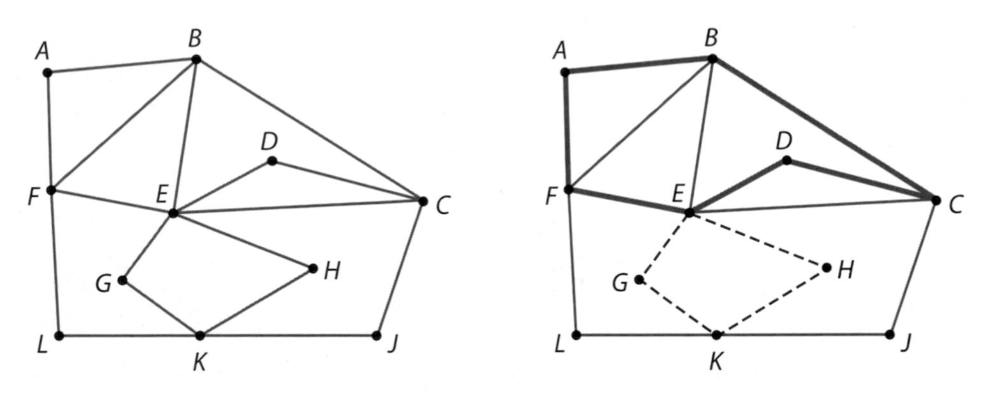

Stage I: Alicia began by drawing a circuit that begins and ends at *A*. The circuit she drew, shown in the diagram by the heavy edges, was *A-B-C-D-E-F-A*. But this does not trace all edges.

Stage II: George added another circuit shown by the dashed edges starting at *E*: *E-G-K-H-E*.

a. Alicia's and George's circuits can be combined to form a single circuit beginning and ending at *A*. List the order of vertices for that combined circuit.

Stage III: Since this circuit still does not trace each edge, a third stage is required.

b. Trace a third circuit which covers the rest of the edges.

c. Combine all the circuits to form an Euler circuit that begins and ends at A. List the vertices of your Euler circuit in order.

d. Use this method to find an Euler circuit in Graph I of Problem 5.

7 Choose your preferred method for finding Euler circuits from Problems 5 and 6. Write specific step-by-step instructions that describe the method you chose. Your instructions should be written so that they apply to *any* graph, not just the one that you may be working on at the moment. Such a list of step-by-step instructions is called an **algorithm**.

Creating algorithms is an important aspect of mathematics. Algorithms are especially important when programming computers to solve problems. Two questions you should ask about any algorithm are *Does it always work?* and *Is it efficient?* You will consider these questions in more detail in the next lesson.

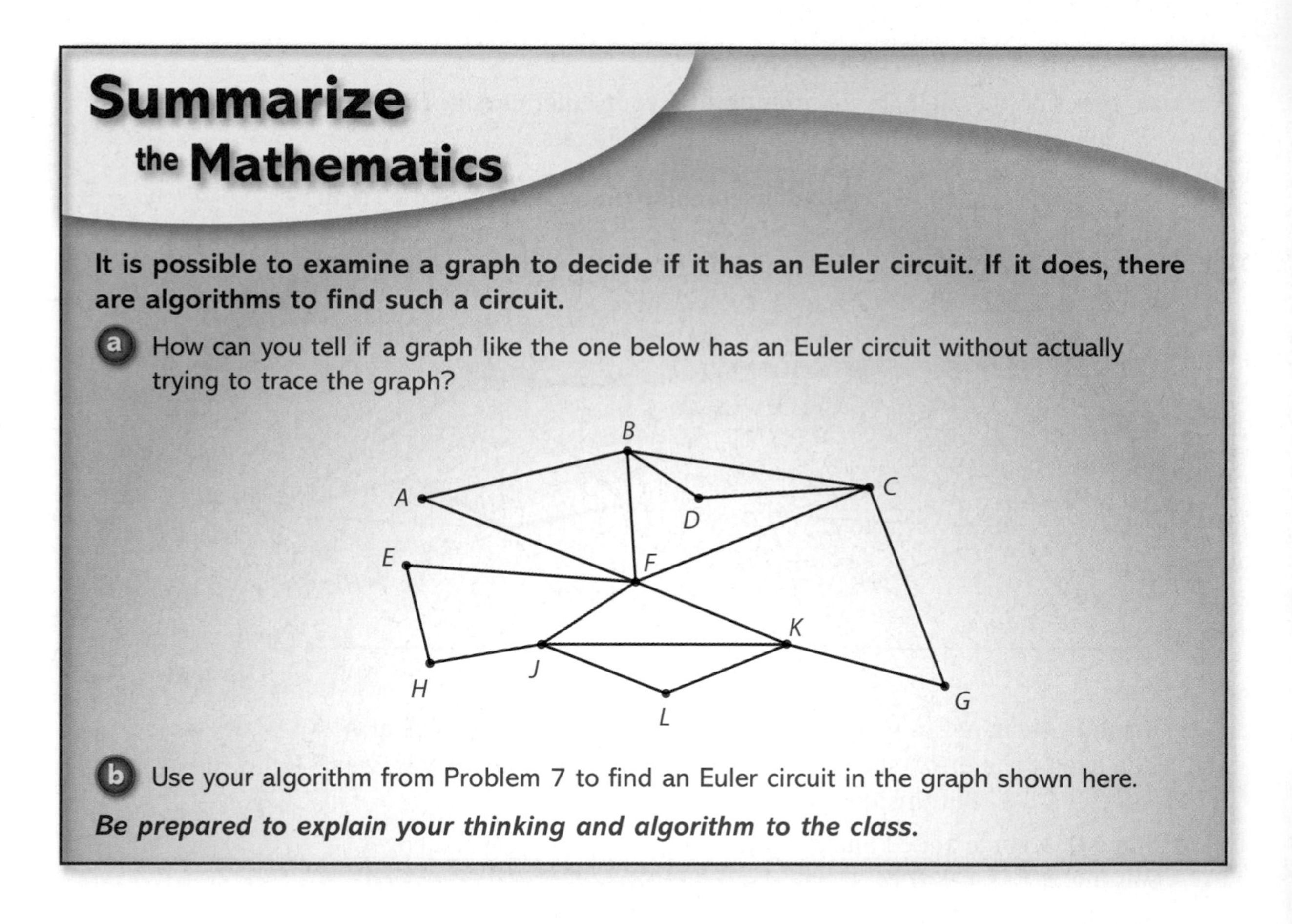

Summarize the Mathematics

It is possible to examine a graph to decide if it has an Euler circuit. If it does, there are algorithms to find such a circuit.

a How can you tell if a graph like the one below has an Euler circuit without actually trying to trace the graph?

b Use your algorithm from Problem 7 to find an Euler circuit in the graph shown here.

Be prepared to explain your thinking and algorithm to the class.

Check Your Understanding

For each of the graphs below, decide if the graph has an Euler circuit. If there is an Euler circuit, use your algorithm to find it. If not, explain how you know that no Euler circuit exists.

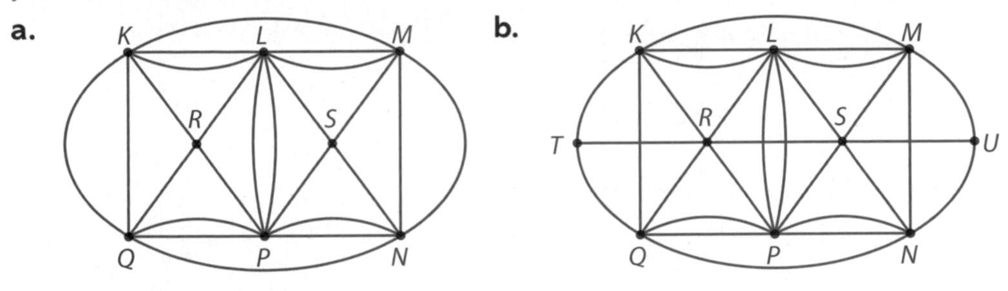

Investigation 3 Graphs and Matrices

Information is often organized and displayed in tables. The use of tables to summarize information can be seen in almost every section of most newspapers. As you work on the problems in this investigation, look for answers to this question:

How can table-like arrays be used to represent vertex-edge graphs and help reason about information contained in the graphs?

1 Examine this information on gold (1st place), silver (2nd place), and bronze (3rd place) medals awarded in the 2006 Winter Olympics.

Medal Count

Country	G	S	B	Total
Germany	11	12	6	29
United States	9	9	7	25
Austria	9	7	7	23
Russian Fed.	8	6	8	22
Canada	7	10	7	24
Sweden	7	2	5	14

Source: www.torino2006.org

a. What do each of the numbers in the row labeled Germany represent?

b. What is the meaning of the number in the fifth row and second column? (Don't count the row and column headings.) In the third row and third column?

c. Finland did not win any gold medals. However, the Finnish team did take home 6 silver medals and 3 bronze medals. How could you modify this chart to include this additional information?

2 Rectangular arrays of numbers, like the one below, are sometimes called **matrices**. Matrices can be used to represent graphs. One way in which a graph can be represented by a **matrix** is shown by the partially completed matrix below.

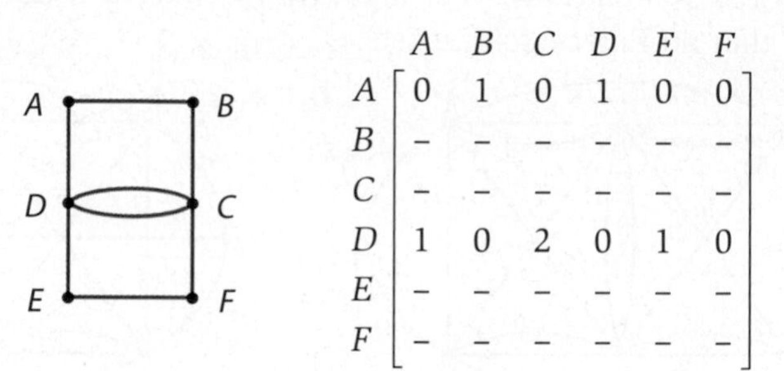

a. Study the first and fourth rows of the matrix. Explain what each entry means in terms of the graph.

b. Copy the matrix and then fill in the missing entries.

c. Vertices in a graph that are connected by an edge are said to be **adjacent vertices**. The matrix you constructed in Part b is called an **adjacency matrix** for the graph, since it contains information about vertices that are adjacent. Each entry in an adjacency matrix is the number of direct connections (edges) between the corresponding pair of vertices. Construct an adjacency matrix for each of the three graphs below.

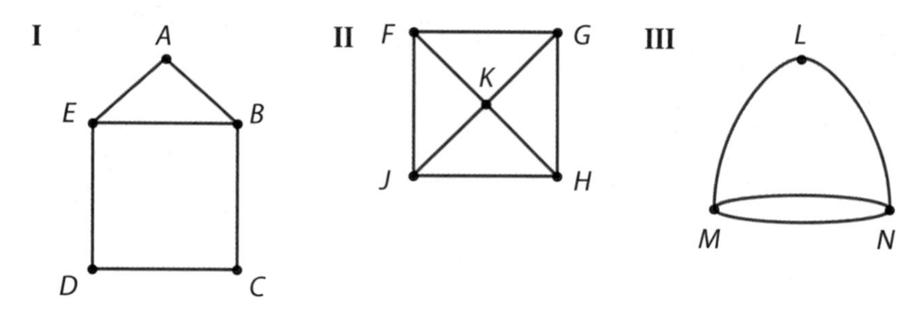

d. It is time-consuming to construct an adjacency matrix for a large graph. If you have access to vertex-edge graph software, construct an adjacency matrix for several graphs (without loops) that contain at least 6 vertices. Click on the entries of the matrix to see the corresponding edge(s) in the graph. Construct a circuit in the graph (not necessarily an Euler circuit) by clicking on entries in the adjacency matrix.

Now examine some common properties of the adjacency matrices you have constructed.

a. The **main diagonal** of a matrix like these consists of the entries in the diagonal running from the top-left corner of the matrix to the bottom-right corner. What do you notice about the main diagonal in these adjacency matrices? Explain this pattern. (The graphs you have worked with so far do not contain loops. To see how loops affect an adjacency matrix, see Extensions Task 28 on page 262.)

b. Describe and explain any symmetry you see in these adjacency matrices.

4 The sums of the numbers in each row of a matrix are called the **row sums** of the matrix.

a. Find the row sums of each of the adjacency matrices in Problem 2 Part c. What do these row sums represent in the graphs?

b. Is it possible to tell by looking at the adjacency matrix for a graph whether the graph has an Euler circuit? Justify your response.

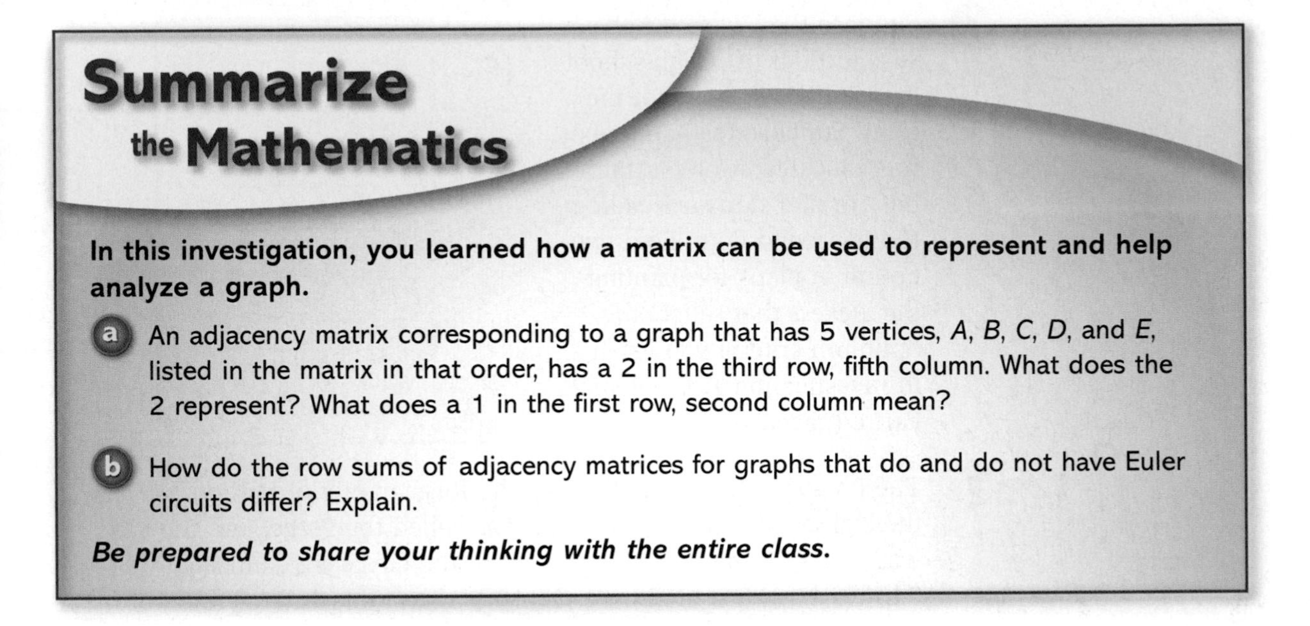

Summarize the Mathematics

In this investigation, you learned how a matrix can be used to represent and help analyze a graph.

a An adjacency matrix corresponding to a graph that has 5 vertices, *A*, *B*, *C*, *D*, and *E*, listed in the matrix in that order, has a 2 in the third row, fifth column. What does the 2 represent? What does a 1 in the first row, second column mean?

b How do the row sums of adjacency matrices for graphs that do and do not have Euler circuits differ? Explain.

Be prepared to share your thinking with the entire class.

Check Your Understanding

Examine the adjacency matrices below, and answer the following questions.

I

$$\begin{array}{c} \\ A \\ B \\ C \end{array}\begin{array}{c} \begin{array}{ccc} A & B & C \end{array} \\ \begin{bmatrix} 0 & 2 & 0 \\ 2 & 0 & 1 \\ 0 & 1 & 0 \end{bmatrix} \end{array}$$

II

$$\begin{array}{c} \\ P \\ Q \\ R \\ S \end{array}\begin{array}{c} \begin{array}{cccc} P & Q & R & S \end{array} \\ \begin{bmatrix} 0 & 1 & 2 & 1 \\ 1 & 0 & 1 & 2 \\ 2 & 1 & 0 & 2 \\ 1 & 2 & 2 & 0 \end{bmatrix} \end{array}$$

a. Does each of the graphs with an adjacency matrix given above have an Euler circuit? How can you tell without drawing the graphs?

b. Draw and label a graph corresponding to each adjacency matrix. Find an Euler circuit if there is one.

On Your Own

Applications

1 Suppose the lockers on the second floor of a high school are located as shown at the right. Suppose the equipment room located at G is at the bottom of a stairway leading to the second floor. Find two optimum plans for painting the lockers that satisfy the optimum criteria you listed in Investigation 1, Problem 2 Part c (page 240).

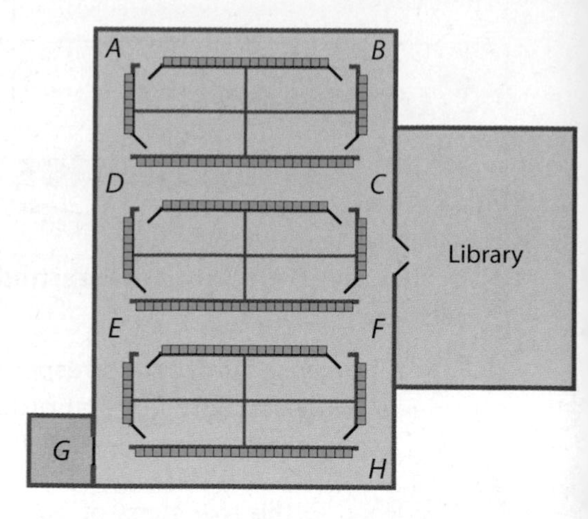

2 The Pregolya River runs through the Russian city of Kaliningrad. In the eighteenth century, the river was called the Pregel and the city was named Königsberg. Four parts of the city were connected by 7 bridges as illustrated here. Citizens often took walking tours of the city by crossing over the bridges. Some people wondered whether it was possible to tour the city by beginning at a point on land, walking across each bridge exactly once, and returning to the same point. This problem, called the *Königsberg bridges problem*, intrigued the mathematician Leonhard Euler, who lived at that time. The paper Euler wrote in 1736 containing the solution to this problem is widely considered to be the first paper on the theory of vertex-edge graphs. Ironically, Euler did not use an actual vertex-edge graph diagram as part of his solution, although his results clearly apply to such graphs. (Euler considered his work on this problem to be part of an area of mathematics informally discussed then as "the geometry of position," since it dealt with relative position, but not distance.)

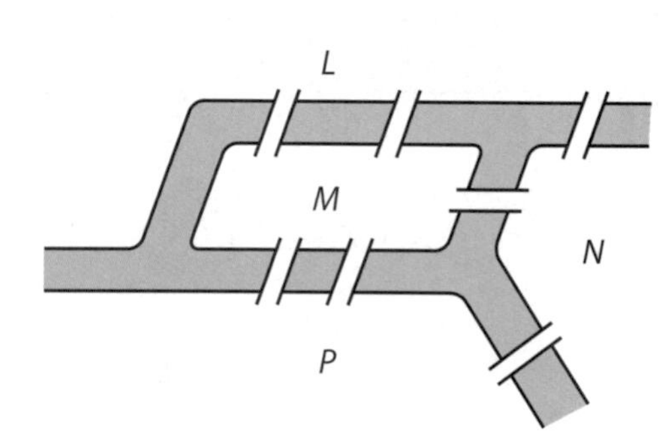

a. Draw a graph in which the vertices represent the 4 land areas (lettered in the figure) and the edges represent bridges.

b. What is the solution to the Königsberg bridges problem? Explain your response.

c. In the time since Euler solved the problem, two more bridges were built. One bridge was added at the left to connect areas labeled L and P. Another bridge was added to connect areas labeled N and P.

i. Draw a graph that models this new situation of land areas and bridges.

ii. Use your graph to determine if it is possible to take a tour of the city that crosses each of the 9 bridges exactly once and allows you to return to the point where you started.

3 The Bushoong are a subgroup of the Kuba chiefdom in the Democratic Republic of Congo (changed from Zaire in 1997). Bushoong children have a long tradition of playing games that involve tracing figures in the sand using a stick. The challenge is to trace each line once and only once without lifting the stick from the sand. Two such figures are given below. (Problem adapted from *Ethnomathematics: A Multicultural View of Mathematical Ideas*, Brooks/Cole Publishing Company, 1991.)

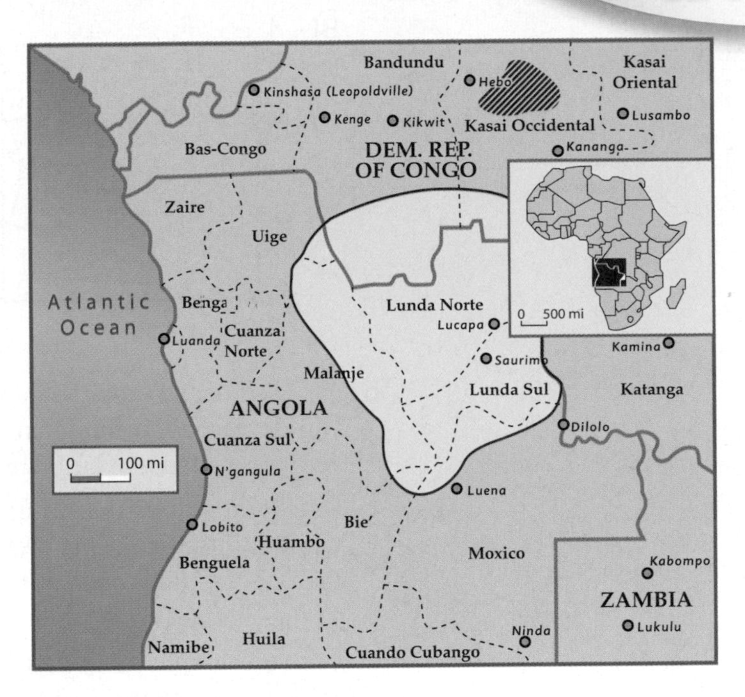

Figure I

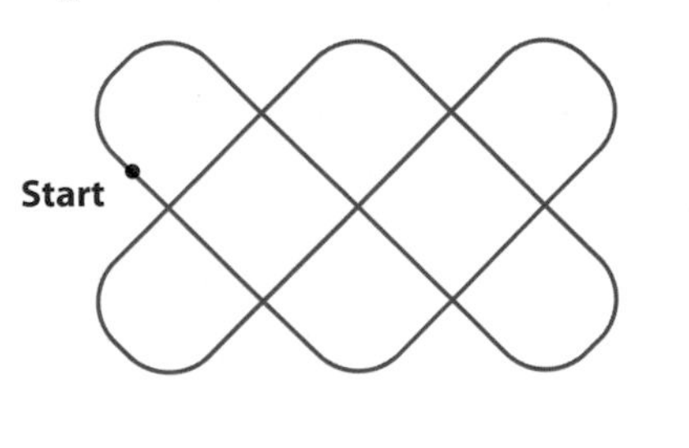

Figure II

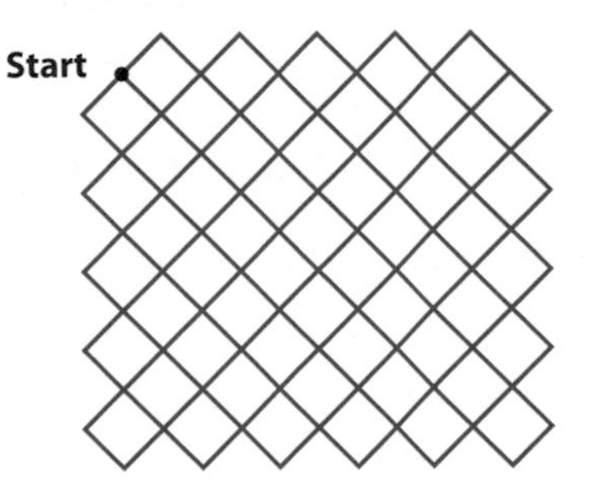

Place a sheet of paper over the figures.

a. Trace each figure without lifting your pencil and without any retracing. Your tracing does not need to end at the same place it started.

b. Try tracing each figure using different "start" points. Summarize your findings.

4 Some popular puzzles involve trying to trace a figure starting and ending at the same vertex without lifting your pencil or tracing an edge more than once. That is, you try to find an Euler circuit.

a. Identify which of the following graphs do not have an Euler circuit. Explain why they do not.

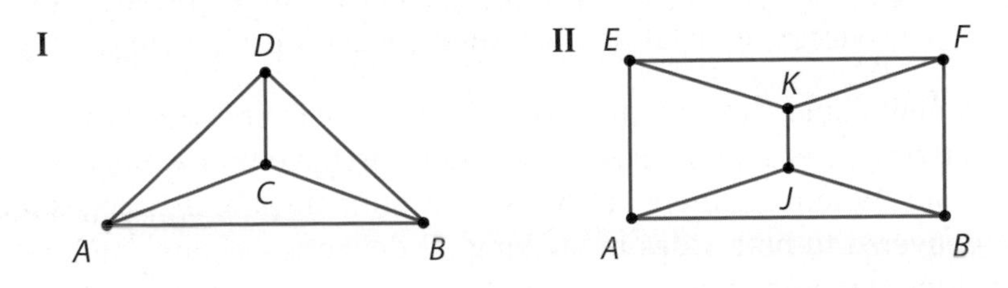

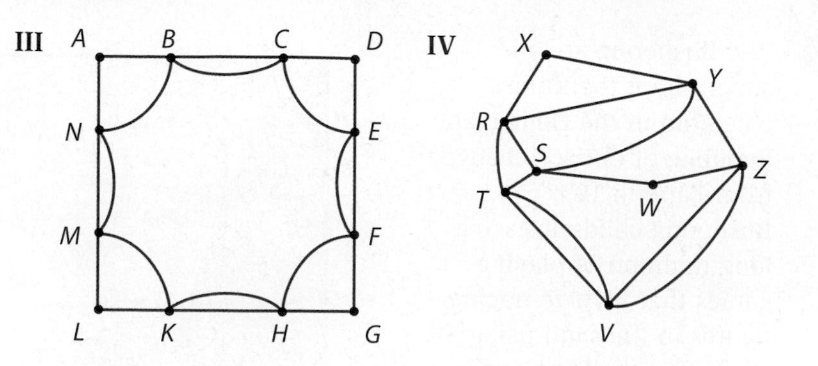

b. For each of the graphs that has an Euler circuit, use the algorithm you developed to find a circuit. Write down the sequence of vertices visited as you trace the circuit.

c. Draw two graphs that would be difficult or impossible to trace without lifting your pencil from the page or tracing an edge more than once—draw one so that it has an Euler circuit and the other so that it does not.

d. Use your graphs from Part c to amaze and teach someone outside of class, as follows. Challenge some people to trace your graphs, without lifting their pencil or tracing any edge more than once and starting and ending at the same point. Then ask them to challenge you in the same way with any graph they draw. See if you can amaze them with how quickly you can tell whether or not it is possible to trace the graph. Then teach them about Euler circuits so they will know the secret too.

5 Suppose the lockers on the second floor of the high school in the locker-painting problem on page 239 are located as shown here.

a. Draw a graph that represents this situation. Be sure to describe what the vertices and edges of your graph represent.

b. Is there a way to paint the lockers by starting and ending at the equipment room and never moving equipment down a hall without painting lockers on one side? Explain.

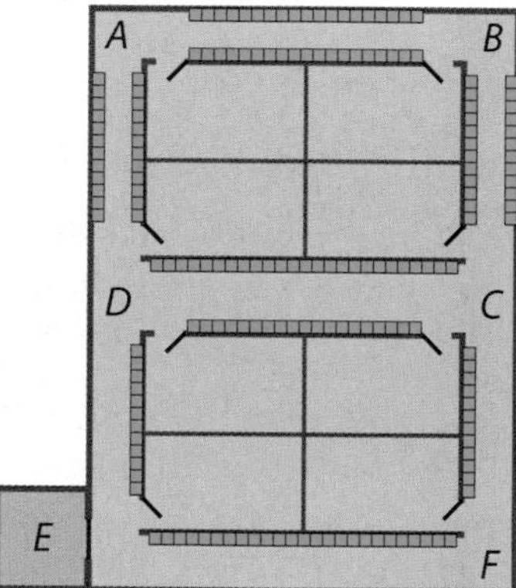

Second-Floor Lockers

c. Is there a way to paint the lockers by starting at D and ending at C and never moving equipment down a hall without painting lockers on one side? Compare the degree of vertices D and C to the degrees of the other vertices. Make a conjecture about graphs in which there is a route through the graph that starts at one vertex, ends at another, and traverses each edge exactly once.

6 A newspaper carrier wants to complete a delivery route without retracing steps. Some streets on the route have houses facing each other. Whenever there are houses on both sides of a street, papers are delivered to both sides by making all deliveries to one side and then along the other side.

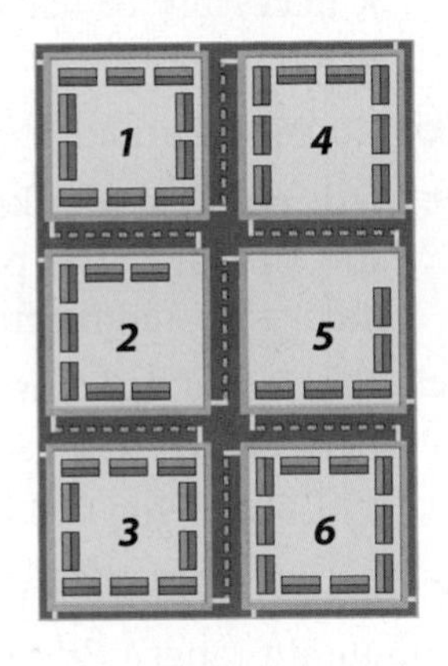

a. Suppose the paper carrier only delivers to the houses on blocks 1, 2, and 3. Construct a vertex-edge graph model for this situation. What do the edges and vertices represent? Find an optimum delivery route.

b. Suppose the paper carrier delivers to the houses on all 6 blocks. Construct a vertex-edge graph model for this situation. Find an optimum delivery route.

c. Now assume that *all* blocks have houses on all 4 sides and all streets continue in both directions.

   i. Add 3 more blocks that are adjacent to the given blocks on the street map. Find an optimum delivery route.

   ii. Can you find an Euler circuit no matter where the 3 new blocks are placed on the route? Explain your response.

   iii. Is it possible to place any number of new blocks on the route and still have an Euler circuit? Explain your reasoning.

7 The map below shows the trails in Tongas State Park. The labeled dots represent rest areas scattered throughout the park.

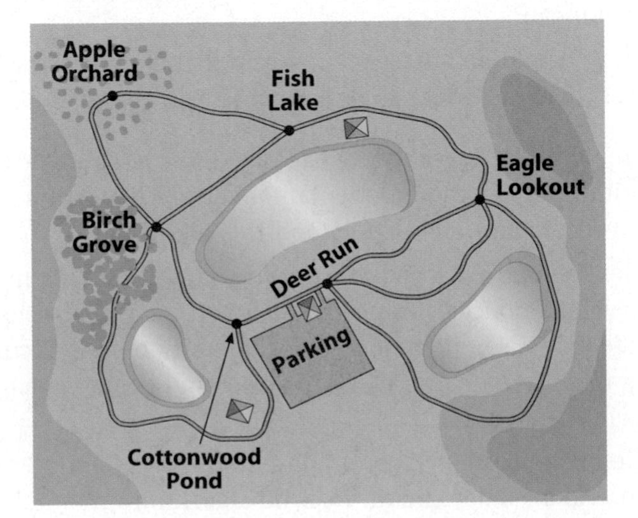

a. How would a graph model of this situation differ from the map? Is it necessary or useful to draw a graph model in this situation? Why or why not?

b. Construct an adjacency matrix related to the park map.

c. Is it possible to hike each of the trails in the park once and return to your car in the parking lot? Explain your answer by using the adjacency matrix from Part b and your knowledge of Euler circuits.

d. The Park Department has received money to build additional trails. Between which rest stops should they build a new trail (or trails) so that people can hike each trail once and return to their cars?

8 Certain towns in southern Alaska are on islands or isolated by mountain ranges. When traveling between these communities, you must take a boat or a plane. Listed below are the routes provided by a local airline.

Routes between:

Anchorage and Cordova
Anchorage and Juneau
Cordova and Yakutat
Juneau and Ketchikan
Juneau and Petersburg
Juneau and Sitka
Petersburg and Wrangell
Sitka and Ketchikan
Wrangell and Ketchikan
Yakutat and Juneau

a. Draw a vertex-edge graph that models this situation.

b. In what ways is your graph model like the map? In what ways is it different?

c. An airline inspector wants to evaluate the airline's operations by flying each route. It is sufficient to fly each route one-way. Can the inspector start in Juneau, fly all the routes exactly once, and end in Juneau?

d. How would an adjacency matrix for the graph show whether or not there is a route as described in Part c?

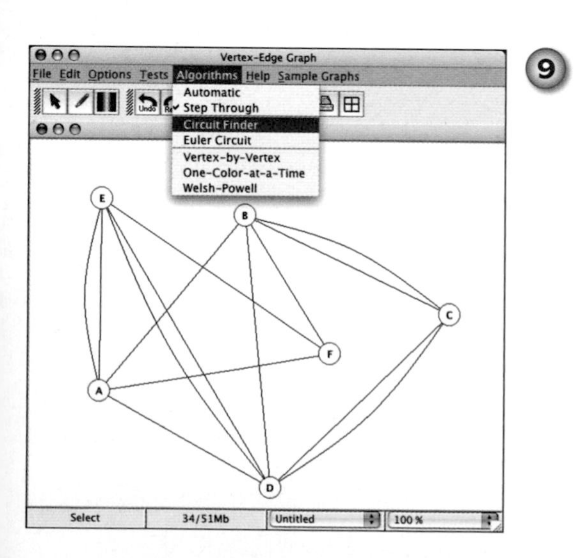

9 In Problem 6 of Investigation 2 (page 245), you learned about an algorithm for finding Euler circuits. Vertex-edge graph software can help you practice and understand this algorithm.

a. Generate an Euler graph from the Sample Graphs menu. (Create several Duplicate Graphs so they will be available to work with as needed.) Use different colors to identify partial circuits in the graph. Then choose black to select and color the edges of the final combined circuit.

b. Generate another Euler graph and use the Circuit Finder algorithm to find an Euler circuit. Use the algorithm in both Automatic and Step Through modes.

Connections

10 The following graphs separate the plane into several *regions*. The exterior of the graph is an infinite region. The interior regions are enclosed by the edges. For example, Graph I separates the plane into four regions (three are enclosed by the graph and the fourth is outside the graph).

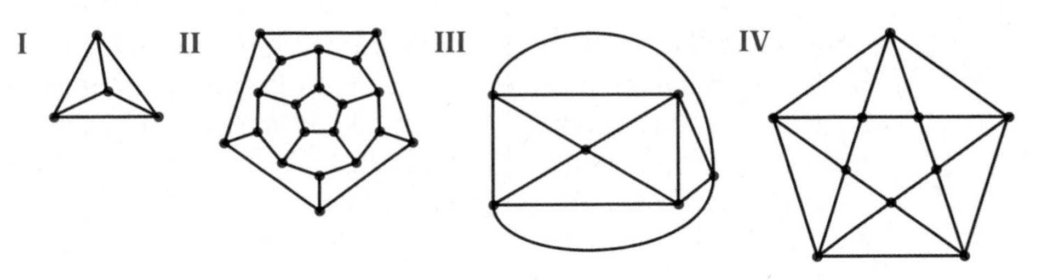

a. Complete a table like the one that follows using the graphs above. Be sure to count the exterior of the graph as one region.

Graph	Number of Vertices (V)	Number of Regions (R)	Number of Edges (E)
I			
II			
III			
IV			

b. Find a formula relating the numbers of vertices V, regions R, and edges E, by using addition or subtraction to combine V, R, and E.

c. Draw several more graphs, and count V, R, and E. Does your formula also work for these graphs?

d. Use your formula to predict how many regions would be formed by an appropriate graph with 5 vertices and 12 edges. Draw such a graph to verify your answer.

11 In this lesson, you discovered that some graphs do not have an Euler circuit.

a. If you are *not* required to start and end at the same vertex, do you think every graph has a path that traces every edge of the graph exactly once? Why or why not?

b. Place a sheet of paper over the graphs below. Try to copy the graphs by tracing each edge exactly once. You don't have to start and end at the same vertex.

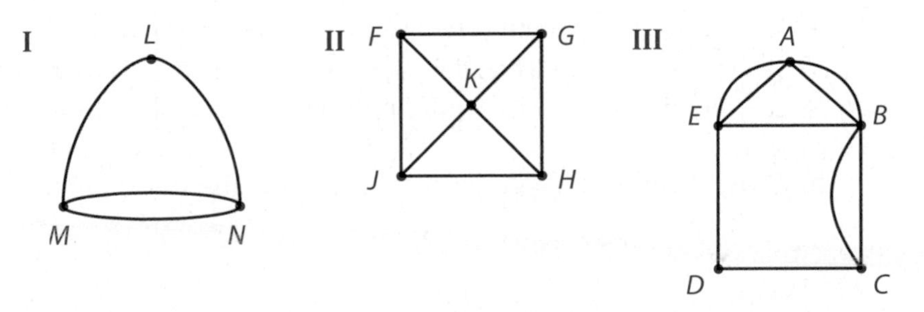

c. For those graphs that can be traced in this manner, how do the starting and ending vertices differ from the other vertices?

d. An **Euler path** is a route through a connected graph that traces each edge of the graph exactly once. Thus an Euler circuit is a special type of Euler path, one in which the starting and ending vertices must be the same. State a rule for determining whether or not a graph has an Euler path in which the starting vertex is different from the ending vertex.

12 Tracing continuous figures is exhibited in cultures around the world. The Malekula live on an island in the South Pacific chain of some eighty islands that comprise the Republic of Vanuatu. As with the Bushoong in Africa (see Applications Task 3), the Malekula also have figures that represent objects or symbols of the culture. For example, Figure I below represents a yam. Figure II is called "the stone of Ambat." (Problem adapted from *Ethnomathematics: A Multicultural View of Mathematical Ideas*, Brooks/Cole Publishing Company, 1991.)

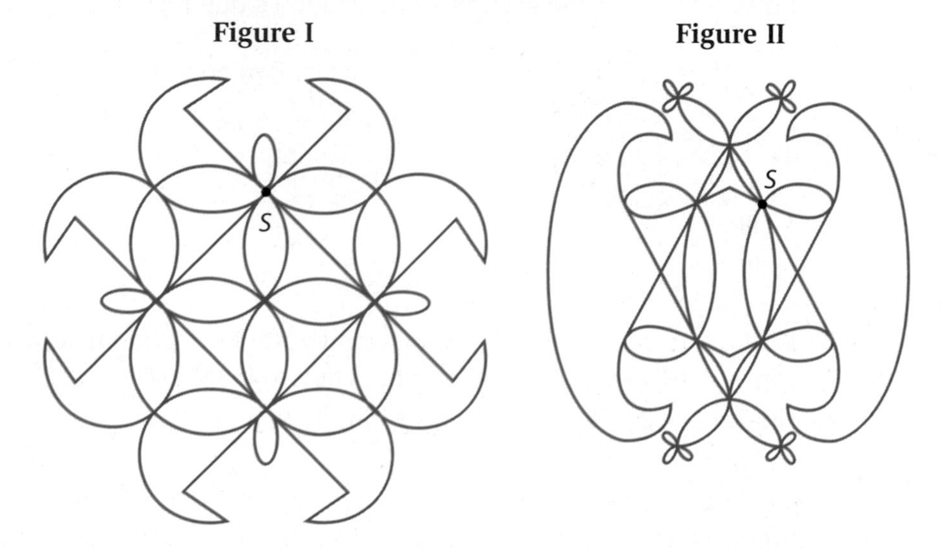

a. Can you trace each of these figures without lifting your pencil or tracing any edges more than once?

b. Describe any *symmetry* you see in each figure.

13 For vertex-edge graphs, the position of the vertices and the length and straightness of the edges are not critical. What is important is the way in which the edges connect the vertices. Consider the following matrix. Each entry shows the shortest distance, in miles, between two corresponding towns.

	W	R	T
Woebegone (W)	–	60	100
Rivendell (R)	60	–	80
Troy (T)	100	80	–

a. Draw a vertex-edge graph that represents the information in the matrix.

b. Use a compass and ruler to draw a scale diagram showing the distances between the towns. Assume straight-line roads between the towns.

c. State a question involving these three towns that is best answered using the scale diagram.

d. State a question that could be answered using either the scale diagram or the graph.

14 The figure below is a *pentagon*. It has 5 vertices and 5 edges. Think about the figure as a vertex-edge graph.

a. Write the adjacency matrix for this graph.

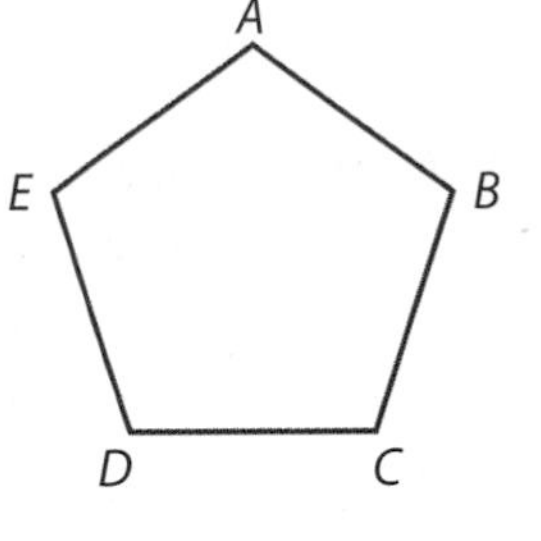

b. Modify a copy of this graph by adding all the *diagonals*. (A diagonal is a line segment connecting 2 vertices that are not adjacent.) Write the adjacency matrix for this modified graph.

c. Write a description of the adjacency matrix for a graph in the shape of a *polygon* with n sides. (A polygon with n sides is similar to the pentagon shown here, but there are n vertices and n edges.) How would you modify the description of the adjacency matrix if the graph consisted of the polygon *and* its diagonals?

Reflections

15 Recall the original 1933 London Underground map on page 238. This is an example of a geometric diagram with vertices and edges for which precise size and shape is not crucial. The cartoon shown here accompanied the introduction of the map. The guide for the London Transport Museum describes the map as follows.

"Producing a map showing the different lines of the Underground system was particularly complicated. At first the Underground lines were shown geographically. A draughtsman, Harry Beck, devised his diagram in 1931. It uses only vertical, horizontal, or 45° diagonals and bears no relation to the real geography of London. At first the publicity department rejected it as too radical, finally publishing it in 1933 to instant enthusiasm from passengers. It was so simple and easy to use that the design is still used today and has been adapted by other cities around the world. It is a design classic." (London Transport Museum Guide)

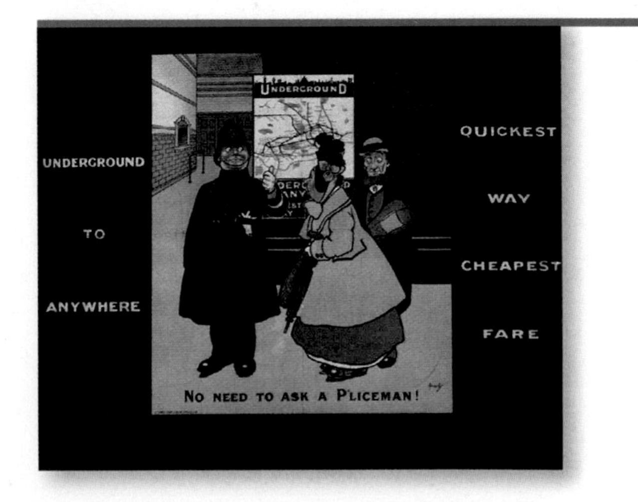

a. Why do you think the publicity department thought the map was "too radical"?

b. Why do you think the map was received with "instant enthusiasm" by the passengers?

On Your Own

16 Two students have each drawn a graph that represents the combined routes of several school buses, as shown below. The vertices represent bus stops. Explain why the two graphs represent the same information. Label corresponding vertices with the same letter. If you have access to vertex-edge graph software, draw one of the graphs, then drag the vertices until it looks like the other graph.

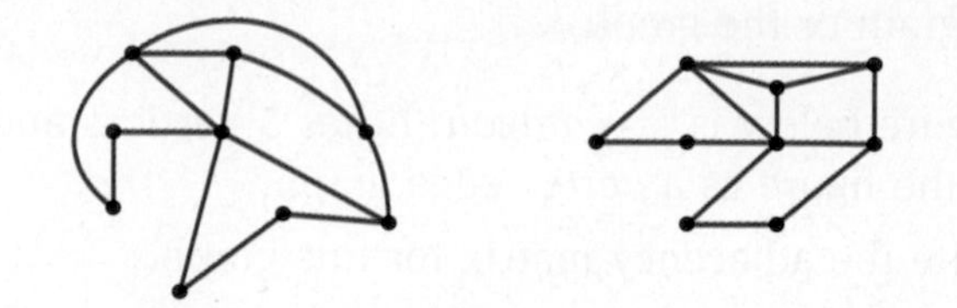

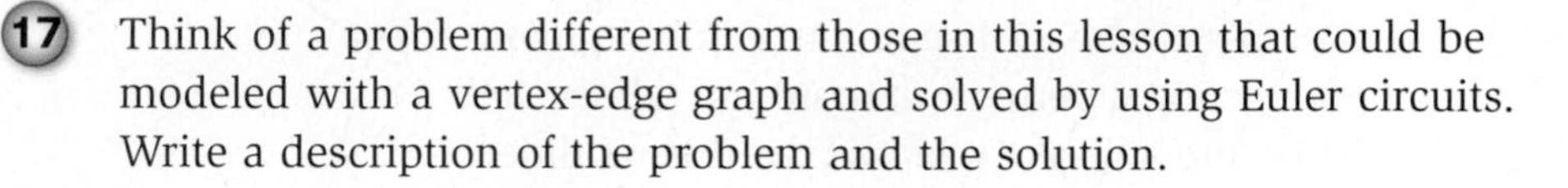

17 Think of a problem different from those in this lesson that could be modeled with a vertex-edge graph and solved by using Euler circuits. Write a description of the problem and the solution.

18 Decide whether you agree with the following statement, and then write an argument to support your position: If a graph has an Euler circuit that begins and ends at a particular vertex, then it will have an Euler circuit that begins and ends at any vertex of the graph.

19 The algorithm for finding an Euler circuit in Problem 6 of Investigation 2 on page 245 is sometimes called the *onionskin algorithm*. Explain how this name describes what the algorithm does.

20 You might think that every matrix can be the adjacency matrix for some graph.

a. Try to draw graphs that have the following adjacency matrices.

i. $\begin{bmatrix} 0 & 3 \\ 3 & 0 \end{bmatrix}$ **ii.** $\begin{bmatrix} 0 & 1 & 2 \\ 1 & 0 & 1 \\ 2 & 1 & 0 \end{bmatrix}$ **iii.** $\begin{bmatrix} 0 & 2 & 1 \\ 2 & 0 & 2 \\ 1 & 1 & 0 \end{bmatrix}$

b. Some matrices cannot be adjacency matrices for graphs. Write a description of the characteristics of a matrix that *could* be the adjacency matrix for a graph.

Extensions

21 Graphs have interesting properties that can be discovered by collecting data and looking for patterns.

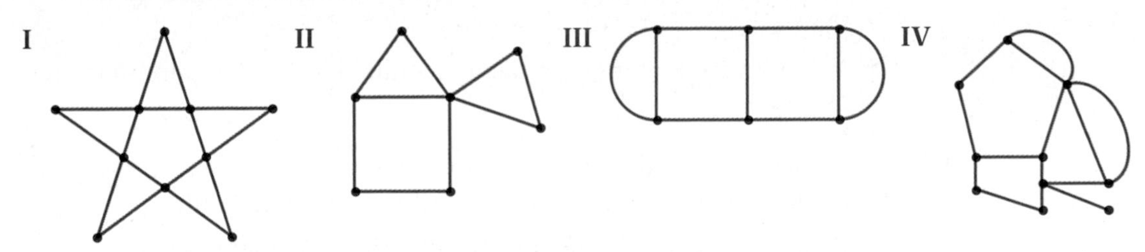

a. Complete a table like the one that follows using the graphs above.

Graph	Sum of the Degrees of All Vertices	Number of Vertices of Odd Degree
I	30	
II		2
III		
IV		

b. Write down any patterns you see in the table.

c. See if the patterns continue when you collect more data. That is, draw a few more graphs, enter the information into the table, and check to see if the patterns you described in Part b are still valid. You might use vertex-edge graph software to help you generate graphs and information quickly.

d. Explain why the sum of the degrees of all the vertices in *any* graph is an even number.

e. Explain why *every* graph has an even number of vertices with odd degree.

22 Most of the graphs you have studied in this lesson have the key property of being *connected*. That is, they are all in one piece. Another way of thinking about connected graphs is that in a connected graph, there is a path from any vertex to any other vertex. Sometimes the way a graph is presented makes it difficult to tell whether or not it is connected.

a. Determine if the following graphs are connected.

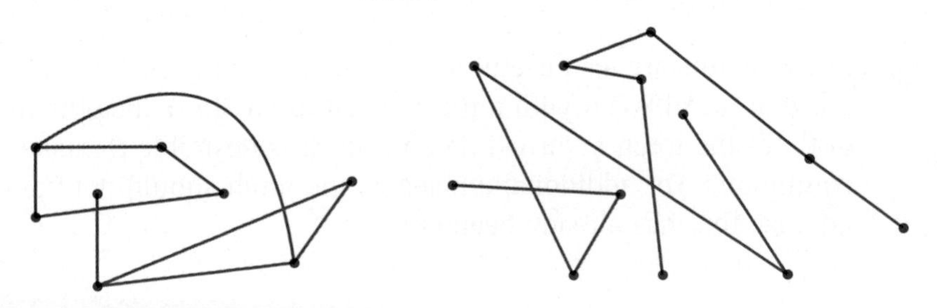

b. Describe a systematic method you could use to check to see if a given graph is connected.

23 Decide whether each of the following statements is true (always true) or false (sometimes false). If a statement is true, explain as precisely as you can why it is true. If a statement is false, draw a *counterexample* that illustrates why it is false.

a. Every vertex of a graph with an Euler circuit has degree greater than 1.

b. If every vertex of a graph has the same degree, the graph has an Euler circuit.

24 In this lesson, you discovered and used an important result about Euler circuits. Now think more carefully about that result.

a. Explain as precisely as you can why this statement is true.

If a graph has an Euler circuit, then all of its vertices have even degree.

b. Explain why or why not the following statement is true.

If all the vertices of a connected graph have even degree, then the graph has an Euler circuit.

25 Dominoes are rectangular tiles used to play a game. Each tile is divided into 2 squares with a number of dots in each square, as in the figure below.

The standard set of dominoes has from 0 to 6 dots in each square. A deluxe set of dominoes has from 0 to 9 dots in each square. In each set, there is exactly one tile representing each possible number-pair combination. To play the game of dominoes, you take turns trying to place dominoes end-to-end by matching the number of dots. For example, for the three dominoes pictured below, the 3-5 domino can be placed next to the 1-3 domino, but the 0-2 domino cannot be placed next to either of the other dominoes.

Is it possible to form a ring of all the dominoes in a standard set placed end-to-end according to the rule above? How about for a deluxe set of dominoes? Explain your answers by reasoning about Euler circuits.

26 Euler circuits are also useful in manufacturing processes where a piece of metal is cut with a mechanical torch. To reduce the number of times the torch is turned on and off, it is desirable to make the cut continuous. For additional efficiency, the torch should not pass along an edge that has already been cut.

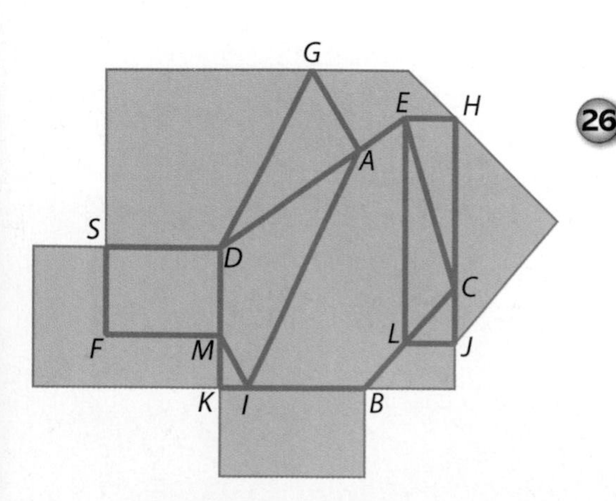

The metal piece must be clamped in air so that the torch does not burn the surface of the workbench. This leads to another condition; namely, any piece that falls off should not require additional cutting. Otherwise, it would have to be picked up and reclamped, a time-consuming process. Find a way to make all the cuts indicated on the pictured piece of metal, so that you begin and end at point S and the above conditions are satisfied.

27 RNA (ribonucleic acid) is a messenger molecule associated with DNA (deoxyribonucleic acid). RNA molecules consist of a chain of bases. Each base is one of 4 chemicals: U (uracil), C (cytosine), A (adenine), and G (guanine). It is difficult to observe exactly what an entire RNA chain looks like, but it is sometimes possible to observe fragments of a chain by breaking up the chain with certain enzymes. Armed with knowledge about the fragments, you can sometimes determine the makeup of the entire chain. One type of enzyme that breaks up an RNA chain is a "G-enzyme." The G-enzyme will break an RNA chain after each G link. For example, consider the following chain.

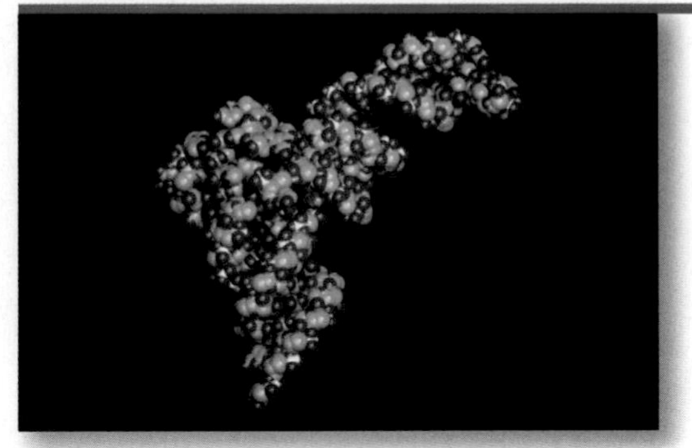

Computer-rendered molecular model of t-RNA

AUUGCGAUC

A G-enzyme will break up this chain into the following fragments.

AUUG CG AUC

Unfortunately, the fragments of a broken-up chain are usually mixed up and in the wrong order. In this task, you will figure out how to reconstruct the chain when given some mixed-up fragments.

a. Suppose a different RNA chain is broken by a G-enzyme into the following fragments (although not necessarily in this order).

AUG AAC CG AG

Explain why the AAC fragment must be the end of the chain.

b. There is another enzyme, called a U-C enzyme, that breaks an RNA chain after each U or C link. For the unknown RNA chain in Part a, the U-C enzyme breaks the chain into the following fragments.

GC GAAC AGAU

As the final step in this chain-breaking process, the fragments are now further broken up using the other enzyme. That is, the fragments formed by the G-enzyme are now broken again, if possible, using the U-C enzyme, and vice versa. The resulting fragments from this process are shown in the table at the right. Each row of the table shows the break-up of each of the 7 fragments above. Complete the table by finding the rest of the final split fragments.

Original Fragment	Final Split Fragments
AUG	AU G
AAC	Not possible to split
CG	
AG	
GC	
GAAC	
AGAU	AG AU

c. Mathematicians and biologists have discovered an amazing technique using Euler paths to reconstruct the unknown RNA chain. Carry out this technique, as follows:

Step 1: Draw vertices for each of the different final split fragments.

Step 2: Draw a **directed edge** (an arrow) from one vertex to another if the two split fragments are part of the same original fragment. The arrow should indicate how the two split fragments are recombined to get the original fragment.

Step 3: You now have a **directed graph**, that is, a graph where the edges have a direction. Find an Euler path through this graph (the start and end are not the same). Keep in mind that as you trace an Euler path, you must move in the direction shown by the directed edges.

Step 4: Put the fragments together as you traverse the Euler path. This will give you the original RNA chain.

28 A **loop** is an edge connecting a vertex to itself. When constructing an adjacency matrix for a graph with loops, a 1 is placed in the position in the matrix that corresponds with an edge joining a vertex to itself. An example of such a graph and its adjacency matrix is shown at the right.

$$\begin{array}{c} \\ A \\ B \end{array} \begin{array}{c} \begin{array}{cc} A & B \end{array} \\ \begin{bmatrix} 1 & 1 \\ 1 & 0 \end{bmatrix} \end{array}$$

a. Recall that the degree of a vertex is the number of edges touching the vertex, except that a loop counts for 2 edge touchings. What is the degree of vertex A?

b. What is the row sum of the first row of the adjacency matrix above? In Investigation 3 of this lesson, you found a connection between row sums of an adjacency matrix and the degree of the corresponding vertex. Does this connection still hold for graphs with loops like the one above?

Some housing developments have houses built on a street that is a "cul-de-sac" so that traffic passing the houses is minimized.

a. Suppose a cul-de-sac is located at the end of the street between blocks 5 and 6 as shown here. Draw a vertex-edge graph that represents this housing development.

b. Find an optimum path for delivering papers to houses in this development.

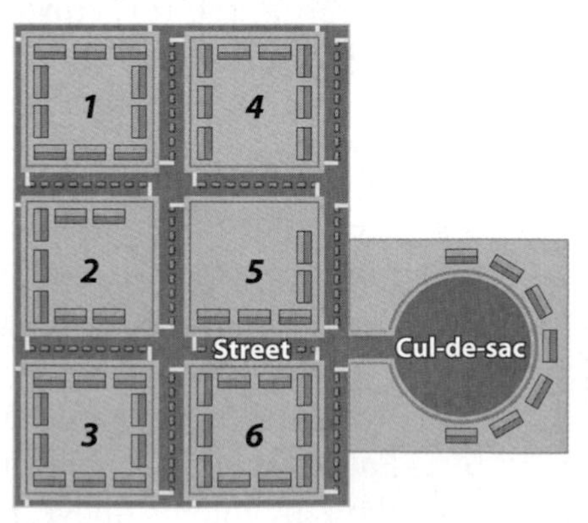

c. You know from this lesson that the degree of a vertex is the number of edges that touch it, except that loops count as two edge touchings. Find the degree of each vertex in your graph.

d. Repeat Parts a, b, and c with a second cul-de-sac constructed at the end of blocks 1 and 4.

e. How does adding a cul-de-sac affect the graph? How does adding a cul-de-sac affect the optimum path for delivering papers to houses in the development?

f. Does the condition about degrees of vertices for graphs with Euler circuits still hold for graphs with loops?

Review

30 Although distance and position are not crucial features of vertex-edge graphs, these features are important in many geometric settings. Consider points in the grid below.

a. What is the distance between points *C* and *B*? Between points *C* and *A*? Explain how to find those distances using the Pythagorean Theorem.

A
B
C

b. Starting at point *C*, move 3 units to the right and 4 units down. Mark the point at this location. How far is this point from point *C*?

c. Find a point that is exactly 10 units from point *A*, but is not directly above or below or directly to the right or left. What geometric shape is formed by all the points that are exactly 10 units from point *A*?

31 Place the following quantities in increasing order without using your calculator. Explain your method.

$$5\%, \frac{1}{10}, 0.5, \frac{1}{9}, \frac{4}{9}, 49\%$$

32 Study the relations represented in the following tables. If a relation is linear, find the slope of the graph representing the relationship.

a.

x	4	6	8	10	12
y	1	0	−1	−2	−3

b.

x	0	1	3	8
y	2	4	6	8

c.

x	0	0.1	0.2	0.3
y	3	2.95	2.9	2.85

33 In a recent poll, 630 students were asked if they like Chinese food. The circle graph below shows the results of the poll. Determine as precisely as possible how many people gave each response.

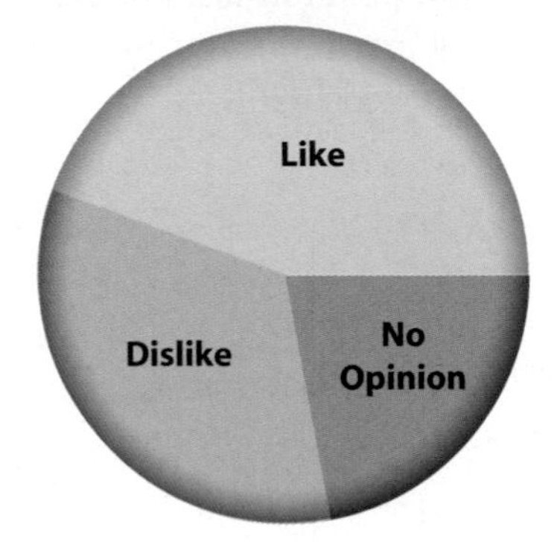

34 Complete a table like the one below showing some possible lengths, widths, and perimeters for a rectangle with an area of 24 square units.

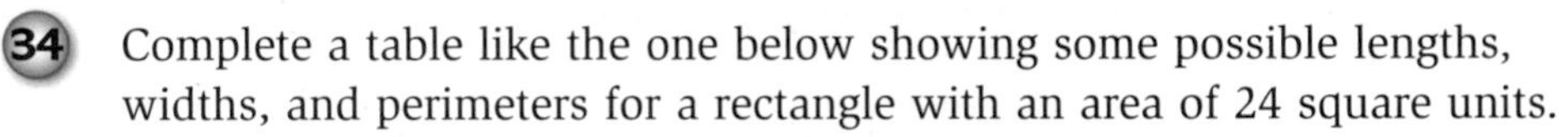

L	1	2	3	4	5	6	7	8	9	10	11	12	24
W	24	12											
P	50	28											

a. Describe the pattern of change in W as L changes.

b. Describe the pattern of change in P as L changes.

c. Find formulas to represent W as a function of L, and P as a function of L.

d. Are either of the patterns of change linear?

35 Solve each equation or inequality.

a. $5x - 6 = 20$

b. $4.85 = 1.25x + 6.1$

c. $80 - \frac{3}{4}x < 20$

d. $75 \leq 15x + 100$

36 Without using your calculator, match each equation with a possible graph of the equation.

a. $y = x - 5$

b. $y = -x + 5$

c. $y = x$

d. $y = -x - 5$

e. $y = x + 5$

f. $y = 5$

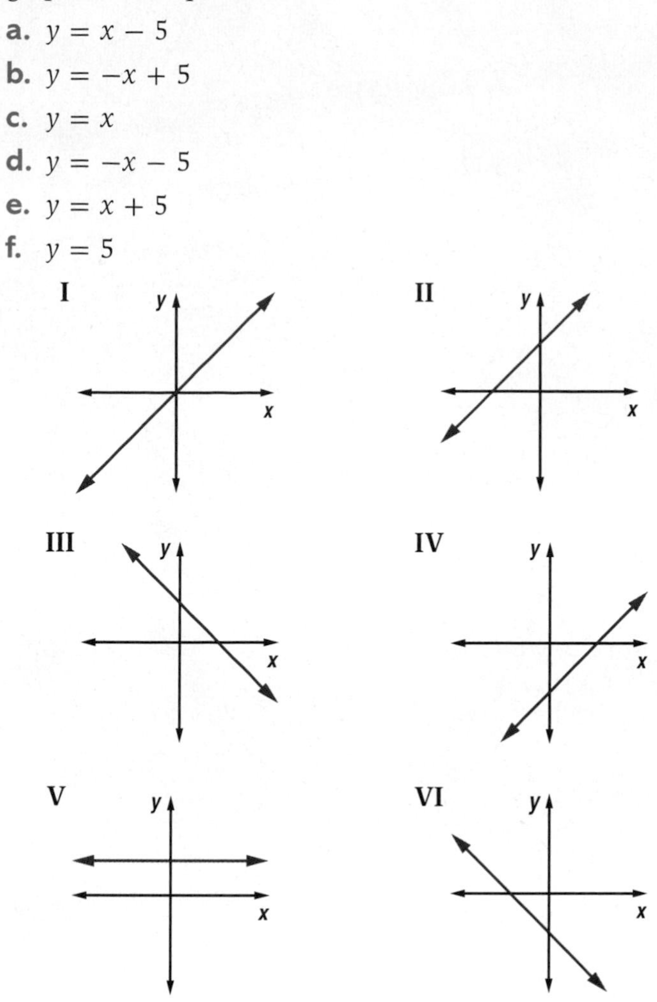

37 Determine if the expressions in each pair are equivalent.

a. 6.02×10^{21} and 602×10^{19}

b. 980×10^{10} and 9.8×10^{8}

c. 0.034×10^{12} and 340×10^{8}

38 Draw sketches of a cone and a cylinder.

LESSON 2

Vertex Coloring: Avoiding Conflict

In the last lesson, you learned about vertex-edge graphs. You used these graphs to model and solve problems related to paths and circuits. They can also be used to solve many other types of problems. In this lesson, you will investigate how graphs can be used to avoid possible conflict among a finite number of objects, people, or other things.

To begin, consider the problem of assigning radio frequencies to stations serving the same region. In cities and towns, you can listen to many different radio stations. Each radio station has its own transmitter that broadcasts on a particular channel, or frequency. The Federal Communications Commission (FCC) assigns the frequencies to the radio stations. The frequencies are assigned so that no two stations interfere with each other. Otherwise, you might tune into "Rock 101.7" and get Mozart instead!

Think About This Situation

Suppose 7 new radio stations have applied for permits to start broadcasting in the same region of the country. Some stations may interfere with each other, others may not.

a What are some factors that may determine whether or not stations interfere with each other?

b How does this situation involve "conflict"?

c How do you think the FCC should assign frequencies to the 7 stations?

d Why do you think the FCC might like to assign the fewest possible number of new frequencies for the 7 stations?

In this lesson, you will learn how to use vertex-edge graphs to solve problems about avoiding conflicts, such as assigning noninterfering radio frequencies, using a technique called *vertex coloring.*

Investigation 1 Building a Model

Suppose the 7 new radio stations that have applied for broadcast permits are located as shown on the grid below. A side of each small square on the grid represents 100 miles. The FCC wants to assign a frequency to each station so that no 2 stations interfere with each other. The FCC also wants to assign the fewest possible number of new frequencies. Suppose that because of geographic conditions and the strength of each station's transmitter, the FCC determines that stations within 500 miles of each other will interfere with each other.

Your work on the problems of this investigation will help you answer the question:

How can vertex-edge graphs be used to assign frequencies to these 7 radio stations so that as few frequencies as possible are used and none of the stations interfere with each other?

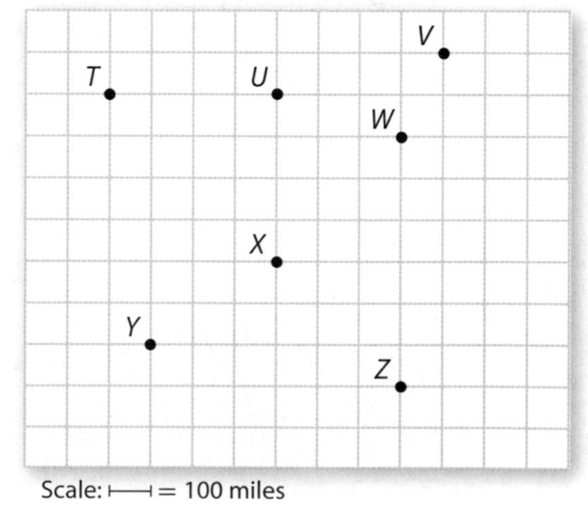

1 For a small problem like this, you could solve it by trial and error. However, a more systematic method is needed for more complicated situations. Working on your own, begin modeling this problem with a graph. Remember, to model a problem with a graph, you must first decide what the vertices and edges represent.

a. What should the vertices represent?

b. How will you decide whether or not to connect 2 vertices with an edge? Complete this statement:

Two vertices are connected by an edge if ... .

c. Now that you have specified the vertices and edges, draw a graph for this problem.

2 Compare your graph with those of your classmates.

a. Did everyone define the vertices and edges in the same way? Discuss any differences.

b. For a given situation, suppose two people define the vertices and edges in two different ways. Is it possible that both ways accurately represent the situation? Explain your reasoning.

c. For a given situation, suppose two people define the vertices and edges in the same way. Is it possible that their graphs have different shapes but both are correct? Explain your reasoning.

3 A common choice for the vertices is to let them represent the radio stations. Edges might be thought of in two ways, as described in Parts a and b below.

a. You might connect 2 vertices by an edge whenever the stations they represent are 500 miles or *less* apart. Did you represent the situation this way? If not, draw a graph where two vertices are connected by an edge whenever the stations they represent are 500 miles or *less* apart.

b. You might connect 2 vertices by an edge whenever the stations they represent are *more* than 500 miles apart. Did you represent the situation this way? If not, draw a graph where 2 vertices are connected by an edge whenever the stations they represent are *more* than 500 miles apart.

c. Compare the graphs from Parts a and b.

i. Are both graphs accurate ways of representing the situation?

ii. Which graph do you think will be more useful and easier to use as a mathematical model for this situation? Why?

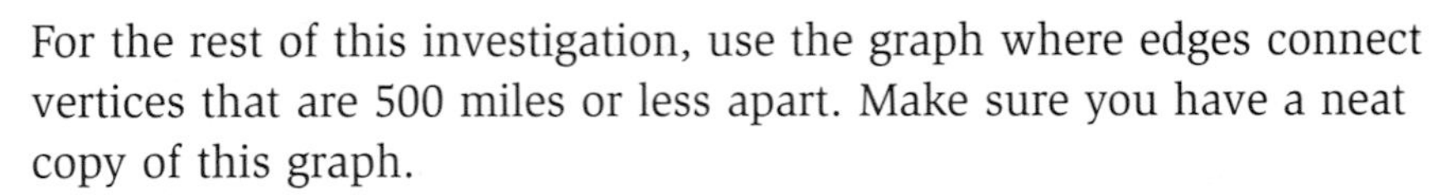

4 For the rest of this investigation, use the graph where edges connect vertices that are 500 miles or less apart. Make sure you have a neat copy of this graph.

a. Are vertices (stations) X and W connected by an edge? Are they 500 miles or less apart? Will their broadcasts interfere with each other?

b. Are vertices (stations) Y and Z connected by an edge? Will their broadcasts interfere with each other?

c. Compare your graph to the graph at the left.

i. Explain why this graph also accurately represents the radio-station problem.

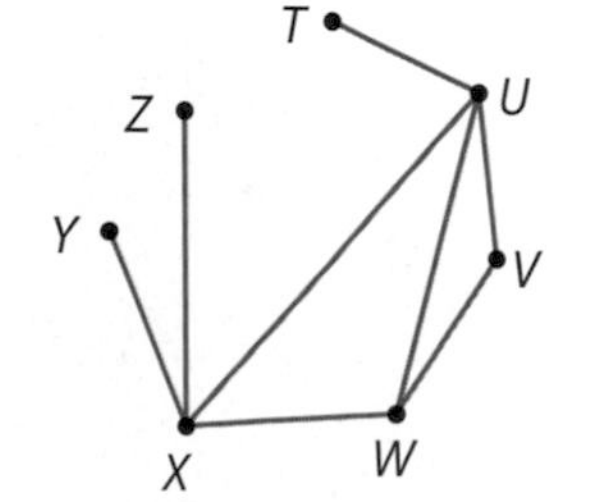

ii. Describe what it means for two graphs to be "the same" even if their appearances are different.

iii. If you have access to vertex-edge graph software, draw this graph and then drag its vertices so it looks like your graph in Problem 3, Part a.

5 Remember that the problem is to assign frequencies so that there will be no interference between radio stations. So far, your graph models this problem as follows. Vertices represent the radio stations. Two vertices are connected by an edge if the corresponding radio stations are within 500 miles of each other. Here's the last step in building the graph model—represent the frequencies as *colors*. So now, assigning frequencies to radio stations means to assign colors to the vertices.

Examine the statements in the following partially completed table. Translate each statement about stations and frequencies into a statement about vertices and colors. (The first one is already done for you.)

Statements about Stations and Frequencies	Statements about Vertices and Colors
Two stations have different frequencies.	Two vertices have different colors.
Find a way to assign frequencies so that stations within 500 miles of each other get different frequencies.	
Use the fewest number of frequencies.	

6 Now use as few colors as possible to **color the graph** for the radio-station problem. That is, assign a color to each vertex so that any 2 vertices that are connected by an edge have different colors. You can use colored pencils or just the names of some colors to do the coloring. Color or write a color code next to each vertex. Try to use the smallest number of colors possible.

7 Compare your coloring with another student's coloring.

a. Do both colorings satisfy the condition that vertices connected by an edge must have different colors?

b. Do both colorings use the same number of colors to color the vertices of the graph? Reach agreement about the minimum number of colors needed.

c. Explain, in writing, why the graph cannot be colored with fewer colors.

d. For 2 particular vertices, suppose one student colors both vertices red while another student colors 1 vertex red and the other blue. Is it possible that both colorings are acceptable? Explain your reasoning.

e. Describe the connection between graph coloring and assigning frequencies to radio stations.

Summarize the Mathematics

Some problems can be solved by coloring the vertices of an appropriate graph.

a What do the vertices, edges, and colors represent in the graph that you used to solve the radio-station problem?

b How did "coloring a graph" help solve the radio-station problem?

c In what ways can two graphs differ and yet still both accurately represent a situation?

Be prepared to share your ideas with the class.

Check Your Understanding

Consider the graph at the right.

a. On a copy of the graph, color the vertices using as few colors as possible.

b. If possible, find a second coloring of the graph in which some of the vertices colored the same in Part a are no longer colored the same. Again use as few colors as possible.

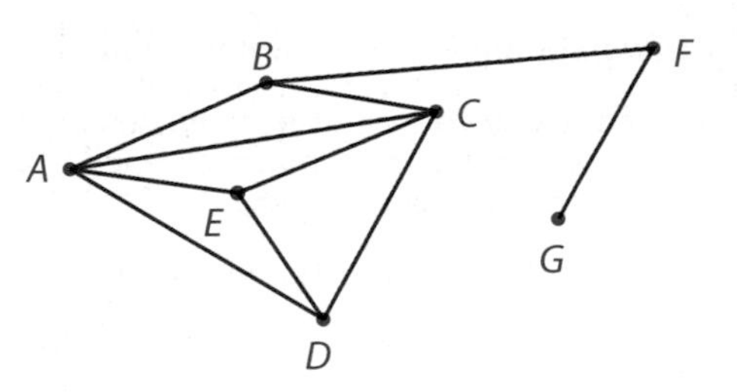

Investigation 2 Scheduling, Mapmaking, and Algorithms

Now that you know how to color a graph, you can use graph coloring to solve many other types of problems. As you work on the problems in this investigation, look for answers to this question:

What are the basic steps of modeling and solving conflict problems using vertex-edge graphs?

Scheduling Meetings There are 6 clubs at King High School that want to meet once a week for one hour, right after school lets out. The problem is that several students belong to more than one of the clubs, so not all the clubs can meet on the same day. Also, the school wants to schedule as few days per week for after-school club meetings as possible. Below is the list of the clubs and the club members who also belong to more than one club.

Clubs and Members

Club	Students Belonging to More Than One Club
Varsity Club	Christina, Shanda, Carlos
Math Club	Christina, Carlos, Wendy
French Club	Shanda
Drama Club	Carlos, Vikas, Wendy
Computer Club	Vikas, Shanda
Art Club	Shanda

1. Consider the club-scheduling problem as a graph-coloring problem.

    a. Your goal is to assign a meeting day (Monday–Friday) to each club in such a way that no 2 clubs that share a member meet on the same day. Also, you want to use as few days as possible. Working on your own, decide what you think the vertices, edges, and colors should represent.

    b. Compare your representations with others. Decide as a group which representations are best. Complete these three statements.

    The vertices represent ... .

    Two vertices are connected by an edge if ... .

    The colors represent ... .

    c. Draw a graph that models the problem.

    d. Color the club-scheduling graph using as few colors as possible.

2. Use your graph coloring in Problem 1 to answer these questions.

    a. Is it possible for every club to meet once per week?

    b. What is the fewest number of days needed to schedule all the club meetings?

    c. On what day should each club meeting be scheduled?

    d. Explain how your coloring of the graph helps you answer each of the questions above.

Coloring Maps Another class of problems for which graph coloring is useful involves coloring maps. You may have noticed in your geography or social-studies course that maps are always colored so that neighboring countries do not have the same color. This is done so that the countries are easily distinguished and don't blend into each other. In the following problems, you will explore the number of different colors necessary to color *any* map in such a way that no 2 countries that share a border have the same color. This is a problem that mathematicians worked on for many years, resulting in a lot of new and useful mathematics. For this problem, *countries* are assumed to be regions that are contiguous (not broken up into separate parts), and *border* means a common boundary of some length (touching at points doesn't matter).

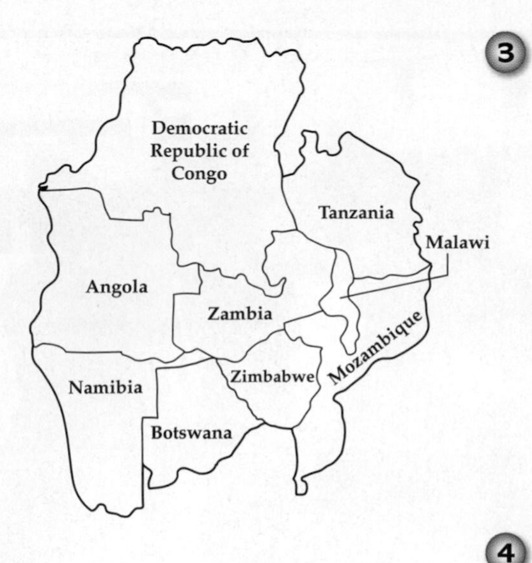

3 Shown here is an uncolored map of a portion of southern Africa.

a. Using a copy of this map, color the map so that no 2 countries that share a border have the same color. (For this problem, you may assume that Botswana and Zambia intersect only at a point, though in fact they share a border about 10 miles long.)

b. How many colors did you use? Try to color the map with fewer colors.

c. Compare your map coloring with that of other classmates.

i. Are the colorings different?

ii. Are the colorings legitimate; that is, do neighboring countries have different colors? If a coloring is not legitimate, fix it.

d. What is the fewest number of colors needed to color this map?

4 In Problem 3, you found the fewest number of colors needed to color the Africa map. Now think about the fewest number of colors needed to color *any* map.

a. Do you think you can color *any* map with at most 5 different colors? Can the map of Africa be colored with 5 colors?

b. The map below has been colored with 5 colors. Is it possible to color the map with fewer than 5 colors? If so, make a copy of the map and color it with as few colors as possible.

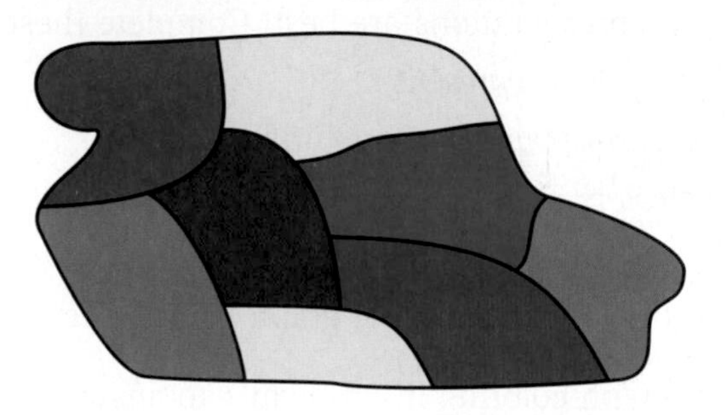

c. What do you think is the *fewest* number of colors needed to color *any* map? Make a conjecture.

i. Compare your conjecture to the conjectures of your classmates. Briefly discuss any differences. Revise your conjecture if you think you should. Test your conjecture as follows, in part ii.

ii. Over the next few days, test your conjecture on other maps outside of class. Revise your conjecture as necessary. Compare and discuss again with your classmates and teacher.

Maps can be colored by working directly with the maps, as you have been doing. But it is also possible to turn a map-coloring problem into a graph-coloring problem. This can be helpful since it allows you to use all the properties and techniques for graphs to help you understand and solve map-coloring problems.

5 To find a graph that models a map-coloring problem, first think about what you did with the radio-station and club-scheduling problems. In both of those problems, the edges were used to indicate some kind of *conflict* between the vertices. The vertices in conflict were connected by an edge and colored different colors. A crucial step in building a graph-coloring model is to decide what the conflict is. Once you know the conflict, you can figure out what the vertices, edges, and colors should represent.

a. What was the conflict in the club-scheduling problem? What was the conflict in the radio-station problem?

b. Make and complete a table like the one below.

Modeling Conflicts

Problem	Conflict if:	Vertices	Connect with an Edge if:	Colors
Radio-station problem	*2 radio stations are 500 miles or less apart*	*radio stations*		*frequencies*
Club-scheduling problem	*two clubs* _______			
Map-coloring problem	*two countries* _______			

6 Consider the map of a portion of southern Africa in Problem 3.

a. Use the information in the table above to create a vertex-edge graph that represents the map.

b. Color the vertices of the graph. Remember that coloring always means that vertices connected by an edge must have different colors. Also, as usual, use as few colors as possible.

c. Compare your coloring with those of other classmates.

i. Are all the colorings legitimate?

ii. Reach agreement on the fewest number of colors needed to color the graph.

iii. Is the minimum number of colors for this *graph*-coloring problem the same as the minimum number of colors for the *map*-coloring problem in Part d of Problem 3? Explain.

Algorithms You have now used vertex coloring to solve several problems. In each problem, you used some method to color the vertices of a graph using as few colors as possible. Recall that a systematic step-by-step method is called an *algorithm*. Finding good graph-coloring algorithms is an active area of mathematical research with many applications. It has proven quite difficult to find an algorithm that colors the vertices of any graph using as few colors as possible. You often can figure out how to do this for a given small graph, as you have done in this lesson. However, no one knows an efficient algorithm that will color *any* graph with the *fewest* number of colors! This is a famous unsolved problem in mathematics. Think about methods you have used to color a graph.

Describe some strategies or algorithms you have used to color the vertices of a graph using the fewest number of colors. Compare and discuss algorithms with your classmates.

One commonly used algorithm is sometimes called the **Welsh-Powell algorithm**. Here's how it works:

Step 1: Begin by making a list of all the vertices starting with the ones of highest degree and ending with those of lowest degree.

Step 2: Color the first uncolored vertex on your list with an unused color.

Step 3: Go down the list coloring as many uncolored vertices with the current color as you can, following the rule that vertices connected by an edge must be different colors.

Step 4: If all the vertices are now colored, you're done. If not, go back to Step 2.

a. Follow the Welsh-Powell algorithm, step by step, to color the two graphs below.

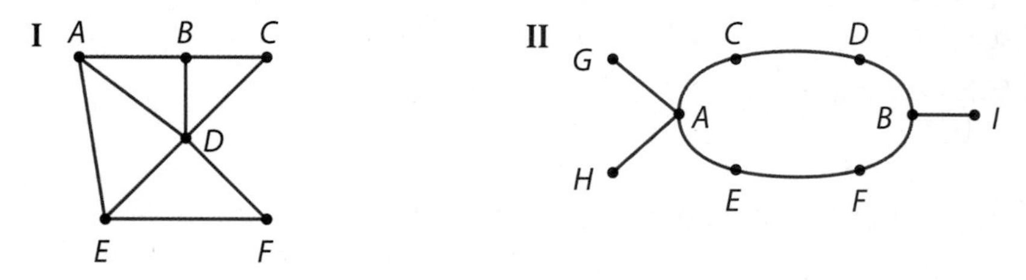

b. Does the Welsh-Powell algorithm always yield a coloring that uses the fewest number of colors possible? Explain your reasoning.

c. Use the Welsh-Powell algorithm to color each graph below and compare your coloring with your previous results.

i. radio-station graph (Investigation 1, Problem 6)

ii. club-scheduling graph (Investigation 2, Problem 1, Parts c and d)

Summarize the Mathematics

In this lesson, you used graphs to avoid conflict in three seemingly different problems.

a Explain how each of the three main problems you solved in this lesson—assigning radio frequencies, club scheduling, and map coloring—involved "conflict among a finite number of objects."

b Describe the basic steps of modeling and solving a conflict problem with a graph.

c The **chromatic number** of a graph is the fewest number of colors needed to color all its vertices so that 2 vertices connected by an edge have different colors. What is the chromatic number of the graphs for each of the three problems in this lesson (radio stations, club meetings, and map coloring)? How is this number related to the solution for each problem?

Be prepared to share your responses with the entire class.

Check Your Understanding

Hospitals must have comprehensive and up-to-date evacuation plans in case of an emergency. A combination of buses and ambulances can be used to evacuate most patients. Of particular concern are patients under quarantine in the contagious disease wards. These patients cannot ride in buses with nonquarantine patients. However, some quarantine patients can be transported together. The records of who can be bused together and who cannot are updated daily.

Suppose that on a given day there are 6 patients in the contagious disease wards. The patients are identified by letters. Here is the list of who cannot ride with whom:

A cannot ride with B, C, or D.
B cannot ride with A, C, or E.
C cannot ride with A, B, or D.
D cannot ride with A or C.
E cannot ride with F or B.
F cannot ride with E.

The problem is to determine how many vehicles are needed to evacuate these 6 patients. Use graph coloring to solve this problem. Describe the conflict, and state what the vertices, edges, and colors represent.

On Your Own

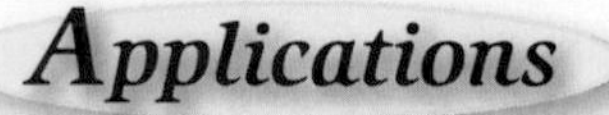

Applications

1 A nursery and garden center plants a certain number of "mix-and-match" flower beds. Each bed contains several different varieties and colors. This allows customers to see possible arrangements of flowers that they might plant.

However, the beds are planted so that no bed contains two colors of the same variety. For example, no bed contains both red roses and coral roses. Also, no bed contains two varieties of the same color. For example, no bed contains both yellow tulips and yellow marigolds. This is done so that the customer can distinguish among and appreciate the different colors and varieties. A list of the varieties and colors that will be planted follows.

Flower Beds

Varieties	Colors
Roses	Red, Coral, White
Tulips	Yellow, Purple, Red
Marigolds	Yellow, Orange

The nursery wants to plant as few mix-and-match beds as possible. In this task, you will determine the minimum number of mix-and-match flower beds.

a. The varieties and colors listed above yield 8 different types of flowers, such as red roses, red tulips, and yellow tulips. List all the other types of flowers that are possible.

b. It is the types of flowers from Part a that will be planted in the mix-and-match beds. The problem is to figure out the minimum number of beds needed to plant these types of flowers so that no bed contains flowers that are the same variety or the same color. First, you need to build a graph-coloring model.

  i. What should the vertices represent?

  ii. What should the edges represent? Why?

  iii. What should the colors of the graph represent?

c. Draw the graph model and color it with as few colors as possible.

d. What is the minimum number of mix-and-match beds needed?

e. Use your graph coloring to recommend to the nursery which types of flowers should go in each of the mix-and-match beds.

f. When using a graph-coloring model, you connect vertices by an edge whenever there is some kind of conflict between the vertices. What was the conflict in this task?

2. A local zoo wants to take visitors on animal-feeding tours. They propose the following tours.

Tour 1 Visit lions, elephants, buffaloes

Tour 2 Visit monkeys, hippos, deer

Tour 3 Visit elephants, zebras, giraffes

Tour 4 Visit hippos, reptiles, bears

Tour 5 Visit kangaroos, monkeys, seals

The animals are fed only once a day. Also, there is only room for 1 tour group at a time at any 1 site. What is the fewest number of days needed to schedule all 5 tours? Explain your answer in terms of graph coloring.

3. You often can color small maps directly from the map, without translating to a graph model. However, using a graph model is essential when the maps are more complicated. The map of South America shown here can be colored either directly or by using a graph-coloring model.

a. Color a copy of the map of South America directly. Use as few colors as possible and make sure that no two bordering countries have the same color.

b. Represent the map as a graph. Then color the vertices of the graph with as few colors as possible.

c. Did you use the same number of colors in Parts a and b?

4. The following figure is part of what is called a *Sierpinski Triangle*. (The complete figure is actually drawn by an infinite process described in Extensions Task 18 on page 283.)

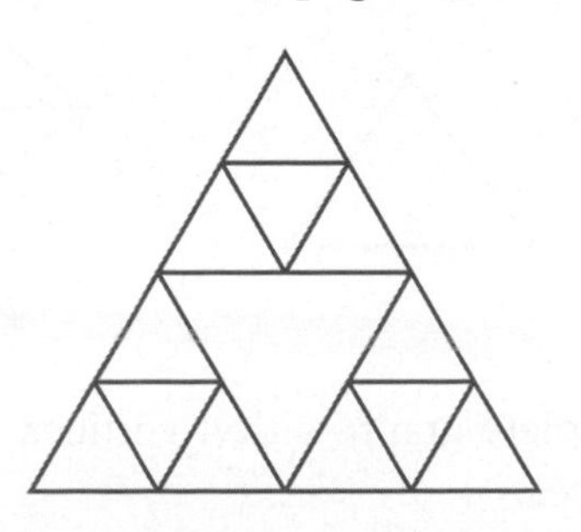

On Your Own

a. Think of this figure as a map in which each triangle *not* containing another triangle is a country. Make and color a copy of the map with as few colors as possible.

b. Construct a graph that represents this map. Color the vertices of the graph with as few colors as possible. Compare the number of colors used with that in Part a.

c. Think of this figure as a map as Sierpinski did: the triangles with points upwards are countries, and the triangles with points downwards are bodies of water, "Sierpinski oceans," that separate the countries. Using this interpretation of countries, color a copy of the map with as few colors as possible. (Leave the Sierpinski oceans uncolored.)

d. Construct a graph for this second map. Color this graph with as few colors as possible. Did you use the same number of colors as in Part c?

Connections

5 Shown here is a student's proposal for a graph that models the radio-station problem from page 267.

a. Is this a legitimate model for the radio-station problem? Explain your reasoning.

b. In this graph model, some edges intersect at places that are not vertices. Can the graph be redrawn without edge-crossings? If so, do so.

c. Graphs that *can* be drawn in the plane with edges intersecting only at the vertices are called **planar graphs**. Which of the graphs below are planar graphs? (Use available software to demonstrate).

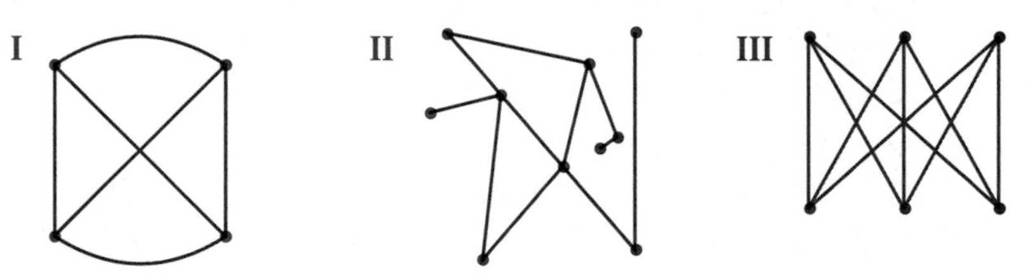

6 This task explores some properties of *complete graphs*. A **complete graph** is a graph that has exactly one edge between every pair of vertices. Complete graphs with 3 and 5 vertices are shown below.

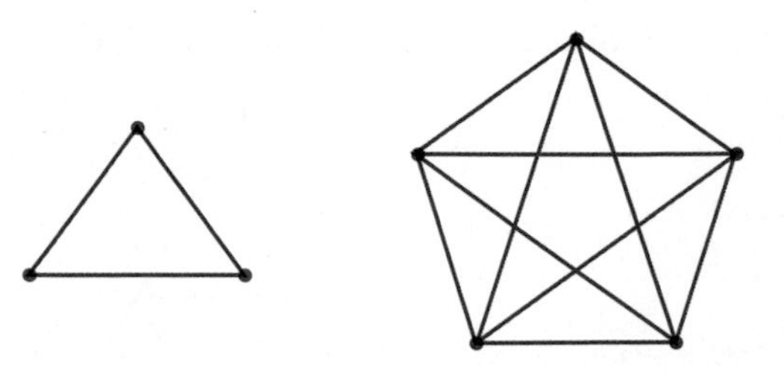

a. Draw a complete graph with 4 vertices. Draw a complete graph with 6 vertices.

b. Make a table that shows the number of edges for complete graphs with 3, 4, 5, and 6 vertices.

c. Look for a pattern in your table. How many edges does a complete graph with 7 vertices have? A complete graph with n vertices?

7 Refer to the definition of a complete graph given in Connections Task 6.

a. What is the minimum number of colors needed to color the vertices of a complete graph with 3 vertices? A complete graph with 4 vertices? A complete graph with 5 vertices?

b. Make a table showing the number of vertices and the corresponding minimum number of colors needed to color a complete graph with that many vertices. Enter your answers from Part a into the table. Find two more entries for the table.

c. Describe any patterns you see in the table.

d. What is the minimum number of colors needed to color a complete graph with 100 vertices? With n vertices?

8 A **cycle graph** is a graph consisting of a single *cycle* (a route that uses each edge and vertex exactly once and ends where it started).

a. Color the vertices of each of the cycle graphs below using as few colors as possible.

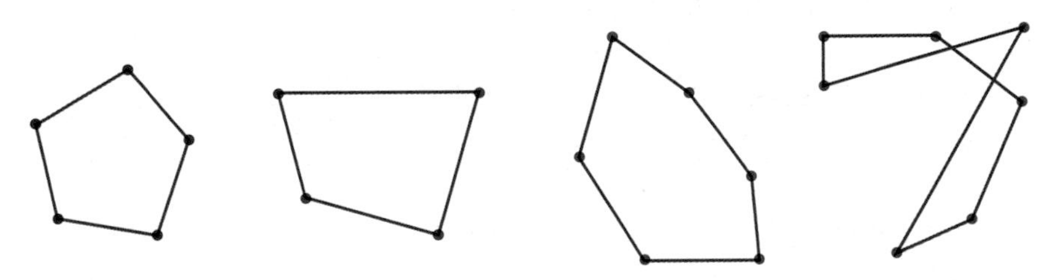

b. Make a conjecture about the minimum number of colors needed to color cycle graphs. Write an argument supporting your conjecture. Test your conjecture by drawing and coloring some large cycle graphs using vertex-edge graph software if available.

9 Besides coloring graphs, it is also possible to color other geometric figures. The three-dimensional figures below are three of the five *regular polyhedra*. You will learn more about regular polyhedra in Unit 6 *Patterns in Shape*. For now, you just need to visualize these three objects.

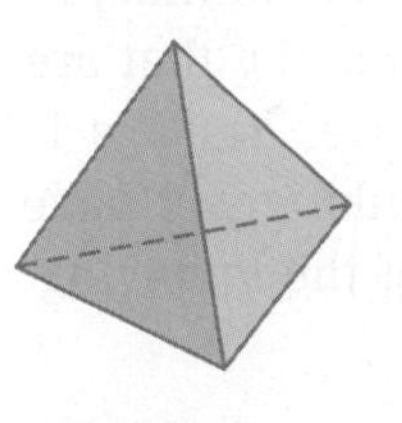

Tetrahedron

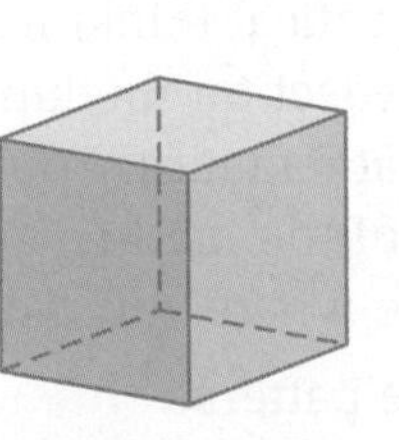

Hexahedron

Octahedron

Complete Parts a, b, and c for each of the above polyhedra. Record your answers for each of these coloring schemes in a table like the one on the next page.

Coloring Polyhedra

Regular Polyhedron	Minimum Number of Colors		
	for Vertices	for Edges	for Faces
Tetrahedron			
Hexahedron			
Octahedron			

a. Color the vertices. Use the minimum number of colors. (Vertices connected by an edge must have different colors.)

b. Color the edges. Use the minimum number of colors. (Edges that share a vertex must have different colors.)

c. Color the faces. Use the minimum number of colors. (Faces that are adjacent must have different colors.)

Reflections

10 Throughout this course, and this unit in particular, you have been doing *mathematical modeling*. Below is a diagram that summarizes the process of mathematical modeling.

Process of Mathematical Modeling

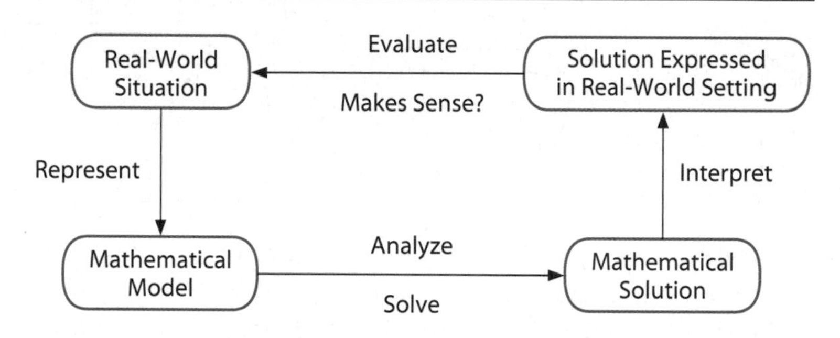

Choose one example of mathematical modeling from this lesson. Use the example to illustrate each part of the diagram.

11 Think of a problem situation different from any in this lesson that could be solved by vertex coloring. Describe the problem and the solution.

12 In this lesson, as well as in previous units, you have engaged in important kinds of mathematical thinking. From time to time, it is helpful to step back and reflect on the kinds of thinking that are broadly useful in doing mathematics. Look back over Lessons 1 and 2 and consider some of the mathematical thinking you have done. Describe an example where you did each of the following.

a. Search for and describe patterns

b. Formulate or find a mathematical model

c. Make and check conjectures

d. Describe and use algorithms

e. Use different representations of the same idea

Extensions

13 Search the Internet or a library for information on mathematicians who have worked on map coloring. Write a one-page report on one mathematician's contribution to the field.

14 In the nineteenth century, mathematicians made a conjecture about the minimum number of colors needed to color any map so that regions with a common boundary have different colors. This conjecture became one of the most famous unsolved problems in mathematics—until 1976 when the problem was solved. Based on your work in this lesson, how many colors do you think are needed to color *any* map? Only consider maps where the regions are connected. So, for example, do not consider a map that has a country that is split into two parts separated by another country.

a. Try to draw a map that requires 3 colors and cannot be colored with fewer colors.

b. Try to draw a map that requires 4 colors and cannot be colored with fewer colors.

c. Try to draw a map that requires 5 colors and cannot be colored with fewer colors.

d. How many colors do you think are necessary to color any map? After you have worked on this problem for a while, search the Internet or a library for recent information on graph theory and map coloring. Find the answer and compare it to your answer. Write a brief report on your findings.

Ken Appel and colleague Wolfgang Haken of the University of Illinois used 1,200 hours of computer time to help solve the map-coloring problem.

15 In Problem 7 on page 274, you described algorithms that you used to color a graph. You may have described one of the following two algorithms (adapted from the description in *Discrete Algorithmic Mathematics, 3rd Edition*, A K Peters, Ltd, 2004, page 294).

Vertex-by-Vertex Algorithm

Step 1: Arbitrarily number all the vertices of the graph: Vertex 1, Vertex 2, Vertex 3, and so on. Also, number the colors: Color 1, Color 2, and so on.

Step 2: Color the first vertex on your list with the first color.

Step 3: Color the next vertex on your list with the lowest-numbered color not already used for an adjacent vertex.

Step 4: Continue vertex-by-vertex until all vertices are colored.

One-Color-at-a-Time Algorithm

Step 1: Arbitrarily number all the vertices of the graph: Vertex 1, Vertex 2, Vertex 3, and so on.

Step 2: Color the lowest-numbered vertex with an unused color.

Step 3: Go down the list of vertices coloring as many uncolored vertices with the current color as you can, following the rule that adjacent vertices must be different colors.

Step 4: If all the vertices are now colored, you're done. If not, go back to Step 2.

a. Color the graph below (from Problem 8 on page 274) using each of the two algorithms above. For this problem, number the vertices in the same way for each algorithm.

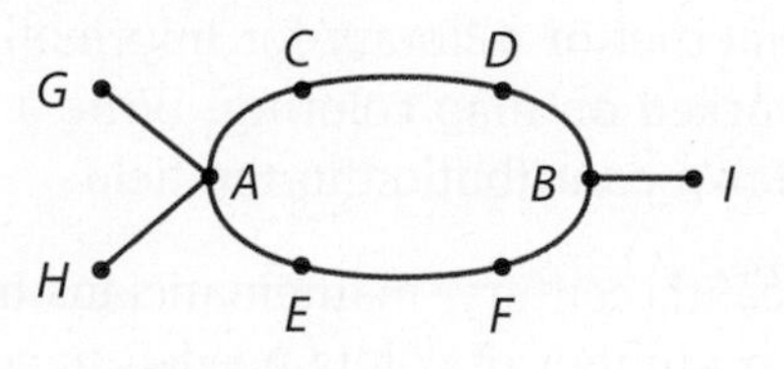

b. Compare the colorings from Part a.

i. Make a conjecture about the colorings produced by these two algorithms for any graph.

ii. Use vertex-edge graph software to test your conjecture with other graphs. For example, load some of the specific graph examples from the Sample Graphs menu. Then color using both algorithms, as found in the Algorithms menu. Each time you apply one of these algorithms, the software creates a random numbering of the vertices of the graph. What happens when the two algorithms are used with the same numbering of the vertices?

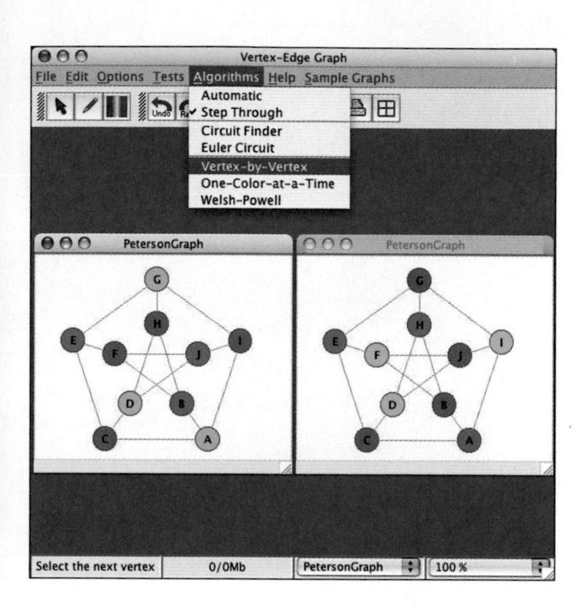

iii. Explain why you think your conjecture is true.

c. Compare the One-Color-at-a-Time algorithm to the Welsh-Powell algorithm in Problem 8 on page 274. Describe similarities and differences.

d. Use vertex-edge graph software to help you further investigate the three algorithms you have considered in this task: the Vertex-by-Vertex algorithm, the One-Color-at-a-Time algorithm, and the Welsh-Powell algorithm. Explore the following questions. Write a brief report summarizing your findings. Include examples, counterexamples, and explanations as needed.

i. What happens when you change the order in which you list the vertices? Do some types of orderings seem to generally be better than others?

ii. Does one algorithm always yield a coloring with fewer colors than the others?

iii. Will any of the algorithms always produce the chromatic number when applied to any graph?

16 In this lesson, coloring a graph has always meant coloring the *vertices* of the graph. It also can be useful to think about **coloring the edges** of a graph. For example, suppose there are 6 teams in a basketball tournament and each team plays every other team exactly once. Games involving different pairs of teams can be played during the same round, that is, at the same time. The problem is to figure out the fewest number of rounds that must be played. One way to solve this problem is to represent it as a graph and then color the *edges*.

a. Represent the teams as vertices. Connect 2 vertices with an edge if the 2 teams will play each other in the tournament. Draw the graph model.

b. Color the edges of the graph so *edges that share a vertex have different colors.* Use as few colors as possible.

c. Think about what the colors mean in terms of the tournament and the number of rounds that must be played. Use the edge coloring to answer these questions.

  i. What is the fewest number of rounds needed for the tournament?

  ii. Which teams play in which rounds?

d. Describe another problem situation that could be solved by edge coloring.

17 Here is an interesting game involving a type of edge coloring that you can play with a friend.

- Place 6 points on a sheet of paper to mark the vertices of a *regular hexagon*, as shown here.
- Each player selects a color different from the other.
- Take turns connecting 2 vertices with an edge. Each player should use his or her color when adding an edge.
- The first player who is forced to form a triangle of his or her own color loses. (Only triangles with vertices among the 6 starting vertices count.)

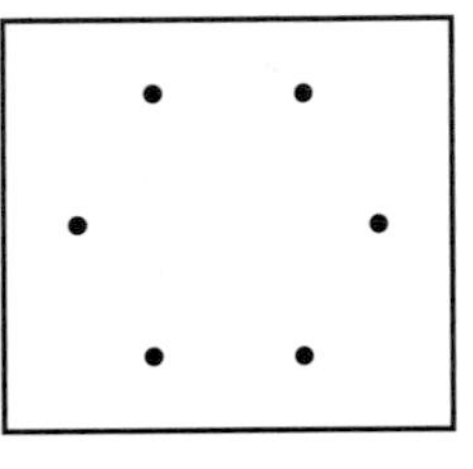

a. Play this game several times and then answer the questions below.

  i. Is there always a winner?

  ii. Which player has the better chance of winning? Explain.

b. Use the results of Part a to help you solve the following problem.

Of any 6 students who are in a room, must there be at least 3 mutual acquaintances or at least 3 mutual strangers?

18 The Sierpinski Triangle is a very interesting geometric figure. If you try to draw it, you will never finish. That's because it is defined by an infinitely repetitive set of instructions. Here are the instructions.

Step 1: Draw an equilateral triangle.

Step 2: Find the midpoint of each side.

Step 3: Connect the midpoints. This will subdivide the triangle into 4 smaller triangles.

Step 4: Remove the center triangle. (Don't actually cut it out, just think about it as being removed. If you wish, you can shade it with a pencil to remind yourself it has been "removed.") Now there are 3 smaller triangles left.

Step 5: Repeat Steps 2–4 with each of the remaining triangles. Continue this process with successively smaller triangles. The first two passes through the instructions are illustrated at the top of the next page.

Sierpinski Triangle quilt made by Diana Venters.

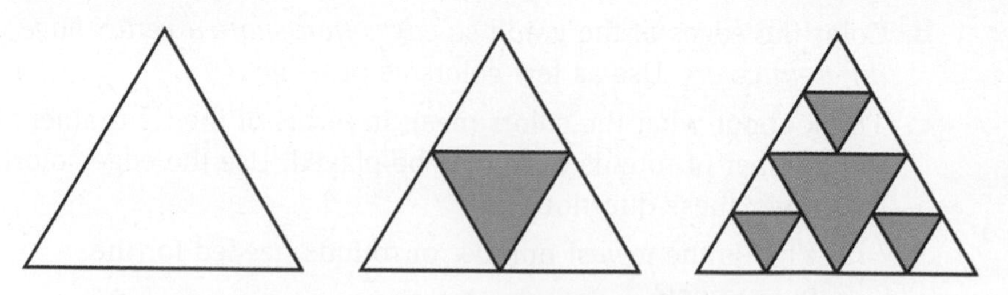

If you continue this process, you never get finished with these instructions because there always will be smaller and smaller triangles to subdivide.

a. On an enlarged copy of the third stage, draw the next stage of the process.

b. Stretch your imagination and think of the Sierpinski Triangle as a map, where the countries are the triangles that don't get removed. What is the minimum number of colors needed to color the map?

Review

19 Each table below represents a linear relation. Complete the table and find the equation of the line.

a.

x	0	1	2	3	4	5	6
y	0			7			

b.

x	0	1	2	3	4	5	6
y	4			7			

c.

x	0	1	2	3	4	5	6
y				7			7

d.

x	0	1	2	3	4	5	6
y				7		3	

20 Try to answer these questions without the use of a calculator. Think about how your answer for one part can help you determine an answer for another part.

a. What percent is 80 of 800?

b. What percent is 8 of 800?

c. What percent is 0.8 of 800?

d. What percent is 4 of 800?

e. What percent is 1 of 800?

f. What percent is 0.5 of 800?

21 Solve $3(x - 1) = 7x + 5$ by any method. Explain your method.

22 Donna wants to buy a painting that regularly sells for a price of \$55 but is on sale for 20% off. If the sales tax is 7%, how much money will Donna need in order to buy the painting?

23 Sketch a graph of each equation.

a. $y = 3x + 4$ b. $y = -\frac{2}{3}x + 6$

c. $y = -2 + \frac{1}{4}x$ d. $y = 3$

24 Without using a calculator, find the value of each expression.

a. $-5^2 + 10$ b. $3^3 - 2^3$

c. $(-2)^2 - 4(-5)$ d. $6(4)^3$

e. $8(0.5)^2$ f. $\frac{3(2^3)}{6}$

g. $\frac{\sqrt{81}}{3}$ h. $(-2)^3 + \sqrt[3]{8}$

25 Calculate the area of each shape.

a. a right triangle that has a base of 7 inches and a height of 4 inches

b. a parallelogram with length 8 cm and height 5 cm

c. a square with perimeter 64 feet

26 Assume that the polygons in each pair shown below are similar with corresponding sides and angles as suggested by the diagrams. Find the unknown side lengths and angle measurements *x, y, z, p, w,* and *t*.

a.

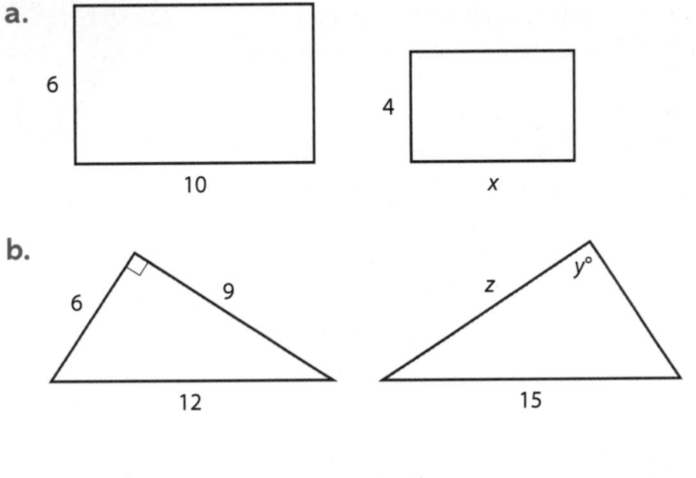

c.

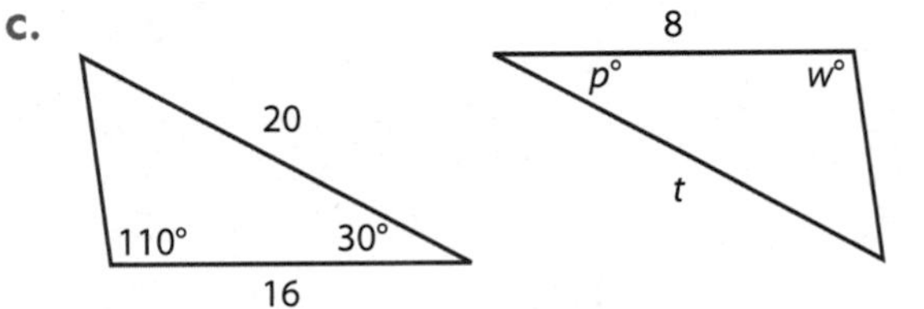

LESSON 3

Looking Back

In this unit, you have studied a type of geometric diagram consisting of vertices and edges called a *graph*, or sometimes *vertex-edge graph*. The essential characteristic of these graphs is the relationship among the vertices, as defined by how the edges connect the vertices. These vertex-edge graphs can be used as models to help understand and solve many interesting types of problems.

You have used Euler circuits and vertex coloring to find optimum circuits and to manage conflicts in a variety of settings. The tasks in this final lesson will help you review and organize your thinking about the use of vertex-edge graphs as mathematical models.

1. One city's Department of Sanitation organizes garbage collection by setting up precise garbage truck routes. Each route takes one day. Some sites that need garbage collection more often are on more than one route. However, if a site is on more than one route, the routes should not visit that site on the same day. Here is a list of routes and the sites on each route that are also on other routes.

   Route 1: Site A, Site C

   Route 2: Site D, Site A, Site F

   Route 3: Site C, Site D, Site G

   Route 4: Site G

   Route 5: Site B, Site F

   Route 6: Site D

   Route 7: Site C, Site F, Site B

   a. Can all 7 routes be scheduled in one week (Monday–Friday)? What is the fewest number of days needed to schedule all 7 routes?

   b. Set up a schedule for the garbage truck routes, showing which routes run on which day of the week.

2 The security guard for an office building must check the building several times throughout the night. The diagrams below are the floor plans for office complexes on two floors of the building. An outer corridor surrounds each office complex. In order to check the electronic security system completely, the guard must pass through each door at least once.

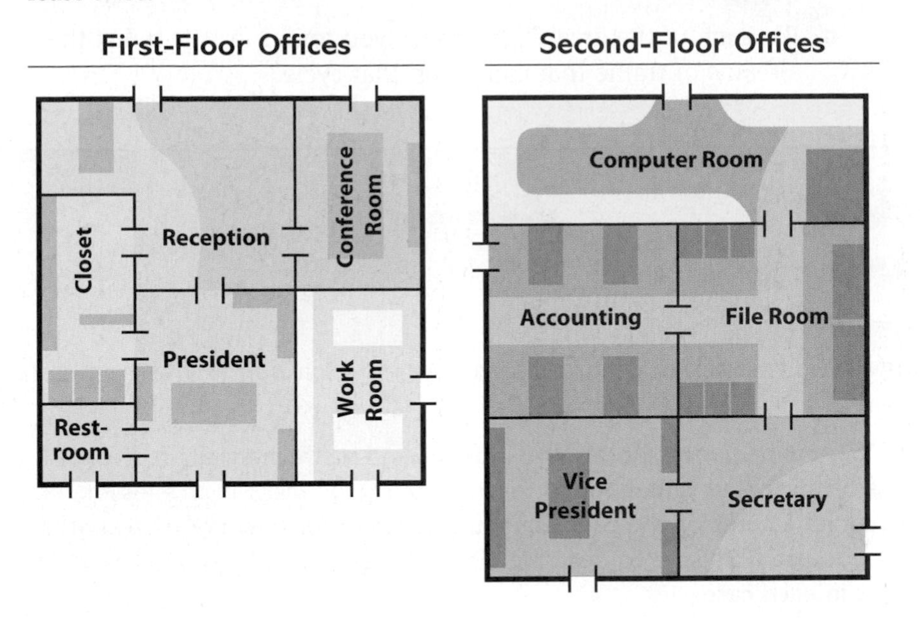

a. For each office complex, can the guard walk through each door exactly once, starting and ending in the outer corridor? If so, show the route the guard could take. If not, explain why not.

b. If it is not possible to walk through each door exactly once starting and ending in the outer corridor, what is the fewest number of doors that need to be passed through more than once? Show a route the guard should take. Indicate the doors that are passed through more than once.

c. Construct an adjacency matrix for the graph modeling the first-floor offices problem. Explain how to use the matrix to solve the problem.

3 Traffic lights are essential for controlling the flow of traffic on city streets, but nobody wants to wait at a light any longer than necessary. Consider the intersection diagrammed below. The arrows show the streams of traffic. There is a set of traffic lights in the center of the intersection.

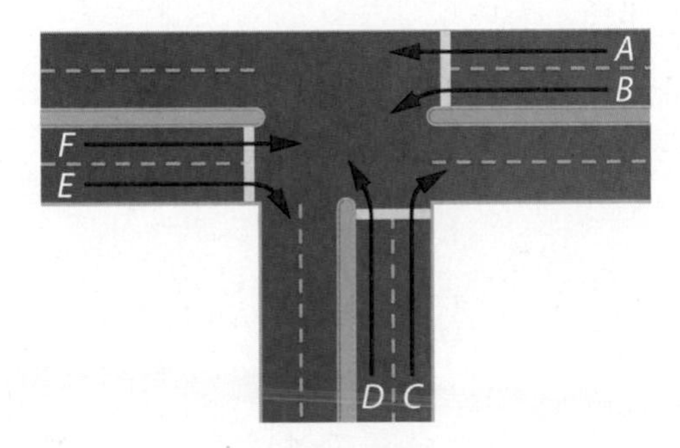

a. Can traffic streams B and D have a green light at the same time? How about B and C? List all the traffic streams that conflict with B.

b. Streams of traffic that have a green light at the same time are said to be on the same green-light cycle. What is the fewest number of green-light cycles necessary to safely accommodate all 6 streams of traffic?

c. For each of the green-light cycles you found in Part b, list the streams of traffic that can be on that cycle.

Summarize the Mathematics

In this unit, you have used vertex-edge graphs as mathematical models to help solve a variety of problems.

a. When constructing a mathematical model, you look for and mathematically represent the essential features of a problem situation. For each of the three tasks in this lesson, describe the essential features of the problem situation and how they are represented in the graph model you used. Be sure to describe what the vertices, edges, and colors (if needed) represent in each case.

b. Key mathematical topics in this unit are Euler circuits and vertex coloring.

   i. What is an Euler circuit?
   ii. How can you tell if a graph has an Euler circuit?
   iii. Describe the types of problems that can be solved with Euler circuits.
   iv. Describe what it means to "color the vertices of a graph."
   v. Describe the types of problems that can be solved by vertex coloring.

Be prepared to share your descriptions and reasoning with the class.

Check Your Understanding

Write, in outline form, a summary of the important mathematical concepts and methods developed in this unit. Organize your summary so that it can be used as a quick reference in future units and courses.

EXPONENTIAL FUNCTIONS

In everyday conversation, the phrase "growing exponentially" is used to describe any situation where some quantity is increasing rapidly with the passage of time. But in mathematics, the terms *exponential growth* and *exponential decay* refer to particular important patterns of change.

For example, when wildlife biologists estimated the population of gray wolves in Michigan, Wisconsin, and Minnesota, they found it growing exponentially—at an annual rate of about 25% from a base of about 170 wolves in 1990 to about 3,100 wolves in 2003.

In this unit, you will develop understanding and skill required to study patterns of change like growth of the midwestern gray wolf population and decay of medicines in the human body.

The key ideas and strategies for studying those patterns will be developed in two lessons.

Lessons

1 *Exponential Growth*

Recognize situations in which variables grow exponentially over time. Write *NOW-NEXT* and "$y = \ldots$" rules that express those patterns of change. Use tables, graphs, and spreadsheets to solve problems related to exponential growth. Use properties of integer exponents to write exponential expressions in useful equivalent forms.

2 *Exponential Decay*

Recognize and solve problems in situations where variables decline exponentially over time. Use properties of fractional exponents to write exponential expressions in useful equivalent forms.

Exponential Growth

In the popular book and movie *Pay It Forward*, 12-year-old Trevor McKinney gets a challenging assignment from his social studies teacher.

Think of an idea for world change, and put it into practice!

Trevor came up with an idea that fascinated his mother, his teacher, and his classmates.

He suggested that he would do something really good for three people. Then when they ask how they can pay him back for the good deeds, he would tell them to "pay it forward"—each doing something good for three other people.

Trevor figured that those three people would do something good for a total of nine others. Those nine would do something good for 27 others, and so on. He was sure that before long there would be good things happening to billions of people all around the world.

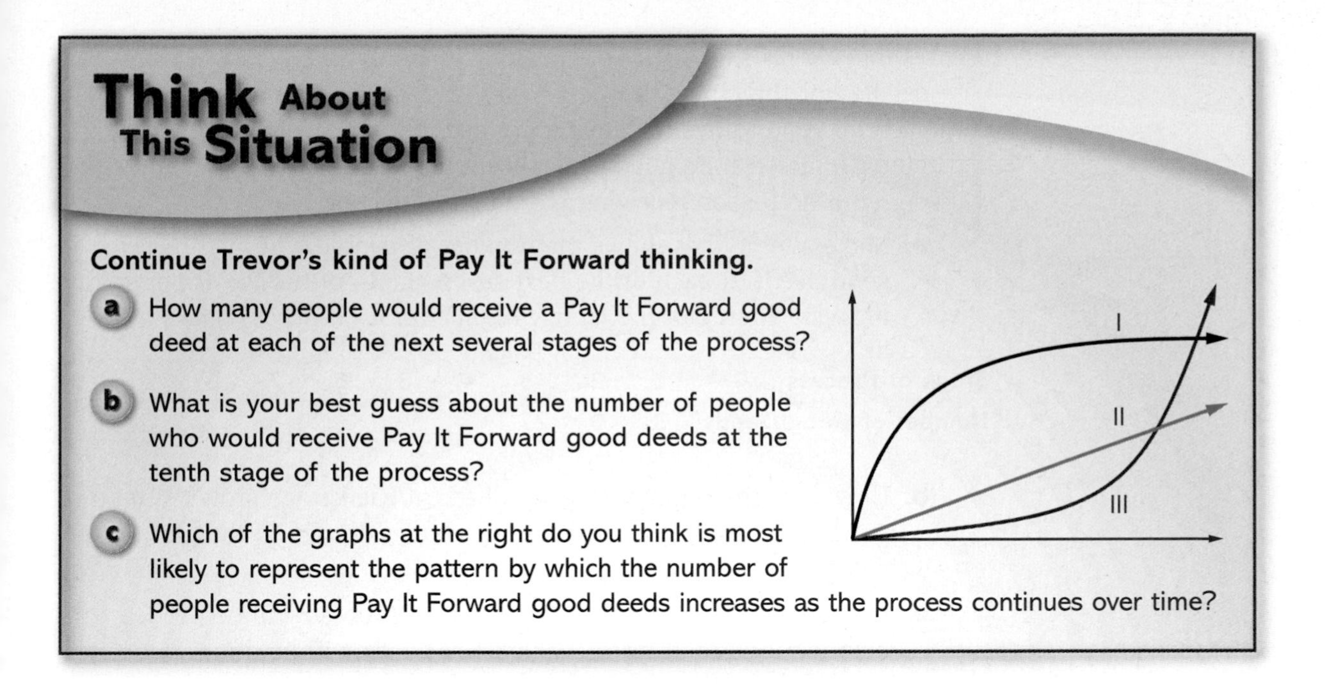

Think About This Situation

Continue Trevor's kind of Pay It Forward thinking.

a How many people would receive a Pay It Forward good deed at each of the next several stages of the process?

b What is your best guess about the number of people who would receive Pay It Forward good deeds at the tenth stage of the process?

c Which of the graphs at the right do you think is most likely to represent the pattern by which the number of people receiving Pay It Forward good deeds increases as the process continues over time?

In this lesson, you will discover answers to questions like these and find strategies for analyzing patterns of change called *exponential growth*. You will also discover some basic properties of exponents that allow you to write exponential expressions in useful equivalent forms.

Investigation 1 Counting in Tree Graphs

The number of good deeds in the Pay It Forward pattern can be represented by a *tree graph* that starts like this:

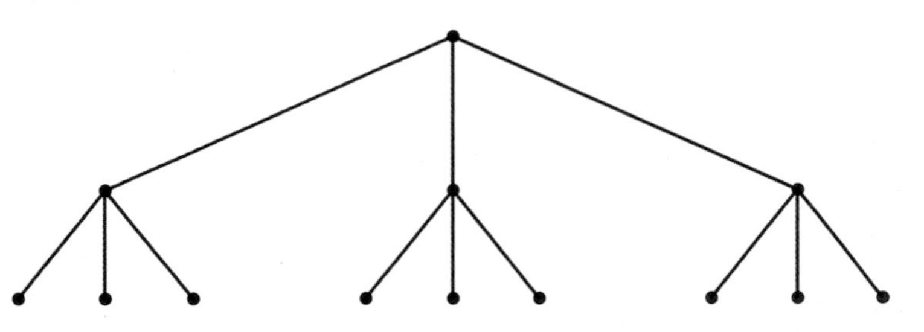

The vertices represent the people who receive and do good deeds. Each edge represents a good deed done by one person for another. As you work on the problems of this investigation, look for answers to these questions:

What are the basic patterns of exponential growth in variations of the Pay It Forward process?

How can those patterns be expressed with symbolic rules?

1. At the start of the Pay It Forward process, only one person does good deeds—for three new people. In the next stage, the three new people each do good things for three more new people. In the next stage, nine people each do good things for three more new people, and so on, with no person receiving more than one good deed.

   a. Make a table that shows the number of people who will receive good deeds at each of the next seven stages of the Pay It Forward process. Then plot the (*stage, number of good deeds*) data.

Stage of Process	1	2	3	4	5	6	7	8	9	10
Number of Good Deeds	3	9	27							

   b. How does the number of good deeds at each stage grow as the tree progresses? How is that pattern of change shown in the plot of the data?

   c. How many stages of the Pay It Forward process will be needed before a total of at least 25,000 good deeds will be done?

2. Consider now how the number of good deeds would grow if each person touched by the Pay It Forward process were to do good deeds for only two other new people, instead of three.

   a. Make a tree graph for several stages of this Pay It Forward process.

   b. Make a table showing the number of good deeds done at each of the first 10 stages of the process and plot those sample (*stage, number of good deeds*) values.

   c. How does the number of good deeds increase as the Pay It Forward process progresses in stages? How is that pattern of change shown in the plot of the data?

   d. How many stages of this process will be needed before a total of 25,000 good deeds will have been done?

3. In the two versions of Pay It Forward that you have studied, you can use the number of good deeds at one stage to calculate the number at the next stage.

   a. Use the words *NOW* and *NEXT* to write rules that express the two patterns.

   b. How do the numbers and calculations indicated in the rules express the patterns of change in tables of (*stage, number of good deeds*) data?

   c. Write a rule relating *NOW* and *NEXT* that could be used to model a Pay It Forward process in which each person does good deeds for four other new people. What pattern of change would you expect to see in a table of (*stage, number of good deeds*) data for this Pay It Forward process?

4. What are the main steps (not keystrokes) required to use a calculator to produce tables of values like those you made in Problems 1 and 2?

5 It is also convenient to have rules that will give the number of good deeds N at any stage x of the Pay It Forward process, without finding all the numbers along the way to stage x. When students in one class were given the task of finding such a rule for the process in which each person does three good deeds for others, they came up with four different ideas:

$$N = 3x$$
$$N = x + 3$$
$$N = 3^x$$
$$N = 3x + 1$$

a. Are any of these rules for predicting the number of good deeds N correct? How do you know?

b. How can you be sure that the numbers and calculations expressed in the correct "$N = ...$" rule will produce the same results as the *NOW-NEXT* rule you developed in Problem 3?

c. Write an "$N = ...$" rule that would show the number of good deeds at stage number x if each person in the process does good deeds for two others.

d. Write an "$N = ...$" rule that gives the number of good deeds at stage x if each person in the process does good deeds for four others.

Summarize the Mathematics

Look back at the patterns of change in the number of good deeds in the different Pay It Forward schemes—three per person and two per person.

a Compare the processes by noting similarities and differences in:

- **i.** Patterns of change in the tables of (*stage*, *number of good deeds*) data;
- **ii.** Patterns in the graphs of (*stage*, *number of good deeds*) data;
- **iii.** The rules relating *NOW* and *NEXT* numbers of good deeds; and
- **iv.** The rules expressing number of good deeds N as a function of stage number x.

b Compare patterns of change in numbers of good deeds at each stage of the Pay It Forward process to those of linear functions that you have studied in earlier work.

- **i.** How are the *NOW-NEXT* rules similar, and how are they different?
- **ii.** How are the "$y = ...$" rules similar, and how are they different?
- **iii.** How are the patterns of change in tables and graphs of linear functions similar to those of the Pay It Forward examples, and how are they different?

Be prepared to share your ideas with the rest of the class.

Check Your Understanding

The patterns in spread of good deeds by the Pay It Forward process occur in other quite different situations. For example, when bacteria infect some part of your body, they often grow and split into pairs of genetically equivalent cells over and over again.

a. Suppose a single bacterium lands in a cut on your hand. It begins spreading an infection by growing and splitting into two bacteria every 20 minutes.

i. Complete a table showing the number of bacteria after each 20-minute period in the first three hours. (Assume none of the bacteria are killed by white blood cells.)

Number of 20-min Periods	1	2	3
Bacteria Count	2	4	

ii. Plot the (*number of time periods, bacteria count*) values.

iii. Describe the pattern of growth of bacteria causing the infection.

b. Use *NOW* and *NEXT* to write a rule relating the number of bacteria at one time to the number 20 minutes later. Then use the rule to find the number of bacteria after fifteen 20-minute periods.

c. Write a rule showing how the number of bacteria N can be calculated from the number of stages x in the growth and division process.

d. How are the table, graph, and symbolic rules describing bacteria growth similar to and different from the Pay It Forward examples? How are they similar to, and different from, typical patterns of linear functions?

Investigation 2 Getting Started

The patterns of change that occur in counting the good deeds of a Pay It Forward scheme and the growing number of bacteria in a cut are examples of *exponential growth*. Exponential functions get their name from the fact that in rules like $N = 2^x$ and $N = 3^x$, the independent variable occurs as an exponent. As you work on the problems in this investigation, look for answers to the following questions:

What are the forms of NOW-NEXT and "y = ..." rules for basic exponential functions?

How can those rules be modified to model other similar patterns of change?

1 Infections seldom start with a single bacterium. Suppose that you cut yourself on a rusty nail that puts 25 bacteria cells into the wound. Suppose also that those bacteria divide in two after every quarter of an hour.

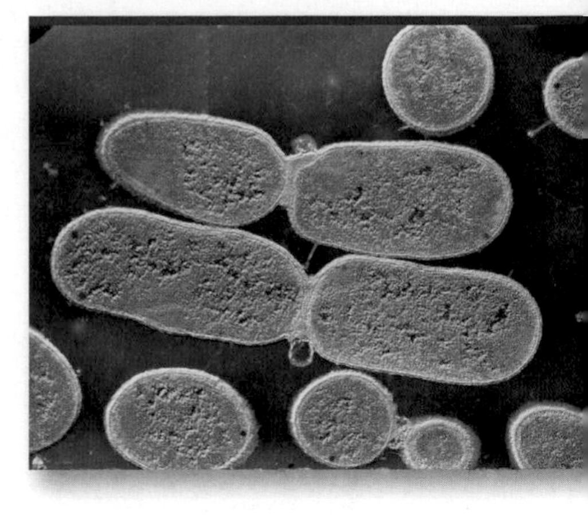

a. Make and record a guess of how many bacteria you think would be present in the cut after 8 hours (32 quarter-hours) if the infection continues to spread as predicted. (Assume that your body does not fight off the infection and you do not apply medication.) Then answer the following questions to check your ability to estimate the rate of exponential growth.

b. Complete a table showing the first several numbers in the bacteria growth pattern:

Number of Quarter-Hour Periods	0	1	2	3	4
Number of Bacteria in the Cut	25	50			

c. Use *NOW* and *NEXT* to write a rule showing how the number of bacteria changes from one quarter-hour to the next, starting from 25 at time 0.

d. Write a rule showing how to calculate the number of bacteria N in the cut after x quarter-hour time periods.

e. Use the rules in Parts c and d to calculate the number of bacteria after 8 hours. Then compare the results to each other and to your initial estimate in Part a.

2 Compare the pattern of change in this situation to the simple case that started from a single bacterium by noting similarities and differences in the:

a. tables of (*number of time periods, bacteria count*) values;

b. graphs of (*number of time periods, bacteria count*) values; and

c. *NOW-NEXT* and "$N = ...$" rules.

3 Investigate the number of bacteria expected after 8 hours if the starting number of bacteria is 30, 40, 60, or 100, instead of 25. For each starting number at time 0, complete Parts a–c. (Divide the work among your classmates.)

a. Make a table of (*number of time periods, bacteria count*) values for 8 quarter-hour time periods.

b. Write two rules that model the bacteria growth—one relating *NOW* and *NEXT* and the other beginning "$N = ...$"

c. Use each rule to find the number of bacteria after 8 hours and check that you get the same results.

d. Now compare results from two of the cases—starting at 30 and starting at 40.

  i. How are the *NOW-NEXT* and "$N = ...$" rules for bacteria counts similar, and how are they different?

  ii. How are patterns in the tables and graphs of (*number of time periods, bacteria count*) data similar, and how are they different?

Just as bacteria growth won't always start with a single cell, other exponential growth processes can start with different initial numbers. Think again about the Pay It Forward scheme in Investigation 1.

4 Suppose that four good friends decide to start their own Pay It Forward tree. To start the tree, they each do good deeds for three different people. Each of those new people in the tree does good deeds for three other new people, and so on.

a. What *NOW-NEXT* rule shows how to calculate the number of good deeds done at each stage of this tree?

b. What "$N = ...$" rule shows how to calculate the number of good deeds done at any stage x of this tree?

c. How would the *NOW-NEXT* and "$N = ...$" rules be different if the group of friends starting the tree had five members instead of four?

d. Which of the Pay It Forward schemes below would most quickly reach a stage in which 1,000 good deeds are done? Why does that make sense?

Scheme 1: Start with a group of four friends and have each person in the tree do good deeds for two different people; or

Scheme 2: Start with only two friends and have each person in the tree do good deeds for three other new people.

In studying exponential growth, it is helpful to know the *initial value* of the growing quantity. For example, the initial value of the growing bacteria population in Problem 1 was 25. You also need to know when the initial value occurs. For example, the bacteria population was 25 after 0 quarter-hour periods.

In Problem 4 on the other hand, 12 good deeds are done at Stage 1. In this context, "Stage 0" does not make much sense, but we can extend the pattern backward to reason that $N = 4$ when $x = 0$.

5 Use your calculator and the [^] key to find each of the following values: 2^0, 3^0, 5^0, 23^0.

a. What seems to be the calculator value for b^0, for any positive value of b?

b. Recall the examples of exponential patterns in bacterial growth. How do the "$N = ...$" rules for those situations make the calculator output for b^0 reasonable?

6 Now use your calculator to make tables of (x, y) values for each of the following functions. Use integer values for x from 0 to 6. Make notes of your observations and discussion of questions in Parts a and b.

i. $y = 5(2^x)$ **ii.** $y = 4(3^x)$

iii. $y = 3(5^x)$ **iv.** $y = 7(2.5^x)$

a. What patterns do you see in the tables? How do the patterns depend on the numbers in the function rule?

b. What differences would you expect to see in tables of values and graphs of the two exponential functions $y = 3(6^x)$ and $y = 6(3^x)$?

7 Suppose you are on a team studying the growth of bacteria in a laboratory experiment. At the start of your work shift in the lab, there are 64 bacteria in one petri dish culture, and the population seems to be doubling every hour.

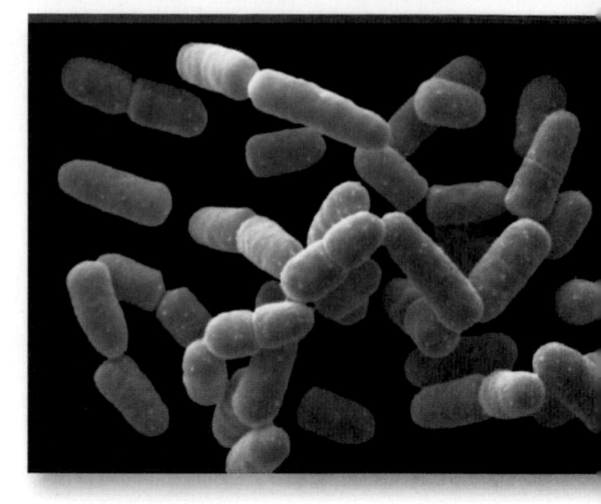

a. What rule should predict the number of bacteria in the culture at a time x hours after the start of your work shift?

b. What would it mean to calculate values of y for negative values of x in this situation?

c. What value of y would you expect for $x = -1$? For $x = -2$? For $x = -3$ and -4?

d. Use your calculator to examine a table of (x, y) values for the function $y = 64(2^x)$ when $x = 0, -1, -2, -3, -4, -5, -6$. Compare results to your expectations in Part c. Then explain how you could think about this problem of bacteria growth in a way so that the calculator results make sense.

8 Study tables and graphs of (x, y) values to estimate solutions for each of the following equations and inequalities. In each case, be prepared to explain what the solution tells about bacteria growth in the experiment of Problem 7.

a. $1{,}024 = 64(2^x)$

b. $8{,}192 = 64(2^x)$

c. $64(2^x) > 25{,}000$

d. $4 = 64(2^x)$

e. $64(2^x) < 5{,}000$

f. $64(2^x) = 32$

Summarize the Mathematics

The exponential functions that you studied in this investigation describe patterns of change in bacteria growth and numbers of people in a Pay It Forward tree. They have some features in common.

a Each *NOW-NEXT* rule fits the pattern *NEXT* = b • *NOW*, starting at a. What do the values of b and a tell about the pattern of change represented by the *NOW-NEXT* rule? How will that pattern be illustrated in a table or a graph of (x, y) values?

b Each "$y = \ldots$" rule fits the pattern $y = a(b^x)$. What do the values of a and b tell about the pattern of change represented by the rule? How will that pattern be illustrated in a table or a graph of (x, y) values?

c What is the value of b^x, when x is 0? What would this result mean in a problem situation where exponential growth is being studied?

d How would you calculate values of b^x when x is a negative number? What would those results mean in a problem situation where exponential growth is being studied?

Be prepared to explain your ideas to the entire class.

Check Your Understanding

Alexander Fleming
Discoverer of Penicillin

The drug penicillin was discovered by observation of mold growing on biology lab dishes. Suppose a mold begins growing on a lab dish. When first observed, the mold covers 7 cm^2 of the dish surface, but it appears to double in area every day.

a. What rules can be used to predict the area of the mold patch 4 days after the first measurement:

  i. using *NOW-NEXT* form?

  ii. using "$y = \ldots$" form?

b. How would each rule in Part a change if the initial mold area was only 3 cm^2?

c. How would each rule in Part a change if the area of the mold patch increased by a factor of 1.5 every day?

d. What mold area would be predicted after 5 days in each set of conditions from Parts a–c?

e. For "$y = \ldots$" rules used in calculating growth of mold area, what would it mean to calculate values of y when x is a negative number?

f. Write and solve equations or inequalities that help to answer these questions.

  i. If the area of a mold patch is first measured to be 5 cm^2 and the area doubles each day, how long will it take that mold sample to grow to an area of 40 cm^2?

  ii. For how many days will the mold patch in part i have an area less than 330 cm^2?

Investigation 3 Compound Interest

Every now and then you may hear about somebody winning a big payoff in a state lottery. The winnings can be 1, 2, 5, or even 100 million dollars. The big money wins are usually paid off in annual installments for about 20 years. But some smaller prizes are paid at once. How would you react if this news report were actually about you?

Kalamazoo Teen Wins Big Lottery Prize

A Kalamazoo teenager has just won the daily lottery from a Michigan lottery ticket that she got as a birthday gift from her uncle. In a new lottery payoff scheme, the teen has two payoff choices.

One option is to receive a single $10,000 payment now.

In the other plan, the lottery promises a single payment of $20,000 ten years from now.

---

1 Imagine that you had just won that Michigan lottery prize.

a. Discuss with others your thinking on which of the two payoff methods to choose.

b. Suppose a local bank called and said you could invest your \$10,000 payment in a special 10-year certificate of deposit (CD), earning 8% interest compounded yearly. How would this affect your choice of payoff method?

As you work on the problems of this investigation, look for answers to the question:

How can you represent and reason about functions involved in investments paying compound interest?

Of the two lottery payoff methods, one has a value of \$20,000 at the end of 10 years. The value (in 10 years) of receiving the \$10,000 payoff now and putting it in a 10-year certificate of deposit paying 8% interest compounded annually is not so obvious.

- After one year, your balance will be:
  $10{,}000 + (0.08 \times 10{,}000) = 1.08 \times 10{,}000 = \$10{,}800$.
- After the second year, your balance will be:
  $10{,}800 + (0.08 \times 10{,}800) = 1.08 \times 10{,}800 = \$11{,}664$.

During the next year, the CD balance will increase in the same way, starting from \$11,664, and so on.

2 Write rules that will allow you to calculate the balance of this certificate of deposit:

a. for the next year, using the balance from the current year.

b. after any number of years x.

3 Use the rules from Problem 2 to determine the value of the certificate of deposit after 10 years. Then decide which 10-year plan will result in more money and how much more money that plan will provide.

4 Look for an explanation of your conclusion in Problem 3 by answering these questions about the potential value of the CD paying 8% interest compounded yearly.

a. Describe the pattern of growth in the CD balance as time passes.

b. Why isn't the change in the CD balance the same each year?

c. How is the pattern of increase in CD balance shown in the shape of a graph for the function relating CD balance to time?

d. How could the pattern of increase have been predicted by thinking about the rules (*NOW-NEXT* and "$y = \ldots$") relating CD balance to time?

5 Suppose that the prize winner decided to leave the money in the CD, earning 8% interest for more than 10 years. Use tables or graphs to estimate solutions for the following equations and inequalities. In each case, be prepared to explain what the solution tells about the growth of a $10,000 investment that earns 8% interest compounded annually.

a. $10{,}000(1.08^x) = 25{,}000$

b. $10{,}000(1.08^x) = 37{,}000$

c. $10{,}000(1.08^x) = 50{,}000$

d. $10{,}000(1.08^x) \geq 25{,}000$

e. $10{,}000(1.08^x) \leq 30{,}000$

f. $10{,}000(1.08^x) = 10{,}000$

6 Compare the pattern of change and the final account balance for the plan that invests $10,000 in a CD that earns 8% interest compounded annually over 10 years to those for the following possible savings plans over 10 years. Write a summary of your findings.

a. Initial investment of $15,000 earning only 4% annual interest compounded yearly

b. Initial investment of $5,000 earning 12% annual interest compounded yearly

Summarize the Mathematics

Most savings accounts operate in a manner similar to the bank's certificate of deposit offer. However, they may have different starting balances, different interest rates, or different periods of investment.

a Describe two ways to find the value of such a savings account at the end of each year from the start to year 10. Use methods based on:

i. a rule relating *NOW* and *NEXT*.

ii. a rule like $y = a(b^x)$.

b What graph patterns would you expect from plots of (*year*, *account balance*) values?

c How would the function rules change if the interest rate changes? If the initial investment changes?

d Why does the dollar increase in the account balance get larger from each year to the next?

e How are the patterns of change that occur with the bank investment similar to and different from those of other functions that you've used while working on problems of Investigations 1 and 2? On problems of previous units?

Be prepared to explain your methods and ideas to the entire class.

Check Your Understanding

In solving change-over-time problems in Unit 1, you discovered that the world population and populations of individual countries grow in much the same pattern as money earning interest in a bank. For example, you used data like the following to predict population growth in two countries.

- Brazil is the most populous country in South America. In 2005, its population was about 186 million. It was growing at a rate of about 1.1% per year.
- Nigeria is the most populous country in Africa. Its 2005 population was about 129 million. It was growing at a rate of about 2.4% per year.

a. Assuming that these growth rates continue, write function rules to predict the populations of these countries for any number of years x in the future.

b. Compare the patterns of growth expected in each country for the next 20 years. Use tables and graphs of (*year since 2005, population*) values to illustrate the similarities and differences you notice.

c. Write and solve equations that give estimates when:

i. Brazil's population might reach 300 million.

ii. Nigeria's population might reach 200 million.

d. Assuming these growth patterns continue, estimate when the population of Nigeria will be greater than the population of Brazil.

Investigation 4 Modeling Data Patterns

In the *Patterns of Change* unit, you used data about wildlife populations to make predictions and to explore effects of protection and hunting policies. For example, you used information from studies of Midwest wolf populations to predict growth over time in that species. You used information about Alaskan bowhead whale populations and hunting rates to make similar projections into the future.

In each case, you began the prediction with information about the current populations and the growth rates as percents. It's not hard to imagine how field biologists might count wolves or whales by patient observation. But they can't observe percent growth rates, and those rates are unlikely to be constant from one year to the next. As you work on the problems of this investigation, look for answers to the following question:

What are some useful strategies for finding functions modeling patterns of change that are only approximately exponential?

1. Suppose that census counts of Midwest wolves began in 1990 and produced these estimates for several different years:

Time Since 1990 (in years)	0	2	5	7	10	13
Estimated Wolf Population	100	300	500	900	1,500	3,100

a. Plot the wolf population data and decide whether a linear or exponential function seems likely to match the pattern of growth well. For the function type of your choice, experiment with different rules to see which rule provides a good model of the growth pattern.

b. Use your calculator or computer software to find both linear and exponential regression models for the given data pattern. Compare the fit of each function to the function you developed by experimentation in Part a.

c. What do the numbers in the linear and exponential function rules from Part b suggest about the pattern of change in the wolf population?

d. Use the model for wolf population growth that you believe to be best to calculate population estimates for the missing years 1994 and 2001 and then for the years 2015 and 2020.

2. Suppose that census counts of Alaskan bowhead whales began in 1970 and produced these estimates for several different years:

Time Since 1970 (in years)	0	5	15	20	26	31
Estimated Whale Population	5,040	5,800	7,900	9,000	11,000	12,600

a. Plot the given whale population data and decide which type of function seems likely to match the pattern of growth well. For the function type of your choice, experiment with different rules to see which provides a good model of the growth pattern.

b. Use your calculator or computer software to find both linear and exponential regression models for the data pattern. Compare the fit of each function to that of the function you developed by experimentation in Part a.

c. What do the numbers in the linear and exponential function rules from Part b suggest about patterns of change in the whale population?

d. Use the model for whale population growth that you believe to be best to calculate population estimates for the years 2002, 2005, and 2010.

Summarize the Mathematics

In the problems of this investigation, you studied ways of finding function models for growth patterns that could only be approximated by one of the familiar types of functions.

a How do you decide whether a data pattern is modeled best by a linear or an exponential function?

b What do the numbers a and b in a linear function $y = a + bx$ tell about patterns in:

- **i.** the graph of the function?
- **ii.** a table of (x, y) values for the function?

c What do the numbers c and d in an exponential function $y = c(d^x)$ tell about patterns in:

- **i.** the graph of the function?
- **ii.** a table of (x, y) values for the function?

d What strategies are available for finding a linear or exponential function that models a linear or exponential data pattern?

Be prepared to share your ideas and reasoning with the class.

Check Your Understanding

Test your ideas about the connections between functions, problem conditions, and data patterns.

a. What *NOW-NEXT* and "$y = \ldots$" rules will express patterns of change in which a variable quantity is increasing:

- **i.** at a rate of 20% per year from a starting value of 750?
- **ii.** at a rate of 4.5% per month from a starting value of 35?
- **iii.** at a rate of 24 per day from a starting value of 18?

b. Write functions that provide good models for the patterns of change that relate p, q, and r to x in the following tables.

i.

x	−10	−5	0	6	15	20	30
p	1	3	5	8	12	15	18

ii.

x	−10	−5	0	6	15	20	30
q	1	8	60	650	25,000	190,000	11,000,000

iii.

x	−10	−5	0	6	15	20	30
r	1.0	1.3	1.6	2.25	3.4	4.4	7.0

Investigation 5 Properties of Exponents I

In solving the problems in Investigations 1–4, you focused on functions modeling exponential growth. You used what you knew about the problem situations to guide development of the function models, to plan calculations that would answer the given questions, and to interpret information in rules, tables, and graphs. For example, you developed and used the rules $B = 2^x$ and $B = 25(2^x)$ to study the pattern of bacteria growth in a cut.

Since doubling occurs so often in questions about exponential growth, it is helpful to know some basic powers of 2, like $2^2 = 4$, $2^3 = 8$, $2^4 = 16$, ... , $2^9 = 512$, and $2^{10} = 1{,}024$. When students in one Wisconsin class had memorized those facts, someone suggested reasoning about even higher powers like this:

Since 2^{10} is about 1,000: the value of 2^{11} should be about 2,000,
the value of 2^{12} should be about 4,000,
the value of 2^{13} should be about 8,000,
⋮
the value of 2^{20} should be about 1,000,000.

How do you suppose the student was thinking about exponents to come up with that estimation strategy?

To develop and test strategies for working with exponential expressions, it helps to know some basic methods of writing these expressions in useful equivalent forms. Remember that the starting point in work with exponents is an expression like b^n, where b is any real number and n is any non-negative integer. The number b is called the **base** of the exponential expression, and n is called the **exponent** or the **power**.

$$b^n = b \cdot b \cdot b \cdot \cdots \cdot b \ (n \text{ factors}) \quad \text{and} \quad b^0 = 1 \ (\text{for } b \neq 0)$$

As you work on the problems in this investigation, look for answers to this question:

How can the above definition of exponent be used to discover and justify other properties of exponents that make useful algebraic manipulations possible?

Products of Powers Work with exponents is often helped by writing products like $b^x \cdot b^y$ in simpler form or by breaking a calculation like b^z into a product of two smaller numbers.

1 Find values for w, x, and y that will make these equations true statements:

a. $2^{10} \cdot 2^3 = 2^y$

b. $5^2 \cdot 5^4 = 5^y$

c. $3 \cdot 3^7 = 3^y$

d. $2^w \cdot 2^4 = 2^7$

e. $b^4 \cdot b^2 = b^y$

f. $9^w \cdot 9^x = 9^5$

2 Examine the results of your work on Problem 1.

a. What pattern seems to relate task and result in every case?

b. How would you use the definition of exponent or other reasoning to convince another student that your answer to Part a is correct?

3 When people work with algebraic expressions that involve exponents, there are some common errors that slip into the calculations. How would you help other students correct their understanding of operations with exponents if their work showed the following errors?

a. $3^5 = 15$

b. $3^4 \cdot 5^2 = 15^8$

c. $3^4 \cdot 5^2 = 15^6$

d. $3^4 + 3^2 = 3^6$

Power of a Power You know that $8 = 2^3$ and $64 = 8^2$, so $64 = (2^3)^2$. As you work on the next problems, look for a pattern suggesting how to write equivalent forms for expressions like $(b^x)^y$ that involve powers of powers.

4 Find values for x and z that will make these equations true statements:

a. $(2^3)^4 = 2^z$

b. $(3^5)^2 = 3^z$

c. $(5^2)^x = 5^6$

d. $(b^2)^5 = b^z$

5 Examine the results of your work on Problem 4.

a. What pattern seems to relate task and result in every case?

b. How would you use the definition of exponent or other reasoning to convince another student that your answer to Part a is correct?

c. What would you expect to see as common errors in evaluating a power of a power like $(4^3)^2$? How would you help someone who made those errors correct their understanding of how exponents work?

Power of a Product The area of a circle can be calculated from its radius r using the formula $A = \pi r^2$. It can be calculated from the diameter using the formula $A = \pi(0.5d)^2$. Next search for a pattern showing how powers of products like $(0.5d)^2$ can be expressed in equivalent forms.

6 Find values for x and y that will make these equations true statements:

a. $(6 \cdot 11)^3 = 6^x \cdot 11^y$

b. $(3\pi)^4 = 3^x \cdot \pi^y$

c. $(2m)^3 = 2^x m^y$

d. $(m^3p)^2 = m^x p^y$

7 Examine the results of your work on Problem 6.

a. What pattern seems to relate task and result in every case?

b. How would you use the definition of exponent or other reasoning to convince another student that your answer to Part a is correct?

c. What would you expect to see as the most common errors in evaluating a power of a product like $(4t)^3$? How would you help someone who made those errors correct their understanding of how exponents work?

Summarize the Mathematics

The problems of this investigation asked you to formulate, test, and justify several principles that allow writing of exponential expressions in convenient equivalent forms.

a For each of the properties of exponents you explored in Problems 1–7, how would you explain the property in words that describe the relationship between two equivalent forms of exponential expressions?

b Summarize the properties of exponents you explored in Problems 1–7 by completing these statements to show equivalent forms for exponential expressions:

i. $b^m \cdot b^n = \ldots$ **ii.** $(b^m)^n = \ldots$ **iii.** $(ab)^n = \ldots$

c What examples would you use to illustrate common errors in use of exponents, and how would you explain the errors in each example?

Be prepared to share your explanations and reasoning with the class.

Check Your Understanding

Use properties of exponents to write each of the following expressions in another equivalent form. Be prepared to explain how you know your answers are correct.

a. $(y^3)(y^6)$

b. $(5x^2y^4)(2xy^3)$

c. $(pq)^3$

d. $(p^3)^5$

e. $(7p^3q^2)^2$

On Your Own

Applications

1 Imagine a tree that each year grows 3 new branches from the end of each existing branch. Assuming that your tree is a single stem when it is planted:

a. How many new branches would you expect to appear in the first year of new growth? How about in the second year of new growth?

b. Write a rule that relates the number of new branches B to the year of growth R.

c. In what year will the number of new branches first be greater than 15,000?

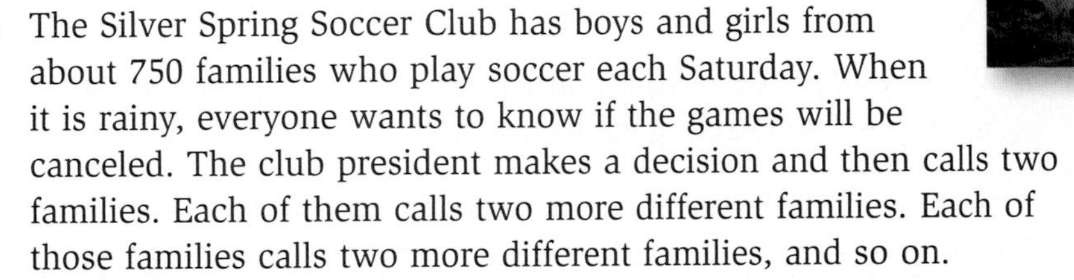

2 The Silver Spring Soccer Club has boys and girls from about 750 families who play soccer each Saturday. When it is rainy, everyone wants to know if the games will be canceled. The club president makes a decision and then calls two families. Each of them calls two more different families. Each of those families calls two more different families, and so on.

a. Sketch a tree graph that shows how the number of people called grows in stages from the first calls by the club president. What do the vertices of the tree graph represent? What do the edges represent?

b. Make a table and a graph showing the number of calls made at each of the first 10 stages of this calling tree.

c. Write two rules that can be used to calculate the number of calls made at various stages of this calling tree—one in *NOW-NEXT* form and another in "$y = ...$" form.

d. How many stages of the calling tree will be needed before all 750 families are contacted?

3 The bacteria *E. coli* often cause illness among people who eat the infected food. Suppose a single *E. coli* bacterium in a batch of ground beef begins doubling every 10 minutes.

a. How many bacteria will there be after 10, 20, 30, 40, and 50 minutes have elapsed? Assume no bacteria die.

b. Write two rules that can be used to calculate the number of bacteria in the food after any number of 10-minute periods—one using *NOW* and *NEXT*, and another beginning "$y = ...$"

c. Use your rules to make a table showing the number of *E. coli* bacteria in the batch of ground beef at the end of each 10-minute period over 2 hours. Then describe the pattern of change in the number of bacteria from each time period to the next.

d. Find the predicted number of bacteria after 4, 5, and 6 hours.

On Your Own

4 The left figure shown below is called a "chair." It can be subdivided into four congruent, smaller "chairs" as shown at the right. Each of the smaller chairs can be subdivided into four congruent, still smaller chairs, and this process can be continued.

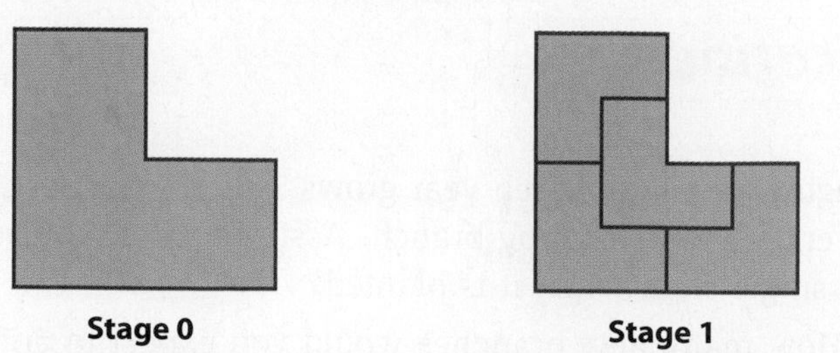

Stage 0 Stage 1

a. Draw a picture of Stage 2 in the process that creates smaller "chairs" and count the number of small chairs at this stage.

b. Make a table that shows the number of small chairs at each stage of the process.

Stage	0	1	2	3	4	5	...	n
Number of "Chairs"	1	4						

c. Write a *NOW-NEXT* rule that shows how the number of chairs increases from each stage to the next.

5 Suppose that the Silver Spring Soccer Club has a meeting of the four club directors to decide on whether or not to cancel a scheduled game. Then the directors each start a branch of a calling tree by calling three families, and each of those families then calls three more families. This process continues until all 750 families are contacted.

a. Sketch a tree graph that shows how the number of people called grows in stages from the first calls by the club directors.

b. Make a table and a graph showing the number of calls made at each of the first 4 stages of this calling tree.

c. Write two rules that can be used to calculate the number of calls made at various stages of this calling tree—one in *NOW-NEXT* form and another in "$y = ...$" form.

d. How many stages of the calling tree will be needed before all 750 families are contacted?

6 Suppose 50 *E. coli* bacteria are introduced into some food as it's being processed, and the bacteria begin doubling every 10 minutes.

a. Make a table and a graph showing the number of bacteria from Stage 0 to Stage 6 of the infection process.

b. Write two rules that can be used to calculate the number of bacteria infecting the food at various stages of this process—one in *NOW-NEXT* form and another in "$y = \ldots$" form.

c. Predict the number of bacteria present after 3 hours. Explain how you made your prediction.

7 Suppose that a local benefactor wants to offer college scholarships to every child entering first grade at an elementary school in her community. For each student, the benefactor puts $5,000 in a separate savings fund that earns 5% interest compounded annually.

a. Make a table and a graph to show growth in the value of each account over the 12 years leading up to college entry.

b. Compare the pattern of growth of the account in Part a to one in which the initial deposit is $10,000. Compare values of each account after 12 years.

c. Compare the pattern of growth of the account in Part a to one in which the interest rate is 10% and the initial deposit is $5,000. Compare values of each account after 12 years.

d. Compare values of the accounts in Parts b and c after 12 years. What does this suggest about the relative importance of interest rate and initial balance in producing growth of an investment earning compound interest?

8 In 2000, the number of people worldwide living with HIV/AIDS was estimated at more than 36 million. That number was growing at an annual rate of about 15%.

a. Make a table showing the projected number of people around the world living with HIV/AIDS in each of the ten years after 2000, assuming the growth rate remains 15% per year.

b. Write two different kinds of rules that could be used to estimate the number of people living with HIV/AIDS at any time in the future.

c. Use the rules from Part b to estimate the number of people living with HIV/AIDS in 2015.

d. What factors might make the estimate of Part c an inaccurate forecast?

On Your Own

9 Studies in 2001 gave a low estimate of 7,700 for the population of Arctic bowhead whales. The natural annual growth rate was estimated to be about 3%. The harvest by Inuit people is very small in relation to the total population. Disregard the harvest for this task.

a. If the growth rate continued at 3%, what populations would be predicted for each year to 2010, using the low 2001 population estimate?

b. Which change of assumptions will lead to a greater 2010 whale population estimate

i. increasing the assumed population annual growth rate to 6%, with the 2001 low population estimate of 7,700, or

ii. increasing the 2001 population estimate to 14,400, but maintaining the 3% growth rate?

c. Find the time it takes for the whale population to double under each of the three sets of assumptions in Parts a and b.

10 The Dow Jones Industrial Average provides one measure of the "health" of the U.S. economy. It is a weighted average of the stock prices for 30 major American corporations. The following table shows the low point of the Dow Jones Industrial Average in selected years from 1965 to 2005.

Year	DJIA Low
1965	840
1970	631
1975	632
1980	759
1985	1,185
1990	2,365
1995	3,832
2000	9,796
2005	10,012

Source: www.analyzeindices.com/dow-jones-history.shtml

a. Find what you believe are the best possible linear and exponential models for the pattern of change in the low value of the Dow Jones Industrial Average over the time period shown in the table (use $t = 0$ to represent 1965). Then decide which you think is the better of the two models and explain your choice.

b. Use your chosen predictive model from Part a to estimate the low value of this stock market average in 2010 and 2015. Explain why you might or might not have confidence in those estimates.

c. Some stockbrokers who encourage people to invest in common stocks claim that one can expect an average return of 10% per year on that investment. Does the rule you chose to model increase in the Dow Jones average support that claim? Why or why not?

11 The following table shows the number of votes cast in a sample of U.S. Presidential elections between 1840 and 2004.

Year of Election	Major Party Candidates	Total Votes Cast
1840	Harrison vs. Van Buren	2,411,118
1860	Lincoln vs. Douglas	4,685,030
1880	Garfield vs. Hancock	9,218,951
1900	McKinley vs. Bryan	14,001,733
1920	Harding vs. Cox	26,757,946
1940	Roosevelt vs. Wilkie	49,752,978
1960	Kennedy vs. Nixon	68,836,385
1980	Reagan vs. Carter	86,515,221
2000	Bush vs. Gore	105,405,100
2004	Bush vs. Kerry	122,267,553

Source: en.wikipedia.org

a. Find rules for what you think are the best possible linear and exponential models of the trend relating votes cast to time (use $t = 0$ to represent the year 1840).

b. Which type of model—linear or exponential—seems to better fit the data pattern? Why do you think that choice is reasonable?

c. In what ways is neither the linear nor the exponential model a good fit for the data pattern relating presidential election votes to time? Why do you think that modeling problem occurs?

12 In 1958, Walter O'Malley paid about $700,000 to buy the Brooklyn Dodgers baseball team. He moved the team to Los Angeles, and in 1998 his son and daughter sold the team for $350,000,000. Assume that the team's value increased exponentially in annual increments according to a rule like $v = a(b^t)$, where $t = 0$ represents the year 1958.

a. What value of a is suggested by the given information?

b. Experiment to find a value of b that seems to give a rule that matches growth in team value prescribed by the given information.

c. What annual percent growth rate does your answer to Part b suggest for the value of the Dodgers team business?

d. According to the model derived in Parts a and b, when did the value of the Dodgers team first reach $1,000,000? $10,000,000? $100,000,000?

13 Find values for w, x, and y that will make these equations true statements.

a. $5^4 \cdot 5^5 = 5^y$ **b.** $3^6 \cdot 3^4 = 3^y$ **c.** $5^3 \cdot 5 = 5^y$

d. $7^w \cdot 7^6 = 7^{11}$ **e.** $1.5^w \cdot 1.5^x = 1.5^6$ **f.** $c^3 \cdot c^5 = c^y$

14 Write each of the following expressions in a simpler equivalent exponential form.

a. $7^4 \cdot 7^9$ b. $4.2^2 \cdot 4.2^5$ c. $x \cdot x^4$

d. $(c^2)(c^5)$ e. $(5x^3y^4)(4x^2y)$ f. $(7a^3bm^5)(b^4m^2)$

g. $(4x^3y^5)(10x)$ h. $(-2c^4d^2)(-cd)$

15 Find values for x, y, and z that will make these equations true statements.

a. $(7^5)^2 = 7^z$ b. $(4.5^2)^3 = x^6$ c. $(9^3)^x = 9^{12}$

d. $(t^3)^7 = t^z$ e. $(7 \cdot 5)^4 = 7^x \cdot 5^y$ f. $(3t)^4 = 3^x t^y$

g. $(5n^3)^2 = 5^x n^y$ h. $(c^5d^3)^2 = c^x d^y$

16 Write each of the following expressions in a simpler equivalent exponential form.

a. $(x^2)^3$ b. $(5a^3c^4)^2$ c. $(3xy^4z^2)^4$ d. $(-5x^3)^2$

Connections

17 Partially completed tables for four relations between variables are given below. In each case, decide if the table shows an exponential or a linear pattern of change. Based on that decision, complete a copy of the table as the pattern suggests. Then write rules for the patterns in two ways: using rules relating *NOW* and *NEXT* y values and using rules beginning "$y = \ldots$" for any given x value.

a.

x	0	1	2	3	4	5	6	7	8
y				8	16	32			

b.

x	0	1	2	3	4	5	6	7	8
y				40	80	160			

c.

x	0	1	2	3	4	5	6	7	8
y				48	56	64			

d.

x	0	1	2	3	4	5	6	7	8
y				125	625	3,125			

18 The diagram below shows the first stages in the formation of a geometric figure called a Koch curve. This figure is an example of a *fractal*. At each stage in the growth of the figure, the middle third of every segment is replaced by a "tent" formed by two equal-length segments. The new figure is made up of more, but shorter, segments of equal length.

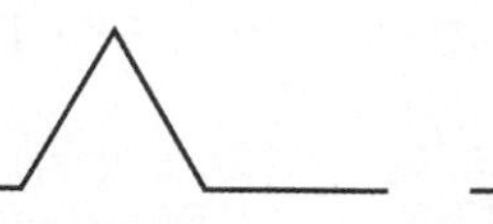

a. Make a sketch showing at least one more stage in the growth of this fractal. Describe any symmetries that the fractal has at *each* stage.

b. Continue the pattern begun in this table:

Stage of Growth	0	1	2	3	4	5	6	7
Segments in Design	1	4						

c. Write a rule showing how the number of segments at any stage of the fractal can be used to find the number of segments at the next stage.

d. Write a rule that can be used to find the number of segments in the pattern at any stage x, without finding the numbers at each stage along the way. Begin your rule, "$y = \ldots$."

e. Use the rule from Part d to produce a table and a graph showing the number of segments in the fractal pattern at each of the first 15 stages of growth. At what stage will the number of segments in the fractal first reach or pass 1 million?

19 News stories spread rapidly in modern society. With broadcasts over television, radio, and the Internet, millions of people hear about important events within hours. The major news providers try hard to report only stories that they know are true. But quite often rumors get started and spread around a community by word of mouth alone.

Suppose that to study the spread of information through rumors, two students started this rumor at 5 P.M. one evening: "Because of the threat of a huge snowstorm, there will be no school tomorrow and probably for the rest of the week." The next day they surveyed students at the school to find out how many heard the rumor and when they heard it.

a. What pattern of rumor spread is suggested by each of the graphs below?

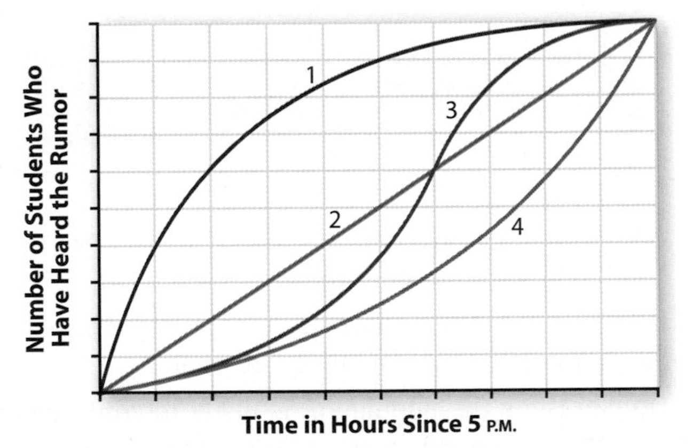

b. Which pattern of change in number of students who have heard the rumor is most likely to match experimental results in case:

i. the rumor is spread by word of mouth from one student to another?

ii. the rumor is mentioned on radio and television broadcasts between 5 and 6 P.M.?

20 On one rainy day, the Silver Spring Soccer Club president decides that she will call all 750 member families herself to tell them about cancellation of play. She figures that she can make 3 calls per minute.

a. How long will it take her to notify all families in the club?

b. Look back at your results from work on Applications Task 2 and estimate the time it would take to inform all families of the cancellation if that calling tree was used instead.

21 Exponential functions, like linear functions, can be expressed by rules relating x and y values and by rules relating *NOW* and *NEXT* y values when x increases in steps of 1. Compare the patterns of (x, y) values produced by these functions: $y = 2(3^x)$ and $y = 2 + 3x$ by completing these tasks.

a. For each function, write another rule using *NOW* and *NEXT* that could be used to produce the same pattern of (x, y) values.

b. How would you describe the similarities and differences in the relationships of x and y in terms of their function graphs, tables, and rules?

22 The population of our world was about 6.5 billion in 2005. At the present rate of growth, that population will double approximately every 60 years. (Source: *The World Factbook 2005*. CIA.)

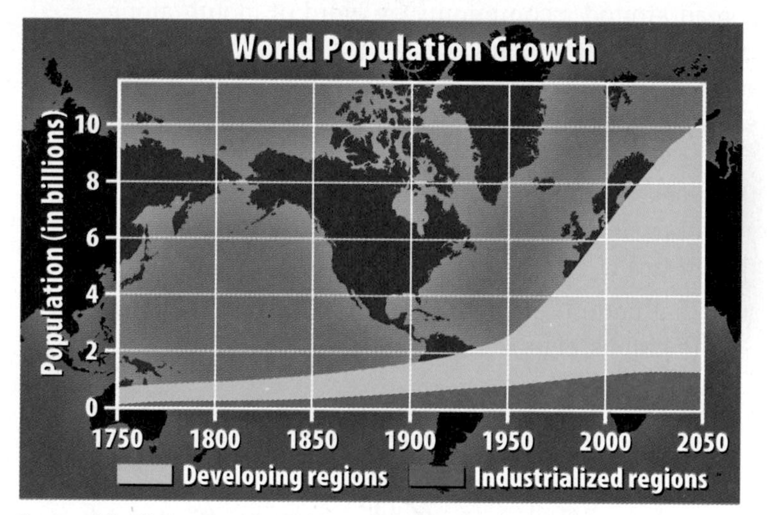

Sources: United Nations Population Division and Population Reference Bureau, 1993.

a. Assuming this rate continues, what will be the population 60, 120, 180, and 240 years from now?

b. How would that growth pattern compare to a pattern that simply added 6.5 billion people every 60 years?

c. Do you think the population is likely to continue growing in the "doubling every 60 years" pattern? Explain your reasoning.

d. How might rapid population growth affect your life in the next 60 years?

23 One way to think about rates of growth is to calculate the time it will take for a quantity to double in value. For example, it is common to ask how long it will take a bank investment or a country's population to double.

a. If the U.S. population in 2000 was about 276 million and growing exponentially at a rate of 0.9% per year, how long will it take for the U.S. population to double?

b. One year's growth is 0.9% of 276 million, or about 2.5 million. How long would it take the U.S. population to double if it increased *linearly* at the rate of 2.5 million per year?

c. How long does it take a bank deposit of $5,000 to double if it earns interest compounded annually at a rate of 2%? At a 4% rate? At a 6% rate? At an 8% rate? At a 12% rate?

d. Examine your (*rate, time to double*) data in Part c. What pattern suggests a way to predict the doubling time for an investment of $5,000 at an interest rate of 3% compounded annually? Check your conjecture. If your prediction was not close, search for another pattern for predicting doubling time and check it.

24 The sketch below shows a small circle of radius 10 millimeters and the results of four different size transformations—each with scale factor 1.5 applied to the previous circle.

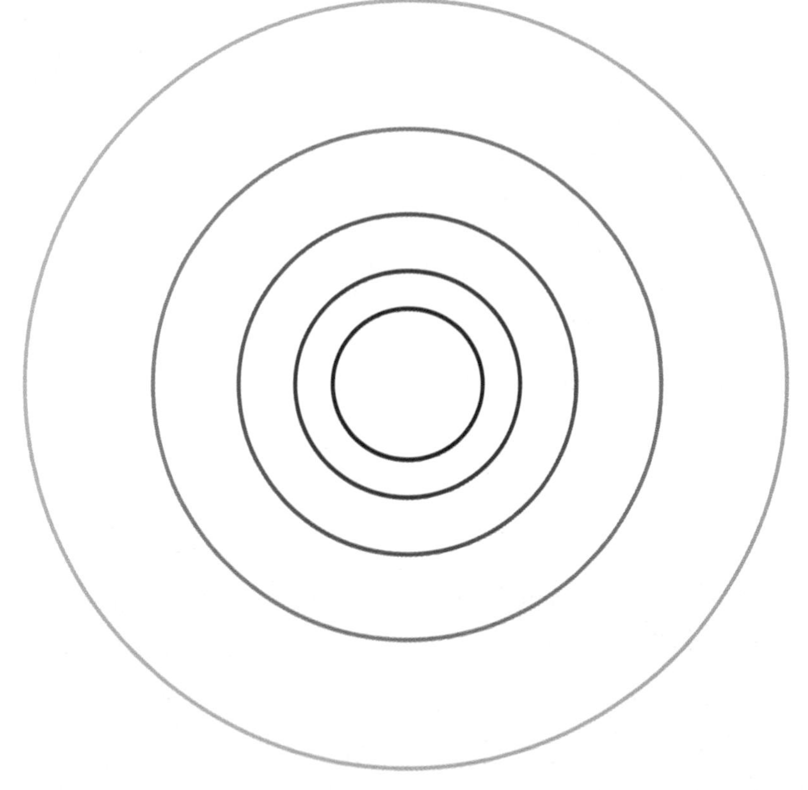

a. Make a table of (*radius, circumference*) values for the 5 circles and plot a graph of the resulting ordered pairs.

b. Write a *NOW-NEXT* rule showing how the circumference of each circle is related to the next larger circle.

c. If the pattern of expanding circles with the same center were continued, what rule would show how to calculate the circumference of the nth circle?

d. Enter the areas of the first 5 circles in the pattern in a new row of your table for Part a. Then write a rule that would show how to calculate the area enclosed by the nth circle in that pattern.

Reflections

25 One common illness in young people is *strep throat*. This bacterial infection can cause painful sore throats. Have you or anyone you know ever had strep throat? How does what you have learned about exponential growth explain the way strep throat seems to develop very quickly?

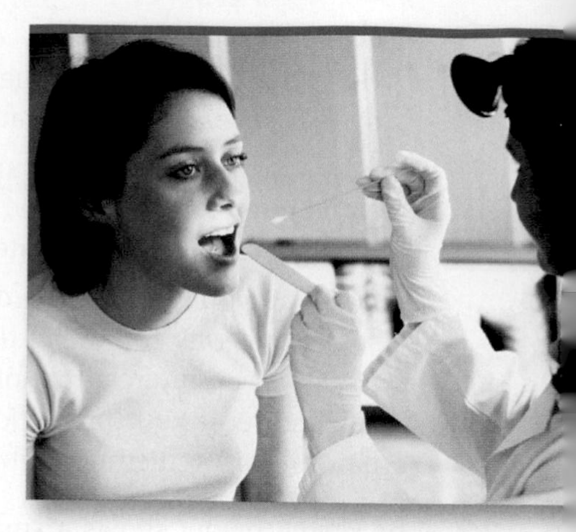

26 Which of the rules for exponential growth by doubling do you prefer: $NEXT = 2 \cdot NOW$ or $y = 2^x$? Give reasons for your preference and explain how the two types of rules are related to each other.

27 Exponential functions, like linear functions, can be expressed by rules relating x and y values.

a. In exponential functions with rules $y = a(b^x)$:

  i. how does the value of a affect the graph?

  ii. how does the value of b affect the graph?

b. In linear functions with rules $y = a + bx$:

  i. how does the value of a affect the graph?

  ii. how does the value of b affect the graph?

28 In each of the calculations in Parts a–d, a student has made an error. Correct the errors. Then describe how you might remind the student how to correctly calculate the answer.

a. $2^3 = 6$ **b.** $-2^2 = 4$

c. $3^0 = 0$ **d.** $(-2)^4 = -16$

29 Make a table to compare the values of 2^b and b^2 for several different positive integer values of b.

b								
2^b								
b^2								

a. For what values of b is $2^b = b^2$?

b. For what values of b is $2^b > b^2$? Explain the reasoning that supports your answer.

Extensions

30 The drawings below show five stages in growth of a design called the *dragon fractal*.

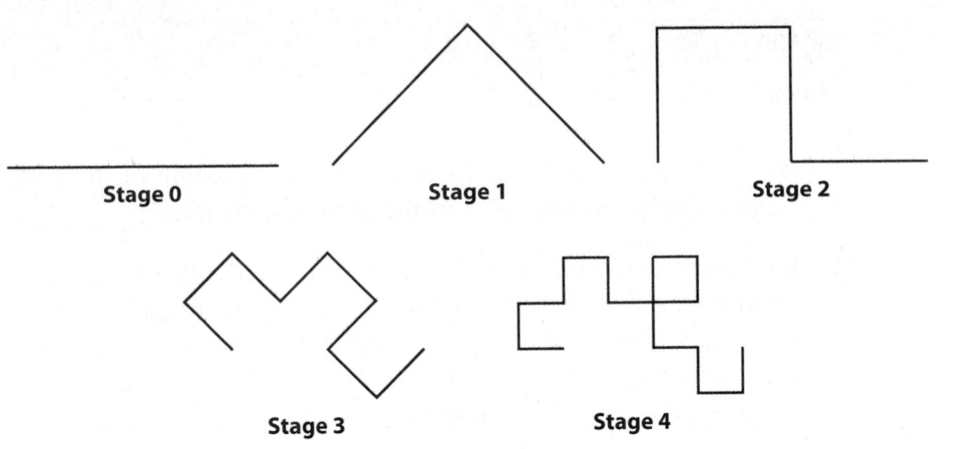

a. Draw Stage 5 of growth in the dragon fractal.

b. What pattern of change do you see in the number of segments of the growing fractal?

c. Make a table and a plot of the data showing that pattern of change.

d. Write a rule relating *NOW* and *NEXT* and a rule beginning "$y = \ldots$" for finding the number of segments in the figure at each stage of growth.

e. How many segments will there be in the fractal design at Stage 16?

f. At what stage will the fractal design have more than 1,000 segments of equal length?

31 In this task, you will examine more closely the Koch curve fractal from Connections Task 18.

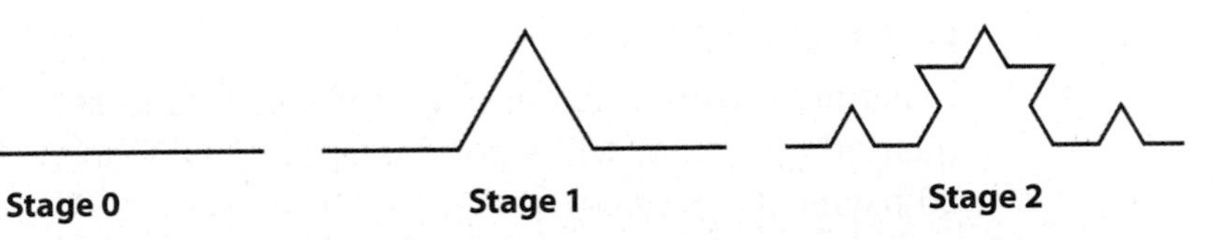

Recall that in moving from one stage to the next, each segment is divided into three equal-length parts. A tent is raised over the middle section with sides equal in length to the parts on each side.

a. If the original line segment is 1 inch long, how long is each segment of the pattern in Stage 1? How long is each segment of the pattern in Stage 2?

b. Complete the following table showing the length of segments in the first 10 stages.

Stage	0	1	2	3	4	5	6	7	8	9
Length	1	$\frac{1}{3}$	$\frac{1}{9}$							

c. Look back at Parts c and d of Connections Task 18 where you wrote rules giving the number of short segments at each stage of the pattern. Then use that information and the results of Part b to complete the following table giving the total length of the pattern at each stage.

Stage	0	1	2	3	4	5	6	7	8	9
Length	1	$\frac{4}{3}$	$\frac{16}{9}$							

d. What appears to be happening to the total length of the pattern as the number of segments in the pattern increases?

32 Banks frequently pay interest more often than once each year. Suppose your bank pays interest compounded *quarterly*. If the annual percentage rate is 4%, then the bank adds 4% ÷ 4 = 1% interest to the account balance at the end of each 3-month period.

a. Explore the growth of a $1,000 deposit in such a bank over 5 years.

b. Compare the quarterly compounding with annual compounding at 4%.

c. Repeat the calculations and comparisons if the annual rate is 8%.

33 Many people borrow money from a bank to buy a car, a home, or to pay for a college education. However, they have to pay back the amount borrowed plus interest. To consider a simple case, suppose that for a car loan of $9,000 a bank charges 6% annual rate of interest compounded quarterly and the repayment is done in quarterly installments. One way to figure the balance on this loan at any time is to use the rule:

$$\textit{new balance} = 1.015 \times \textit{old balance} - \textit{payment}.$$

a. Use this rule to find the balance due on this loan for each quarterly period from 0 to 20, assuming that the quarterly payments are all $250.

b. Experiment with different payment amounts to see what quarterly payment will repay the entire $9,000 loan in 20 payments (5 years).

c. Create a spreadsheet you can use in experiments to find the effects of different quarterly payment amounts, interest rates, and car loan amounts. Use the spreadsheet to look for patterns relating those variables for 5-year loans. Write a brief report of your findings. (You might want to have cells in the spreadsheet for each of the variables, *original loan amount, interest rate,* and *quarterly payment amount,* and then a column that tracks the *outstanding loan balance* over a period of 20 quarters.)

34 The Wheaton Boys and Girls Club has 511 members and a calling tree in which, starting with the president, members are asked to pass on news to two other members.

a. What function rule shows how to calculate the total number T of members informed after x stages of the process have been completed? It might help to begin by finding these sums:

1

$1 + 2$

$1 + 2 + 4$

$1 + 2 + 4 + 8$

$1 + 2 + 4 + 8 + 16$

$\vdots$

Use the function rule you came up with to make a table and a graph showing values of T for $x = 0, 1, 2, 3, 4, 5, 6, 7, 8$.

b. What function rule gives the number of people N in the club who have not yet been called after stage x of the calling tree process? Use the function rule you came up with to make a table and graph showing values of N for $x = 0, 1, 2, 3, 4, 5, 6, 7, 8$.

c. Describe the patterns of change shown in tables and graphs of the rules that express T and N as functions of x and compare them to each other and to other functions you've encountered.

Review

35 Write each of the following calculations in more compact form by using exponents.

a. $5 \times 5 \times 5 \times 5$

b. $3 \times 3 \times 3 \times 3 \times 3 \times 3 \times 3 \times 3$

c. $1.5 \times 1.5 \times 1.5 \times 1.5 \times 1.5 \times 1.5$

d. $(-10) \times (-10) \times (-10) \times (-10) \times (-10) \times (-10) \times (-10) \times (-10)$

e. $\underbrace{6 \times 6 \times \cdots \times 6}_{n \text{ factors}}$

f. $\underbrace{a \times a \times \cdots \times a}_{n \text{ factors}}$

36 Do these calculations *without* use of the exponent key (^ or y^x) on your calculator.

a. 5^4 **b.** $(-7)^2$ **c.** 10^0

d. $(-8)^3$ **e.** 2^8 **f.** 2^{10}

37 Draw and label each triangle as accurately as you can. After you draw the triangle, use a ruler to determine the lengths of the remaining sides and a protractor to find the measures of the remaining angles.

	AB	*BC*	*CA*	m∠*ABC*	m∠*BCA*	m∠*CAB*
a.	2.5 in.	4 in.		125°		
b.	12 cm			45°		45°

38 Given that $\frac{1}{4}$ of 160 is 40, and that 10% of 160 is 16, find the values of the following without using a calculator. Explain how you obtained your answers.

a. $\frac{3}{4}$ of 160, $\frac{5}{4}$ of 160, and $1\frac{3}{4}$ of 160

b. 5% of 160, 95% of 160, and 105% of 160

39 Anthony took an inventory of the colors of the shirts that he owns and made a table of his findings.

Shirt Color	Green	Blue	Black	Red	Other
Number	5	2	4	3	6

a. Identify two types of graphs that would be appropriate for Anthony to make to display this data. Choose one of them and make it.

b. What percent of Anthony's shirts are red?

c. What percent of Anthony's shirts are not green?

d. If Anthony were to randomly choose a shirt to wear, what is the probability that he would choose a black shirt?

40 A line passes through the points (1, 1) and (5, −7). Determine whether or not each point below is also on the line. Explain your reasoning.

a. (8, −12) **b.** (0, 3) **c.** (3, −4) **d.** (−3, 9)

41 Consider the two solids shown below.

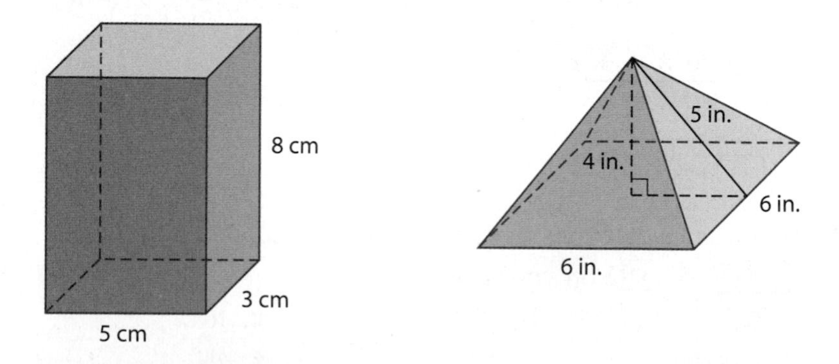

For each solid, complete the following.

a. How many faces does the solid have? For each face, describe the shape and give its dimensions.

b. Identify all faces that appear to be parallel to each other.

c. Identify all faces that appear to be perpendicular to each other.

d. Find the surface area.

e. Find the volume. (Recall that the formula for the volume of a pyramid is $V = \frac{1}{3}Bh$, where B is the area of the base and h is the height of the pyramid.)

42 Evaluate each of these algebraic expressions when $x = 3$ and be prepared to explain why you believe you've produced the correct values.

a. $3x + 7$

b. $7 + 3x$

c. $5 - 4x + 3x^2$

d. $8(4x - 9)^2$

e. $\frac{5x - 2}{5 + 2x}$

f. $-2x^2$

g. $(-2x)^3$

h. $\sqrt{25 - x^2}$

i. $\frac{x + 12}{x + 4}$

43 When you use formulas to find values of dependent variables associated with specific values of independent variables, you need to be careful to "read" the directions of the formula the way they are intended. For example, the formula for surface area of a circular cylinder is $A = 2\pi r^2 + 2\pi rh$.

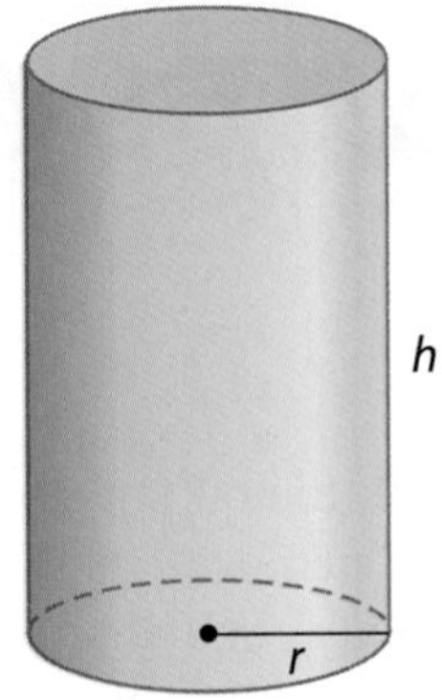

a. What is the surface area of a cylinder with radius 5 inches and height 8 inches? How do you know your calculation used the formula correctly?

b. Sketch a cylinder for which the height is the same length as the radius r. Show that the surface area of the cylinder is $A = 4\pi r^2$. What algebraic properties did you use in your reasoning?

44 Graph each of the following lines on grid paper. Then, for Parts a and b, write an equation of the line.

a. A line with slope of $\frac{2}{3}$ and y-intercept at $(0, 6)$

b. A line with slope of 0.5 and x-intercept at $(-2, 0)$

c. A line with equation $y = -3x - 5$

Exponential Decay

In 1989, the oil tanker Exxon Valdez ran aground in waters near the Kenai peninsula of Alaska. Over 10 million gallons of oil spread on the waters and shoreline of the area, endangering wildlife. That oil spill was eventually cleaned up—some of the oil evaporated, some was picked up by specially equipped boats, and some sank to the ocean floor as sludge. But the experience had lasting impact on thinking about environmental protection.

For scientists planning environmental cleanups, it is important to be able to predict the pattern of dispersion in such contaminating spills. Suppose that an accident dropped some pollutant into a large aquarium. It's not practical to remove all water from the aquarium at once, so the cleanup has to take place in smaller steps. A batch of polluted water is removed and replaced by clean water. Then the process is repeated.

Think about the following experiment that simulates pollution and cleanup of the aquarium.

- Mix 20 black checkers (the pollution) with 80 red checkers (the clean water).
- Remove 20 checkers from the mixture (without looking at the colors) and replace them with 20 red checkers (clean water). Record the number of black checkers remaining. Then shake the new mixture. This simulates draining off some of the polluted water and replacing it with clean water.

- In the second step, remove 20 checkers from the new mixture (without looking at the colors) and replace them with 20 red checkers (more clean water). Record the number of black checkers remaining. Then stir the new mixture.
- Repeat the remove-replace-record-mix process for several more steps.

Think About This Situation

The graphs below show two possible outcomes of the pollution and cleanup simulation.

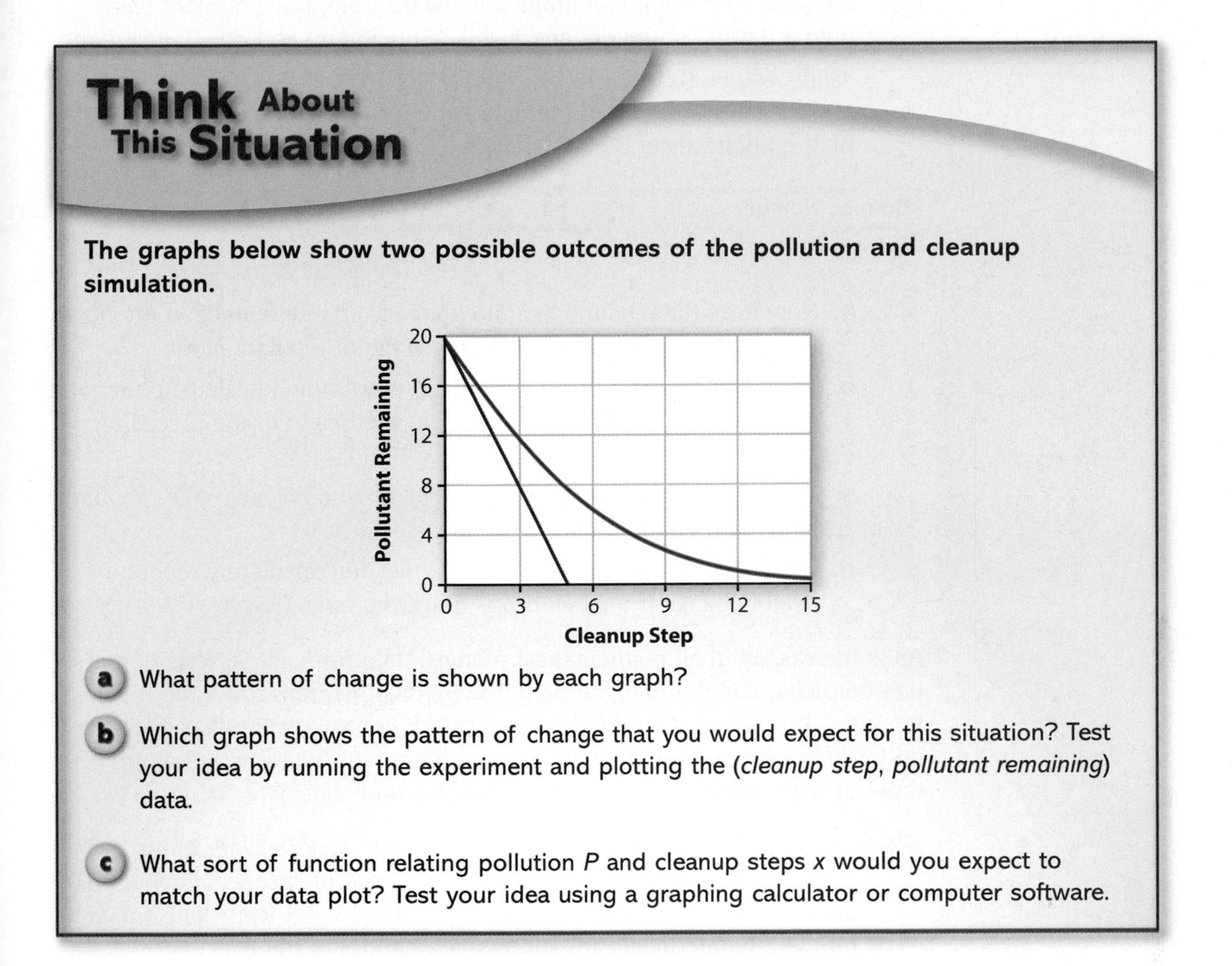

a What pattern of change is shown by each graph?

b Which graph shows the pattern of change that you would expect for this situation? Test your idea by running the experiment and plotting the (*cleanup step*, *pollutant remaining*) data.

c What sort of function relating pollution P and cleanup steps x would you expect to match your data plot? Test your idea using a graphing calculator or computer software.

The pollution cleanup experiment gives data in a pattern that occurs in many familiar and important problem situations. That pattern is called *exponential decay*. Your work on problems of this lesson will reveal important properties and uses of exponential decay functions and fractional exponents.

Investigation 1 More Bounce to the Ounce

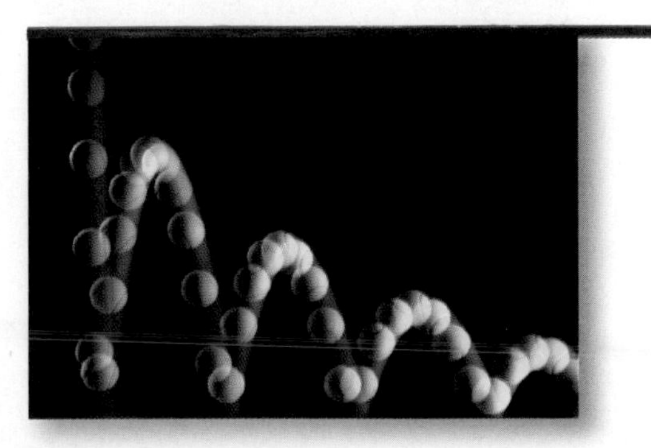

Most popular American sports involve balls of some sort. In designing those balls, one of the most important factors is the bounciness or *elasticity* of the ball. For example, if a new golf ball is dropped onto a hard surface, it should rebound to about $\frac{2}{3}$ of its drop height. The pattern of change in successive rebound heights will be similar to that of the data in the pollution cleanup experiment.

As you work on the problems of this investigation, look for answers to this question:

What mathematical patterns in tables, graphs, and symbolic rules are typical of exponential decay relations?

1 Suppose a new golf ball drops downward from a height of 27 feet onto a paved parking lot and keeps bouncing up and down, again and again. Rebound height of the ball should be $\frac{2}{3}$ of its drop height. Make a table and plot of the data showing expected heights of the first ten bounces of the golf ball.

Bounce Number	0	1	2	3	4	5	6	7	8	9	10
Rebound Height (in feet)	27										

a. How does the rebound height change from one bounce to the next? How is that pattern shown by the shape of the data plot?

b. What rule relating *NOW* and *NEXT* shows how to calculate the rebound height for any bounce from the height of the preceding bounce?

c. What rule beginning "$y = \ldots$" shows how to calculate the rebound height after any number of bounces?

d. How will the data table, plot, and rules for calculating rebound height change if the ball drops first from only 15 feet?

As is the case with all mathematical models, data from actual tests of golf ball bouncing will not match exactly the predictions from rules about ideal bounces. You can simulate the kind of quality control testing that factories do by running some experiments in your classroom. Work with a group of three or four classmates to complete the next problems.

2 Get a golf ball and a tape measure or meter stick for your group. Decide on a method for measuring the height of successive rebounds after the ball is dropped from a height of at least 8 feet. Collect data on the rebound height for successive bounces of the ball.

a. Compare the pattern of your data to that of the model that predicts rebounds which are $\frac{2}{3}$ of the drop height. Would a rebound height factor other than $\frac{2}{3}$ give a better model for your data? Be prepared to explain your reasoning.

b. Write a rule using *NOW* and *NEXT* that relates the rebound height of any bounce of your tested ball to the height of the preceding bounce.

c. Write a rule beginning "$y = \ldots$" to predict the rebound height after any bounce.

3 Repeat the experiment of Problem 2 with some other ball such as a tennis ball or a volleyball.

a. Study the data to find a reasonable estimate of the rebound height factor for your ball.

b. Write a rule using *NOW* and *NEXT* and a rule beginning "$y = \ldots$" to model the rebound height of your ball on successive bounces.

Summarize the Mathematics

Different groups might have used different balls and dropped the balls from different initial heights. However, the patterns of (*bounce number*, *rebound height*) data should have some similar features.

a Look back at the data from your experiments.

  i. How do the rebound heights change from one bounce to the next in each case?

  ii. How is the pattern of change in rebound height shown by the shape of the data plots in each case?

b List the *NOW-NEXT* and the "$y = \ldots$" rules you found for predicting the rebound heights of each ball on successive bounces.

  i. What do the rules relating *NOW* and *NEXT* bounce heights have in common in each case? How, if at all, are those rules different, and what might be causing the differences?

  ii. What do the rules beginning "$y = \ldots$" have in common in each case? How, if at all, are those rules different, and what might be causing the differences?

c What do the tables, graphs, and rules in these examples have in common with those of the exponential growth examples in Lesson 1? How, if at all, are they different?

d How are the exponential decay data patterns, graphs, and rules similar to and different from those of linear functions and other types of functions you've studied in earlier units?

Be prepared to compare your data, models, and ideas with the rest of the class.

Check Your Understanding

When dropped onto a hard surface, a brand new softball should rebound to about $\frac{2}{5}$ the height from which it is dropped.

a. If the softball is dropped 25 feet from a window onto concrete, what pattern of rebound heights can be expected?

  i. Make a table and plot of predicted rebound data for 5 bounces.

  ii. What *NOW-NEXT* and "$y = \ldots$" rules give ways of predicting rebound height after any bounce?

b. Here are some data from bounce tests of a softball dropped from a height of 10 feet.

Bounce Number	1	2	3	4	5
Rebound Height (in feet)	3.8	1.3	0.6	0.2	0.05

  i. What do these data tell you about the quality of the tested softball?

  ii. What bounce heights would you expect from this ball if it were dropped from 20 feet instead of 10 feet?

c. What *NOW-NEXT* and "$y = \ldots$" rules would model rebound height of an ideal softball if the drop were from 20 feet?

d. What rule beginning "$y = \ldots$" shows how to calculate the height y of the rebound when a new softball is dropped from any height x? What connections do you see between this rule and the rule predicting rebound height on successive bounces of the ball?

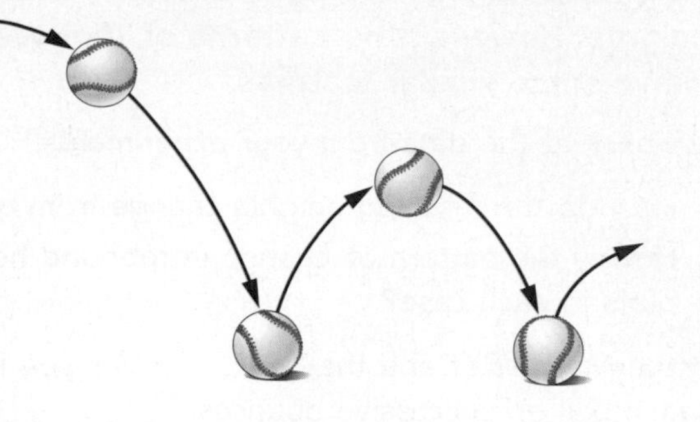

Investigation 2 Medicine and Mathematics

Prescription drugs are a very important part of the human health equation. Many medications are essential in preventing and curing serious physical and mental illnesses.

Diabetes, a disorder in which the body cannot metabolize glucose properly, affects people of all ages. In 2005, there were about 14.6 million diagnosed cases of diabetes in the United States. It was estimated that another 6.2 million cases remained undiagnosed. (Source: diabetes.niddk.nih.gov/dm/pubs/statistics/index.htm)

In 5–10% of the diagnosed cases, the diabetic's body is unable to produce insulin, which is needed to process glucose.

To provide this essential hormone, these diabetics must take injections of a medicine containing insulin. The medications used (called insulin delivery systems) are designed to release insulin slowly. The insulin itself breaks down rather quickly. The rate varies greatly among individuals, but the following graph shows a typical pattern of insulin decrease.

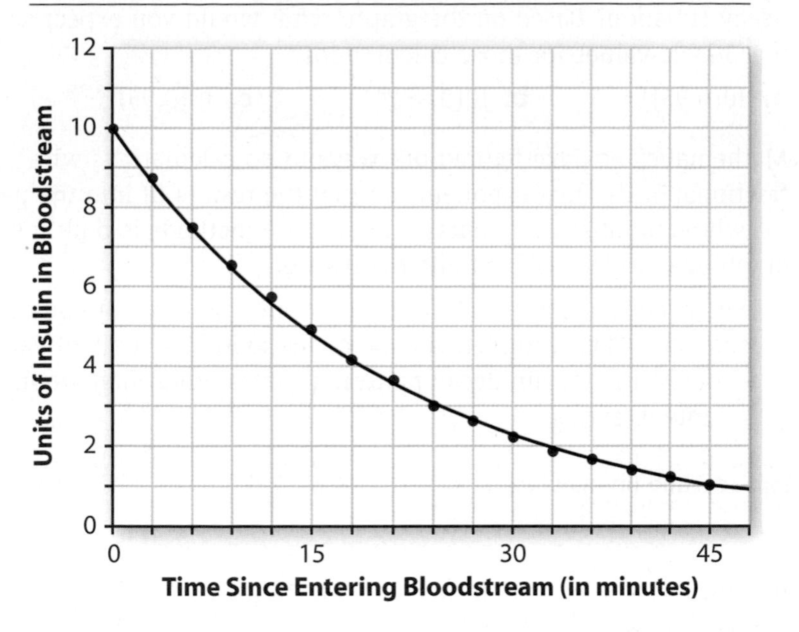

As you work on the problems of this investigation, look for answers to the following questions:

How can you interpret and estimate or calculate values of expressions involving fractional or decimal exponents?

How can you interpret and estimate or calculate the half-life of a substance that decays exponentially?

1. Medical scientists often are interested in the time it takes for a drug to be reduced to one half of the original dose. They call this time the **half-life** of the drug. What appears to be the half-life of insulin in this case?

2. The pattern of decay shown on this graph for insulin can be modeled well by the function $y = 10(0.95^x)$, where x is the number of minutes since the insulin entered the bloodstream.

   a. Use your calculator or computer software to see how well a table of values and graph of this rule matches the pattern in the graph above.

   b. What do the numbers 10 and 0.95 tell about the amount of insulin in the bloodstream?

   c. Based on the function modeling insulin decay, what percent of active insulin is actually used up with each passing minute?

3. What rule relating *NOW* and *NEXT* shows how the amount of insulin in the blood changes from one minute to the next, once 10 units have entered the bloodstream?

4. The insulin decay graph shows data points for three-minute intervals following the original insulin level. But the curve connecting those points reminds us that the insulin breakdown does not occur in sudden bursts at the end of each minute! It occurs *continuously* as time passes.

What would each of the following calculations tell about the insulin decay situation? Based on the graph, what would you expect as reasonable values for those calculations?

a. $10(0.95)^{1.5}$ **b.** $10(0.95)^{4.5}$ **c.** $10(0.95)^{18.75}$

5 Mathematicians have figured out ways to do calculations with fractional or decimal exponents so that the results fit into the pattern for whole number exponents. One of those methods is built into your graphing calculator or computer software.

a. Enter the function $y = 10(0.95^x)$ in your calculator or computer software. Then complete a copy of the following table of values showing the insulin decay pattern at times other than whole-minute intervals.

Elapsed Time (in minutes)	0	1.5	4.5	7.5	10.5	13.5	16.5	19.5
Units of Insulin in Blood	10							

b. Compare the entries in this table with data shown by points on the graph on the preceding page.

c. Study tables and graphs of your function to estimate, to the nearest tenth of a minute, solutions for the following equations and inequality. In each case, be prepared to explain what the solution tells about decay of insulin.

i. $2 = 10(0.95^x)$ **ii.** $8 = 10(0.95^x)$

iii. $10(0.95^x) > 1.6$

6 Use the function $y = 10(0.95^x)$ to estimate the half-life of insulin for an initial dose of 10 units. Then estimate the half-life in cases when the initial dose is 15 units. When it is 20 units. When it is 25 units. Explain the pattern in those results.

Summarize the Mathematics

In this investigation, you have seen another example of the way that patterns of exponential decay can be expressed by function rules like $y = a(b^x)$.

a What *NOW-NEXT* rule describes this pattern of change?

b What do the values of a and b tell about the situation being modeled? About the tables and graphs of the (x, y) values?

c How can you estimate or calculate values of b^x when x is not a whole number?

d What does the half-life tell about a substance that decays exponentially? What strategies can be used to estimate or calculate half-life?

Be prepared to compare your responses with those of your classmates.

Check Your Understanding

The most famous antibiotic drug is penicillin. After its discovery in 1929, it became known as the first *miracle drug*, because it was so effective in fighting serious bacterial infections.

Drugs act somewhat differently on each person. But, on average, a dose of penicillin will be broken down in the blood so that one hour after injection only 60% will remain active. Suppose a patient is given an injection of 300 milligrams of penicillin at noon.

a. Write a rule in the form $y = a(b^x)$ that can be used to calculate the amount of penicillin remaining after any number of hours x.

b. Use your rule to graph the amount of penicillin in the blood from 0 to 10 hours. Explain what the pattern of that graph shows about the rate at which active penicillin decays in the blood.

c. Use the rule from Part a to produce a table showing the amount of active penicillin that will remain at *quarter-hour* intervals from noon to 5 P.M.

- **i.** Estimate the half-life of penicillin.
- **ii.** Estimate the time it takes for an initial 300-mg dose to decay so that only 10 mg remain active.

d. If 60% of a penicillin dose remains active one hour after an injection, what percent has been broken down in the blood?

Investigation 3 Modeling Decay

When you study a situation in which data suggest a dependent variable decreasing in value as a related independent variable increases, there are two strategies for finding a good algebraic model of the relationship. In some cases, it is possible to use the problem conditions and reasoning to determine the type of function that will match dependent to independent variable values. In other cases, some trial-and-error exploration or use of calculator or computer curve-fitting software will be necessary before an appropriate model is apparent.

From a scientific point of view, it is always preferable to have some logical explanation for choice of a model. Then the experimental work is supported by understanding of the relationship being studied. As you work on the following problems, look for answers to these questions:

What clues in problem conditions are helpful in deriving function models for experimental data involving decay?

How can logical analysis of an experiment be used as a check of a function model produced by your calculator or computer curve-fitting software?

 Suppose that you were asked to conduct this experiment:

- Get a collection of 100 coins, shake them well, and drop them on a tabletop.
- Remove all coins that are lying heads up and record the number of coins left.
- Repeat the shake-drop-remove-record process until 5 or fewer coins remain.

a. If you were to record the results of this experiment in a table of (*drop number, coins left*) values, what pattern would you expect in the data? What function rule would probably be the best model relating drop number n to number of coins left c?

b. Conduct the experiment, record the data, and then use your calculator or curve-fitting software to find a function model that seems to fit the data pattern well.

c. Compare the model suggested by logical analysis of the experiment to that found by fitting a function to actual data. Decide which you think is the better model of the experiment and be prepared to explain your choice.

2 Suppose that the experiment in Problem 1 is modified in this way:

- Get a collection of 100 coins and place them on a table top.
- Roll a six-sided die and remove the number of coins equal to the number on the top face of the die. Record the number of coins remaining. For example, if the first roll shows 4 dots on the top of the die, remove four coins, leaving 96 coins still on the table.
- Repeat the roll-remove-record process until 10 or fewer coins remain.

a. If you were to record the results of this experiment in a table of (*roll number, coins left*) values, what pattern would you expect in that data? What function rule would probably be the best model relating roll number n to number of coins left c?

b. Conduct the experiment, record the data, and then use your calculator or curve-fitting software to find a function model that seems to fit the data pattern well.

c. Compare the model suggested by logical analysis of the experiment to that found by fitting a function to actual data. Decide which you think is the better model of the experiment and be prepared to explain your choice.

 How are the data from the experiments in Problems 1 and 2 and the best-fitting function models for those data different? Why are those differences reasonable, in light of differences in the nature of the experiments that were conducted?

Summarize the Mathematics

In this investigation, you compared two strategies for developing models of patterns in experimental data where a dependent variable decreases in value as a related independent variable increases.

a. What differences did you notice between models suggested by logical analysis of the experiments and by curve-fitting based on real data?

b. How were the models for each experiment similar, and how were they different? How are those similarities and differences explained by logical analysis? How are they illustrated by patterns in experimental data plots?

c. What kinds of problem conditions suggest situations in which a linear model is likely to be best? Situations in which an exponential model is likely to be best?

Be prepared to compare your responses with those from other groups.

Check Your Understanding

Consider the following experiment:

- Start with a pile of 90 kernels of unpopped popcorn or dry beans.
- Pour the kernels or beans onto the center of a large paper plate with equal-sized sectors marked as in the diagram below.

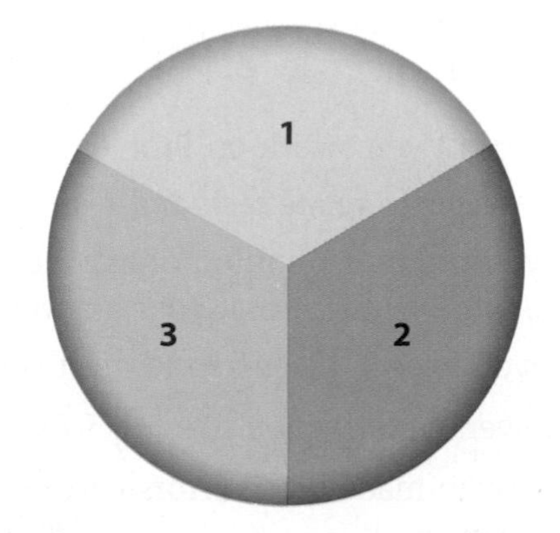

- Shake the plate so that the kernels or beans scatter into the various sectors in a somewhat random pattern.
- Remove all kernels that land on the sector marked "1" and record the trial number and the number of kernels or beans remaining.
- Repeat the shake-remove-record process several times.

a. If you were to record the results of this experiment in a table of (*trial number, kernels left*) values, what pattern would you expect in that data? What function rule would probably be the best model for the relationship between trial number n and kernels left k?

b. Conduct the experiment and record the data. Then use your calculator or curve-fitting software to find the model that seems to fit the data pattern well.

c. Compare the models suggested by logical analysis of the experiment and by fitting of a function to actual data. Decide which is the better model of the experiment and explain your choice.

Investigation 4 Properties of Exponents II

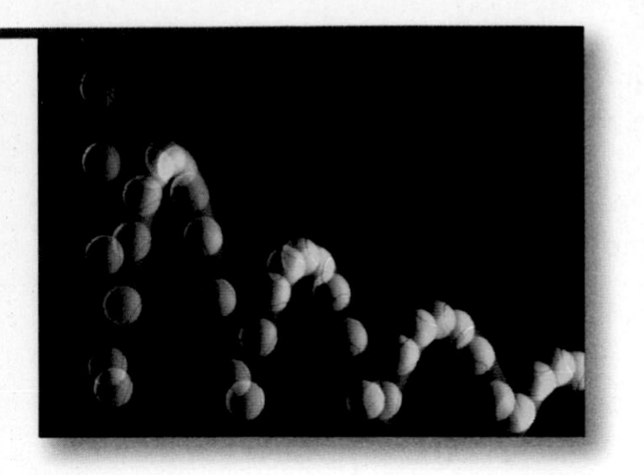

In studying the rebound height of a bouncing ball, you calculated powers of the fraction $\frac{2}{3}$. You can calculate a power like $\left(\frac{2}{3}\right)^4$ by repeated multiplication $\left(\frac{2}{3}\right)\left(\frac{2}{3}\right)\left(\frac{2}{3}\right)\left(\frac{2}{3}\right)$. But there is a shortcut rule for such calculations with exponents.

As you work on the problems in this investigation, make notes of answers to this question:

What exponent properties provide shortcut rules for calculating powers of fractions, quotients of powers, and negative exponents?

Powers of a Fraction As you work on the next calculations, look for a pattern suggesting ways to write powers of fractions in useful equivalent forms.

1 Find values of x and y that will make these equations true statements:

a. $\left(\frac{3}{5}\right)^3 = \frac{3^x}{5^y}$

b. $\left(\frac{c}{5}\right)^2 = \frac{c^x}{5^y}$

c. $\left(\frac{4}{n}\right)^5 = \frac{4^x}{n^y}$ $(n \neq 0)$

d. $\left(\frac{c^2}{n}\right)^3 = \frac{c^x}{n^y}$ $(n \neq 0)$

2 Examine the results of your work on Problem 1.

a. What pattern seems to relate task and result in every case?

b. How would you use the definition of exponent or other reasoning to convince another student that your answer to Part a is correct?

c. What would you expect to see as the most common errors in evaluating powers of a fraction like $\left(\frac{3}{5}\right)^4$? Explain how you would help someone who made those errors correct their understanding of how exponents work.

Quotients of Powers Since many useful algebraic functions require division of quantities, it is helpful to be able to simplify expressions involving quotients of powers like $\frac{b^x}{b^y}$ $(b \neq 0)$.

3 Find values for x, y, and z that will make these equations true statements.

a. $\frac{2^{10}}{2^3} = 2^z$

b. $\frac{3^6}{3^2} = 3^z$

c. $\frac{10^9}{10^3} = 10^z$

d. $\frac{2^x}{2^5} = 2^7$

e. $\frac{7^x}{7^y} = 7^2$

f. $\frac{b^5}{b^3} = b^z$

g. $\frac{3^5}{3^5} = 3^z$

h. $\frac{b^x}{b^x} = b^z$

4. Examine the results of your work on Problem 3.

   a. What pattern seems to relate task and result in every case?

   b. How would you use the definition of exponent or other reasoning to convince another student your answer to Part a is correct?

   c. What would you expect to see as the most common errors in evaluating quotients of powers like $\frac{8^{12}}{8^4}$? Explain how you would help someone who made those errors correct their understanding of how exponents work.

5. Use your answers to Problem 3 Parts g and h and Problem 4 Part a to explain why it is reasonable to define $b^0 = 1$ for any base b ($b \neq 0$).

Negative Exponents Suppose that you were hired as a science lab assistant to monitor an ongoing experiment studying the growth of an insect population. If the population when you took over was 48 and it was expected to double every day, you could estimate the population for any time in the future or the past with the function $p = 48(2^x)$.

Future estimates are easy: One day from now, the population should be about $48(2^1) = 96$; two days from now it should be about $48(2^2) = 48(2)(2) = 192$, and so on.

Estimates of the insect numbers in the population before you took over require division: One day earlier, the population should have been about $48(2^{-1}) = 48 \div 2 = 48\left(\frac{1}{2}\right) = 24$; two days ago, it should have been about:

$$\begin{aligned} 48(2^{-2}) &= (48 \div 2) \div 2 \\ &= 48 \div 2^2 \\ &= 48\left(\frac{1}{2^2}\right) \\ &= 12 \end{aligned}$$

This kind of reasoning about exponential growth suggests a general rule that for any nonzero number b and any integer n, $b^{-n} = \frac{1}{b^n}$.

6. The rule for operating with negative integer exponents also follows logically from the property about quotients of powers and the definition $b^0 = 1$. Justify each step in the reasoning below.

$$\begin{aligned} \frac{1}{b^n} &= \frac{b^0}{b^n} && (1) \\ &= b^{0-n} && (2) \\ &= b^{-n} && (3) \end{aligned}$$

7. Use the relationship between fractions and negative integer exponents to write each of the following expressions in a different but equivalent form. In Parts a–f, write an equivalent fraction that does not use exponents at all.

   a. 5^{-3} **b.** 6^{-1} **c.** 2^{-4} **d.** $\left(\frac{2}{5}\right)^{-1}$

   e. $\left(\frac{1}{2}\right)^{-3}$ **f.** $\left(\frac{2}{5}\right)^{-2}$ **g.** x^{-3} **h.** $\frac{1}{a^4}$

8 Examine the results of your work in Problems 6 and 7.

a. How would you describe the rule defining negative integer exponents in your own words?

b. What would you expect to see as the most common errors in evaluating expressions with negative integer exponents like $\left(\frac{4}{3}\right)^{-2}$? How would you help someone who made those errors correct their understanding of how negative integer exponents work?

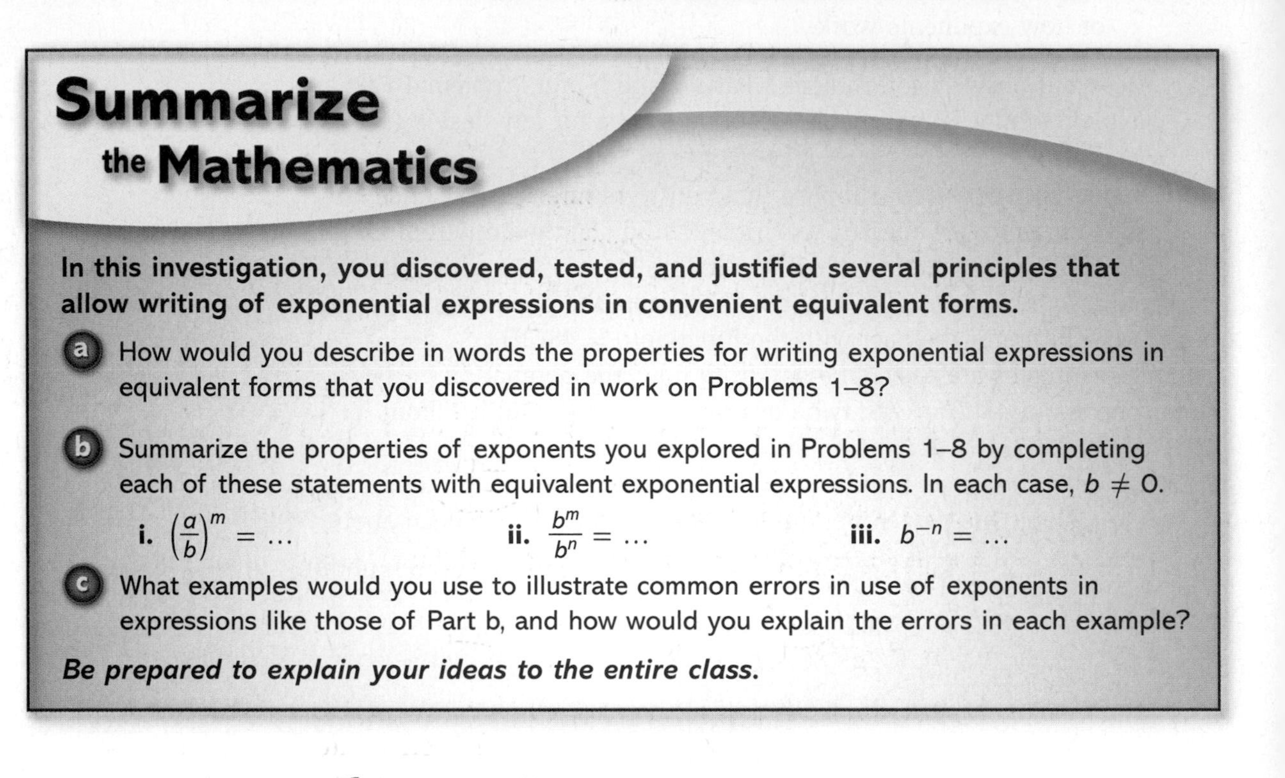

Summarize the Mathematics

In this investigation, you discovered, tested, and justified several principles that allow writing of exponential expressions in convenient equivalent forms.

a How would you describe in words the properties for writing exponential expressions in equivalent forms that you discovered in work on Problems 1–8?

b Summarize the properties of exponents you explored in Problems 1–8 by completing each of these statements with equivalent exponential expressions. In each case, $b \neq 0$.

i. $\left(\frac{a}{b}\right)^m = \ldots$ **ii.** $\frac{b^m}{b^n} = \ldots$ **iii.** $b^{-n} = \ldots$

c What examples would you use to illustrate common errors in use of exponents in expressions like those of Part b, and how would you explain the errors in each example?

Be prepared to explain your ideas to the entire class.

Check Your Understanding

Use properties of exponents to write each of the following expressions in another equivalent form and be prepared to explain how you know your answers are correct.

a. $(y^3)(y^6)$ **b.** $(5x^2y^4)(2xy^3)$ **c.** $\frac{a^7}{a^5}$

d. $\left(\frac{5}{3}\right)^3$ **e.** $(pq)^3$ **f.** $(7p^3q^2)^2$

g. $(T^3)^2$ **h.** $\frac{2}{p^{-4}}$ **i.** -5^2

j. $(-5)^2$ **k.** $2a^0$ **l.** $(2a)^0$

m. $4a^{-2}$ **n.** $(4a)^{-2}$

Investigation 5 Square Roots and Radicals

In your work on problems of insulin decay, you found that some questions required calculation with exponential expressions involving a fractional base and fractional powers. For example, estimating the amount of insulin active in the bloodstream 1.5 minutes after a 10-unit injection required calculating $10(0.95^{1.5})$.

Among the most useful expressions with fractional exponents are those with power one-half. It turns out that one-half powers are connected to the square roots that are so useful in geometric calculations like those involving the Pythagorean Theorem. For any non-negative number b,

$$b^{\frac{1}{2}} = \sqrt{b}.$$

Expressions like $\sqrt{b}$, $\sqrt{5}$, and $\sqrt{9 - x^2}$ are called *radicals*. As you work on the following problems, keep this question in mind:

How can you use your understanding of properties of exponents to guide your thinking about one-half powers, square roots, radical expressions, and rules for operating with them?

1 For integer exponents m and n, you know that $(a^m)^n = a^{mn}$. That property can be extended to work with fractional exponents.

a. Write each of these expressions in standard number form without exponents or radicals.

i. $\left(2^{\frac{1}{2}}\right)^2$ **ii.** $\left(5^{\frac{1}{2}}\right)^2$ **iii.** $\left(12^{\frac{1}{2}}\right)^2$ **iv.** $\left(2.4^{\frac{1}{2}}\right)^2$

b. How do the results of Part a explain why the definition $b^{\frac{1}{2}} = \sqrt{b}$ makes sense?

2 Write each of the following expressions in an equivalent form using radicals and then in simplest number form (without exponents or radicals).

a. $(25)^{\frac{1}{2}}$ **b.** $(9)^{\frac{1}{2}}$ **c.** $\left(\frac{9}{4}\right)^{\frac{1}{2}}$ **d.** $(100)^{\frac{1}{2}}$

3 The diagram below shows a series of squares with side lengths increasing in sequence 1, 2, 3, 4, and one diagonal drawn in each square.

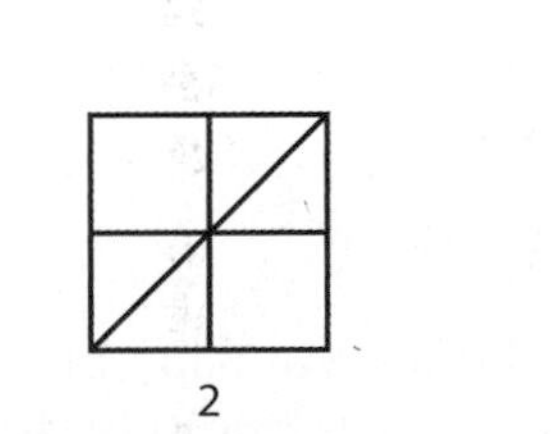

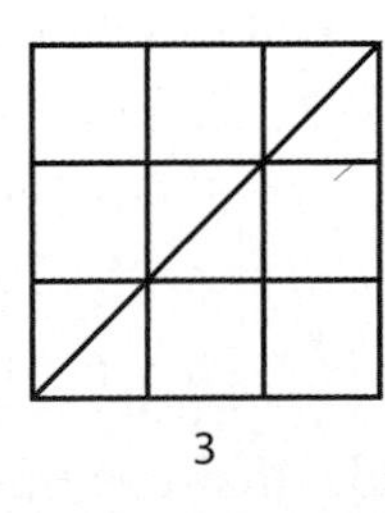

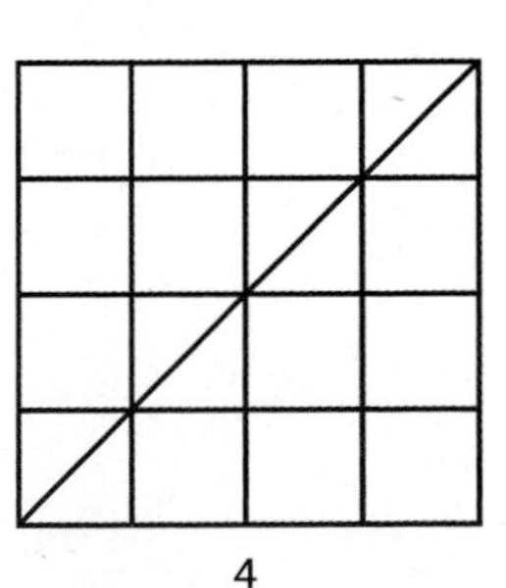

a. Use the Pythagorean Theorem to find the exact length of the diagonal of each square.

b. How are the lengths of the diagonals in the three larger squares related to the length of the diagonal of the unit square?

c. Look for a pattern in the results of Part b to complete the statement beginning:

The length d of each diagonal in a square with sides of length s is given by d = … .

4 The pattern relating side and diagonal lengths in a square illustrates a useful rule for simplifying radical expressions:

For any non-negative numbers a and b: $\sqrt{ab} = \sqrt{a}\,\sqrt{b}$.

a. What properties of square roots and exponents justify the steps in this argument? For any non-negative numbers a and b:

$$\begin{aligned} \sqrt{ab} &= (ab)^{\frac{1}{2}} && (1) \\ &= a^{\frac{1}{2}}b^{\frac{1}{2}} && (2) \\ &= \sqrt{a}\,\sqrt{b} && (3) \end{aligned}$$

b. Modify the argument in Part a to justify this property of radicals:

For any non-negative numbers a and b $(b \neq 0)$, $\sqrt{\frac{a}{b}} = \frac{\sqrt{a}}{\sqrt{b}}$.

5 Use the properties of square roots in Problem 4 to write expressions a–h in several equivalent forms. In each case, try to find the simplest equivalent form—one that involves only one radical and the smallest possible number inside that radical. Check your ideas with calculator estimates of each form. For example,

$$\begin{aligned} \sqrt{48} &= \sqrt{4}\,\sqrt{12} \\ &= 2\sqrt{12} \\ &= 2\sqrt{4}\,\sqrt{3} \\ &= 2 \cdot 2\sqrt{3} \\ &= 4\sqrt{3} \end{aligned}$$

Calculator estimates show that $\sqrt{48} \approx 6.93$ and $4\sqrt{3} \approx 6.93$.

a. $\sqrt{9 \cdot 5}$ **b.** $\sqrt{18}\,\sqrt{8}$ **c.** $\sqrt{45}$ **d.** $\sqrt{4 \cdot 9}$

e. $\sqrt{4 \cdot \frac{1}{9}}$ **f.** $\sqrt{\frac{9}{4}}$ **g.** $\sqrt{12}$ **h.** $\sqrt{96}$

6 The properties of square roots in Problem 4 are like distributive properties—taking the square root distributes over the product or the quotient of two (or more) numbers. One of the most common errors in working with square roots is distributing the square root sign over addition. However,

$$\sqrt{a + b} \neq \sqrt{a} + \sqrt{b}$$

except in some very special cases. Use several pairs of positive values for a and b to show that taking square roots *does not* distribute over addition (or subtraction).

Summarize the Mathematics

In this investigation, you explored the relationship between fractional exponents and square roots and important properties of radical expressions.

a For $n \geq 0$, what does $\sqrt{n}$ mean, and why does it make sense that $\sqrt{n} = n^{\frac{1}{2}}$?

b What property of square roots can be used to express $\sqrt{n}$ in equivalent, often simpler, forms?

c What formula gives the length of each diagonal in a square with sides of length s?

Be prepared to share your thinking with the entire class.

Check Your Understanding

Use your understanding of fractional exponents and radical expressions to help complete the following tasks.

a. How could you use a calculator with only +, −, ×, and ÷ keys to check these claims about values of expressions involving fractional exponents?

i. $225^{\frac{1}{2}} = 15$ **ii.** $7^{\frac{1}{2}} \approx 2.65$

b. Find the values of these expressions, without use of a calculator.

i. $36^{\frac{1}{2}}$ **ii.** $\sqrt{81}$

iii. $\left(\frac{25}{16}\right)^{\frac{1}{2}}$ **iv.** $\sqrt{\frac{49}{81}}$

c. Use the property that for non-negative numbers a and b, $\sqrt{ab} = \sqrt{a}\sqrt{b}$ to help write each of these radical expressions in at least two equivalent forms.

i. $\sqrt{30}$ **ii.** $\sqrt{10}\sqrt{40}$

iii. $\sqrt{\frac{7}{25}}$ **iv.** $\sqrt{\frac{2}{3}}\sqrt{\frac{3}{2}}$

d. What is the length of each diagonal in a square with sides of length 7 centimeters?

e. Give a counterexample to show that for nonnegative numbers a and b, $\sqrt{a - b}$ is *not* equal to $\sqrt{a} - \sqrt{b}$

On Your Own

Applications

1 If a basketball is properly inflated, it should rebound to about $\frac{1}{2}$ the height from which it is dropped.

a. Make a table and plot showing the pattern to be expected in the first 5 bounces after a ball is dropped from a height of 10 feet.

b. At which bounce will the ball first rebound less than 1 foot? Show how the answer to this question can be found in the table and on the graph.

c. Write a rule using *NOW* and *NEXT* and a rule beginning "$y = ...$" that can be used to calculate the rebound height after many bounces.

d. How will the data table, plot, and rules change for predicting rebound height if the ball is dropped from a height of 20 feet?

e. How will the data table, plot, and rules change for predicting rebound height if the ball is somewhat over-inflated and rebounds to $\frac{3}{5}$ of the height from which it is dropped?

Records at the Universal Video store show that sales of new DVDs are greatest in the first month after the release date. In the second month, sales are usually only about one-third of sales in the first month. Sales in the third month are usually only about one-third of sales in the second month, and so on.

a. If Universal Video sells 180 copies of one particular DVD in the first month after its release, how many copies are likely to be sold in the second month? In the third month?

b. What *NOW-NEXT* and "$y = ...$" rules predict the sales in the following months?

c. How many sales are predicted in the 12th month?

d. In what month are sales likely to first be fewer than 5 copies?

e. How would your answers to Parts a–d change for a different DVD that has first-month sales of 450 copies?

3 You may have heard of athletes being disqualified from competitions because they have used anabolic steroid drugs to increase their weight and strength. These steroids can have very damaging side effects for the user. The danger is compounded by the fact that these drugs leave the human body slowly. With an injection of the steroid *cyprionate*, about 90% of the drug and its by-products will remain in the body one day later. Then 90% of that amount will remain after a second day, and so on. Suppose that an athlete tries steroids and injects a dose of 100 milligrams of cyprionate. Analyze the pattern of that drug in the athlete's body by completing the next tasks.

a. Make a table showing the amount of the drug remaining at various times.

Time Since Use (in days)	0	1	2	3	4	5	6	7
Steroid Present (in mg)	100	90	81					

b. Make a plot of the data in Part a and write a short description of the pattern shown in the table and the plot.

c. Write two rules that describe the pattern of amount of steroid in the blood.

i. Write a *NOW-NEXT* rule showing how the amount of steroid present changes from one day to the next.

ii. Write a "$y = \ldots$" rule that shows how one could calculate the amount of steroid present after any number of days.

d. Use one of the rules in Part c to estimate the amount of steroid left after 0.5 and 8.5 days.

e. Estimate, to the nearest tenth of a day, the half-life of cyprionate.

f. How long will it take the steroid to be reduced to only 1% of its original level in the body? That is, how many days will it take until 1 milligram of the original dose is left in the body?

4 When people suffer head injuries in accidents, emergency medical personnel sometimes administer a paralytic drug to keep the patient immobile. If the patient is found to need surgery, it's important that the immobilizing drug decay quickly.

For one typical paralytic drug, the standard dose is 50 micrograms. One hour after the injection, half the original dose has decayed into other chemicals. The halving process continues the next hour, and so on.

a. How much of the drug will remain in the patient's system after 1 hour? After 2 hours? After 3 hours?

b. Write a rule that shows how to calculate the amount of drug that will remain x hours after the initial dose.

c. Use your rule to make a table showing the amount of drug left at half-hour intervals from 0 to 5 hours.

d. Make a plot of the data from Part c and a continuous graph of the function on the same axes.

e. How long will it take the 50-microgram dose to decay to less than 0.05 microgram?

5 Radioactive materials have many important uses in the modern world, from fuel for power plants to medical x-rays and cancer treatments. But the radioactivity that produces energy and tools for "seeing" inside our bodies can have some dangerous effects too; for example, it can cause cancer in humans.

The radioactive chemical *strontium-90* is produced in many nuclear reactions. Extreme care must be taken in transport and disposal of this substance. It decays slowly—if an amount is stored at the beginning of a year, 98% of that amount will still be present at the end of the year.

a. If 100 grams (about 0.22 pound) of strontium-90 are released by accident, how much of that radioactive substance will still be around after 1 year? After 2 years? After 3 years?

b. Write two different rules that can be used to calculate the amount of strontium-90 remaining from an initial amount of 100 grams at any year in the future.

c. Make a table and a graph showing the amount of strontium-90 that will remain from an initial amount of 100 grams at the end of every 10-year period during a century.

Years Elapsed	0	10	20	30	40	50	...
Amount Left (in g)	100						

d. Find the amount of strontium-90 left from an initial amount of 100 grams after 15.5 years.

e. Find the number of years that must pass until only 10 grams remain.

f. Estimate, to the nearest tenth of a year, the half-life of strontium-90.

6 The values of expensive products like automobiles *depreciate* from year to year. One common method for calculating the depreciation of automobile values assumes that a car loses 20% of its value every year. For example, suppose a new pickup truck costs \$20,000. The value of that truck one year later will be only $20{,}000 - 0.2(20{,}000) = \$16{,}000$.

a. Why is it true that for any value of x, $x - 20\%x = 80\%x$? How does this fact provide two different ways of calculating depreciated values?

b. Write *NOW-NEXT* and "$y = \ldots$" rules that can be used to calculate the value of the truck in any year.

c. Estimate the time when the truck's value is only \$1,000. Show how the answer to this question can be found in a table and on a graph.

d. How would the rules in Part b change if the truck's purchase price was only \$15,000? What if the purchase price was \$25,000?

7 In Applications Task 4 of Lesson 1, you counted the number of "chairs" at each stage in a design process that begin like this:

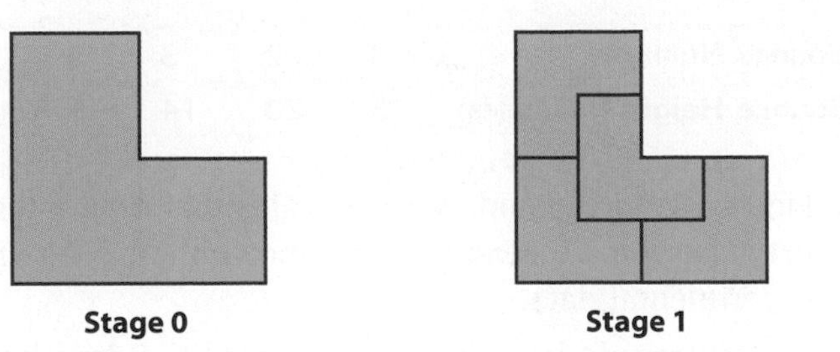

The chair at Stage 0 can be made by placing three square tiles in an "L" pattern. Suppose that the tiles used to make the chair design at Stage 0 are each one-centimeter squares. Then the left side and the bottom of that chair are each two centimeters long.

a. Complete a table like this that shows the lengths of those chair sides in smaller chairs used at later stages of the subdivision process.

Subdivision Stage	0	1	2	3	4	5	...	*n*
Side Length (in cm)	2	...						

b. Write two rules that show how to calculate the side length (in cm) of the smaller chair at any stage—one using *NOW* and *NEXT*, and another beginning "$L = \ldots$."

c. The area of the chair at Stage 0 is 3 square centimeters. What is the area of each small chair at Stage 1? At Stage 2? At Stage 3? At Stage n?

d. Write two rules that show how to calculate the area (in cm^2) of the smaller chairs at any stage—one using *NOW* and *NEXT*, and another beginning "$A = \ldots$."

8 Fleas are one of the most common pests for dogs. If your dog has fleas, you can buy many different kinds of treatments, but they wear off over time. Suppose the half-life of one such treatment is 10 days.

a. Make a table showing the fraction of an initial treatment that will be active after 10, 20, 30, and 40 days.

b. Experiment with your calculator or computer software to find a function of the form $y = b^x$ (where x is time in days) that matches the pattern in your table.

On Your Own

9 Suppose that an experiment to test the bounce of a tennis ball gave the data in the following table.

Bounce Number	1	2	3	4	5	6
Bounce Height (in inches)	35	20	14	9	5	3

a. Find *NOW-NEXT* and "$y = ...$" rules that model the relationship between bounce height and bounce number shown in the experimental data.

b. Use either rule from Part a to estimate the drop height of the ball.

c. Modify the rules from Part a to provide models for the relationship between bounce height and bounce number in case the drop height was 100 inches. Then make a table and plot of estimates for the heights of the first 6 bounces in this case.

d. What percent seems to describe well the relationship between drop height and bounce height of the tennis ball used in the experiment?

Consider the following experiment:

- Start with a pile of 100 kernels of popcorn or dry beans.
- Pour the kernels or beans onto the center of a large paper plate with equal-sized sectors marked as in the following diagram. Shake the plate so that the kernels or beans scatter into the various sectors in a somewhat random pattern.
- Remove all kernels that land on the sectors marked "1" *and* "2" and record the trial number and the number of kernels or beans remaining.
- Repeat the shake-remove-count process several times.

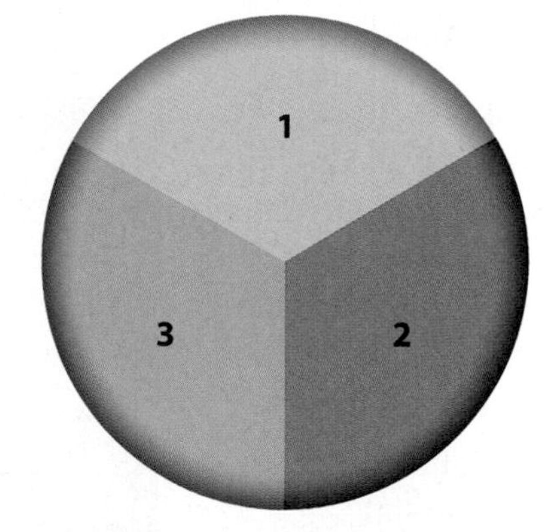

a. If you were to record the results of this experiment in a table of (*trial number, kernels left*) values, what pattern would you expect in that data? What function rule would probably be the best model for the relationship between trial number n and kernels left k?

b. Which of the following data patterns seems most likely to result from performing the experiment and why?

Table I

Trial Number	1	2	3	4	5
Kernels Left	65	40	25	15	10

Table II

Trial Number	1	2	3	4	5
Kernels Left	80	60	40	20	0

Table III

Trial Number	1	2	3	4	5
Kernels Left	35	15	5	2	0

11 Suppose that you performed the following experiment:

- Roll 100 fair dice and remove all that show 2, 4, or 6 dots on the top face.
- Roll the remaining dice and remove all that show 2, 4, or 6 dots on the top face.
- Repeat the roll-and-remove process, recording the number of dice left at each roll.

a. Complete a table like this showing your prediction of the number of dice remaining after each roll-and-remove stage of the experiment.

Roll Number	0	1	2	3	4	5	6	7
Estimated Dice Left	100							

b. Write *NOW-NEXT* and "$y = \ldots$" rules that model the relationship between roll number and dice left shown in your table.

c. Suppose that your teacher claimed to have done a similar experiment, starting with only 30 dice, and got the results shown in the next table.

Roll Number	0	1	2	3	4	5	6
Dice Left	30	17	10	4	3	1	1

Is the teacher's claim reasonable? What evidence supports your judgment?

12 Find values of x and y that will make these equations true statements.

a. $\left(\frac{5}{4}\right)^3 = x$

b. $\left(\frac{5}{d}\right)^2 = \frac{5^x}{d^y}$ $(d \neq 0)$

c. $\left(\frac{n}{4}\right)^3 = \frac{n^x}{y}$

d. $\left(\frac{t^3}{s}\right)^4 = \frac{t^x}{s^y}$ $(s \neq 0)$

13 Write each of the following expressions in a simpler equivalent exponential form.

a. $\left(\frac{4x}{n}\right)^2$

b. $\left(\frac{32x^2y^5}{8x^3y}\right)^2$

c. $\left(\frac{5x}{4y^3}\right)^0$

14 Find values for x and y that will make these equations true statements.

a. $\frac{5^7}{5^5} = 5^y$ b. $\frac{3^x}{3^5} = 3^6$ c. $\frac{t^5}{t^2} = t^y$ d. $\frac{6.4^9}{6.4^9} = 6.4^y$

15 Write each of the following expressions in a simpler equivalent exponential form.

a. $\frac{7^{11}}{7^4}$ b. $\frac{25x^3}{5x}$ c. $\frac{30x^3y^2}{6xy}$ d. $\frac{a^3b^4}{ab^4}$

16 Write each of the following expressions in equivalent exponential form. For those involving negative exponents, write an equivalent form without using negative exponents. For those involving positive exponents, write an equivalent form using negative exponents.

a. 4.5^{-2} b. $(7x)^{-1}$ c. $\left(\frac{2}{5}\right)^{-1}$ d. $\left(\frac{1}{5}\right)^{-4}$

e. $5x^{-3}$ f. $\left(\frac{2}{5}\right)^2$ g. $(4ax)^{-2}$ h. $\frac{5}{t^3}$

17 In Parts a–h below, write the number in integer or common fraction form, where possible. Where not possible, write an expression in simplest form using radicals.

a. $\sqrt{49}$ b. $\sqrt{28}$ c. $98^{\frac{1}{2}}$ d. $\sqrt{\frac{64}{25}}$

e. $\sqrt{6}\sqrt{24}$ f. $\sqrt{9 + 16}$ g. $\sqrt{\frac{12}{49}}$ h. $(\sqrt{49})^2$

18 Answer these questions about the side and diagonal lengths of squares.

a. How long is the diagonal of a square if each side is 12 inches long?

b. How long is each side of a square if the diagonal is $5\sqrt{2}$ inches long?

c. How long is each side of a square if the diagonal is 12 inches long?

d. What is the area of a square with a diagonal $5\sqrt{2}$ inches long?

e. What is the area of a square with a diagonal length d units?

Connections

19 One of the most interesting and famous fractal patterns is named after the Polish mathematician Waclaw Sierpinski. The start and first two stages in making a triangular *Sierpinski carpet* are shown below. Assume that the area of the original equilateral triangle is 12 square meters.

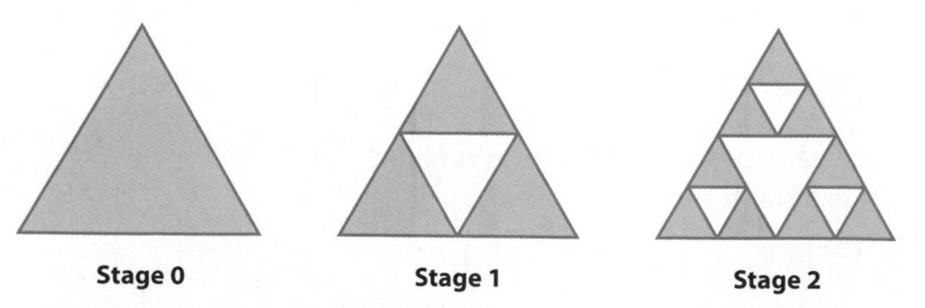

Stage 0 **Stage 1** **Stage 2**

a. Sketch the next stage in the pattern. Note how, in typical fractal style, small pieces of the design are similar to the design of the whole.

b. Make a table showing (*cutout stage, area remaining*) data for cutout stages 0 to 5 of this process.

c. Make a plot of the data in Part b.

d. Write two different rules that can be used to calculate the area of the remaining carpet at different stages. One rule should show change from one stage to the next. The other should be in the form "$y = \ldots$."

e. How many stages are required to reach the point where there is:

i. more hole than carpet remaining?

ii. less than 0.1 square meters of carpet remaining?

20 For each of the following rules, decide whether the function represented is an example of:

- An increasing linear function
- A decreasing linear function
- An exponential growth function
- An exponential decay function
- Neither linear nor exponential function

In each case, explain how the form of the rule was used in making your decision.

a. $y = 5(0.4^x)$

b. $y = 5 + 0.4x$

c. $y = 0.4(5^x)$

d. $y = 0.4 + 5x$

e. $y = \frac{5}{x}$

f. $y = \frac{0.4}{x}$

g. $y = 5 - 0.4x$

h. $y = 0.4 - 5x$

i. $NEXT = 0.4 \cdot NOW$

j. $NEXT = NOW + 0.4$

k. $NEXT = NOW - 5$

l. $NEXT = 5 \cdot NOW$

21 The graphs, tables, and rules below model four exponential growth and decay situations. For each graph, there is a matching table and a matching rule. Use what you know about the patterns of exponential relations to match each graph with its corresponding table and rule. In each case, explain the clues that can be used to match the items without any use of a graphing calculator or computer.

Graphs

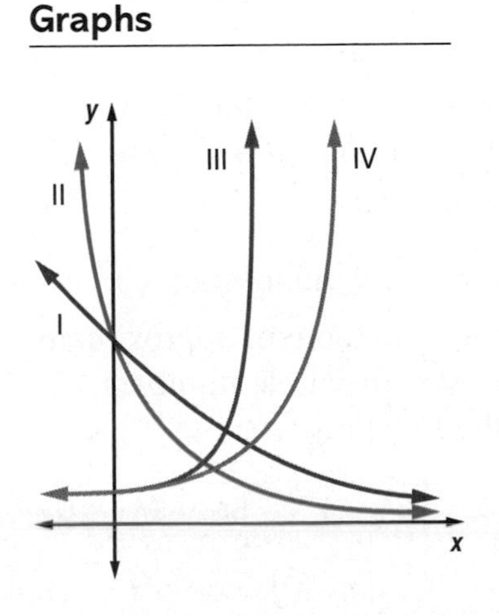

Tables

A

x	1	2	3	4
y	40	16	6.4	2.56

B

x	1	2	3	4
y	30	90	270	810

C

x	1	2	3	4
y	60	36	21.6	12.96

D

x	1	2	3	4
y	20	40	80	160

Rules

(1) $y = 100(0.6^x)$

(2) $y = 100(0.4^x)$

(3) $y = 10(2^x)$

(4) $y = 10(3^x)$

On Your Own

22 When very large numbers are used in scientific work, they are usually written in what is called *scientific notation*—that is, as the product of a decimal between 1 and 10 (usually rounded to three decimal places) with some power of 10. For example, basic measurements of the Earth are often given in scientific notation like this:

Measurement	Standard Form	Scientific Notation
Land Area (in m^2)	58,969,045,000,000	5.897×10^{13}
Volume (in km^3)	1,083,000,000,000,000,000	1.083×10^{18}
Population	6,214,891,000	6.215×10^{9}
Mass (in kg)	5,976,000,000,000,000,000,000,000	5.976×10^{24}

a. Write each of these large numbers in scientific notation rounded to three decimal places.

i. 234,567,890 **ii.** 54,987 **iii.** 1,024,456,981,876

b. Use negative exponents to write each of these numbers in scientific notation.

i. 0.0234 **ii.** 0.00002056 **iii.** 0.000000000008

c. Translate each of these numbers, given in scientific notation, to standard numeral form.

i. 7.82×10^8 **ii.** 5.032×10^6 **iii.** 8.1×10^{-3}

d. Express the results of these calculations in scientific notation, without using a calculator. Be prepared to explain your reasoning and how you use properties of exponents to reach the results.

i. $(4 \times 10^{12}) \times (3 \times 10^5)$

ii. $(40 \times 10^{12}) \div (5 \times 10^5)$

iii. $(4 \times 10^{12}) \times (3 \times 10^{-5})$

e. Use the Earth measurement data in the table to answer these questions. Express your answers in both scientific and standard notation.

i. How much land surface is there for each person living today?

ii. Each kilogram of mass is equal to 1,000 grams. What is the mass of the Earth in grams?

In 2001, the U.S. national public debt was 5.807×10^{12} dollars, and the U.S. population was about 2.85×10^8 people. What does this imply in terms of national public debt per person?

Every non-negative number x, has a non-negative square root $\sqrt{x}$.

a. Use your calculator to complete the following table of approximate values for the square root function $y = \sqrt{x}$ for whole numbers from 0 to 10. Then sketch a graph of the function.

x	0	1	2	3	4	5	6	7	8	9	10
$y = \sqrt{x}$											

b. How is the pattern of change shown in the table and graph of the square root function similar to and different from those of exponential growth and decay functions? How about linear functions?

25 The shell of the chambered nautilus is one of the most beautiful designs in nature. The outside image is a spiral, and segments of the spiral match chambers within the shell that increase in size as the spiral unfolds.

The spiral diagram to the left of the nautilus picture below is similar to the shell. Each outside segment is 1 centimeter long. The individual "chambers" are right triangles.

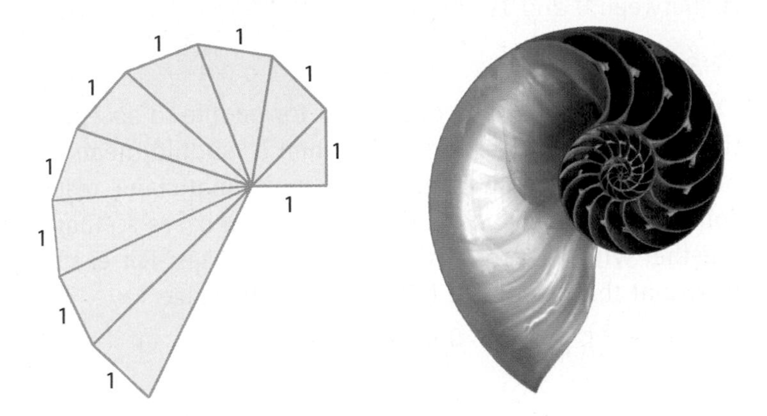

a. Make a table showing the pattern of lengths for the segments that divide the "shell chambers." Report each segment length in radical form.

b. What rule tells the length of the hypotenuse in the *n*th "chamber"?

Reflections

26 When some students were discussing the ball bounce experiment, one said that he thought the ball might rebound less on each bounce, but it would never actually stop bouncing just a little bit. His partners disagreed. They said that because the rebound height decreases on successive bounces, the rebound *time* also decreases. They said that the sum of rebound times would be:

$$\frac{1}{2} + \frac{1}{4} + \frac{1}{8} + \frac{1}{16} + \cdots = 1$$

What do you think?

On Your Own

27 Suppose a person taking steroid drugs is hospitalized due to a side effect from the drug. Tests taken upon admittance show a steroid concentration of 1.0. The next test one day later shows a concentration of 0.75. Based on these results, the person's family and friends assume that in three more days the drug will be out of the person's system.

a. What pattern of change are the family and friends assuming?

b. What might be a more accurate pattern prediction? Why is that pattern more reasonable?

28 For a function with rule $y = a(b^x)$ where $a > 0$, what conclusions can you draw about the tables and the graphs of (x, y) values when b is

a. between 0 and 1?

b. greater than 1?

29 The definition $b^0 = 1$ is often hard for people to accept. They argue that if b^5 means "5 factors of b," then b^0 should mean "zero factors of b" and this should be zero. The mathematician's response is that sometimes we make definitions for special cases so that they fit together with patterns covering all other cases. For example, they point out that we define $(-3)(-3) = 9$ because

$$\begin{aligned} 0 &= (-3)(0) \\ &= (-3)(3 + (-3)) \\ &= (-3)(3) + (-3)(-3) \\ &= -9 + (-3)(-3) \end{aligned}$$

a. What do you think of this argument for $(-3)(-3) = 9$?

b. What do you think of the general practice of making definitions in special cases so that they fit the rules that apply to all other cases?

30 A student at Sam Houston High School in Texas wrote the following equations to show equivalence of exponential forms. Decide which of the statements are correct. Correct those that are incorrect by revising the right side of the given equations. Give the student advice for correct thinking about the indicated calculations.

a. $\frac{x^2}{x^3} = x$ b. $(2x)(x) = 3x^2$ c. $\frac{30x^3y^3}{6xy} = 24x^3y^3$

Extensions

31 The African Black Rhinoceros is the second largest of all land mammals. The black rhino has walked the earth for 40 million years, and prior to the 19th century over 1,000,000 of the species roamed the plains of Africa. However, that number has been drastically reduced by hunting and loss of natural habitat.

The next table shows the very sharp decline in black rhino numbers between 1970 and 1993.

Year	1970	1980	1984	1986	1993
Population (in 1,000s)	65	15	9	3.8	2.3

Source: *Mammals of the World,* fifth ed., vol. 2. Johns Hopkins University Press: Baltimore, 1991; www.rhinos-irf.org/information/blackrhino/index.htm

a. Experiment with exponential and linear models for the data pattern shown in the table (use 0 for 1970). Decide on a model that seems to be a good fit for the data pattern.

b. Use your model of choice to predict the black rhino population for 2000 and 2005.

c. Since 1996, intense anti-poaching efforts have had encouraging results. Black rhino population estimates for 2000 rose to 2,700 and for 2005 rose to 3,600. Include this additional data and experiment with possible models for the data pattern. Use your model of choice to predict the black rhino population for 2010.

d. Suppose that black rhinos are not poached and their natural habitat is left intact. Assume also that the population would increase at a natural rate of 4% each year after 1993. How would the African population change by 2010 under those conditions?

e. As is the case with populations of Alaskan bowhead whales, native Africans might be allowed an annual hunting quota. Suppose that in 1993, the quota was set at 50 per year. What *NOW-NEXT* rule shows how to explore the effect of this hunting and a 4% natural population growth rate? What 2010 population is predicted under those conditions?

f. Construct a spreadsheet to explore the effects of different natural growth rates and hunting quotas and summarize what you learn in a report. In particular, find hunting quotas that would lead to stable black rhino populations if natural population growth rates were 2%, 5%, 7%, and 10% and the current population is 4,000.

32 Cigarette smoke contains nicotine, an addictive and harmful chemical that affects the brain, nervous system, and lungs. It leads to very high annual health care costs for our country.

Suppose an individual smokes one cigarette every 40 minutes over a period of 6 hours and that each cigarette introduces 100 units of nicotine into the bloodstream. The half-life of nicotine is 20 minutes.

a. Make a chart that tracks the amount of nicotine in that smoker's body over the 6-hour period, making entries in the chart for every 20-minute period. Describe the pattern of nicotine build-up.

b. Compare the pattern in Part a with the pattern resulting from smoking a cigarette every 20 minutes.

c. Write *NOW-NEXT* rules showing how the amount of nicotine in the body changes over time (in 40-minute intervals for Part a and 20-minute intervals for Part b). Compare these rules to those of simple exponential growth and decay and explain the differences.

d. Because nicotine is strongly addictive, it is difficult for smokers to break the habit. Suppose that a long-time smoker decides to quit "cold turkey." That is, rather than reducing the number of cigarettes smoked, the smoker resolves never to pick up another cigarette. How would the level of nicotine in that smoker's body change over time?

e. How do the results of your analysis in Part d suggest that addiction to nicotine is psychological as well as physical?

33 Driving after drinking alcohol is both dangerous and illegal. The National Highway Traffic Safety Administration reported 2,400 youth (15 to 20 years old) alcohol-related traffic fatalities in 2002—an average of about 6 per day. (Source: *Traffic Safety Facts 2002: Young Drivers*; www.nhtsa.dot.gov)

Many factors affect a person's Blood Alcohol Concentration (BAC), including body weight, gender, and amount drunk. American Medical Association guidelines suggest that a BAC of 0.05 is the maximum safe level for activities like driving a car.

The following chart gives typical data relating body weight and number of drinks consumed to BAC for people of various weights.

Approximate Blood Alcohol Concentrations

Weight (in pounds)	1 drink	2 drinks	3 drinks	4 drinks	5 drinks
100	0.05	0.09	0.14	0.18	0.23
120	0.04	0.08	0.11	0.15	0.19
140	0.03	0.07	0.10	0.13	0.16
160	0.03	0.06	0.09	0.11	0.14
180	0.03	0.05	0.08	0.10	0.13

a. Study the data in the table and decide how BAC for each weight seems to be related to number of drinks consumed. Find *NOW-NEXT* and "$y = \ldots$" rules for the function that seems the best model at each weight. Explain what each rule tells about the effects of additional drinks on BAC.

b. The next table shows how BAC changes over time after drinking stops for a 100-pound person who has had 3 drinks.

Time (in hours)	0	2	4	6	8
BAC	0.14	0.12	0.10	0.08	0.06

i. What type of function seems to model that pattern of change well?

ii. Find *NOW-NEXT* and "$y = \ldots$" rules for the function that seems the best model.

iii. Explain what those rules tell about the way BAC declines over time.

c. Suppose that the pattern in Part b relating BAC to time since last drink applies for the situation in which a 100-pound person has 5 drinks. What is the prediction of time required for the blood alcohol of that 100-pound person to return to a "safe" level of 0.05?

34 To study behavior of exponential functions for fractional values of the independent variable, consider several numbers between 0 and 1, like 0.25, 0.5, and 0.75.

a. How would you expect the values of 5^0, $5^{0.25}$, $5^{0.5}$, $5^{0.75}$, and 5^1 to be related to each other?

b. What are your best estimates for the values of $5^{0.25}$, $5^{0.5}$, and $5^{0.75}$?

c. Use a calculator to check your ideas in Parts a and b.

d. Graph the function $y = 5^x$ for $-1 \leq x \leq 2$ and explain how it shows the observed pattern of change in values for 5^0, $5^{0.25}$, $5^{0.5}$, $5^{0.75}$, and 5^1.

35 Use what you know about properties of exponents to evaluate these expressions.

i. $\left(3^{\frac{1}{4}}\right)^4$ **ii.** $\left(5^{\frac{1}{4}}\right)^4$ **iii.** $\left(16^{\frac{1}{4}}\right)^4$

iv. $16^{\frac{1}{4}}$ **v.** $\left(16^{\frac{1}{4}}\right)^3$ **vi.** $(16)^{\frac{3}{4}}$

a. Look for a pattern to help you explain what $b^{\frac{1}{4}}$ and $b^{\frac{3}{4}}$ must mean for any positive value of b.

b. Based on your explorations, what meanings are suggested for the expressions $b^{\frac{1}{n}}$ and $b^{\frac{m}{n}}$ when b is a positive number and m and n are positive integers? Use your calculator to check your ideas in the case of some specific examples.

36 Any number that can be expressed as an integer or a common fraction is called a *rational number*. For example, 12, $\frac{3}{5}$, and $\frac{7}{5}$ are rational numbers. If a number cannot be expressed as an integer or common fraction, it is called an *irrational number*.

When you use a computer algebra system for arithmetic and algebraic calculations, it will generally report any numerical results as integers or common fractions whenever that is possible.

a. Use a computer algebra system to see which of the numbers $\sqrt{2}, \sqrt{3}, \sqrt{4}, \sqrt{5}, \ldots, \sqrt{50}$ are rational numbers and which are irrational.

b. Based on your work in Part a, what seems to be a way to decide when $\sqrt{n}$ is rational and when it is irrational?

c. Use a computer algebra system to evaluate $\sqrt{\frac{a}{b}}$ for several different fractions like $\sqrt{\frac{4}{9}}, \sqrt{\frac{4}{7}}, \sqrt{\frac{5}{9}}$, and $\sqrt{\frac{6}{35}}$.

d. Based on your work in Part c, what seems to be a way to decide when $\sqrt{\frac{a}{b}}$ is rational and when it is irrational?

Review

37 Work on problems that involve exponential growth and decay often requires skill in use of percents to express rates of increase or decrease. Suppose that you are asked to figure new prices for items in a sporting goods store. Show two ways to calculate each of the following price changes—one that involves two operations (either a multiplication and an addition or a multiplication and a subtraction) and another that involves only one operation (multiplication).

a. Reduce the price of a \$90 warm-up suit by 20%.

b. Increase the price of a \$25 basketball by 30%.

c. Reduce the price of a \$75 skateboard by 60%.

d. Increase the price of a \$29 sweatshirt by 15%.

e. Reduce the price of a \$15 baseball cap by $33\frac{1}{3}\%$.

f. Increase the price of a \$60 tennis racket by 100%.

38 Sketch and label a diagram for each situation. Then find the measure of the indicated segment or angle.

a. $m\angle XYZ = 140°$ and $\overrightarrow{YB}$ bisects $\angle XYZ$. Find $m\angle BYZ$.

b. $AB = 5$ cm and C is the midpoint of $\overline{AB}$. Find AC.

c. M is the midpoint of $\overline{XY}$, $XM = 3$ cm. Find XY.

d. $\overrightarrow{PQ}$ bisects $\angle RPT$. $m\angle RPQ + m\angle QPT = 82°$. Find $m\angle RPT$ and $m\angle RPQ$.

39 Write *NOW-NEXT* rules that match each of the following *linear decay* functions. Then explain how the "decay" in all three cases is different from exponential decay and how the difference(s) would appear in tables and graphs of (x, y) values for the functions.

a. $y = 5 - 2x$

b. $y = -0.5x + 1$

c. $y = -\frac{x}{3} - 6$

40 In Unit 4, *Vertex-Edge Graphs*, you learned about coloring maps and coloring vertex-edge graphs. Suppose the sketch at the right represents a simple map, where country *A* shares a border with country *B*. (Rules forbid having countries meet at a point.)

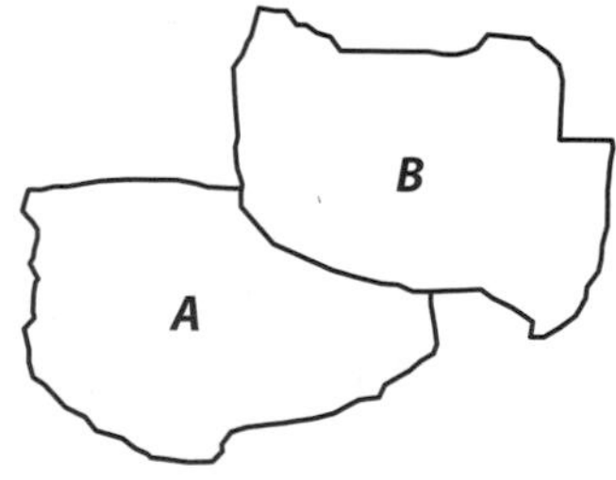

a. Copy this map. Add a country *C* that shares a border with both countries *A* and *B*. Now add a country that shares a border with each of countries *A*, *B*, and *C*. Continue to add a country, one at a time, each country sharing a border with *each* of the previous countries. Try different placements of the countries. What was the maximum number of countries possible?

b. The vertex-edge graph shown represents the same situation as the map on the previous page—the edge represents a shared border. As before, add vertices (countries) and edges (borders) one at a time. What is the maximum number of vertices on the graph?

A • ——— • B

c. Conjecture: If every country shares a border with every other country, then the largest number of countries possible is ________ .

41 Find the missing side lengths in each triangle. Express your answers in both radical and decimal approximation form.

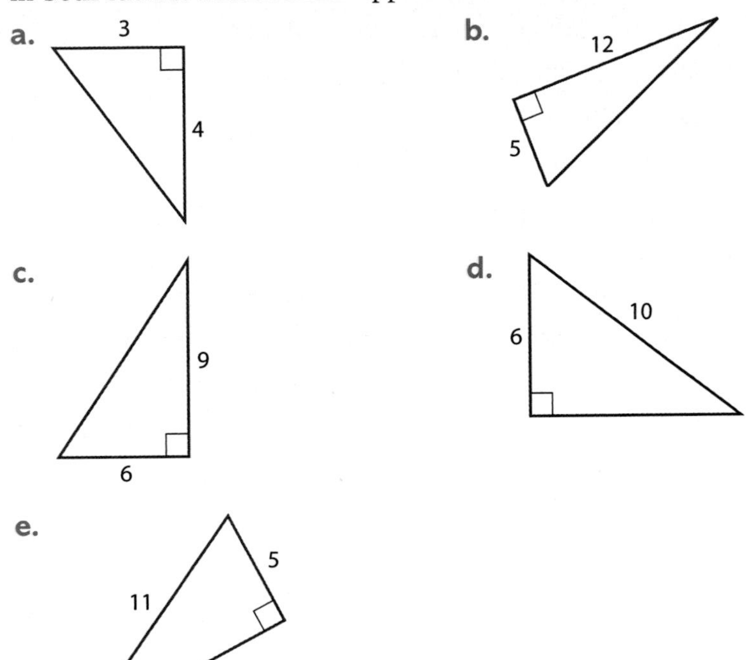

e. 11, 5

42 Without using a calculator, decide if the following statements are true or false. If the statement is false, explain why.

a. $2 \cdot 2^5 = 4^5$

b. $2 \cdot 2^5 = 2^5$

c. $2 \cdot 2^5 = 2^6$

d. $2 \cdot 3^x + 3^x = (2 + 1)3^x$

e. $3 \cdot 3^x = 3^{x+1}$

43 Make a copy of the diagram at the right. Shade the part of the large square that represents $\frac{1}{2}$ of $\frac{1}{2}$ of $\frac{1}{2}$, and give the answer as a fraction.

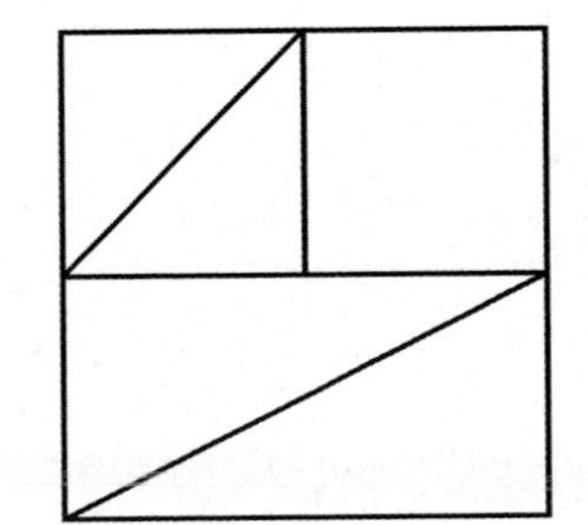

44 Examine each equation below and decide on the method you would use to solve it: use a table, use a graph, or reason with the symbols themselves. Then solve each equation using your preferred method and check your answer.

a. $2^x = 100$

b. $2x + 5 = 100$

c. $100(x + 2) = 800$

d. $100(1.5^x) = 200$

e. $100(b^4) = 200$

45 Write an equation for the line that matches each description.

a. Has slope of −0.5 and y-intercept at (0, 2)

b. Contains the points (0, 5) and (−4, −10)

c. Is horizontal and contains the point (7, 12)

d. Contains the points (−2, 7.5) and (1, 3)

46 A basketball team is selling sweatshirts in order to raise money for new uniforms. The rule that gives their profit p in dollars based on the number of sweatshirts they sell n is $p = 5n - 175$.

a. Explain the meaning of the 175 and the 5 in terms of the situation.

b. If they sell 265 sweatshirts, how much profit will they make?

c. How many sweatshirts must they sell in order to make a profit of $2,000?

Looking Back

In this unit, you studied patterns of change in variables that can be modeled well by exponential functions. These functions can all be expressed with rules like

$$NEXT = b \cdot NOW, \text{ starting at } a$$
or
$$y = a(b^x).$$

As a result of your work on Lessons 1 and 2, you should be better able to recognize situations in which variables are related by exponential functions, to use data tables and graphs to display patterns in those relationships, to use symbolic rules to describe and reason about the patterns, and to use graphing calculators, spreadsheets, and computer algebra systems to answer questions that involve exponential relationships. You should also be able to write exponential expressions in useful equivalent forms.

The tasks in this final lesson will help you review your understanding of exponential functions and apply that understanding in solving several new problems.

1. **Counting Codes** Code numbers are used in hundreds of ways every day—from student and social security numbers to product codes in stores and membership numbers in clubs.

   a. How many different 2-digit codes can be created using the digits 0, 1, 2, 3, 4, 5, 6, 7, 8, and 9 (for example, 33, 54, 72, or 02)?

   b. How many different 3-digit codes can be created using those digits?

   c. How many different 4-digit codes can be created using those digits?

   d. Using any patterns you may see, complete a table like the one below showing the relation between number of digits and number of different possible codes.

Number of Digits	1	2	3	4	5	6	7	8	9
Number of Codes									

e. Write a rule using *NOW* and *NEXT* to describe the pattern in the table of Part d.

f. Write a rule that shows how to calculate the number of codes C for any number of digits D used.

g. Music and video stores stock thousands of different items. How many digits would you need in order to have code numbers for up to 8,500 different items?

h. How will your answers to Parts d–f change if the codes were to begin with a single letter of the alphabet (A, B, C, ... , or Z) as in A23 or S75?

2 **Eyes on the Prizes** In one women's professional golf tournament, the money a player wins depends on her finishing place in the standings. The first-place finisher wins $\frac{1}{2}$ of the \$1,048,576 in total prize money. The second-place finisher wins $\frac{1}{2}$ of what is left; then the third-place finisher wins $\frac{1}{2}$ of what is left, and so on.

U.S. Women's Open Champion Annika Sorenstam

a. What fraction of the *total* prize is won

i. by the second-place finisher?

ii. by the third-place finisher?

iii. by the fourth-place finisher?

b. Write a rule showing how to calculate the fraction of the total prize money won by the player finishing in nth place, for any positive integer n.

c. Make a table showing the actual prize money in dollars (not fraction of the total prize money) won by each of the first five place finishers.

Place	1	2	3	4	5
Prize (in dollars)					

d. Write a rule showing how to calculate the actual prize money in dollars won by the player finishing in place n. How much money would be won by the 10th-place finisher?

e. How would your answers to Parts a–d change if

i. the total prize money were reduced to \$500,000?

ii. the fraction used was $\frac{1}{4}$ instead of $\frac{1}{2}$?

f. When prize monies are awarded using either fraction $\frac{1}{2}$ or $\frac{1}{4}$, could the tournament organizers end up giving away more than the stated total prize amount? Explain your reasoning.

3 **Cold Surgery** Hypothermia is a life-threatening condition in which body temperature falls well below the norm of 98.6°F. However, because chilling causes normal body functions to slow down, doctors are exploring ways to use hypothermia as a technique for extending time of delicate operations like brain surgery.

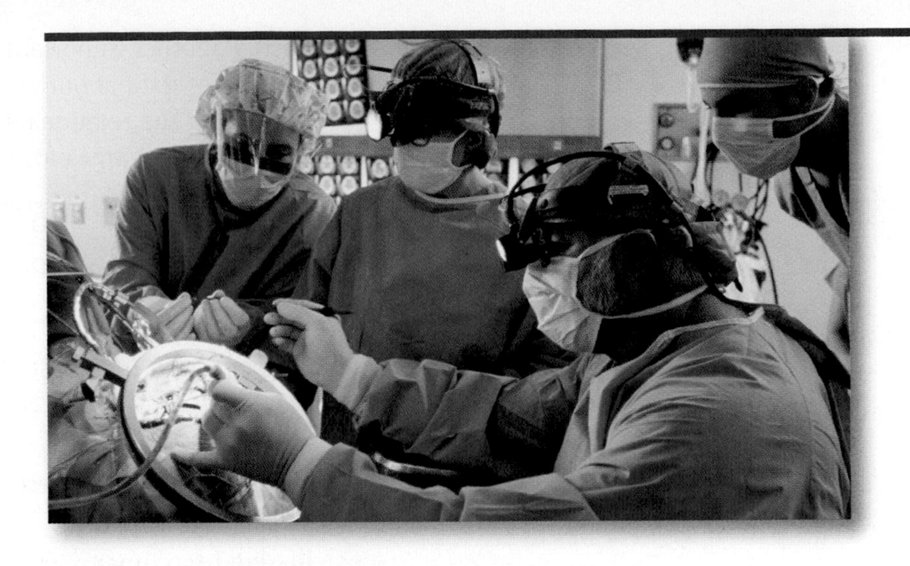

a. The following table gives experimental data illustrating the relationship between body temperature and brain activity.

Body Temperature (in °F)	50	59	68	77	86	98.6
Brain Activity (% Normal)	11	16	24	37	52	100

Source: *USA Today*, August 1, 2001, "Surgery's Chilling Future Will Put Fragile Lives on Ice."

i. Plot the table data and find a "$y = \ldots$" rule that models the pattern in these data relating brain activity level to body temperature. Then express the same relationship with an equivalent *NOW-NEXT* rule.

ii. Use your rules to estimate the level of brain activity at a body temperature of 39°F, the lowest temperature used in surgery experiments on pigs, dogs, and baboons.

iii. Find the range of body temperatures at which brain activity is predicted to be about 75% of normal levels.

b. The next table gives experimental data illustrating the relationship between body temperature and safe operating time for brain surgery.

Body Temperature (in °F)	50	59	68	77	86	98.6
Safe Operating Time (in minutes)	45	31	21	14	9	5

i. Plot the table data and find a "$y = \ldots$" rule that models the pattern in these data relating safe operating time to body temperature. Then express the same relationship with an equivalent *NOW-NEXT* rule.

ii. Use your rules to estimate the safe operating time at a body temperature of 39°F.

iii. Find the body temperature at which safe operating time is predicted to be at least 25 minutes.

c. Cost is another important variable in medical practice. The next table gives data about charges for a sample of routine surgeries, illustrating the relationship between time required for the operation and hospital charges for use of the operating room.

Time (in minutes)	30	60	90	120	150	180
Cost (in $)	950	1,400	1,850	2,300	2,750	3,200

i. Plot the table data and find a "$y = \ldots$" rule that models the pattern in these data relating surgery cost to time. Then express the same relationship with an equivalent *NOW-NEXT* rule.

ii. Use your rules to estimate the cost of an operation that takes 45 minutes.

iii. Find the time of an operation for which cost is predicted to be $5,000.

d. Compare the "$y = \ldots$" rules for the three functions in Parts a, b, and c. In each case, explain how the rules alone can be used to predict the pattern of change in the dependent variable as the independent variable increases.

4 Exponent Properties In Lessons 1 and 2, you discovered and practiced several principles for writing exponential expressions in equivalent (often simpler) forms. Use those principles to find values of x and y that make the following equations true statements.

a. $(2.3^5)(2.3^3) = 2.3^x$

b. $2.3^x = 1$

c. $(3.5^x)^y = 3.5^{12}$

d. $\frac{7^9}{7^4} = 7^x$

e. $\frac{7^x}{7^4} = 7^2$

f. $(7^3)^x = 7^6$

g. $\left(\frac{3}{5}\right)^4 = \frac{3^x}{5^y}$

h. $(4a)^3 = 4^x a^y$

i. $\frac{1}{7^4} = 7^x$

5 Fractional Powers and Radicals In Lesson 2, you also discovered and practiced use of expressions in which fractional powers occur. Special attention was paid to square roots, using the exponent one-half. Use what you learned to answer these questions.

a. The value of $3^2 = 9$ and $3^3 = 27$. What does this information tell about the approximate values of $3^{2.4}$ and $3^{2.7}$?

b. For each of these equations, find two different pairs of integer values for a and b that make the equation true.

i. $\sqrt{48} = a\sqrt{b}$

ii. $\sqrt{a}\sqrt{b} = \sqrt{36}$

Summarize the Mathematics

When two variables are related by an exponential function, that relationship can be recognized from key features of the problem situations, from patterns in tables and graphs of (x, y) data, and from the rules that show how to calculate values of one variable from given values of the other.

a In deciding whether an exponential function describes the relationship between two variables, what hints do you get from

i. the nature of the situation and the variables involved?

ii. the patterns in graphs or scatterplots?

iii. the patterns in data tables?

b Exponential functions, like linear functions, can be expressed by a rule relating x and y values and by a rule relating *NOW* and *NEXT* y values.

i. Write a general rule for an exponential function, "$y = \ldots$."

ii. Write a general rule relating *NOW* and *NEXT* for an exponential function.

iii. What do the parts of the rules tell you about the problem situation?

iv. How do you decide whether a given exponential function rule will describe growth or decay, and why does your decision rule make sense?

c Suppose that you develop or discover a rule (*NOW-NEXT* or "$y = \ldots$") that shows how a variable y is an exponential function of another variable x. Describe the different strategies you could use to complete tasks like these:

i. Find the value of y associated with a specific given value of x.

ii. Find the value of x that gives a specific target value of y.

iii. Describe the way that the value of y changes as the value of x increases or decreases.

d Complete each equality to give a useful equivalent form of the first expression.

i. $a^m a^n = \ldots$ **ii.** $(a^m)^n = \ldots$ **iii.** $a^0 = \ldots$

iv. $(ab)^n = \ldots$ **v.** $\frac{1}{a^n} = \ldots$ **vi.** $\left(\frac{a}{b}\right)^n = \ldots$

vii. $\frac{a^m}{a^n} = \ldots$ **viii.** $\sqrt{ab} = \ldots$ **ix.** $\sqrt{\frac{a}{b}} = \ldots$

Be prepared to share your responses and thinking with the class.

Check Your Understanding

Write, in outline form, a summary of the important mathematical concepts and methods developed in this unit. Organize your summary so that it can be used as a quick reference in future units and courses.

Patterns in Shape

Shape is an important and fascinating aspect of the world in which you live. You see shapes in nature, in art and design, in architecture and mechanical devices. Some shapes, like the Rock and Roll Hall of Fame building, are three-dimensional. Others, like the architect's plans for the building are two-dimensional.

In this unit, your study will focus on describing and classifying two-dimensional and three-dimensional shapes, on visualizing and representing them with drawings, and on analyzing and applying their properties. You will develop understanding and skill in use of the geometry of shape through work on problems in three lessons.

Lessons

1 *Two-Dimensional Shapes*

Use combinations of side lengths and angle measures to create congruent triangles and quadrilaterals. Investigate properties of these figures by experimentation and by careful reasoning. Use those properties to study the design of structures and mechanisms and to solve problems.

2 *Polygons and Their Properties*

Recognize and use symmetry and other properties of polygons and of combinations of polygons that tile a plane.

3 *Three-Dimensional Shapes*

Recognize, visualize, and develop drawing methods for representing three-dimensional shapes. Analyze and apply properties of polyhedra.

Two-Dimensional Shapes

In previous units, you used the shape of graphs to aid in understanding patterns of linear and nonlinear change. In this unit, you will study properties of some special geometric shapes in the plane and in space. In particular, you will study some of the geometry of two-dimensional figures called *polygons* and three-dimensional figures called *polyhedra*, formed by them.

The geometry of shape is among some of the earliest mathematics. It was used in ancient Egypt to construct the pyramids and to measure land. For example, when the yearly floods of the Nile River receded, the river often followed a different path. As a result, the shape and size of fields along the river changed from year to year. It is believed that the Egyptians used ropes tied with equally-spaced knots to re-establish land boundaries. To see how a knotted rope might be used in building design and measuring, think about how you could use a piece of rope tied into a 24-meter loop with knots at one-meter intervals.

Think About This Situation

Suppose that you and two or three friends each grabbed the rope at a different knot and pulled outward until the loop formed a particular shape.

a) How could you position yourselves so that the resulting shape was an equilateral triangle? An isosceles triangle? A right triangle?

b) How are the perimeters of the three triangles related? How do you think the areas are related?

c) How could you position yourselves so that the resulting shape was a square? A rectangle? A parallelogram that is not a rectangle?

d) How are the perimeters of the three quadrilaterals related? How do you think the areas are related?

As you complete the investigations in the following three lessons, you will discover why some shapes are used so frequently in building and in design. You will also discover how knowledge of a few basic properties of geometric figures can be used to reason to many additional properties of those shapes.

Investigation 1 Shape and Function

Buildings and bridges, like most objects around you, are three-dimensional. They have length, height, and depth (or width). To better understand the design of these objects, it is often helpful to examine the two-dimensional shapes of their components. Triangles and special quadrilaterals such as rectangles are among the most commonly occurring two-dimensional shapes in structural designs.

In this first investigation, you will explore conditions on the sides of triangles and quadrilaterals that affect their shape. In the process, you will discover some physical properties of these shapes that have important applications. As you work on the following problems, look for answers to these questions:

What conditions on side lengths are needed to build triangles and quadrilaterals? What additional constraints are needed to build special quadrilaterals?

Why and how are triangles used in the design of structures like bridge trusses?

Why and how are quadrilaterals used in the design of devices like windshield wipers?

Using strands of uncooked thin spaghetti, conduct the following experiment at least three times. Keep a record of your findings, including sketches of the shapes you make.

- Mark any two points along the length of a strand of spaghetti and break the spaghetti at those two points.
- Try to build a triangle with the pieces end-to-end.
- If a triangle can be built, try to build a differently shaped triangle with the same side lengths.

a. Was it possible to build a triangle in each case? If a triangle could be built, could you build a differently shaped triangle using the same three segments? Compare your findings with those of others.

b. If a triangle can be built from three segments, how do the segment lengths appear to be related? Use a ruler and compass to test your conjecture for segments of length 3 cm, 4 cm, and 5 cm. For segments of length 5 cm, 6 cm, and 12 cm. Revise your conjecture if needed.

c. Suppose a, b, and c are side lengths of *any* triangle. Write an equation or inequality relating a, b, and c. How many different equations or inequalities can you write relating a, b, and c?

d. Write in words the relationship that must be satisfied by the side lengths of any triangle (do not use letters to name the side lengths). This relationship is called the **Triangle Inequality**.

You may recall from your prior mathematical study that triangles can be classified in terms of their sides as *scalene*, *isosceles*, or *equilateral*.

a. What type of triangle were most of the triangles that you built in your experiment? Explain as carefully as you can why you might expect that result.

b. Draw an isosceles triangle that is not equilateral. Suppose a is the length of two of the sides and b is the length of the *base* of your triangle. How must a and b be related? Explain your reasoning using the Triangle Inequality.

c. Can you build an equilateral triangle of any side length a? Explain your reasoning using the Triangle Inequality.

3 Now use strands of spaghetti to conduct this quadrilateral-building experiment at least three times. Keep a record of your findings, including sketches of the shapes you make.

- Mark any three points along the length of a strand of spaghetti and break the spaghetti at those three points.
- Try to build a quadrilateral with the pieces end-to-end.
- If a quadrilateral can be built, try to build another, differently shaped quadrilateral with the same side lengths.

a. Was it possible to build a quadrilateral in each case? If a quadrilateral could be built, could you build a differently shaped quadrilateral using the same four segments? Compare your findings with those of others.

b. If a quadrilateral can be built from four side lengths, how are the side lengths related? Use a ruler and compass to test your conjecture for segments of length 3 cm, 5 cm, 8 cm, and 10 cm. For segments of length 4 cm, 4 cm, 7 cm, and 15 cm. For segments of length 2 cm, 4 cm, 8 cm, and 16 cm.

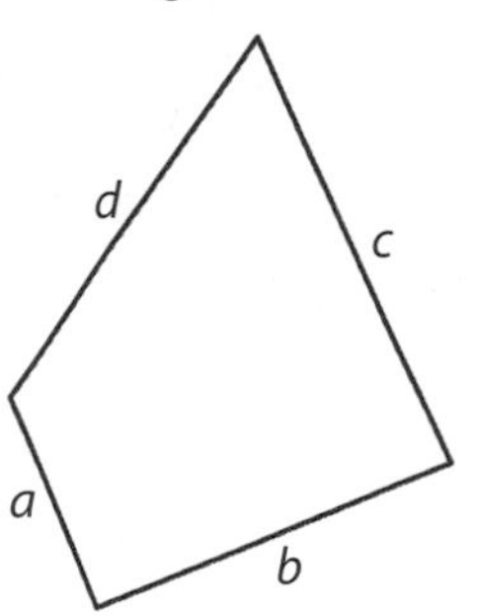

c. Suppose a, b, c, and d are consecutive side lengths of any quadrilateral. Write an equation or inequality relating a, b, c, and d. How many different equations or inequalities can you write relating a, b, c, and d?

d. Write in words the relationship that must be satisfied by the four side lengths of any quadrilateral (do not use letters to name side lengths).

Quadrilaterals are more complicated than triangles. They have more sides and more angles. In Problem 3, you discovered that using the same four side lengths of a quadrilateral, you could build quite different shapes. Quadrilaterals are classified as *convex*—as in the case of the quadrilateral below on the left—or *nonconvex*—as in the case of the quadrilateral on the right.

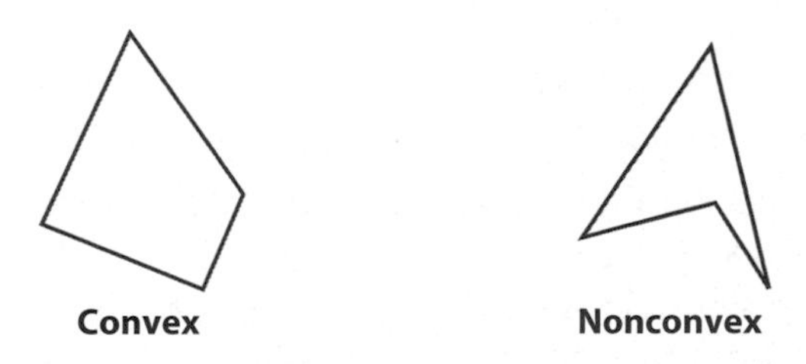

4 Some special convex quadrilaterals can be characterized in terms of side lengths. For example, in completing Part c of the Think About This Situation, you likely created a **parallelogram** by forming a quadrilateral with opposite sides the same length.

a. Show how you can build a parallelogram using four segments cut from a strand of spaghetti and placed end-to-end.

- **i.** How many differently shaped parallelograms can you build with those four segments?
- **ii.** What additional constraint(s) would you have to build into the shape for it to be a rectangle?

b. A **kite** is a quadrilateral with two distinct pairs of consecutive sides the same length.

i. Build a kite using the same four segments of spaghetti, in Part a, placed end-to-end.

ii. How many differently shaped kites can you build with those four pieces?

c. A **rhombus** is a quadrilateral with all four sides the same length.

i. Build a rhombus using four segments from a strand of spaghetti placed end-to-end.

ii. How many differently shaped rhombi can you build with those four pieces?

iii. Explain why a rhombus is a parallelogram.

iv. What additional constraint(s) would you have to build into a rhombus for it to be a square?

The results of your experiments in building triangles and quadrilaterals lead to important physical applications.

Working with a partner, use plastic or cardboard strips and paper fasteners to make each of the models shown below.

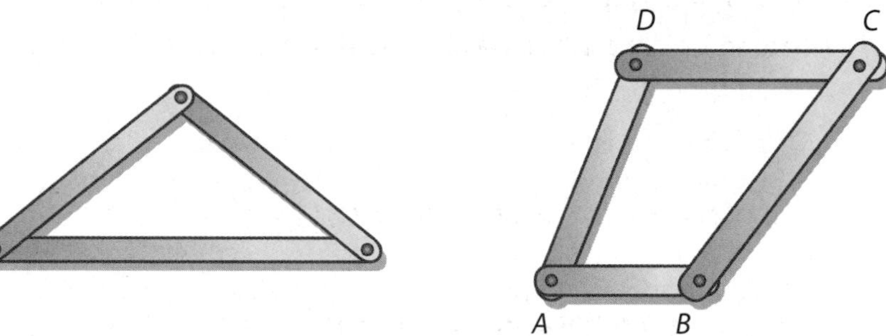

a. Can you change the shape of the triangle model? Can you change any of the features of the model? Explain.

b. What features of the quadrilateral model can you change? What features of the model cannot change?

c. Now add a *diagonal* strip $\overline{BD}$ to your quadrilateral model. What features of this model can change?

d. Triangles are **rigid**. They retain their shape when pressure is applied. Quadrilaterals are rigid when *triangulated* with a diagonal. The process of triangulating is often called *bracing*. How are these facts utilized in the design of the bridge truss shown on page 363?

e. Describe two structures or objects in your community or home that employ the rigidity of triangles in their design.

The nonrigidity of quadrilaterals has important physical applications. For example, mechanical engineers use the flexibility of quadrilaterals in the design of *linkages*.

6 An important feature of a quadrilateral or 4-bar linkage is that if any side is held fixed so it does not move and another side is moved, then the movement of the remaining sides is completely determined. The side that is fixed is called the *frame*. The two sides attached to the frame are called *cranks*. The crank most directly affected by the user is called the *driver* crank; the other is called the *follower* crank. The side opposite the frame is called the *coupler*.

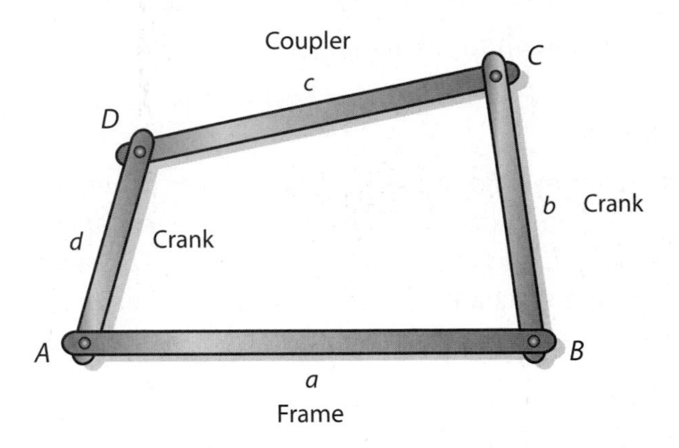

Quadrilateral linkages have different characteristics depending on the lengths a, b, c, d of the sides and which side is used as a crank. Explore some of those characteristics using linkage strips or computer software like the "Design a Linkage" custom tool.

a. Working with a partner, make several different quadrilateral linkages so that strip $\overline{AB}$ is the longest side and fixed; strip $\overline{AD}$ is the shortest side and acts as one of the cranks. Investigate how lengths a and d are related to lengths b and c when $\overline{AD}$ can rotate completely. In this case, how does the follower crank move? The coupler? Write a summary of your findings.

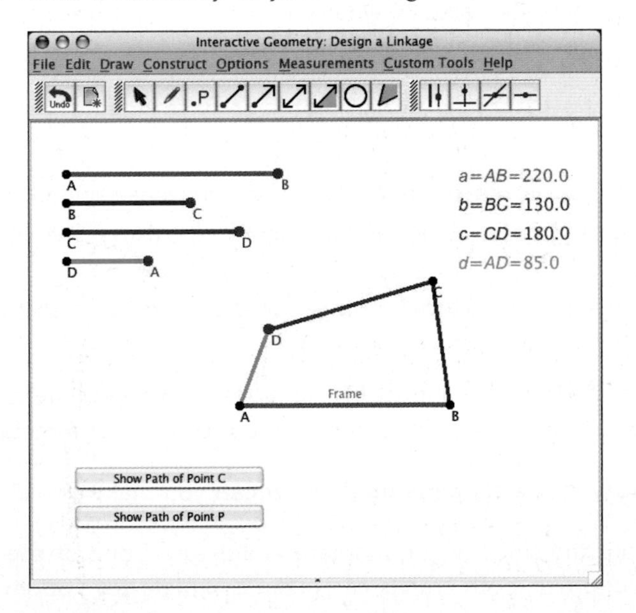

b. The principle you discovered in Part a is called **Grashof's Principle**. How could you use a mechanism satisfying Grashof's Principle to drive the agitator in a washing machine or an automotive windshield wiper?

c. Use a quadrilateral linkage satisfying Grashof's Principle to investigate the mechanics of the linkage if the shortest side is used as the frame. Summarize your findings.

7 Examine the bus windshield wiper mechanism shown at the left. The wiper blade is attached to the mechanism in a fixed position.

a. Make a sketch of this mechanism. Label the frame, cranks, and coupler.

b. Explain why this is a parallelogram linkage.

c. As the linkage moves, what paths do the ends of the wiper blade follow?

d. If the wiper blade is vertical (as shown) when the mechanism is at the beginning of a cycle, describe the positions of the blade when the mechanism is one quarter of the way through its cycle and when the mechanism is halfway through its cycle.

e. Sketch the region of the windshield that the blade can keep clean.

Summarize the Mathematics

In this investigation, you experimented with building triangles and quadrilaterals with different side lengths. You also investigated how the rigidity of triangles and the nonrigidity of quadrilaterals influence their uses in the design of structures and devices.

a Describe the similarities and differences in what you discovered in your triangle-building and quadrilateral-building experiments.

b Suppose you are told that a triangular garden plot is to have sides of length 5 m, 12 m, and 13 m.

i. Explain why it is possible to have a triangular plot with these dimensions.

ii. Explain how you and a partner could lay out such a plot using only a 15-meter tape measure.

iii. How many differently-shaped triangular plots could be laid out with these dimensions? Why?

c What constraints are needed on the lengths of the sides of a quadrilateral for it to be a parallelogram? What additional constraint(s) are needed for it to be a rectangle?

d What does it mean to say that a shape is rigid? How can you make a quadrilateral rigid?

e What must be true about the sides of a quadrilateral linkage if one of the cranks can make a complete revolution? If both cranks can make complete revolutions?

Be prepared to share your ideas and reasoning with the class.

✓ Check Your Understanding

Four-bar linkages illustrate how geometric shape and function are related.

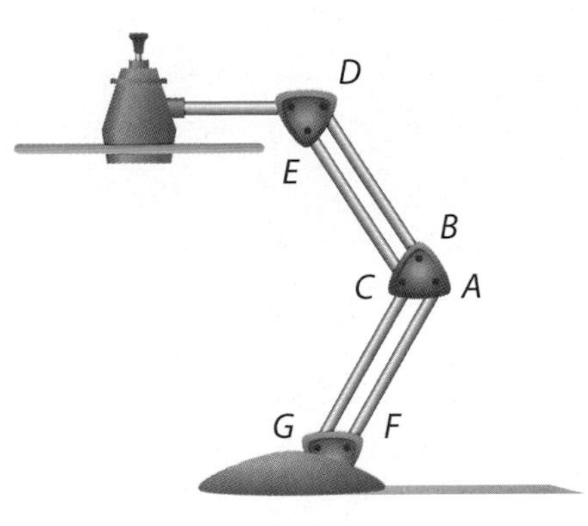

a. Examine the adjustable lamp in the diagram. The pivots at the labeled points are snug, but they will allow pivoting to adjust the lamp.

 i. Explain why the parallelogram linkages used in this lamp remain parallelograms as the position of the lamp is adjusted.

 ii. Visualize and describe how the position of the lamp should change as you make parallelogram *AFGC* vertical.

b. Suppose you are given segments of the following lengths: 7, 8, 24, 25.

 i. If possible, sketch and label several different quadrilaterals that can be formed with these side lengths.

 ii. Suppose you build a quadrilateral linkage with consecutive sides of lengths 7, 24, 8, and 25. What can you say about the length of the shortest possible brace that will make the quadrilateral rigid?

 iii. Can a quadrilateral linkage with a rotating crank be constructed from strips of these lengths? Explain your reasoning.

 iv. Can a quadrilateral linkage be made from these strips with two rotating cranks? Explain.

Investigation 2 Congruent Shapes

Roof trusses are manufactured in different shapes and sizes but they are most often triangular in shape. The "W" or Fink truss shown below is the most widely-used design in building today. The locations of the truss components provide for the most uniform distribution of stresses and forces. The rigidity of triangles is a key element in the design of these trusses. An equally important element is that all trusses for a particular roof are identical or *congruent*.

As you work on the problems of this investigation, look for answers to the following questions:

How can you test whether two shapes are congruent?

What combination of side or angle measures is sufficient to determine if two trusses or other triangular shapes are congruent?

 As a builder at the home site pictured on page 369, how could you test whether the two trusses standing against the garage wall are congruent? Could you use the same method to test if those two trusses are congruent to the ones already placed in position on the double-car garage?

Congruent figures have the same shape and size, regardless of *position* or *orientation*. In congruent figures, corresponding segments have the same length and corresponding angles have the same measure. The marks in the diagrams below indicate corresponding side lengths and angle measures that are identical.

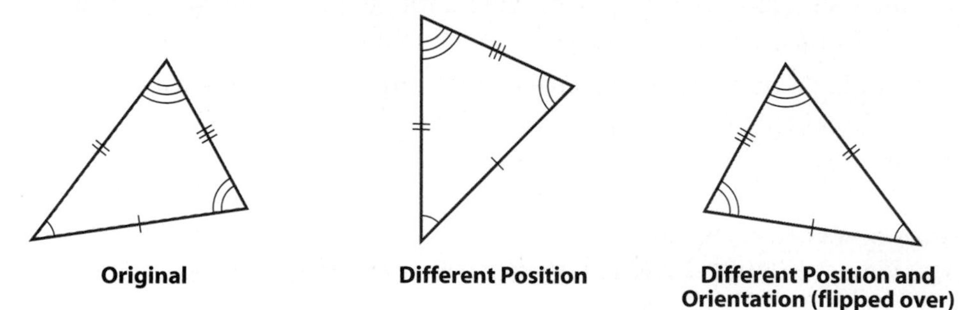
Original | Different Position | Different Position and Orientation (flipped over)

One way to test for congruence of two trusses, or any two figures, is to see if one figure can be made to coincide with the other by sliding, rotating, and perhaps flipping it. This is, of course, very impractical for large trusses. Your work in the previous investigation suggests an easier method.

 In Investigation 1, you found that given three side lengths that satisfy the Triangle Inequality, you could build only one triangle.

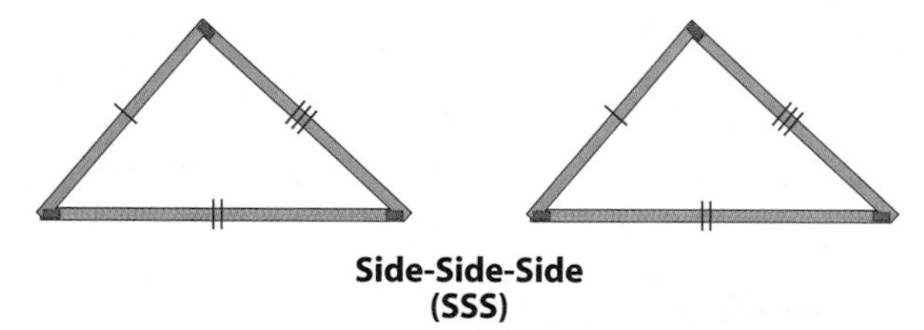
Side-Side-Side (SSS)

a. Explain as carefully as you can why simply measuring the lengths of the three corresponding sides of two triangular roof trusses is sufficient to determine if the trusses are congruent.

b. Could you test if the two trusses are congruent by measuring the lengths of just two corresponding sides? Explain.

In the following problems, you will explore other combinations of side lengths and angle measures that would provide a simple test of whether two triangular roof trusses are congruent.

Use strands of spaghetti along with a ruler and a compass, or geometry software like the "Triangle Congruence" custom tool to conduct the following triangle-building experiments.

For each condition in Parts a–c:

- Try to build a triangle satisfying the given condition. You choose segment lengths. Use one or two of the angles below as a template for the angle(s) of your triangle.

- If a triangle can be built, try to build another with the same three parts.
- Make a note if the condition could be used to test for congruence of two triangles.

For each experiment, compare your findings with your classmates and resolve any differences. Keep a record of your agreed-upon findings. Include sketches of the shapes you make.

a. *Side-Angle-Side (SAS) Condition*: You know the lengths of two sides and the measure of the angle between the two sides.

b. *Side-Side-Angle (SSA) Condition*: You know the lengths of two sides and the measure of an angle not between the two sides.

c. *Angle-Side-Angle (ASA) Condition*: You know the measures of two angles and the length of the side between the two angles.

4 You may recall from your prior mathematics study that the sum of the measures of the angles of a triangle is 180°.

a. How is this **Triangle Angle Sum Property** demonstrated by folding a paper model of a triangle as shown below?

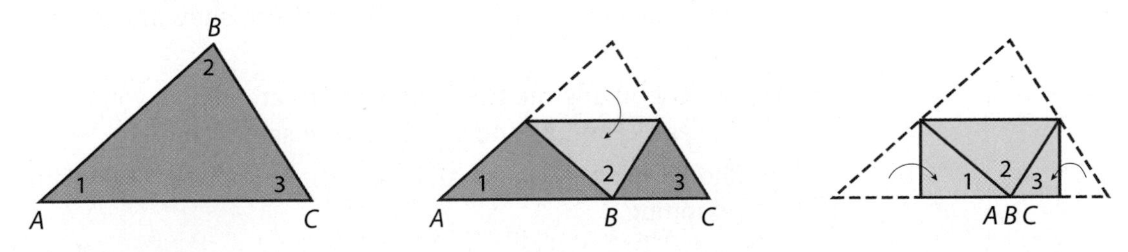

b. Using a protractor and ruler, carefully draw a triangle with angle measures 40°, 60°, and 80°.

c. Could a building contractor test whether two triangular roof trusses are congruent by measuring only the corresponding angles? Explain your reasoning.

The Kingpost truss shown below is used primarily for support of single-car garages or short spans of residential construction. The shape of the truss is an isosceles triangle. The support brace $\overline{BD}$ connects the peak of the truss to the midpoint of the opposite side.

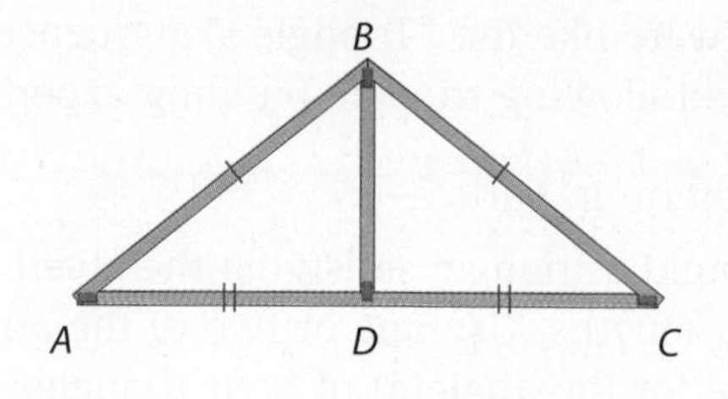

a. How are the specifications (given information) for this truss shown in the diagram?

b. Based on the specifications for this truss and the results of your experiments, explain as carefully as you can why $\triangle ABD$ is congruent to $\triangle CBD$, written $\triangle ABD \cong \triangle CBD$. (The congruence notation always lists the letters for corresponding vertices in the same order.)

c. To properly support the roof, it is important that the brace $\overline{BD}$ is perpendicular to side $\overline{AC}$. Based on your work in Part b, explain why the placement of brace $\overline{BD}$ guarantees that $\overline{BD}$ is perpendicular to $\overline{AC}$ (in symbols, $\overline{BD} \perp \overline{AC}$).

d. An important property of the Kingpost truss, and *any* isosceles triangle, is that the angles opposite the congruent sides (called **base angles**) are congruent. How does your work in Part b guarantee that $\angle A \cong \angle C$?

Study the diagram below of a "W" truss. $\triangle ABC$ is an isosceles triangle. Points D, E, F, and G are marked on the truss so that $\overline{CG} \cong \overline{BF}$ and $\overline{CD} \cong \overline{BE}$.

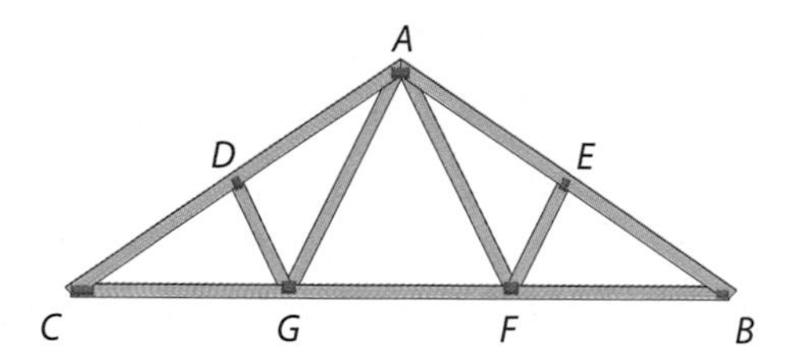

a. On a copy of the truss, use tick marks to show the given information.

b. When building the truss, explain as carefully as you can why braces $\overline{DG}$ and $\overline{EF}$ should be cut the same length.

c. Should braces $\overline{AG}$ and $\overline{AF}$ be cut the same length? Explain your reasoning.

Summarize the Mathematics

In this investigation, you discovered combinations of side lengths or angle measures that were sufficient to determine if two triangles were congruent. You also explored how you could use congruent triangles to reason about properties of an isosceles triangle.

a Which sets of conditions—SSS, SAS, SSA, ASA, and AAA—can be used to test if two triangles are congruent?

b Write each *Triangle Congruence Condition* in words and illustrate with a diagram.

c If $\triangle PQR \cong \triangle XYZ$, what segments are congruent? What angles are congruent?

d Describe properties of an isosceles triangle that you know by definition or by reasoning.

Be prepared to share your ideas and reasoning with the class.

Check Your Understanding

Wood trusses commonly employ two or more triangular components in their construction. For each truss below, examine the two labeled triangular components. Is enough information provided for you to conclude that the triangles are congruent? Explain your reasoning.

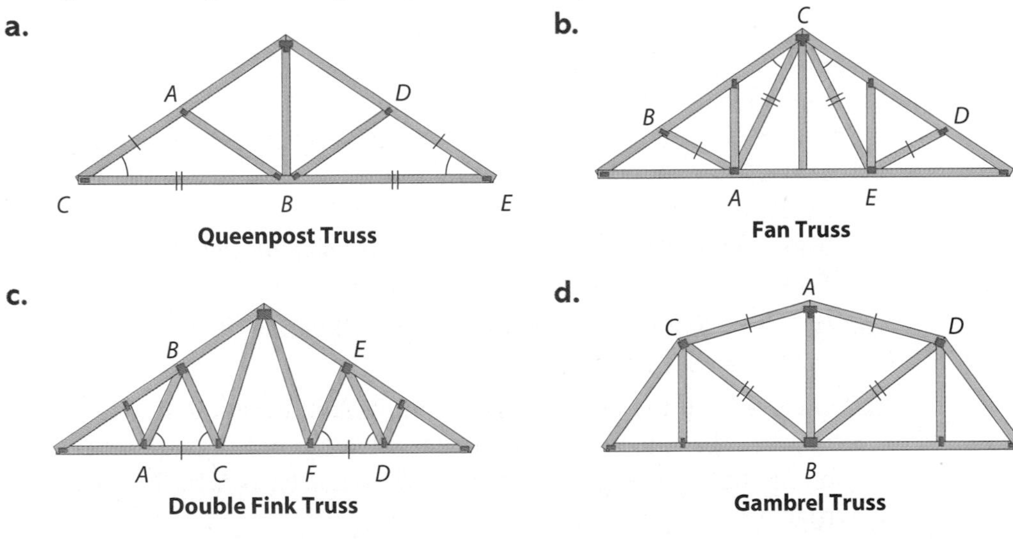

Investigation 3 Reasoning with Shapes

In your work with the Kingpost truss, you discovered some important properties of isosceles triangles—not by conducting experiments and looking for patterns but by careful reasoning from statements of facts that you and your classmates already understand and agree on. As you work on problems of this investigation, look for answers to the following questions:

What strategies are useful in reasoning about properties of shapes?

What are some additional properties of triangles and quadrilaterals that have important applications?

As you may recall, the support brace $\overline{BD}$ of a Kingpost truss as shown below on the left connects the peak of the truss to the midpoint of the opposite side. You used congruent triangles to show that $\overline{BD} \perp \overline{AC}$. In this case, $\overline{BD}$ is said to be a **perpendicular bisector** of $\overline{AC}$, that is $\overline{BD} \perp \overline{AC}$ at the midpoint D of $\overline{AC}$.

To design a Kingpost truss that has the same *span* $\overline{AC}$, but less *pitch* (slope), Beth located point E on the perpendicular bisector of $\overline{AC}$ as shown below on the right.

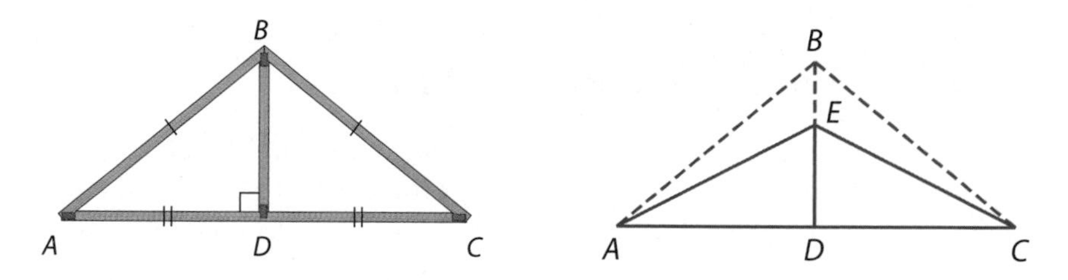

She was confident that the new truss would still be an isosceles triangle. She reasoned as follows:

I need to show that $\overline{EA} \cong \overline{EC}$. Consider $\triangle ADE$ and $\triangle CDE$.
Since $\overline{ED}$ is the $\perp$ bisector of $\overline{AC}$, $\angle ADE$ and $\angle CDE$ are right angles and $\overline{AD} \cong \overline{CD}$. The triangles share $\overline{ED}$.
So, $\triangle ADE \cong \triangle CDE$.
Since corresponding parts of congruent triangles are congruent, $\overline{EA} \cong \overline{EC}$.

a. Is Beth's reasoning correct? How does she know that $\triangle ADE \cong \triangle CDE$?

b. On a copy of the diagram above on the left, design a new truss that has the same span but greater pitch by locating a point F on the line $\overleftrightarrow{BD}$. Explain carefully why your truss is an isosceles triangle.

c. Explain why *any* point on the perpendicular bisector of a segment will be equally distant from the endpoints of the segment.

2 The truss shown at the right is often used for portions of a house in which a sloped interior ceiling is desired. It is designed so that $\overline{AB} \cong \overline{CB}$ and $\overline{AD} \cong \overline{CD}$.

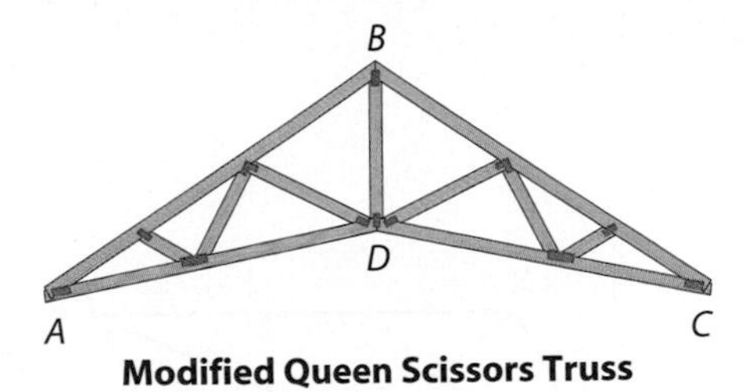

Modified Queen Scissors Truss

a. How could you reason with congruent triangles to explain why $\angle ABD \cong \angle CBD$?

b. What other pairs of angles in the truss must also be congruent? Why?

In Investigation 1, you found that you could make a quadrilateral linkage rigid by adding a diagonal brace. Diagonals are also helpful in reasoning about properties of quadrilaterals.

3 Recall that by definition of a parallelogram, opposite sides are the same length, or congruent.

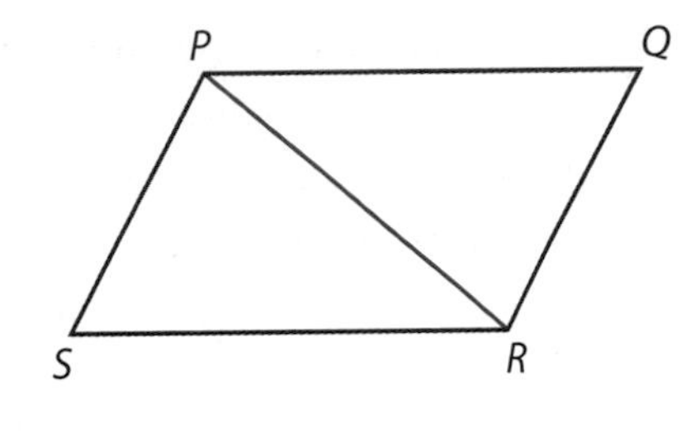

a. On a copy of parallelogram *PQRS*, use tick marks to indicate segments that are congruent.

b. Provide an argument to justify the statement:

A diagonal of a parallelogram divides the parallelogram into two congruent triangles.

c. Angles in a parallelogram like $\angle Q$ and $\angle S$ are called **opposite angles**.

i. Explain why $\angle Q \cong \angle S$.

ii. What reasoning would you use to show that the other pair of opposite angles, $\angle P$ and $\angle R$, are congruent? Compare your argument with others.

d. What is the sum of the measures of the angles of ▱*PQRS*? Give reasons that support your answer.

e. Would your answer and reasons in Part d change if the figure were a quadrilateral but *not* a parallelogram? Explain your reasoning.

4 Information on diagonal lengths can be used to test whether a quadrilateral is a special quadrilateral. The diagram below shows results of three trials of an experiment with two linkage strips fastened at their midpoints.

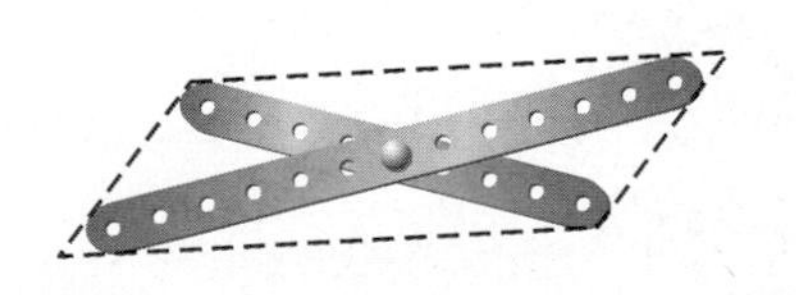 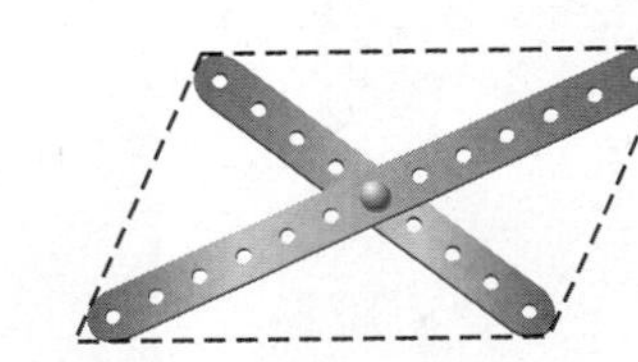 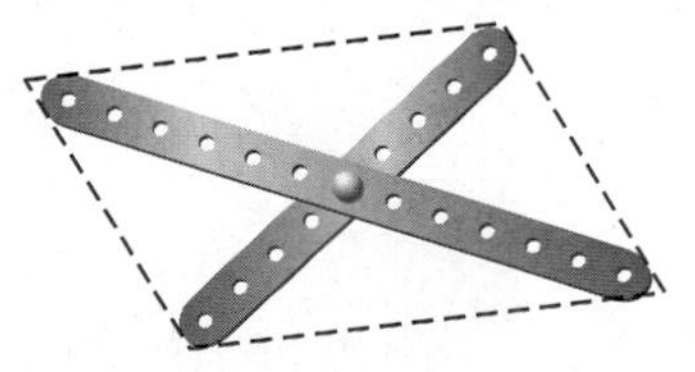

a. In each case, what appears to be true about the quadrilateral that has the given strips as its diagonals? Do you think the same conclusion would hold if you conducted additional trials of the experiment?

Parts b–e will provide you a guide to preparing a supporting argument for the statement:

If the diagonals of a quadrilateral bisect each other, then the quadrilateral is a parallelogram.

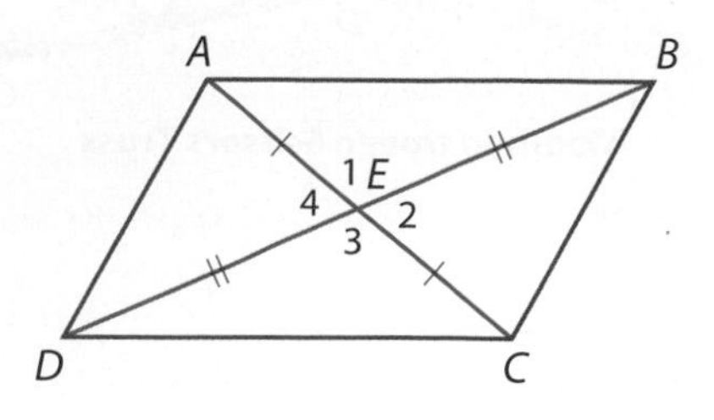

b. Study this diagram of a quadrilateral with diagonals that bisect each other.

i. Are pairs of segments given as congruent properly marked? Explain.

ii. To show quadrilateral *ABCD* is a parallelogram, you must show that opposite sides are the same length. To show that opposite sides $\overline{AB}$ and $\overline{CD}$ are congruent, what triangles would you try to show are congruent? What additional information would you need?

c. Angles positioned like $\angle 1$ and $\angle 3$, and $\angle 2$ and $\angle 4$, are called **vertical angles**. Each pair of vertical angles *appears* to be congruent. A student at Bellevue High School in Washington gave the following argument to justify that $\angle 1 \cong \angle 3$.

i. Give a reason to support each statement.

$m\angle 1 + m\angle 2 = 180°$	($m\angle 1$ is read "measure of $\angle 1$")	(1)
$m\angle 2 + m\angle 3 = 180°$		(2)
$m\angle 1 + m\angle 2 = m\angle 2 + m\angle 3$		(3)
$m\angle 1 = m\angle 3$		(4)
So, $\angle 1 \cong \angle 3$.		(5)

ii. Use similar reasoning to write an argument justifying that $\angle 2 \cong \angle 4$.

d. Explain why it follows that $\triangle AEB \cong \triangle CED$ and $\triangle AED \cong \triangle CEB$.

e. Why can you conclude that $\overline{AB} \cong \overline{CD}$ and $\overline{AD} \cong \overline{CB}$? That quadrilateral *ABCD* must be a parallelogram?

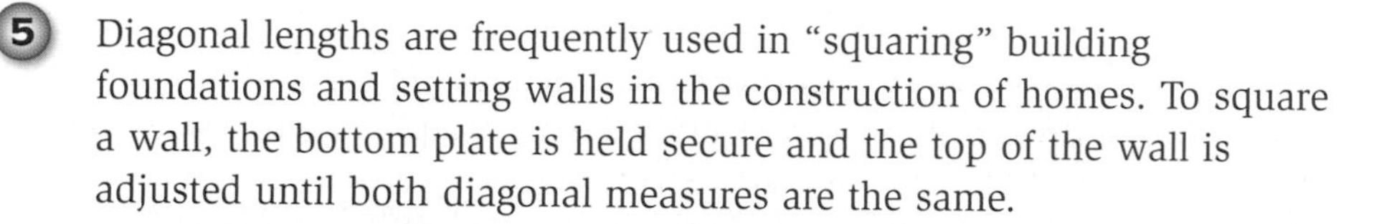

5 Diagonal lengths are frequently used in "squaring" building foundations and setting walls in the construction of homes. To square a wall, the bottom plate is held secure and the top of the wall is adjusted until both diagonal measures are the same.

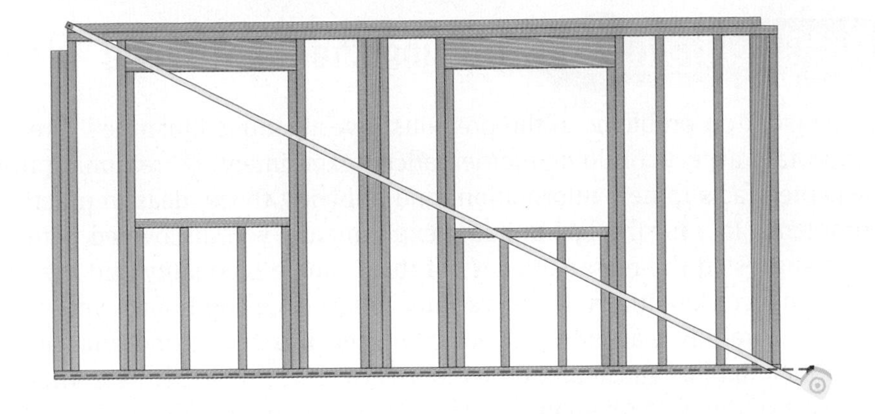

a. Assuming the top and bottom plates are the same length and the two wall studs at the ends are the same length, explain as carefully as you can why the statement, "If the diagonals are the same length, then the wall frame is a rectangle," is true. Your explanation should include a labeled diagram, a statement of what information is given in terms of the diagram, and supporting reasons for your statements.

b. Compare your argument with others. Correct any errors in reasoning.

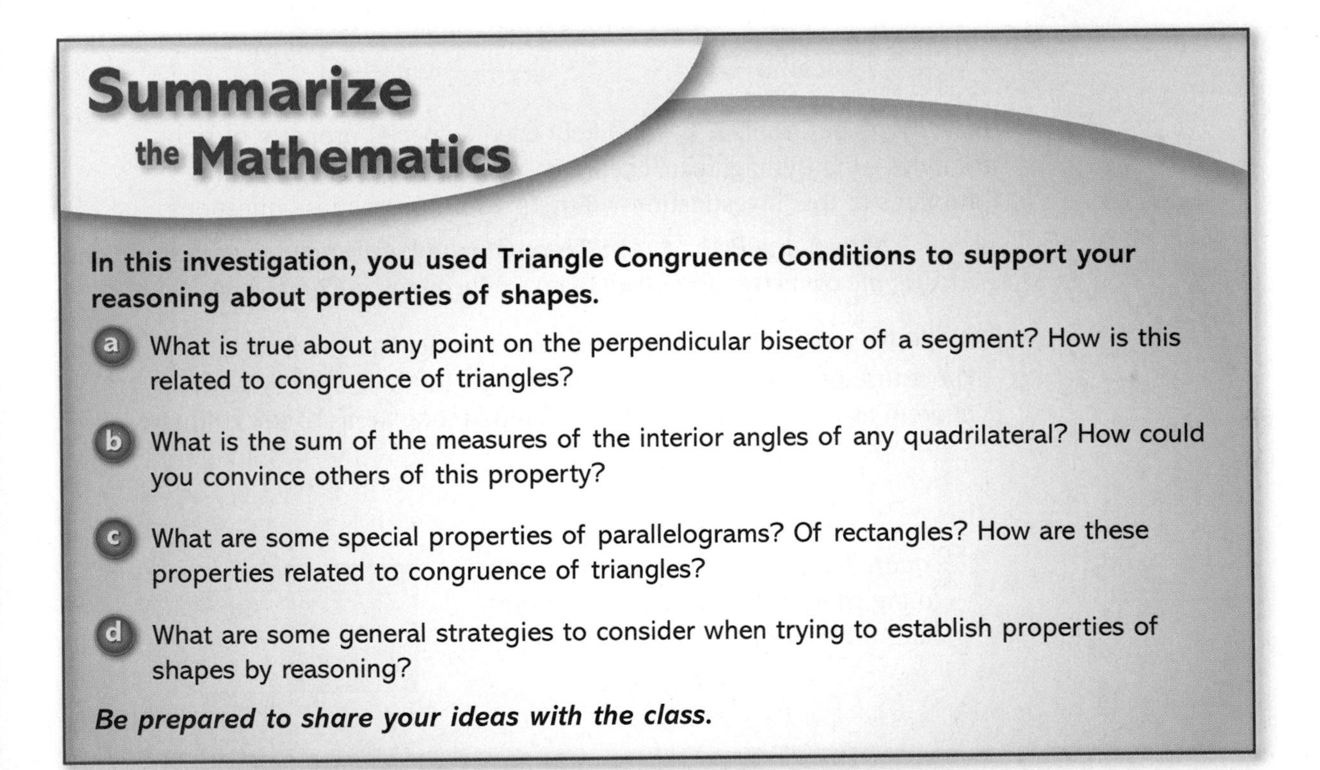

Summarize the Mathematics

In this investigation, you used Triangle Congruence Conditions to support your reasoning about properties of shapes.

a. What is true about any point on the perpendicular bisector of a segment? How is this related to congruence of triangles?

b. What is the sum of the measures of the interior angles of any quadrilateral? How could you convince others of this property?

c. What are some special properties of parallelograms? Of rectangles? How are these properties related to congruence of triangles?

d. What are some general strategies to consider when trying to establish properties of shapes by reasoning?

Be prepared to share your ideas with the class.

Check Your Understanding

Refer to kite *ABCD* with diagonal $\overline{AC}$ shown at the right.

a. Use careful reasoning to explain why $\angle 1 \cong \angle 2$ and $\angle 3 \cong \angle 4$.

b. What must be true about the shorter diagonal $\overline{DB}$? Why?

Investigation 4 Getting the Right Angle

Your work on problems in the previous investigations illustrated three important aspects of doing mathematics—experimenting, reasoning from accepted facts to new information, and applying those ideas to practical problems. In your triangle-building experiments, you discovered patterns that suggested the reasonableness of the Triangle Inequality and the Triangle Congruence Conditions. Using various congruence conditions, you were able to carefully reason to properties of special triangles and quadrilaterals. You then applied those properties to a variety of problems. Keep these aspects of doing mathematics in mind as you complete this investigation.

1 Bridging, shown in the diagram below, provides stability between adjacent floor joists. It is generally used when floor spans are greater than 8 feet. If the floor joists are set approximately 16 inches apart, to what length should the bridging be cut? Why should all pieces be cut the same length?

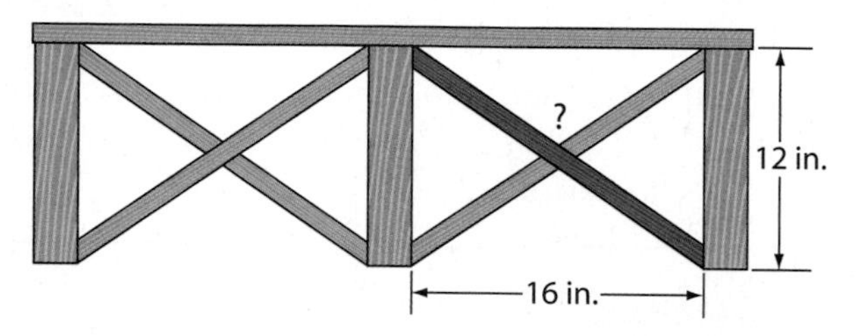

In working on Problem 1, you likely used a special property of right triangles—the Pythagorean Theorem. Your work on the remaining problems of this investigation will help you answer these questions:

Why is the Pythagorean Theorem true for all right triangles?
Is the converse of the Pythagorean Theorem true and, if so, why?

The Pythagorean Theorem is often used to calculate the length of the hypotenuse of a right triangle. You can also think of the Pythagorean Theorem as a statement of a relationship among areas of three squares.

For any right triangle, the area of the square built on the hypotenuse is equal to the sum of the areas of the squares built on the two legs.

$$a^2 + b^2 = c^2$$

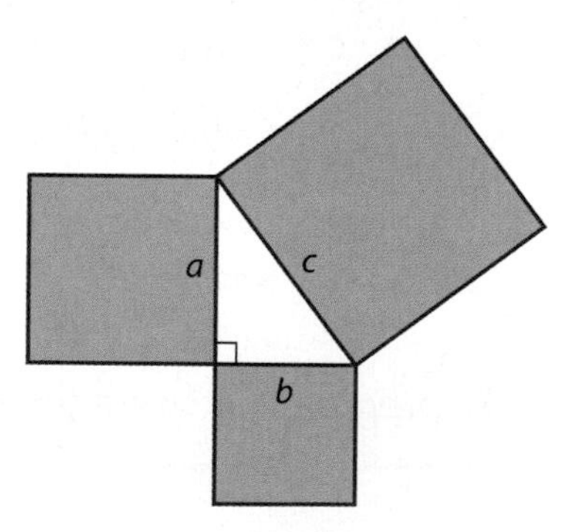

The Greek philosopher Pythagoras (572–497 B.C.) is sometimes credited with first providing a general argument for why this relationship is true for all right triangles. However, the oldest recorded justification is found in an ancient Chinese manuscript written more than 500 years before Pythagoras. The ancient Babylonians and Egyptians also discovered special cases of the relationship.

Since there are infinitely many right triangles, it would be impossible to check that $a^2 + b^2 = c^2$ for all of them. Pythagoras's argument, like that outlined in Problem 2, involves reasoning from known facts rather than relying on patterns in specific right triangles. In Problem 2, the argument involves finding the area of the same square *PQRS* in two different ways.

2 Study the diagrams below of a right triangle, a square built on the hypotenuse of the triangle, and an arrangement of congruent copies of the triangle around the square.

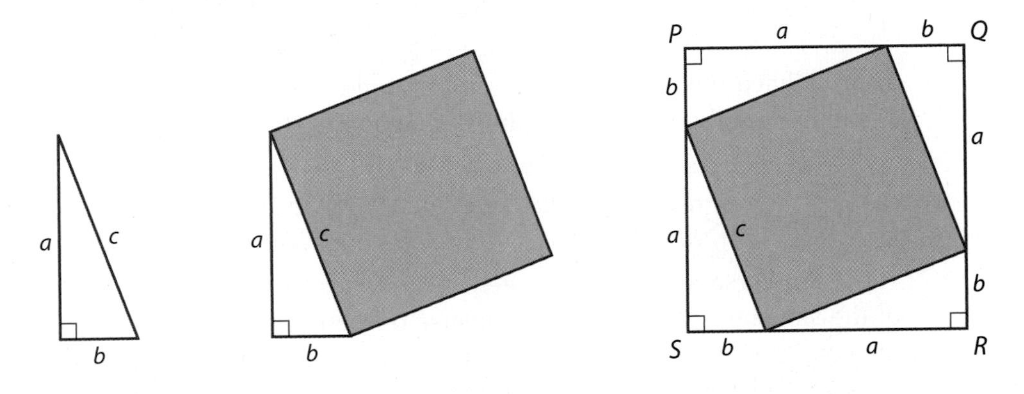

a. Explain as carefully as you can why quadrilateral *PQRS* is a square. Your explanation should include how you know that the sides are straight line segments.

b. Describe two ways to calculate the area of square *PQRS*.

c. Now study this diagram, which shows another way of thinking about the area of square *PQRS*.

On a copy of this diagram, add two line segments to create four right triangles congruent to the original right triangle. Explain how you know that the triangles are congruent.

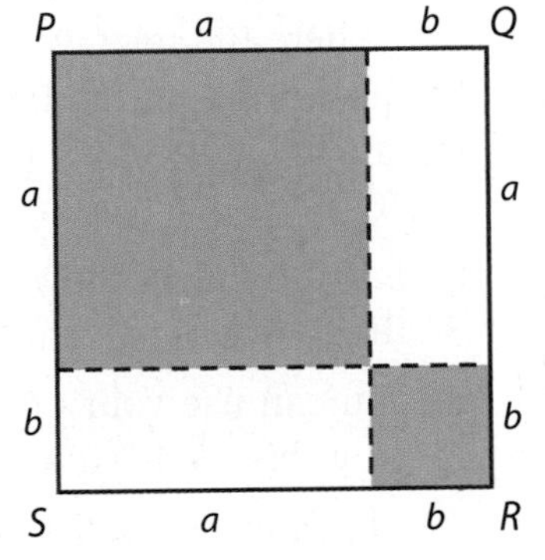

d. Place a copy of the right-most diagram above Part a side-by-side with your modification of the diagram in Part c.

i. How do the areas of the two large squares compare?

ii. Suppose you remove the four congruent triangles from each of the diagrams. What can you say about the areas of the remaining pieces?

iii. Explain as precisely as you can what you have shown.

Now look back at the rope-stretching problem at the beginning of this lesson (page 363). In an attempt to form a right triangle, one group of students at Washington High School stretched the knotted rope as shown below.

They claimed the triangle was a right triangle since $8^2 + 6^2 = 10^2$. These students used the **converse of the Pythagorean Theorem** in their reasoning:

If the sum of the squares of the lengths of two sides of a triangle equals the square of the length of the third side, then the triangle is a right triangle.

The **converse** of an *if-then* statement reverses the order of the two parts of the statement. Although the converse of the Pythagorean Theorem *is* true, the converse of a true statement may not necessarily be true. For example, consider the statement, "If I'm in math class, then I'm in school," and the converse, "If I'm in school, then I'm in math class." Is the converse necessarily true?

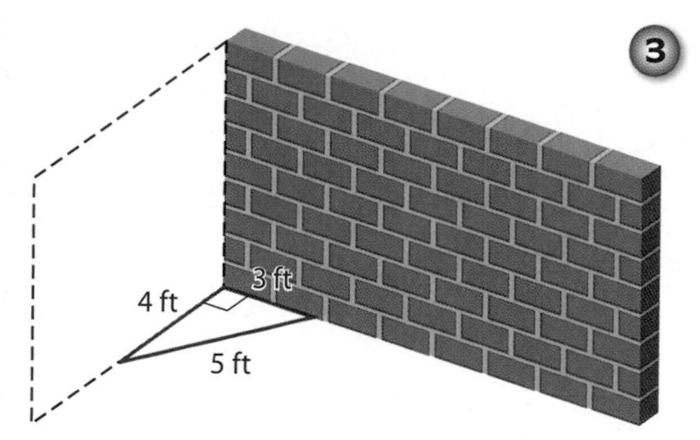

3 To lay out a wall perpendicular to an existing wall, a builder measures 3 feet along the base of the existing wall and 4 feet along the floor line where the new wall is to be placed. The builder then checks if the distance between these two points is 5 feet. If so, she knows that the angle between the existing wall and the wall to be constructed is 90°.

a. Is the builder using the Pythagorean Theorem or the converse of the Pythagorean Theorem? Explain.

b. You can use your understanding of triangle congruence to explain why this "3-4-5 triangle" method guarantees a right angle.

i. Draw segments of length 3 cm, 4 cm, 5 cm. Then, using a ruler and compass, construct a triangle with these side lengths.

ii. Use a ruler and protractor to draw a separate 90° angle. From the vertex of the angle, mark off a segment of length 3 cm on a side and of 4 cm on the other side. Connect the two sides to form a right triangle. According to the Pythagorean Theorem, what should be the length of the hypotenuse?

iii. Explain why the 3-4-5 triangle in part i is congruent to the triangle in part ii.

iv. Why must the 3-4-5 triangle have a right angle? Where is it located?

c. You can use similar reasoning to show, in general, that if you start with a $\triangle ABC$ where the lengths of its sides a, b, and c satisfy $a^2 + b^2 = c^2$, then you can conclude that $\triangle ABC$ is a right triangle with right angle at C.

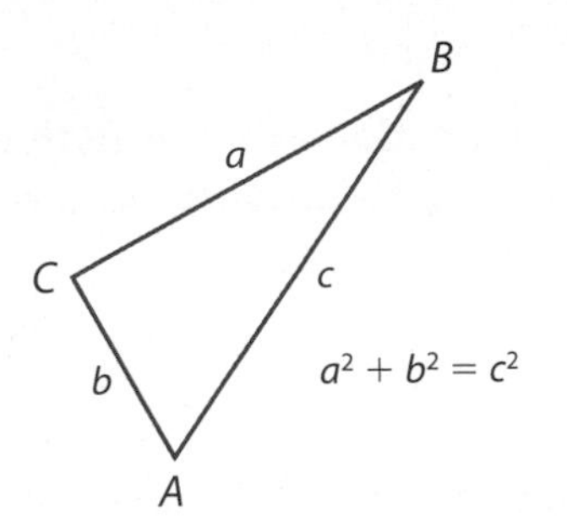

To prove that $\triangle ABC$ is a right triangle, you can reason like you did in Part b.

i. On a separate sheet of paper, draw and label a *right* triangle with sides (other than the hypotenuse) of the given lengths a and b.

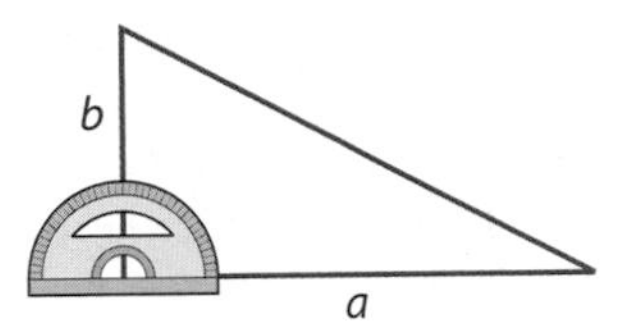

ii. Write an expression for the length of the hypotenuse of this triangle.

iii. Why is the triangle you created congruent to the given triangle, $\triangle ABC$?

iv. Why must the given triangle, $\triangle ABC$, be a right triangle? Why is $\angle C$ the right angle?

4 In preparing an architectural drawing of right triangular components of a building, is it possible to draw a triangle congruent to a given *right* triangle under each of the following conditions? In each case, explain your reasoning.

Tennessee Aquarium, Chattanooga, TN

a. You measure the lengths of the two legs of the given right triangle.

b. You measure the lengths of a leg and the hypotenuse of the given right triangle.

Summarize the Mathematics

In this investigation, you examined applications of the Pythagorean Theorem and its converse. You also used careful reasoning to provide arguments for why these statements are true.

a. Describe the general idea behind your argument that the Pythagorean Theorem is true for all right triangles.

b. Describe the general idea behind your argument that the converse of the Pythagorean Theorem is true.

c. Give two examples, one mathematical and one not involving mathematics, to illustrate that if a statement is true, its converse may not be true.

d. What is the smallest number of side lengths you need to compare in order to test if two right triangles are congruent? Does it make a difference which side lengths you use? Explain.

Be prepared to share ideas and examples with the class.

Check Your Understanding

In the Think About This Situation (page 363), you were asked to consider whether four students could form various shapes using a 24-meter loop of knotted rope with knots one meter apart. Reconsider some of those questions using the mathematics you learned in this investigation.

a. Explain how you could use the 24-meter knotted rope to form a right triangle and how you know the shape is a right triangle.

b. Now explain how you could use the 24-meter knotted rope to form a rectangle and how you know that the shape is a rectangle.

c. Look back at your work in Part b. Could you form a second differently shaped rectangle? Explain.

d. Suppose you and two classmates were given a 30-meter loop of rope with knots tied one meter apart. Could you position yourselves so that the resulting triangle is a right triangle? Explain your reasoning.

On Your Own

Applications

1. Suppose you are given four segments with lengths 5 cm, 5 cm, 12 cm, 12 cm. Think about building shapes using three or four of these lengths.
   a. How many different triangles can you build? Identify any special triangles.
   b. Can you build a parallelogram? If so, how many different ones can you build?
   c. Can you build a kite? If so, how many different kites can you build?
   d. How many different quadrilaterals can you build that are not parallelograms?

2. Four large oil fields are located at the vertices of a quadrilateral *ABCD* as shown. Oil from each of the four fields is to be pumped to a central refinery. To minimize costs, the refinery is to be located so that the amount of piping required is as small as possible.

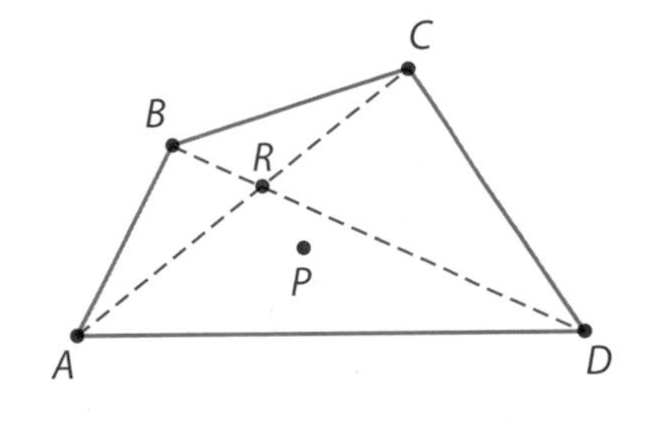

   a. If the refinery is located at position *R*, write an algebraic expression that shows the amount of piping required.
   b. For oil fields *A* and *C*, explain why position *R* is a better location for the refinery than position *P*.
   c. Explain why position *R* is a better location for the refinery than position *P* in terms of all four oil fields.
   d. Is there a better location for the refinery than position *R*? Explain your reasoning.

3. Understanding the body mechanics involved in various physical activities is important to sports physicians and trainers. The diagram at the right shows a person pedaling a bicycle. Key points in the pedaling motion are labeled.

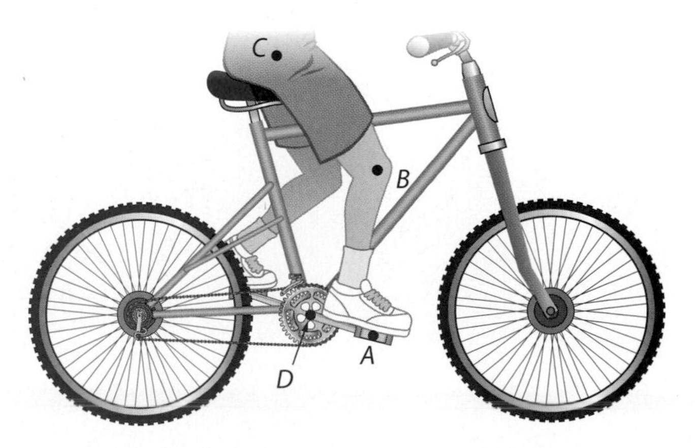

   a. What kind of linkage is represented by *ABCD*?
   b. Identify the frame, the coupler, the drive crank, and the follower crank.
   c. What modifications to the situation would allow it to be modeled by a parallelogram linkage? Should a sports trainer recommend these modifications? Explain your reasoning.

On Your Own

4 A Double Pitch truss, with side lengths and angle measures, is shown below.

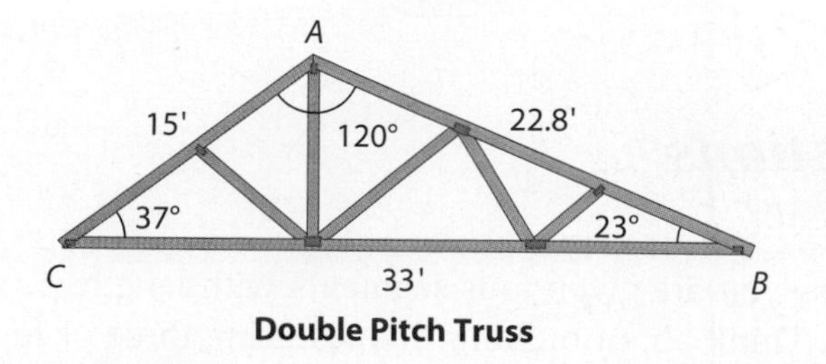

Double Pitch Truss

Which sets of measurements below would be sufficient to test whether a truss *PQR* is congruent to the given truss *ABC*? Explain your reasoning in each case.

a. $PQ = 22.8'$, $PR = 15'$, and $m\angle P = 120°$

b. $PQ = 22.8'$, $PR = 15'$, and $m\angle R = 23°$

c. $RQ = 33'$, $m\angle Q = 23°$, and $m\angle R = 37°$

d. $m\angle P = 120°$, $m\angle R = 37°$, and $m\angle Q = 23°$

e. $PQ = 22.8'$, $RQ = 33'$, and $PR = 15'$

5 Examine each of the following pairs of triangles and the markings that indicate congruence of corresponding angles and sides. In each case, decide whether the information given by the markings ensures that the triangles are congruent. If the triangles are congruent, write the congruence relation and cite an appropriate congruence condition to support your conclusion.

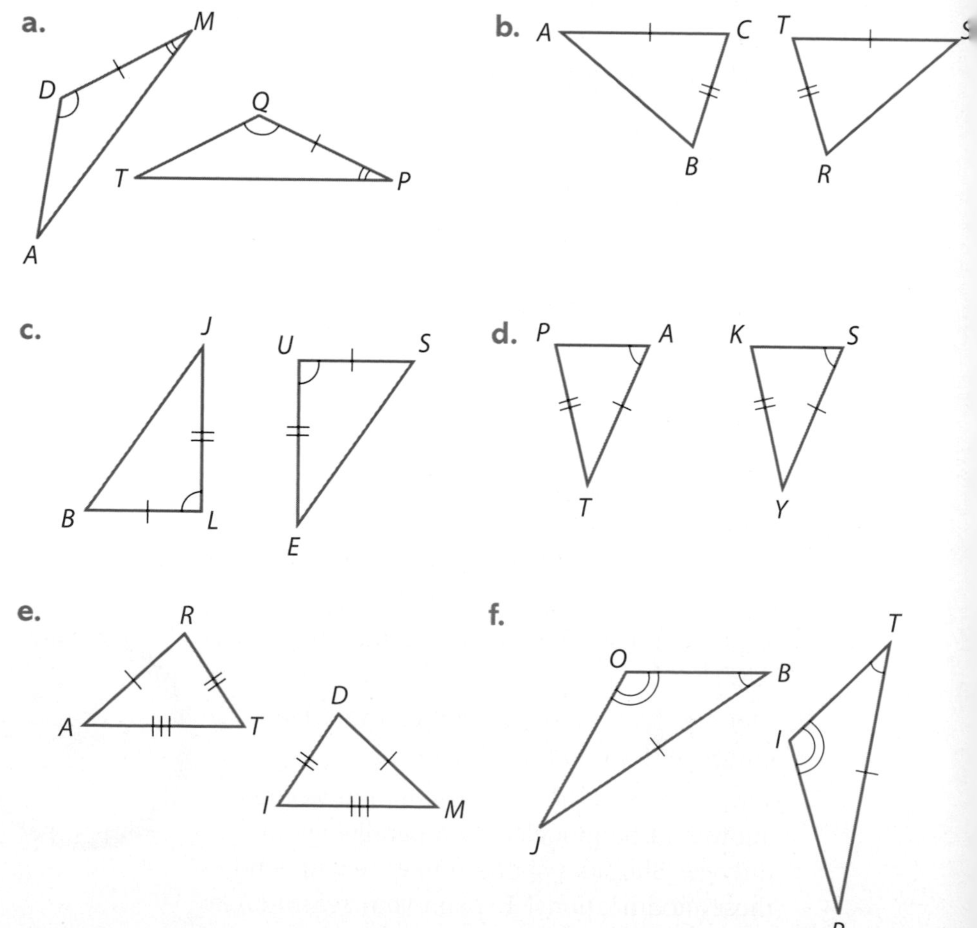

6. Modified Queenpost trusses are often used for roofs that have wide spans and low pitch.

In manufacturing this particular isosceles triangular truss, the bracing is positioned according to specifications in the diagram below.

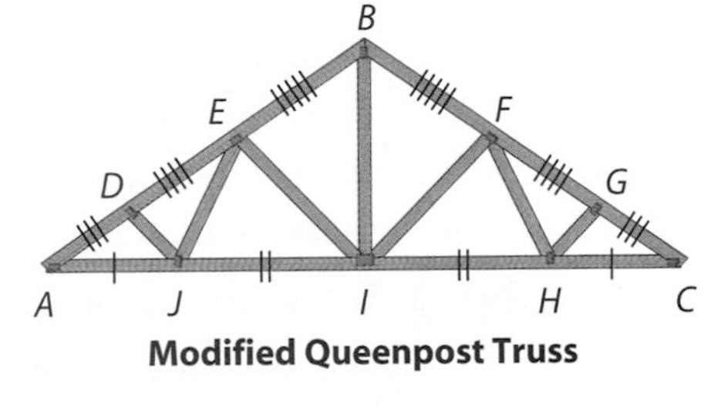

Modified Queenpost Truss

a. Explain carefully why braces $\overline{DJ}$ and $\overline{GH}$ should be cut the same length.

b. Explain why braces $\overline{EJ}$ and $\overline{FH}$ should be cut the same length.

c. Give reasons why $\overline{EI}$ and $\overline{FI}$ should be cut the same length.

d. Is quadrilateral *EBFI* a special quadrilateral? If so, name it and explain how you know.

7. The diagram below illustrates how a carpenter's square is often used to bisect an angle. (A **bisector of an angle** is a ray that begins at the vertex of the angle and divides the angle into two angles of equal measure.) The square is positioned as shown so that $PQ = RQ$ and $PS = RS$.

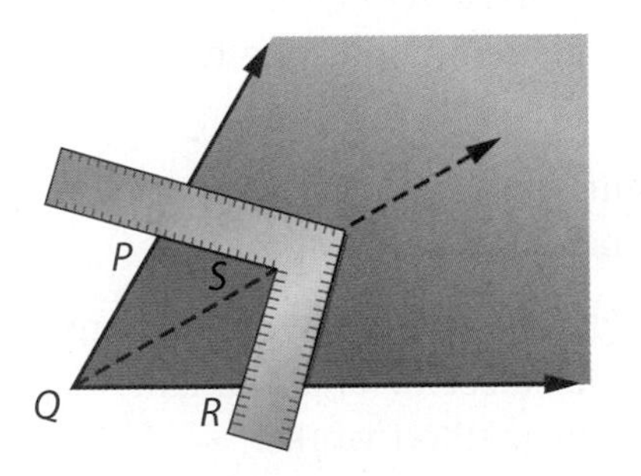

a. Explain why this information is sufficient to conclude that $\triangle PQS \cong \triangle RQS$.

b. Why does ray *QS* (written $\overrightarrow{QS}$) bisect $\angle PQR$?

8 Draftsmen and industrial designers use a variety of tools in their work. Depending on the nature of the task, these tools vary from sophisticated CAD (computer-assisted design) software to compasses and *straightedges* (rulers with no marks for measuring).

a. Draw an acute angle, $\angle ABC$. Using a compass, a straightedge, and the algorithm below, construct the bisector of $\angle ABC$.

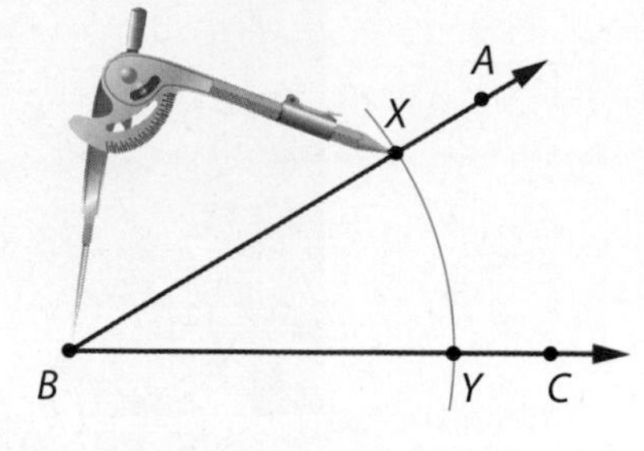

Angle Bisector Algorithm: To bisect $\angle ABC$, do the following.

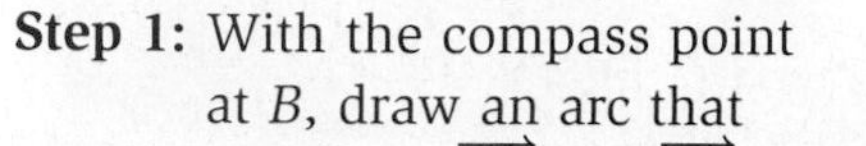

Step 1: With the compass point at B, draw an arc that intersects $\overrightarrow{BA}$ and $\overrightarrow{BC}$; call the intersection points X and Y, respectively.

Step 2: With the compass point at point X and using a radius greater than $\frac{1}{2}XY$, draw an arc in the interior of $\angle ABC$. Then, keeping the same radius, place the compass point at Y and draw a second arc that intersects the first. Label the point of intersection D.

Step 3: Draw the ray $\overrightarrow{BD}$. $\overrightarrow{BD}$ bisects $\angle ABC$.

b. Explain why this algorithm produces the bisector of $\angle ABC$. That is, explain how you know that $\overrightarrow{BD}$ bisects $\angle ABC$. In what way(s) is this algorithm similar to the technique in Applications Task 7?

c. Can this algorithm be used to construct the bisector of a right angle and an *obtuse angle* (an angle with measure greater than 90°)? Explain your reasoning.

d. Think of a line as a "straight" angle. Add steps to the Angle Bisector Algorithm to produce an algorithm for constructing a perpendicular to a given point P on a line.

i. Draw a line $\overleftrightarrow{AB}$ containing point P. Use your algorithm and a compass and straightedge to construct a perpendicular to $\overleftrightarrow{AB}$ at P.

ii. Explain how you know that the line you constructed is perpendicular to $\overleftrightarrow{AB}$ at P.

e. How would you modify your algorithm to construct a perpendicular bisector of a segment? Explain as carefully as you can why your method works.

9 Use a ruler to carefully draw a triangle, $\triangle XYZ$. Design and test an algorithm for using a compass and a straightedge to construct $\triangle ABC$ so that $\triangle ABC \cong \triangle XYZ$. Provide an argument that your algorithm will always work.

10 In Investigation 3, you were able to provide an argument for why opposite angles of any parallelogram are congruent. Experimenting with a parallelogram linkage should convince you that **consecutive angles** of a parallelogram like $\angle 1$ and $\angle 2$ may not always be congruent.

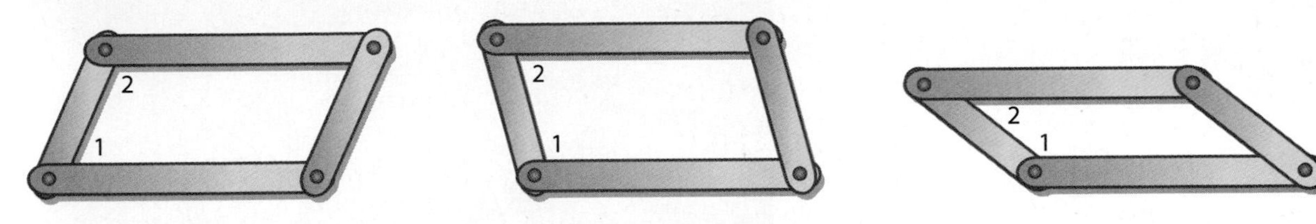

In the first diagram, $m\angle 1 < m\angle 2$. In the next two diagrams, as $m\angle 1$ increases, $m\angle 2$ decreases. Thinking that there might be some relationship between the angles, students in a class at Columbia-Hickman High School measured the angles and in each case found that $m\angle 1 + m\angle 2$ was about 180°. They tried to find reasons that might explain this relationship.

Examine the reasoning of each student below.

- Give a reason that would support each statement made by the students.
- Then decide if the conclusion follows logically from knowing that quadrilateral *ABCD* is a parallelogram.

a. Anna drew ▱*ABCD* at the right and set out to show that $m\angle A + m\angle B = 180°$. She reasoned as follows.

Since *ABCD* is a quadrilateral, I know that
$m\angle A + m\angle B + m\angle C + m\angle D = 360°$.
Since *ABCD* is a parallelogram, I know that
$\angle A \cong \angle C$ and $\angle B \cong \angle D$.
It follows that $m\angle A + m\angle B + m\angle A + m\angle B = 360°$.
So, $2m\angle A + 2m\angle B = 360°$.
Therefore, $m\angle A + m\angle B = 180°$.

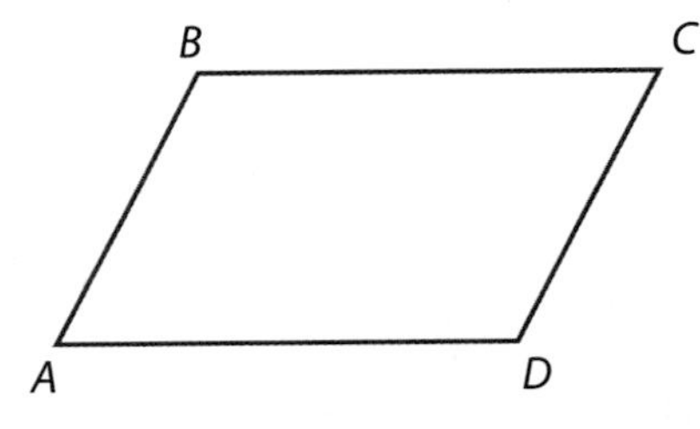

b. Andy drew ▱*ABCD* with diagonal $\overline{BD}$ and then reasoned to show that $m\angle A + m\angle B = 180°$.

I know that $\triangle ABD \cong \triangle CDB$.
So, $\angle BDA \cong \angle DBC$.
I know that $m\angle A + m\angle ABD + m\angle BDA = 180°$.
So, $m\angle A + m\angle ABD + m\angle DBC = 180°$.
Therefore, $m\angle A + m\angle B = 180°$.

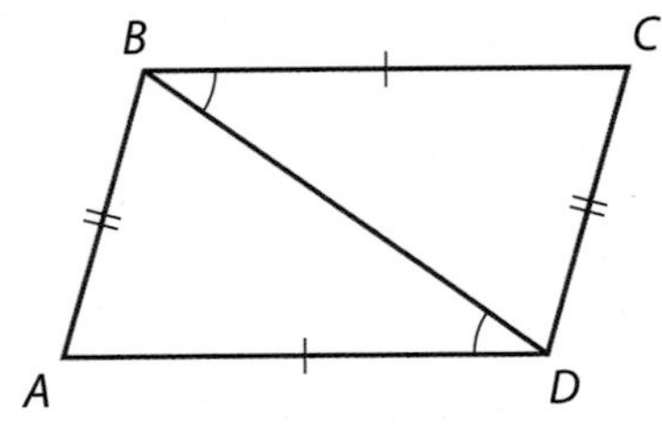

On Your Own

11 Materials tend to expand when heated. This expansion needs to be considered carefully when building roads and railroad tracks.

In the case of a railroad track, each 220-foot-long rail is anchored solidly at both ends. Suppose that on a very hot day a rail's length expands by 1.2 inches, causing it to buckle as shown below.

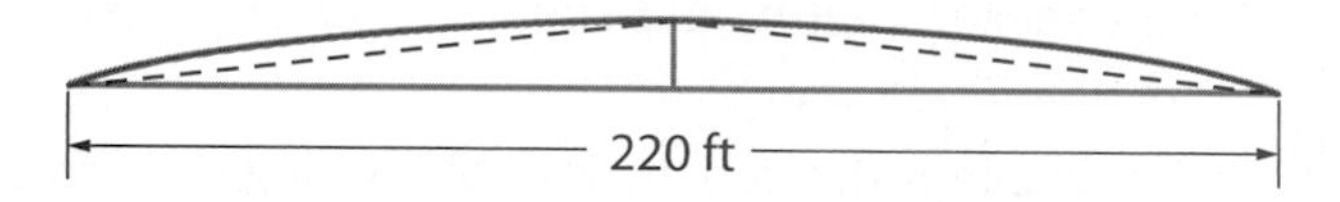

a. At what point along the rail do you think the buckling will occur?

b. Do you think you could slide a gym bag between the raised rail and the track bed?

c. Approximate this situation using right triangles, and then calculate an estimate of the height of the buckle.

d. Would you expect your estimate of the height of the buckle to be more or less than the actual value? Explain your reasoning.

e. Research *expansion joints*. How does the use of these joints in railroad tracks and concrete highways minimize the problem you modeled in Part c?

12 You can represent the diagonals of a quadrilateral with two linkage strips attached at a point.

a. What must be true about the diagonal strips, and how should you attach them so that the quadrilateral is a parallelogram?

b. What must be true about the diagonal strips, and how should you attach them so that the quadrilateral is a rectangle?

c. What constraint(s) must be placed on the diagonal strips and their placement if the quadrilateral is to be a kite? Give reasons to justify that the shape with your arrangement of diagonals is a kite.

d. What constraints must be placed on the diagonal strips and how they are attached in Part a if the quadrilateral is to be a square? Give reasons to justify that the shape with your arrangement of diagonals is a square.

13 When a ball with no spin and medium speed is banked off a flat surface, the angles at which it strikes and leaves the cushion are congruent. You can use this fact and knowledge of congruent triangles to your advantage in games of miniature golf and pool.

To make a hole-in-one on the miniature golf green to the right, visualize a point H' so that the side ℓ is the perpendicular bisector of H and H'. Aim for the point P where $\overline{BH}$ intersects ℓ. If you aim for point P, give reasons to justify that the ball will follow the indicated path to the hole. That is, show that $\angle 3 \cong \angle 1$.

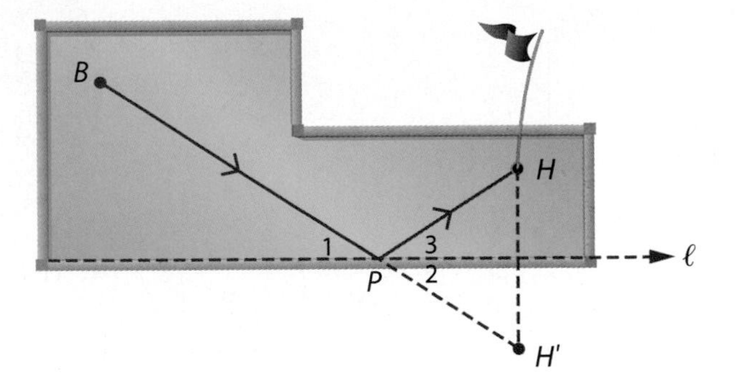

Connections

14 Examine the 5-bar linkage at the right.

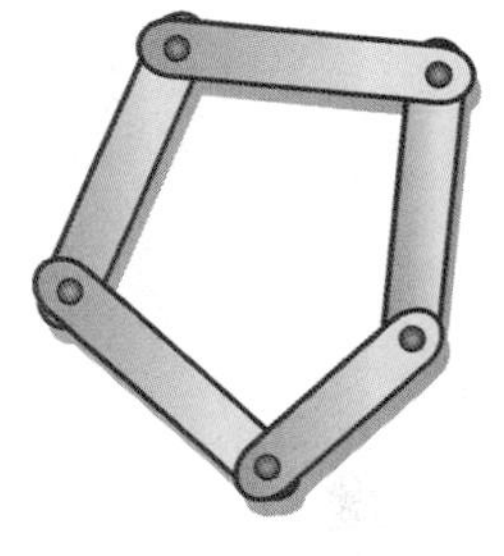

a. Explain why this linkage is not rigid.

b. Make a sketch of the linkage showing how you could make it rigid. How many braces did you use? Is that the fewest number possible?

c. What is the fewest number of braces required to make a 6-bar linkage rigid? To make an 8-bar linkage rigid? Draw a sketch illustrating your answers.

d. Try to generalize your reasoning. What is the fewest number of braces required to make an n-bar linkage rigid? How many triangles are formed?

15 *Perimeter* and *area* are important characteristics of two-dimensional shapes. By recalling the formula for the area of a rectangle $A = base \times height$ and using visual thinking, you can develop and easily recall formulas for the areas of parallelograms and triangles. Study each pair of diagrams below in which b is the length of a *base* and h is the corresponding *height* of the shape. Write a formula for the area A of the shape and then explain how the diagrams helped you reason to the formula.

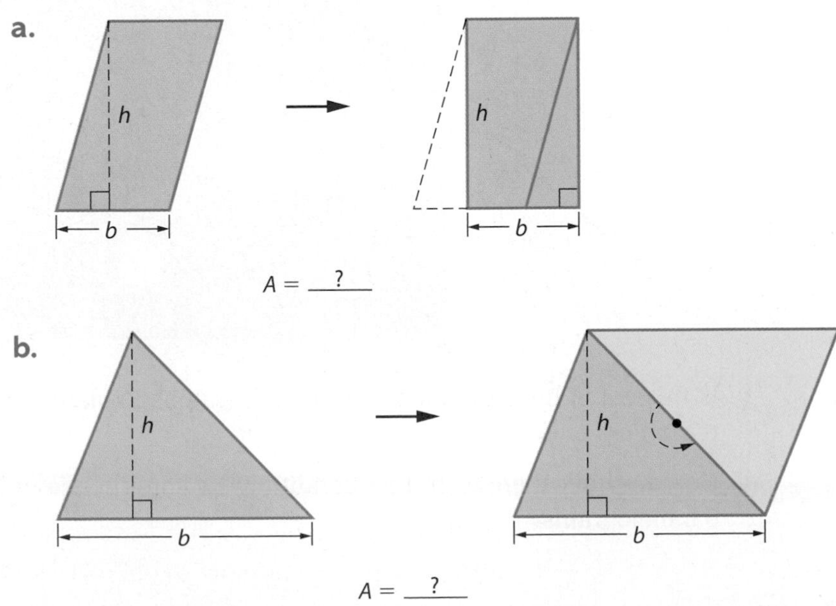

16 The circle below has been dissected into eight sections. These sections can be reassembled to form an "approximate" parallelogram.

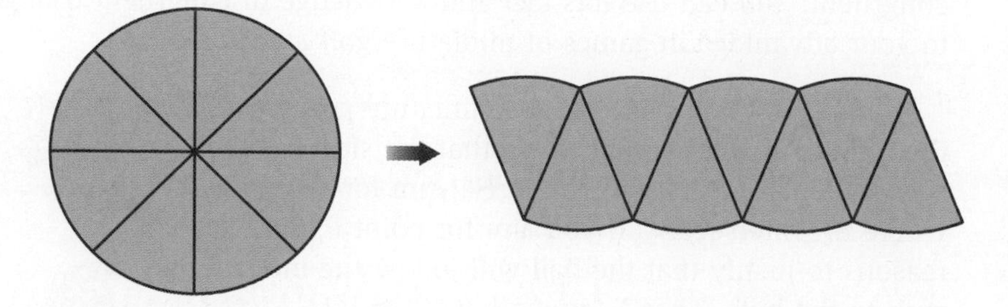

a. How is the base of this "approximate" parallelogram related to the circle?

b. What is the height of the "approximate" parallelogram?

c. How could you dissect the circle into sections to get a better approximation to a parallelogram?

d. Use the above information to produce the formula for the area of a circle.

17 In Applications Task 10, you gave reasons why the sum of the measures of two consecutive angles of a parallelogram is 180°. In the case of ▱*PQRS*, this means that $m\angle P + m\angle Q = 180°$ and $m\angle P + m\angle S = 180°$.

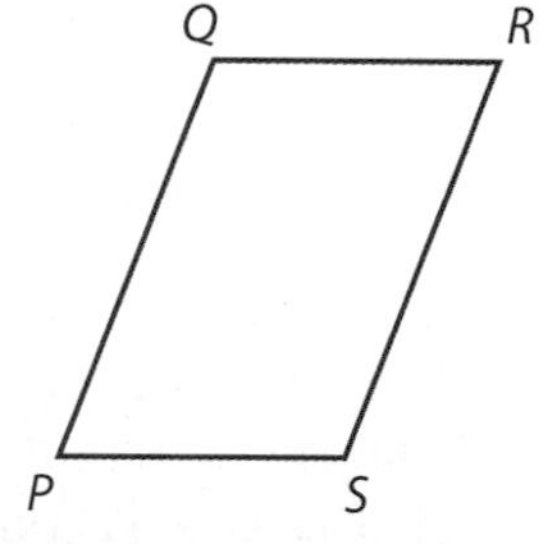

a. Write two similar statements involving other pairs of angles of ▱*PQRS*.

b. Suppose you are given the indicated specifications for a parallelogram window frame *ABCD*.

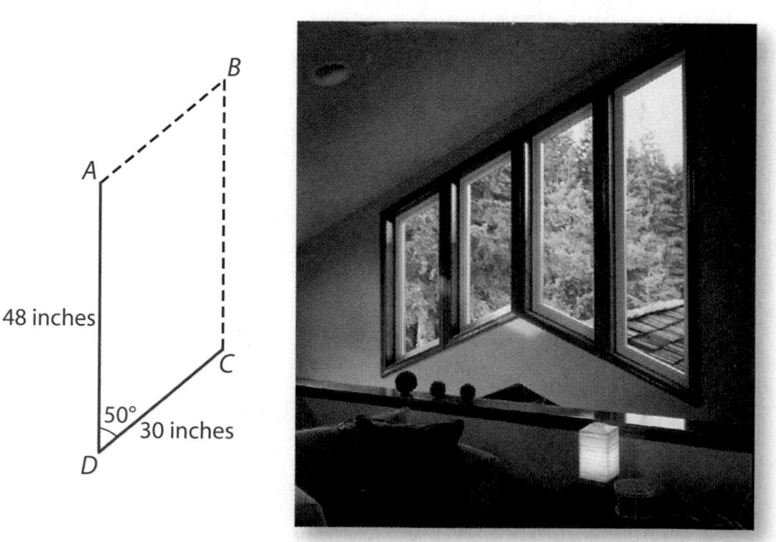

i. Is this enough information to build the frame? Explain your reasoning.

ii. Do you think there is an SAS condition for congruence of parallelograms? Explain.

18 The sum of the measures of consecutive angles of a parallelogram is 180°. (See Applications Task 10.) That property helps to explain the use of the term "parallel" in parallelogram.

In a parallelogram, opposite sides are parallel.

That is, if opposite sides of a parallelogram are extended, the lines will not intersect. Parts a–c provide an outline of why, in ▱$ABCD$, $\overline{BC}$ must be parallel to $\overline{AD}$. The reasoning depends on you.

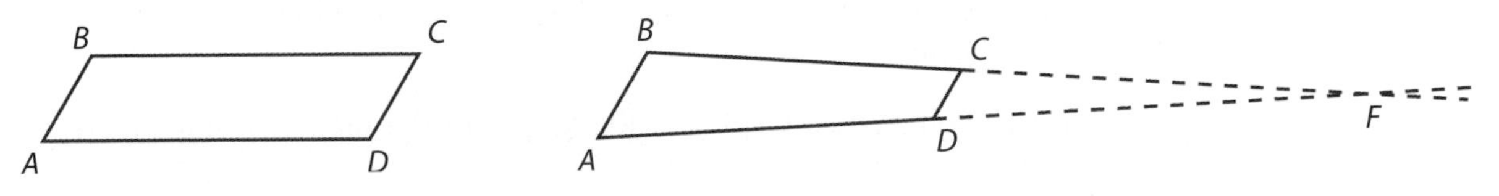

a. What is true about m∠A and m∠B? Why?

b. Now either $\overline{BC}$ is parallel to $\overline{AD}$, or $\overline{BC}$ is *not* parallel to $\overline{AD}$. If $\overline{BC}$ is *not* parallel to $\overline{AD}$, then the situation would look something like that in the diagram above on the right. What must be true about m∠A + m∠B + m∠F? Why?

c. Explain why the situation in Part b is impossible. What does this tell you about the assumption that $\overline{BC}$ was not parallel to $\overline{AD}$? What can you conclude?

d. How could you use similar reasoning to show that $\overline{AB}$ must be parallel to $\overline{CD}$?

19 Two diagrams used in your reasoning about the Pythagorean Theorem (page 379) are shown below.

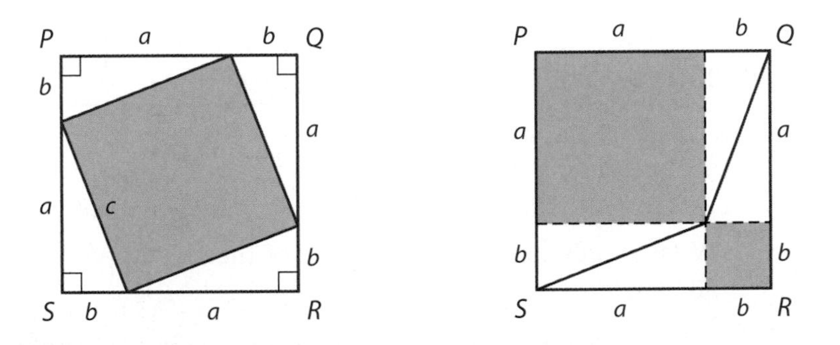

a. Write an expression for the area of square $PQRS$ that involves a, b, and c using the diagram above at the left.

b. Write an algebraic expression for the area of square $PQRS$ that involves only a and b, using the diagram above at the right.

c. Use your results from Parts a and b and algebra to show that $a^2 + b^2 = c^2$.

20 Draw squares of side lengths 2, 4, 7, 8, 10, and 11 centimeters on centimeter grid paper.

a. Measure the diagonals to the nearest 0.1 cm. Record your data in a table.

b. Make a plot of your (*side length, diagonal length*) data. Find a linear model that fits the trend in the data.

  i. What is the slope of the line? What does it mean?

  ii. What is the *y*-intercept? Does it make sense in this context? Explain.

c. Use your model to predict the length of the diagonal of a square with side length of 55 cm.

d. Compare your predicted length to that computed by using the Pythagorean Theorem. Explain any differences.

e. Write a rule that would express *exact* diagonal length D in terms of side length s for any square.

21 The diagram below shows an equilateral triangle, $\triangle ABC$, with an altitude from A to M that forms two smaller triangles.

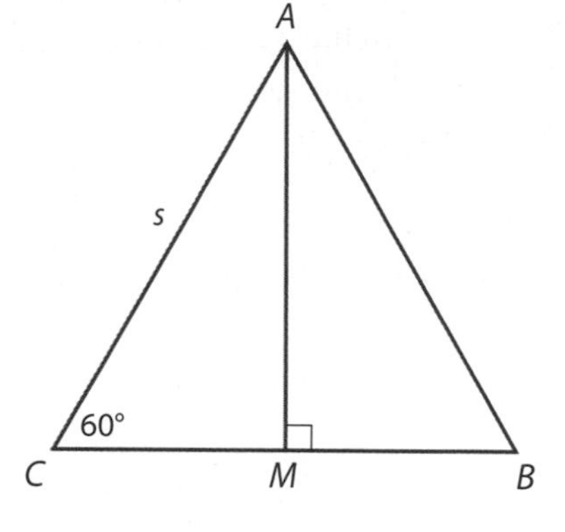

a. What is the measure of $\angle MAC$?

b. Explain as precisely as you can why $\triangle AMC \cong \triangle AMB$.

c. Find the exact length of the altitude $\overline{AM}$ when the sides of the equilateral triangle have length:

  i. 5 cm **ii.** 8 cm

  iii. 10 cm **iv.** 1 cm

d. Now consider the general problem of finding side lengths of a 30°-60°-90° triangle in which the hypotenuse is of length s.

  i. What expression gives the length of the side opposite the 30° angle?

  ii. What expression gives the length of the side opposite the 60° angle?

e. Write in words how the lengths of the sides of a 30°-60°-90° triangle are related.

Reflections

22 In Investigation 3, you were able to provide reasons justifying that the base angles of an isosceles triangle are congruent. Why does it follow logically that an equilateral triangle is *equiangular*; that is, the three angles of an equilateral triangle are congruent? What is the measure of each angle?

23 Why will any parallelogram linkage have two rotating cranks?

24 Explain why there is no Side-Side-Side-Side (SSSS) congruence condition for quadrilaterals.

25 Explain why opposite angles of a rhombus are congruent. Are both pairs of opposite angles of a kite congruent? Explain.

26 Look back at Problem 3 of Investigation 4. Could a builder also lay out a wall perpendicular to an existing wall by measuring the existing wall at 6 feet, the location of the new wall at 8 feet, and then check if the distance between the two points is 10 feet? Explain your reasoning. Which method would likely give greater accuracy? Why?

Extensions

27 The diagram at the right shows a quadrilateral linkage with frame $\overline{AB}$ satisfying Grashof's Principle that you discovered in Problem 6 of Investigation 1 (page 367). When the shortest crank $\overline{AD}$ makes a complete revolution, the other crank $\overline{BC}$ oscillates between two positions moving back and forth in an arc as indicated in the diagram.

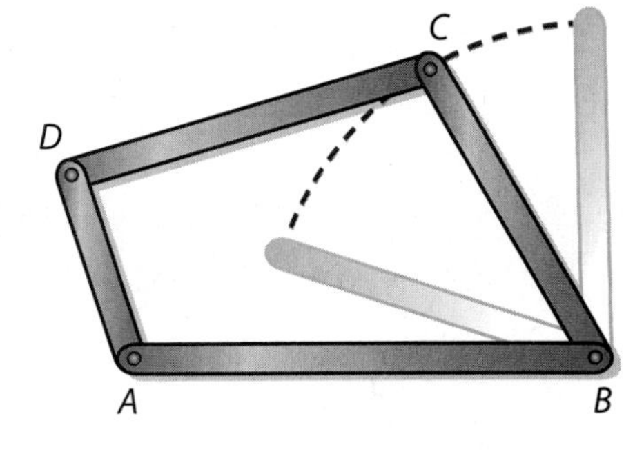

a. Use software like the "Design a Linkage" custom tool to investigate the possible paths of point C under the following two conditions.

i. Quadrilateral $ABCD$ is a kite.

ii. Quadrilateral $ABCD$ is a parallelogram, including special types.

Consider the two cases where the frame is the longest or the shortest side. Write a paragraph summarizing your findings.

b. Repeat Part a for the case of a point $P \neq C$ on the coupler.

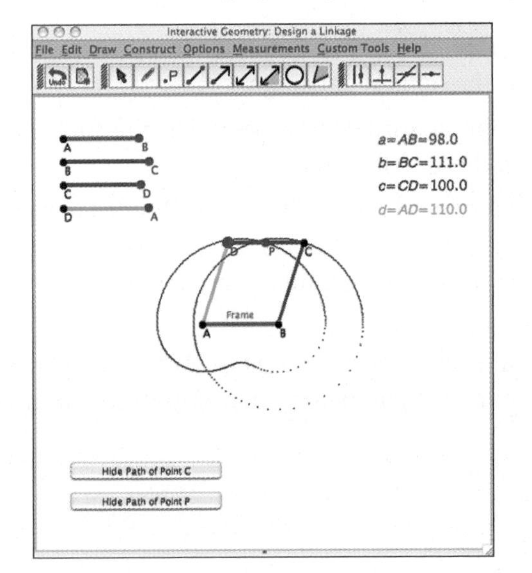

28 In order for kites to fly well, they need to have a high ratio of *lift area* to weight. For two-dimensional kites, the lift area is just the area of the kite.

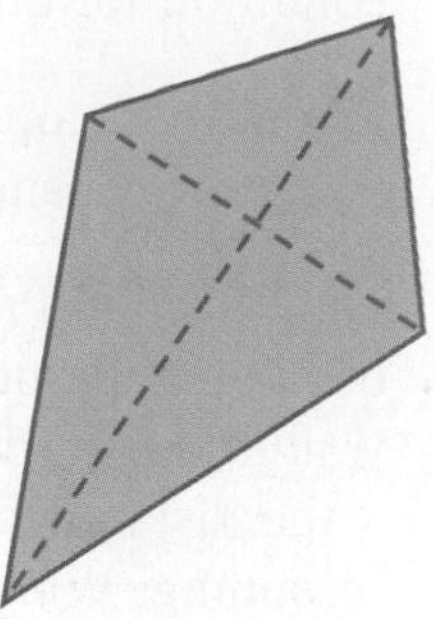

a. On a copy of the kite shown, label its vertices and use markings to show which segments are congruent by definition of a kite.

b. Use congruent triangles to help you find the lift area of the traditional kite shown with cross pieces of lengths 0.8 m and 1.0 m.

c. Suppose the lengths of the diagonals of the kite are a and b where $a < b$. Use the diagram to help develop a formula for the area of a kite.

d. Can you also use your formula to find the area of a rhombus? Explain your reasoning.

e. Could this formula be used to find the area of any other quadrilaterals? Explain.

29 As noted at the beginning of this lesson, ancient Egyptians had to deal with changes in shape and size of fields caused by the annual flooding of the Nile River. Historians have evidence that the Egyptians calculated the areas of quadrilateral-shaped fields using the formula shown below.

$$A = \frac{1}{2}(a + c) \cdot \frac{1}{2}(b + d)$$

where a, b, c, d are the lengths of consecutive sides of the quadrilateral.

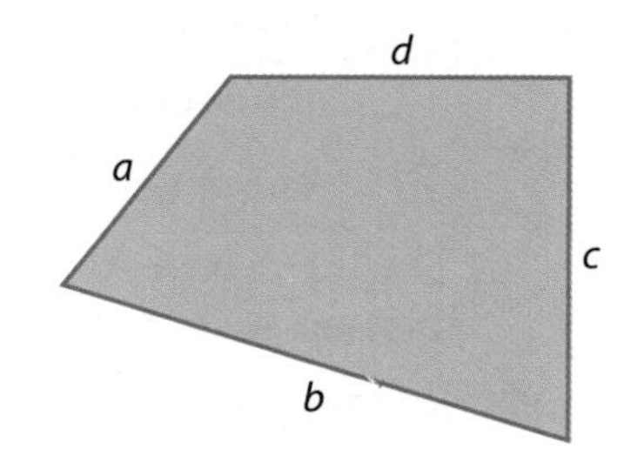

a. State this formula in words (without using the labels a, b, c, d) using the idea of "average."

b. Describe quadrilaterals for which the formula gives an exact calculation of the area.

c. Use software like the "Areas of Quadrilaterals" custom tool to explore cases of other quadrilaterals. For which quadrilateral shapes does the formula overestimate the area? Underestimate the area?

d. Explain why your findings in Part c make sense in terms of area formulas for parallelograms, trapezoids, and kites.

30 The television industry has set standards for the sizing of television screens. The ratio of height h to width w is called the *aspect ratio*. The aspect ratio for a conventional television screen is 3:4. That is $\frac{h}{w} = \frac{3}{4}$.

a. Write a rule expressing h as a function of w.

b. Use the Pythagorean Theorem to write a rule relating h, w, and the diagonal length 27.

c. Use your rules in Parts a and b to find the standard dimensions of a 27-inch diagonal TV screen.

d. Check the dimensions you obtained against actual measurements of a 27-inch TV screen.

w

h

27 in.

Review

31 Write an equation that matches each graph. The scale on both axes is 1.

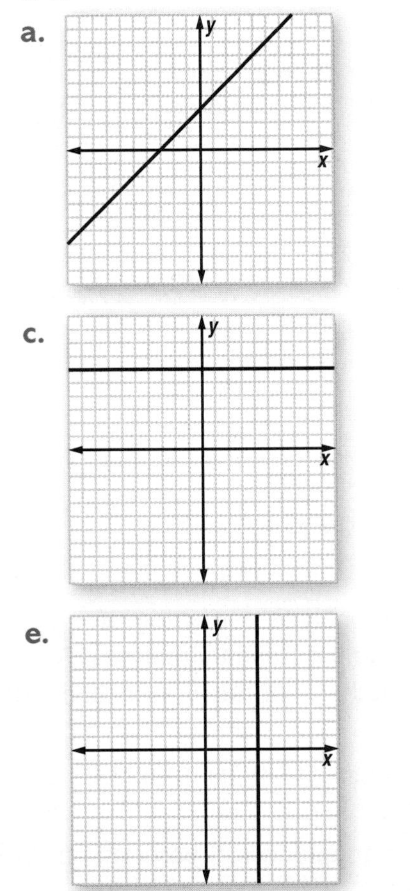

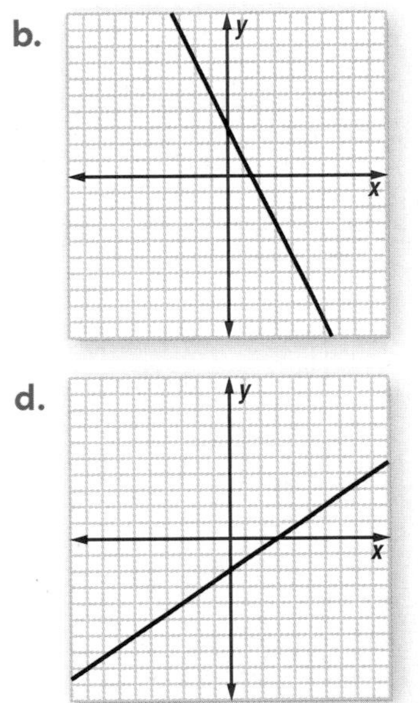

32 You have a number cube with the numbers 2, 3, 4, 5, 6, and 7 on the faces. You roll the cube and look at the number showing on the top face.

a. What is the probability of rolling an even number?

b. What is the probability of rolling a prime number?

c. What is the probability of rolling a number less than 3?

d. What is the probability of rolling an odd number that is greater than 4?

33 Use the fact that $36 \times 15 = 540$ and mental computation to evaluate the following.

a. $\frac{5{,}400}{36}$ **b.** 3.6×15 **c.** 72×15 **d.** $\frac{45 \cdot 36}{3}$

34 Without measuring, find the measure of each indicated angle.

a. m∠*ABC* **b.** m∠*CBD*

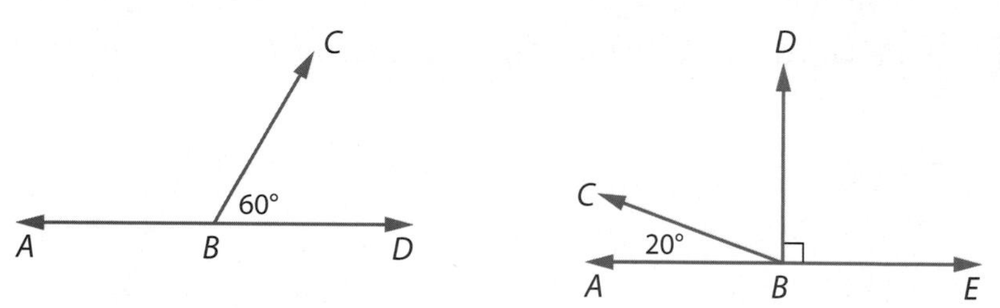

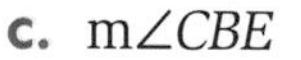
c. m∠*CBE*

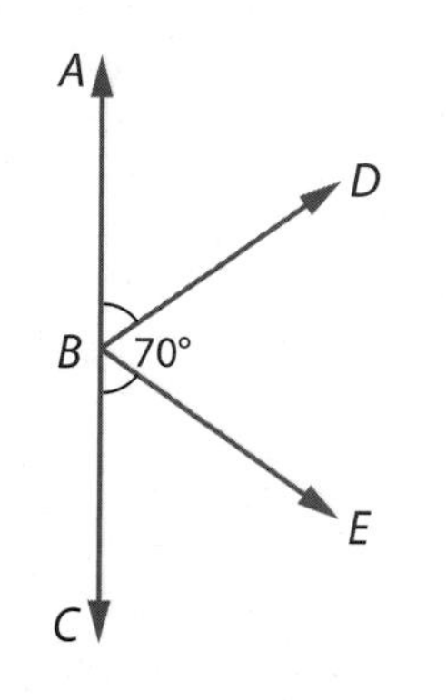

35 Algebraic models can often help you analyze a situation. In this task, you will write algebraic rules for several different situations. Before you write each rule, think about whether a linear or an exponential rule would be better for the situation.

a. Alena's telephone credit card charges \$0.50 just to make a call and then charges \$0.04 for each minute.

i. Write a rule that gives the charge for a call based upon the length of the call in minutes.

ii. Alena wants to be able to make a 40-minute call to her friend. How much will this call cost her?

b. Randy owns a car that is currently worth $8,750. The value of his car decreases by 15% each year.

i. Write a rule that gives the value of Randy's car t years from now.

ii. In how many years will Randy's car first be worth less than $2,000?

36 Solve the following equations by reasoning with the symbols themselves.

a. $3x - 5 = 9x + 4$

b. $9 + \frac{1}{2}x = 14$

c. $3.2 = 5x + 0.7$

d. $2(4x - 8) = 8x + 14$

e. $2(5^x) = 250$

f. $(-2)(-2)^x = 16$

37 Find the value of each expression without using a calculator.

a. -5^2

b. $(-3)^2 - 4(2) + 21$

c. $148 - 3(-5)$

d. $\frac{-15 + 8(-3)}{2}$

e. $-6 + (3 - 5)^3$

38 Rewrite each expression in an equivalent form as an integer or radical expression in simplest form.

a. $\sqrt{\frac{16}{49}}$

b. $\sqrt{44}$

c. $\sqrt{3}\sqrt{15}$

d. $2\sqrt{63}$

e. $81^{\frac{1}{2}}$

f. $\frac{\sqrt{8}}{2}$

g. $\sqrt[3]{27}$

h. $\sqrt[3]{-1}$

39 The table below gives the number of words spelled correctly (out of ten) by a group of students preparing for a spelling competition.

Number of Words Spelled Correctly	6	7	8	9	10
Number of Students	5	4	10	8	3

a. Calculate the mean and standard deviation of the number of words spelled correctly.

b. Colin and Lindsey tried to spell these ten words, and they both spelled all ten of the words correctly. Lindsey then added her score of 10, Colin's score of 10, and the average of the other 30 students and then divided that sum by 3 to get a new mean of 9.33. Is Lindsey's mean the correct mean of all 32 students? If not, explain the problem with her reasoning.

40 Often you will need to convert measures from one unit to another. Use what you know about seconds, minutes, hours, and days to complete each statement.

a. 80 minutes = _______ seconds = _________ hours

b. 3 days = _________ hours = _________ minutes

c. 300,000 seconds = ________ days = _________ years

Polygons and Their Properties

Triangles and quadrilaterals are special classes of **polygons**—closed figures in a plane, formed by connecting line segments endpoint-to-endpoint with each segment meeting exactly two other segments. The segments are the *sides* of the polygons, and the points that they join are the *vertices*. Some other polygonal shapes that can be seen in daily life and with which you may be familiar are shown above.

Think About This Situation

As you examine the photos on the previous page, try to identify the polygon in each case and think about some of its features.

a) How would you describe the shape of each polygon?

b) What features do each of these polygons appear to have in common?

c) The design of most bolts and nuts are based on polygons with an even number of sides. Why do you think this is the case? Why do you think the nuts on many public water mains and fire hydrants have the shape shown?

d) Why do you think a stop sign has the shape it has? Would a square or rectangle work just as well?

e) Why do you think the cells of a honeycomb are shaped as they are? Would other polygons work just as well?

f) Based on your previous work with triangles and quadrilaterals, what are some natural questions you might ask about other polygons?

In this lesson, you will investigate properties of polygons, including relationships among their sides, angles, and diagonals. You also will explore the symmetry of polygons and patterns formed by combinations of polygons. These properties and patterns have important applications in art, design, and manufacturing.

Investigation 1 Patterns in Polygons

As the thin metal sheets that form the aperture of a camera move together or apart, they determine the amount of light that passes through a camera lens. The closing apertures on various cameras also determine polygons that differ in their number of sides.

Polygons can be classified in several different ways. One of the most commonly used classifications is in terms of the number of sides they have.

Number of Sides	Name
3	Triangle
4	Quadrilateral
5	Pentagon
6	Hexagon
7	Septagon
8	Octagon

Number of Sides	Name
9	Nonagon
10	Decagon
11	11-gon
12	Dodecagon
15	15-gon
n	n-gon

1 Name the polygons pictured on pages 398 and 399.

The polygonal shapes in the photos are examples of **regular polygons**. In regular polygons, all sides are congruent and all angles are congruent. These shapes also have a certain balance or regularity of form that can be explained in terms of their *symmetry*. As you work on the following problems, look for answers to these questions:

How can you accurately draw or build a regular polygon?

How can you describe the symmetry of a regular polygon and other shapes?

2 You can discover a method for accurately drawing a regular polygon by conducting the following experiment with a two-mirror kaleidoscope.

Hinge two mirrors together with tape so that you can adjust the angle between them. Draw a line segment on a sheet of paper and place a dot on the segment. Position the mirrors as shown in the photo. Adjust the mirrors so that they make an isosceles triangle with the segment and you can see a regular pentagon. Carefully trace the angle formed by the mirrors. Measure the angle.

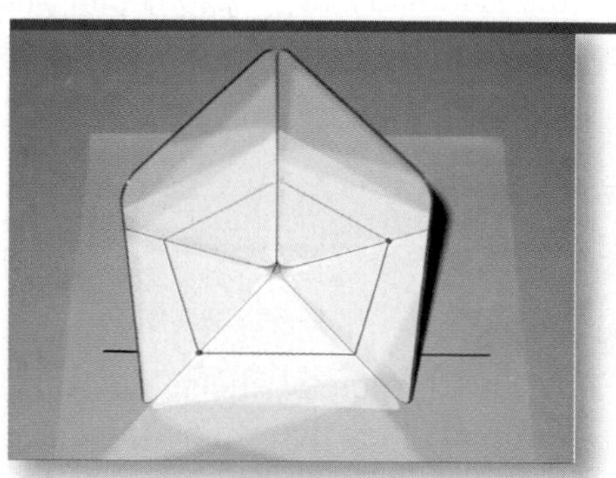

a. Complete a table like the one below by adjusting the mirrors to form a regular polygon with the given number of sides.

Regular Polygon	3	4	5	6	7	8
Angle Between Mirrors	?	?	72°	?	?	?

b. Predict the angle between the mirrors necessary to form a decagon. Check your prediction.

c. Write a rule that gives the measure M of the angle between the mirrors as a function of the number of sides of the regular n-gon produced by the two-mirror kaleidoscope.

The angle you traced in each case is called a **central angle** of the regular polygon.

d. To draw a regular pentagon, first draw a circle with a compass. Next draw a radius and then use a protractor to draw a central angle of 72°. The points where the sides of the angle intersect the circle are two of the vertices of the pentagon. How can you find the remaining three vertices using only a compass? Draw the pentagon.

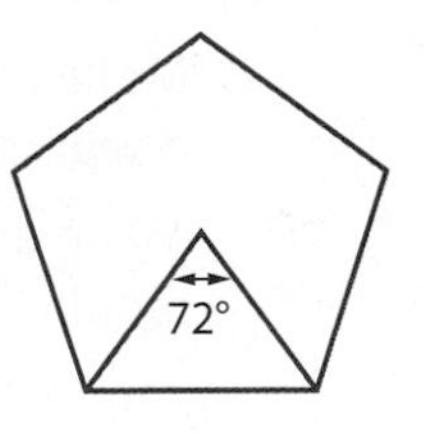

e. Make an accurate drawing of a regular octagon.

f. Write a step-by-step description of a general method for drawing a regular n-gon.

3 If you rotate a tracing of a regular pentagon 72° about the center of the pentagon, the tracing will *coincide* (match) with the original figure. Try it. The regular pentagon has 72° **rotational symmetry**.

a. Explain why a regular pentagon also has 144° rotational symmetry. What other rotational symmetries does a regular pentagon have that are less than 360°? We do not consider a 360° rotation since a tracing of any figure will coincide with the original figure after a rotation of 360°.

b. What are the rotational symmetries of an equilateral triangle? Do other triangles have rotational symmetry? Explain.

c. What are the rotational symmetries of a square? Of a regular hexagon?

d. Make a conjecture about the number of rotational symmetries of a regular n-gon. What angle measures will these symmetries have? Test your conjecture for the case of a regular octagon.

4 A regular pentagon also has **reflection** or **line symmetry**; sometimes called **mirror symmetry**.

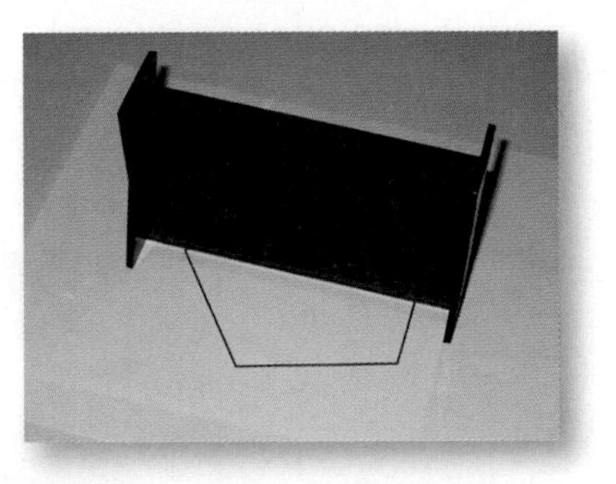

When a mirror or piece of dark-colored Plexiglas is placed on a *line of symmetry*, half of the figure and its reflected image form the entire figure.

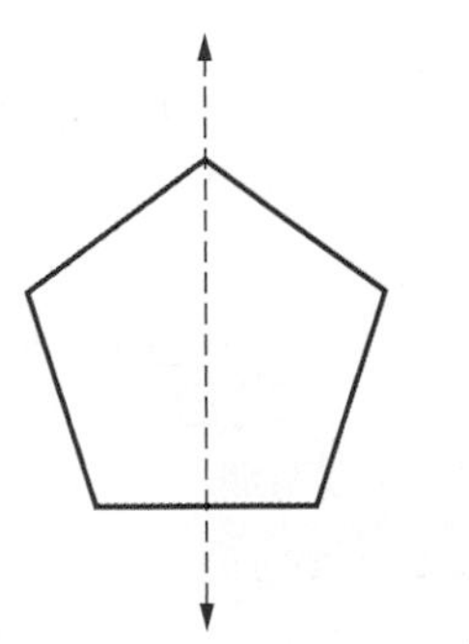

When a tracing of the figure is folded along the *line of symmetry*, one-half of the figure exactly coincides with the other half.

a. On a tracing of the regular pentagon above, draw each line of symmetry.

b. Draw an equilateral triangle and then draw each of its symmetry lines. What other triangles have lines of symmetry? What property or properties do these triangles have in common?

c. Describe the lines of symmetry of a square.

d. Draw a regular hexagon and then draw its lines of symmetry.

e. Make and test a conjecture about the number of symmetry lines of a regular octagon.

f. Make a conjecture about the number of symmetry lines of a regular n-gon.

- **i.** Describe where the symmetry lines cut a regular n-gon when n is an even number.
- **ii.** Describe where the symmetry lines cut a regular n-gon when n is an odd number.

g. How is the line of symmetry of a figure related to the segment connecting a point on the figure with its mirror image?

5 Now consider the symmetries of special quadrilaterals that are not regular polygons—kites, general parallelograms, rhombuses, and rectangles.

a. Which of these other special quadrilaterals have line symmetry?

- **i.** In each case, sketch the shape and each of its symmetry lines.
- **ii.** Which of the quadrilaterals have symmetry lines that join vertices? What do these quadrilaterals have in common?
- **iii.** Which of the quadrilaterals have symmetry lines that do not join vertices? Where are the symmetry lines located? How do such quadrilaterals differ from those in part ii?

b. Which of these other special quadrilaterals have rotational symmetry?

- **i.** In each case, what are the angles of the rotational symmetries?
- **ii.** What property or properties do the quadrilaterals with rotational symmetry have in common?

6 Symmetry is perhaps the most common geometric characteristic of shapes found in nature. Symmetry is also often an integral part of the art forms created by people throughout history and across many cultures of the world. Examine carefully each of the figures shown below.

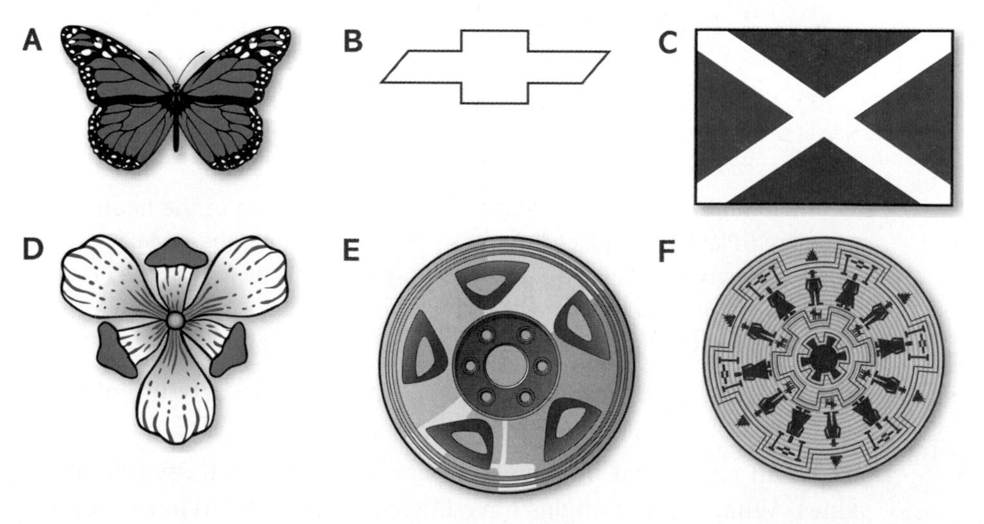

a. Which of these figures have reflection symmetry? Using a copy of these figures, for each figure with reflection symmetry, draw its line(s) of symmetry.

b. Which of the shapes have rotational symmetry? For each shape with rotational symmetry, give the angle(s) through which it can be rotated to coincide with itself.

c. If a figure has reflection symmetry, must it have rotational symmetry? Explain your reasoning.

d. If a figure has rotational symmetry, must it have reflection symmetry? Explain.

Summarize the Mathematics

In this investigation, you learned how to draw regular polygons and discovered special patterns relating the number of sides to the measure of a central angle and to the number and nature of rotational and reflection symmetries.

a. Explain how you would accurately draw a regular *n*-gon.

b. Explain how you can test to see if a figure has line symmetry. Describe the number and positions of the lines of symmetry of a regular *n*-gon.

c. Explain how you can test if a figure has rotational symmetry. Describe the number of rotational symmetries of a regular *n*-gon, and their angles.

Be prepared to share your ideas and thinking with the class.

Check Your Understanding

Make an accurate drawing of a regular nonagon.

a. What is the measure of a central angle?

b. Describe all the rotational symmetries.

c. Sketch all the lines of symmetry. Where do the symmetry lines cut the sides of the nonagon?

Investigation 2 The Triangle Connection

Polygons with the same corresponding side lengths can have quite different shapes. As in the special case of quadrilaterals, polygons can be convex or nonconvex. In a **convex polygon**, no segment connecting two vertices is outside the polygon. Unless otherwise stated, in the remainder of this unit and in future units, polygons will be assumed to be convex.

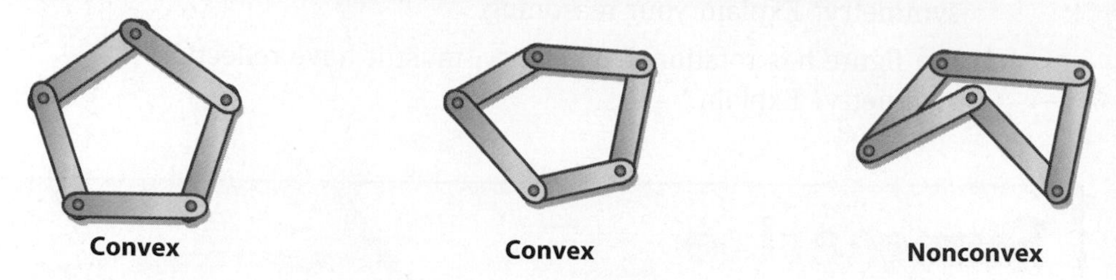

The shape and symmetry of a polygon depend on both side lengths and angle measures. As you work on Problems 1–4, look for answers to the following question:

How are the measures of the angles of any polygon related to the number of sides?

1 In Lesson 1, you learned that polygon shapes of four or more sides are not rigid. They can be made rigid by adding diagonal braces.

a. How could you use this idea of *triangulation* to find the sum of the measures of the interior angles of a pentagon? Compare your method and angle sum with others.

b. Use similar reasoning to find the sum of the measures of the interior angles of a hexagon. Why is it not necessary that the hexagon be a regular hexagon?

c. Complete a table like the one below for polygons having up to 9 sides. Examine your table for patterns relating sides, triangles, and angle sums.

Number of Sides	Number of Triangles	Sum of Interior Angles
4		
5	3	540°
6		
⋮		

d. Predict the sum of the measures of the interior angles of a decagon (10 sides). Check your prediction with a sketch.

e. Suppose a polygon has n sides. Write a rule that gives the sum of the measures of its interior angles S as a function of the number of its sides n.

f. Test your rule for $n = 3$ (a triangle) and $n = 4$ (a quadrilateral).

g. Why is your function in Part e a linear function?

 i. What is the slope of the graph of your function?

 ii. What does the slope mean in terms of the variables? Does the y-intercept make sense? Why or why not?

2 By extending each side of a polygon, you create an *exterior angle* at each vertex.

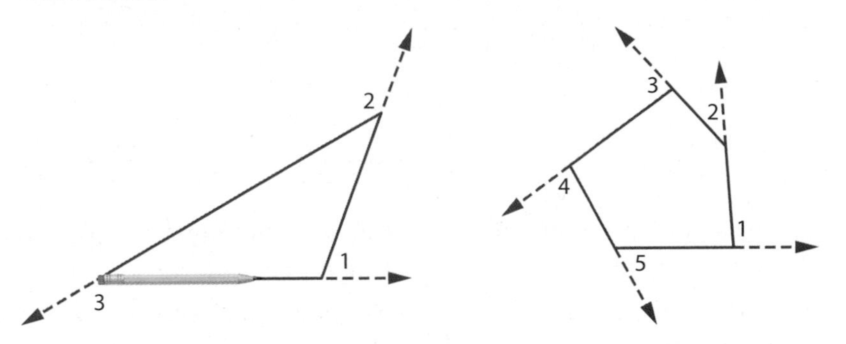

a. Conduct the following exploration.

For each shape shown above:

(1) Place your pencil along the horizontal side of the shape with the tip pointing to the right as shown.

(2) Slide your pencil to the right until the eraser end is at the vertex of $\angle 1$.

(3) Turn your pencil about the vertex so that it aligns with the second side.

(4) Repeat Steps 2 and 3 for each vertex until your pencil has made a complete trip around the polygon and is in its original position.

When the trip is completed, by how much has the pencil turned? What does this suggest about the sum of the measures of the exterior angles of the shape?

b. Repeat the exploration by drawing a different polygon of your choice. Did you find supporting evidence for your conjecture? What do you think is true about the sum of the measures of the exterior angles of any convex polygon?

3 Exterior angles can be created by extending sides in either direction. Suppose a, b, c, and d represent the measures of the exterior angles of quadrilateral $ABCD$.

a. Use the diagram and careful reasoning to write a general argument for why your conjecture in Problem 2 is true for any quadrilateral. Compare your reasoning with others and resolve any differences.

b. Explain how you might use a similar argument in the case of an octagon.

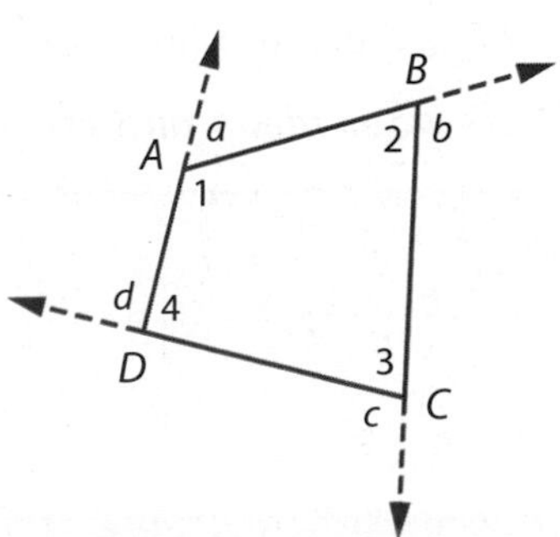

If a polygon is a regular polygon, then you can find a relationship between the number of its sides and the measure of *each* interior angle.

a. What is the measure of an interior angle of a regular hexagon?

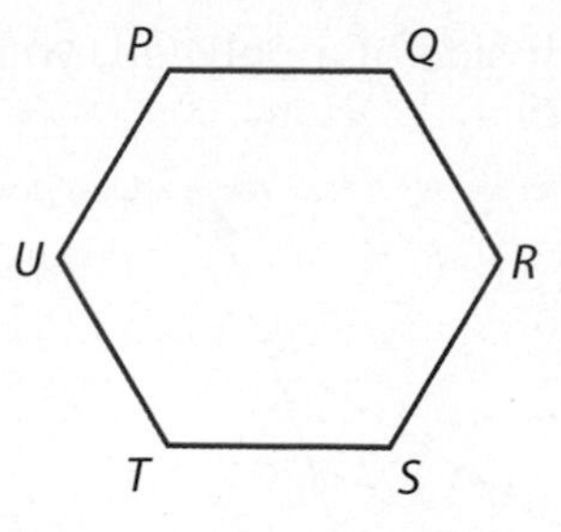

b. Write a rule that gives the measure A of an interior angle of a regular polygon as a function of its number of sides n.

c. What is the measure of an exterior angle of a regular hexagon? Of a regular 15-gon?

d. Write a rule that gives the measure E of an exterior angle of a regular polygon as a function of its number of sides n.

e. Visualize an example of a regular n-gon with an exterior angle drawn. If you added the expressions for your rules in Parts b and d, what should you get? Try it.

f. Are your functions in Parts b and d linear functions? If so, what is the constant rate of change in each case?

Summarize the Mathematics

In this investigation, you used reasoning with triangles to help discover a pattern relating the number of sides of a polygon to the sum of the measures of its interior angles. You also discovered a surprising pattern involving the sum of the measures of the exterior angles of a polygon.

a Explain how you can find the sum of the measures of the interior angles of a polygon. What is the measure of one interior angle of a regular n-gon?

b What is true about the sum of the measures of the exterior angles of a polygon? What is the measure of one exterior angle of a regular n-gon?

Be prepared to share your ideas and reasoning with the class.

Check Your Understanding

Being able to recognize traffic signs by their shape and color is important when driving and is often tested on exams for a driver's license. Examine the school crossing sign at the right.

a. Identify the shape of the sign and describe the symmetries of this shape.

b. Use the design specifications shown and symmetry to find the lengths of the remaining sides of the sign.

c. Find the measures of the remaining interior angles.

d. On a copy of the shape, extend the sides to form an exterior angle at each vertex. Find the measure of each exterior angle.

Investigation 3 Patterns with Polygons

One of the most common and interesting applications of polygon shapes is their use as tiles for floors and walls. The photo below shows portions of two different *tilings* at the Center for Mathematics and Computing at Carleton College in Northfield, Minnesota. The portion of the tiling in the center of the photo is based on special tiles and a procedure for placing them created by Sir Roger Penrose, a British mathematician at the University of Oxford. Can you identify the types of polygons used for the Penrose tiles?

Later in this lesson you will have the opportunity to explore the variety of patterns that can be created with these tiles. Researchers have recently discovered that certain chemicals naturally organize themselves in similar patterns, some of which are used to make nonstick coating for pots and pans.

As you work on the following problems, make notes that will help you answer this question:

Which polygons or combinations of polygons will tile the plane?

 The figures below show portions of **tilings** or **tessellations** of equilateral triangles and squares. The tilings are made of repeated copies of a shape placed edge-to-edge so that they completely cover a region without overlaps or gaps.

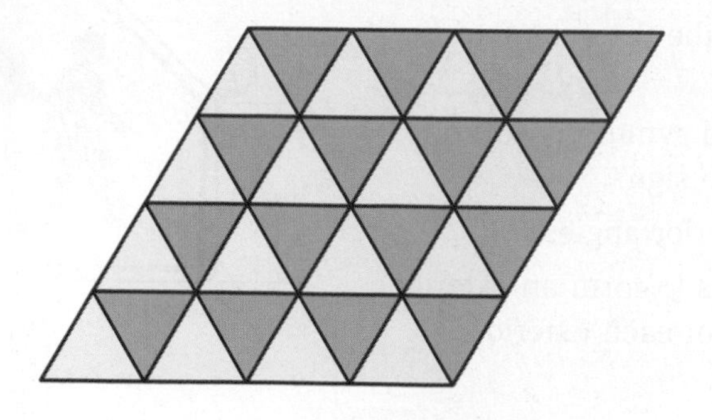

a. Assume that the tilings are extended indefinitely in all directions to cover the plane.

i. Describe the various ways that you can *slide* a tracing of each tiling so that it coincides with the original tiling. These tilings have **translation symmetry**.

ii. How could you describe the translation symmetries using arrows?

iii. Do the extended tilings have any reflection symmetry? If so, describe the lines of symmetry.

iv. Do the extended tilings have any rotational symmetries? If so, describe the centers and angles of rotation.

b. For these two tilings:

i. what is the sum of the measures of the angles at a common vertex?

ii. what is the measure of each angle at a common vertex?

c. In the tiling with equilateral triangles, identify other common polygons formed by two or more adjoined triangles that also produce a tiling. Sketch each and show the equilateral triangles that form the new tile. What does this suggest about other polygons that could be used to tile? Explain your reasoning.

2 Now explore if other triangles can be used as tiles.

a. Working in groups, each member should cut from poster board a small triangle that is *not* equilateral. Each member's triangle should have a different shape. Individually, explore whether a tiling of a plane can be made by using repeated tracings of your triangle. Draw and compare sketches of the tilings you made.

b. Can more than one tiling pattern be made by using copies of one triangle? If so, illustrate with sketches.

c. Do you think any triangle could be used to tile a plane? Explain your reasoning. You may find software like the "Tilings with Triangles or Quadrilaterals" custom tool helpful in exploring this question.

3 The most common tiling is by squares. In this problem, you will explore other quadrilaterals that can be used to make a tiling.

a. Each member of your group should cut a nonsquare quadrilateral from poster board. Again, each of the quadrilaterals should be shaped differently. Individually, investigate whether a tiling of a plane can be made with the different quadrilaterals. Draw sketches of the tilings you made.

b. Can more than one tiling pattern be made using the same quadrilateral shape? If so, illustrate and explain.

c. Make a conjecture about which quadrilaterals can be used to tile a plane.

i. Test your conjecture using software like the "Tilings wih Triangles or Quadrilaterals" custom tool.

ii. Explain why your conjecture makes sense in terms of what you know about angles of quadrilaterals.

4 You have seen two regular polygons that tile the plane. Now explore other regular polygons that could be used to make a tiling.

a. Can a regular pentagon tile the plane? Explain your reasoning.

b. Can a regular hexagon tile the plane? Explain.

c. Will any regular polygon of more than six sides tile the plane? Provide an argument to support your conjecture.

d. Tilings that consist of repeated copies of a single regular polygon with edges that match exactly are called **regular tessellations**. Which regular polygons can be used to make a regular tessellation?

5 As you saw at the beginning of this investigation, tilings can involve more than one type of shape. Another example of such a tiling is shown below. This tiling is from the Taj Mahal mausoleum in India.

a. How many different shapes are used in the tiling? Draw a sketch of each shape.

b. Each shape can be divided into equilateral triangles, so this tiling is related to the tiling in Problem 1. On a copy of the tiling, show the equilateral triangles that make up each shape.

6. Combinations of regular octagons and squares are frequently used to tile hallways and kitchens of homes. They are also often used in the design of outdoor patios.

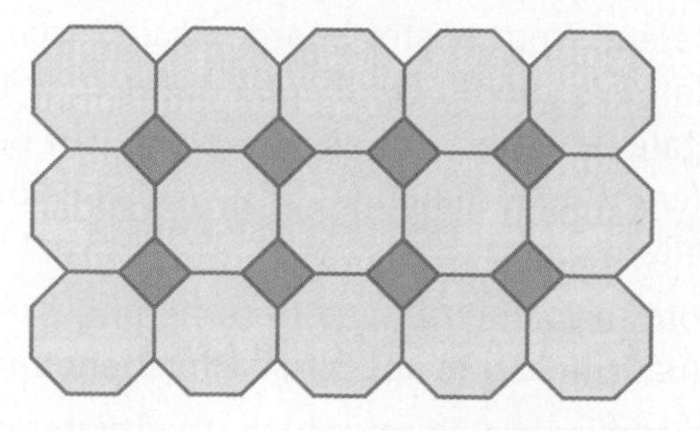

a. Explain why the polygons fit together with no overlaps or gaps. At each vertex, is there the same arrangement of polygons?

b. Tessellations of two or more regular polygons that have the same arrangement of polygons at each vertex are called **semiregular tessellations**. Use tiles made from poster board or software like the "Tilings with Regular Polygons" custom tool to test whether there can be a semiregular tessellation that has at each vertex a regular hexagon, two squares, and an equilateral triangle. If possible, draw a sketch of such a tessellation.

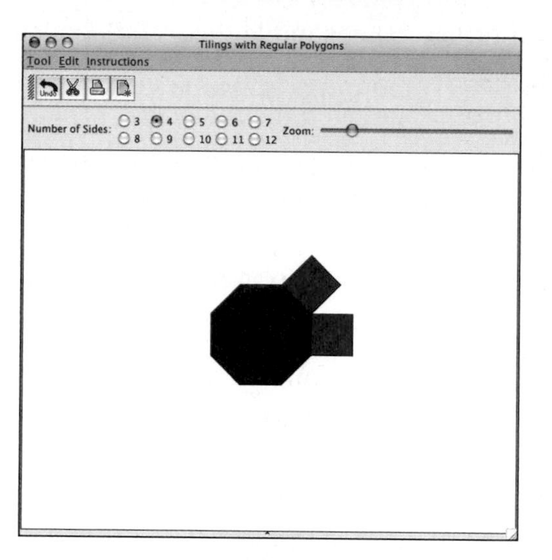

c. Semiregular tessellations are coded by listing the number of sides of the polygons that meet at a single vertex. The numbers are arranged in order with a smallest number first. The tessellation above Part a is coded 4, 8, 8, for square, octagon, octagon.

i. Use this code to describe the tessellation you drew in Part b. Give the code for each of the three possible regular tessellations.

ii. Determine if the vertex arrangement 3, 6, 3, 6 describes a semiregular tessellation.

iii. Is there another semiregular tessellation that can be formed using equilateral triangles and regular hexagons? Explain.

d. There are eight different semiregular tessellations. You have examples of four of them. Find at least one more example and name it using the vertex arrangement code in Part c. Compare your findings with those of other classmates.

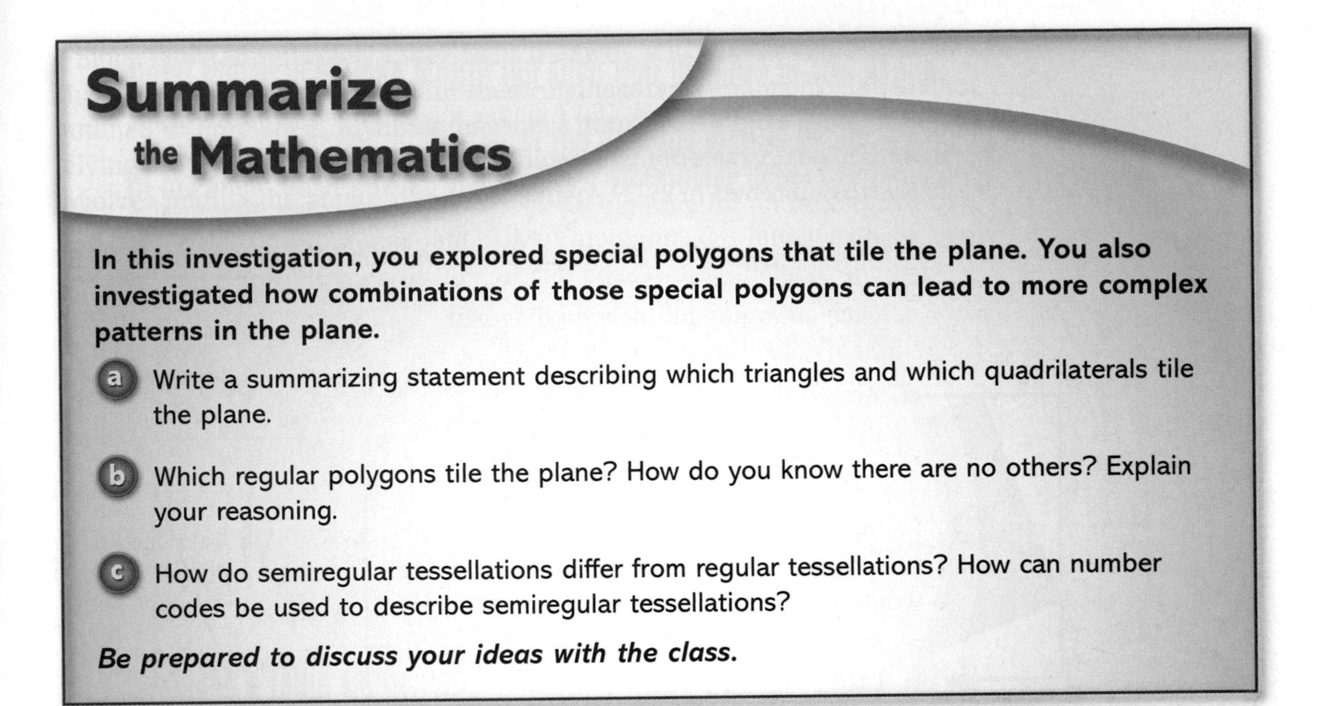

Summarize the Mathematics

In this investigation, you explored special polygons that tile the plane. You also investigated how combinations of those special polygons can lead to more complex patterns in the plane.

a Write a summarizing statement describing which triangles and which quadrilaterals tile the plane.

b Which regular polygons tile the plane? How do you know there are no others? Explain your reasoning.

c How do semiregular tessellations differ from regular tessellations? How can number codes be used to describe semiregular tessellations?

Be prepared to discuss your ideas with the class.

✓ Check Your Understanding

You have seen that a regular pentagon will not tessellate the plane. There are nonregular convex pentagons that will tessellate. But not many. Researchers have identified only 14 types. Whether there are more remains an open question.

Using a copy of the figure at the right, find a pentagon in the figure that will tile the plane. Shade it. Show as many different tiling patterns for your pentagon as you can.

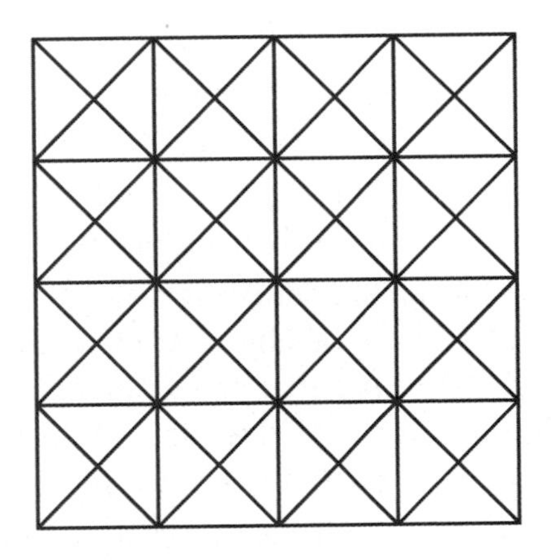

On Your Own

Applications

1 In regular pentagon *ABCDE*, four diagonals have been drawn.

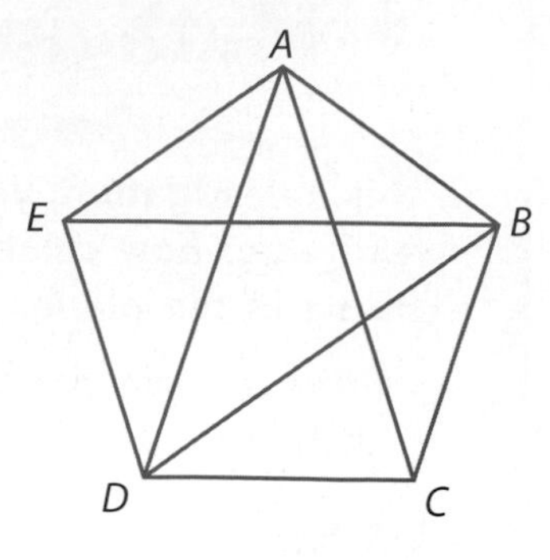

a. Use careful reasoning to explain why $\overline{AD} \cong \overline{AC}$.

b. Is diagonal $\overline{AD}$ congruent to diagonal $\overline{DB}$? Explain your reasoning.

c. Give an argument for why diagonal $\overline{AD}$ is congruent to diagonal $\overline{EB}$.

d. Are all the diagonals of a regular pentagon congruent? Explain your reasoning.

2 Suppose every student in your mathematics class shook hands with each of the other students at the beginning of your next class. What would be the total number of handshakes?

a. Show how you can represent this problem with a polygon for the case of just 4 students.

i. How many handshakes were involved?

ii. How is your answer related to the components of the polygon?

b. Represent and solve the problem for the special case of 5 students. Of 6 students.

c. How did you represent students in your models? How did you represent handshakes between students?

d. Use any numerical or visual pattern in your models to help solve the original problem.

e. In addition to using patterns to solve problems, it is also important to be able to explain and, whenever possible, generalize the patterns you discover. A student in a Wisconsin classroom claimed that a class of n students would involve $n^2 - 2n$ handshakes. The student reasoned as follows.

> I thought of students as vertices of an n-gon and handshakes as the sides and diagonals. An n-gon has n sides and n angles. From each vertex, I can draw a diagonal to $n - 3$ other vertices. So, I can draw $n(n - 3)$ diagonals. So, the number of handshakes is $n + n(n - 3) = n^2 - 2n$.

i. Is this reasoning correct? If not, identify and correct errors in the reasoning.

ii. Write a rule that expresses the number of handshakes as a function of the number of students n.

f. Write an expression in symbols and in words for calculating the number of diagonals of an n-gon.

3. In designing this mall garden, the architect proposed a gazebo in the shape of a regular octagon. The octagonal roof cupola was to have a width of 9 feet 8 inches. What should be the dimensions of each side? (*Hint*: Use the *auxiliary lines* drawn to help guide your thinking.)

4. Objects in nature are often approximately symmetric in form.

   a. The shapes below are single-celled sea plants called *diatoms*.

      i. Identify all of the symmetries of these diatoms. Ignore interior details.

      ii. For those with reflection symmetry, sketch the shape and show the lines of symmetry.

      iii. For those with rotational symmetry, describe the angles of rotation.

A 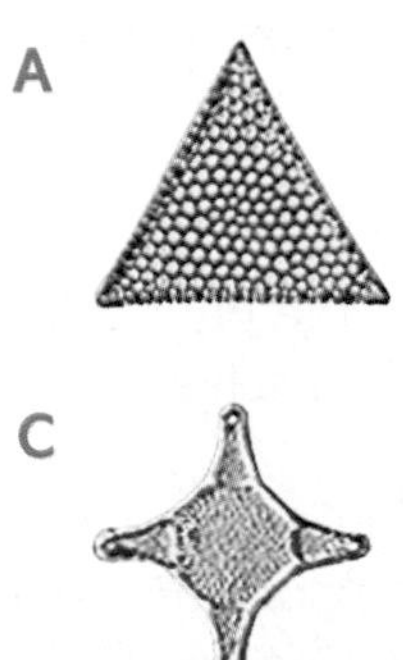

B 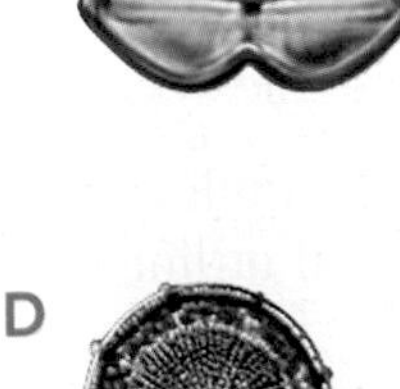

C

D

   b. Identify all of the symmetries of the two flowers shown below.

      i. If the flower has line symmetry, sketch the shape and draw all lines of symmetry.

      ii. If the flower has rotational symmetry, describe the angles of rotation.

A 

Geranium

B 

Periwinkle

c. It has been said that no two snowflakes are identical. Yet every snowflake has some common geometric properties.

  i. Identify the symmetries of the snowflakes below.

  ii. In terms of their symmetry, how are the snowflakes alike?

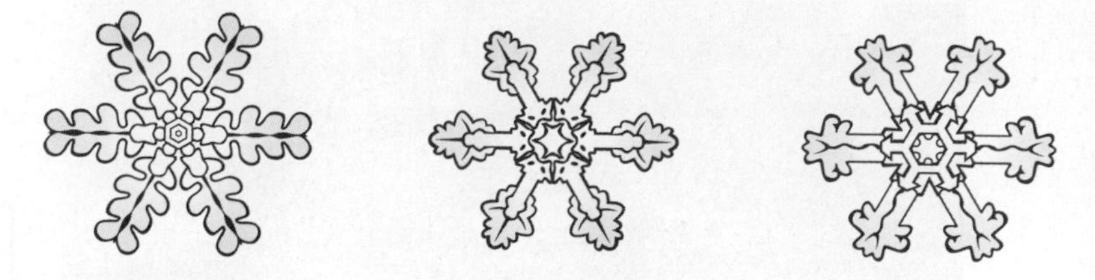

Native American rug

Star of Bethlehem quilt

5 Polygons and symmetry are important components of the arts and crafts of many cultures.

a. Examine the photo of a Native American rug.

  i. Describe the line and rotational symmetries of this rug. Sketch two design elements within the rug that have rotational symmetry. Describe the angles through which each can be turned.

  ii. Sketch two design elements within the rug that have line symmetry. Draw the line(s) of symmetry.

  iii. Are there any design elements which have both rotational and line symmetry? If so, identify them. Where is the center of rotation in relation to the lines of symmetry?

b. The design of the quilt to the left is called "Star of Bethlehem."

  i. What rotational symmetries do you see in the fundamental "pinwheel stars"? Give the angles of rotation for each of these symmetries.

  ii. Is there line symmetry in the "pinwheel stars"? Explain.

  iii. Does the quilt as a whole, including the pinwheel stars, have rotational or line symmetry? Describe each symmetry you find.

6 Here is a two-person game that can be played on any regular polygon. To play, place a penny on each vertex of the polygon. Take turns removing one or two pennies from adjacent vertices. The player who picks up the last coin is the winner.

a. Suppose the game is played on a nonagon, as shown at the right. Try to find a strategy using symmetry that will permit the second player to win always. Write a description of your strategy.

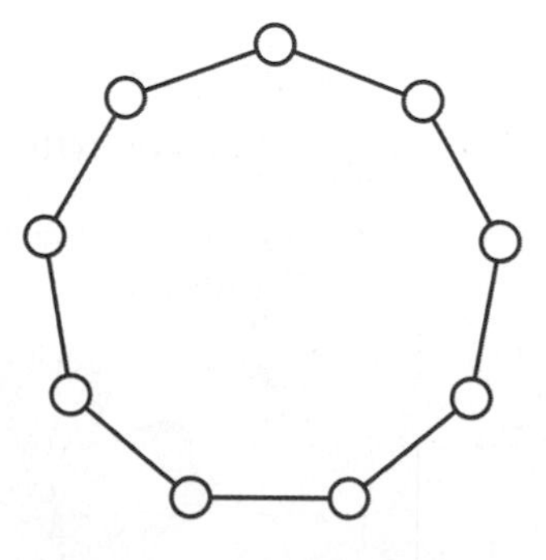

b. Will the strategy you found work if the game is played on any polygon with an odd number of vertices? Explain your reasoning.

c. Suppose the game is played on a polygon with an even number of vertices, say an octagon. Try to find a strategy that will guarantee that the second player still can win always. Write a description of this strategy.

7 Bees produce honeycomb cells with cross sections that are approximately the shape of a regular hexagon.

a. If all the cells are to be congruent, what other regular polygon shapes might they have used?

b. Suppose the perimeter of a cross section of one cell of a honeycomb is 24 mm. Find the area of the cross section, assuming the cell has the following shape.

i. an equilateral triangle

ii. a square

iii. a regular hexagon

c. Which cell has the greatest cross-sectional area for a fixed perimeter of 24 mm? As the number of sides of a regular polygon with fixed perimeter increases, how does the corresponding area change?

d. Write a statement summarizing how shape is an important factor in the design of the cells of a honeycomb.

8 It is possible to create intriguing tessellations by carefully modifying the sides of a polygon. The Dutch artist M.C. Escher was a master of these modifications. He created this tessellation of *Pegasus*, the mythical winged horse.

a. Assuming the tessellation above is extended indefinitely in all directions, describe its symmetries.

b. Study the process below which illustrates how Escher may have created his "flying horse" tessellation from a square.

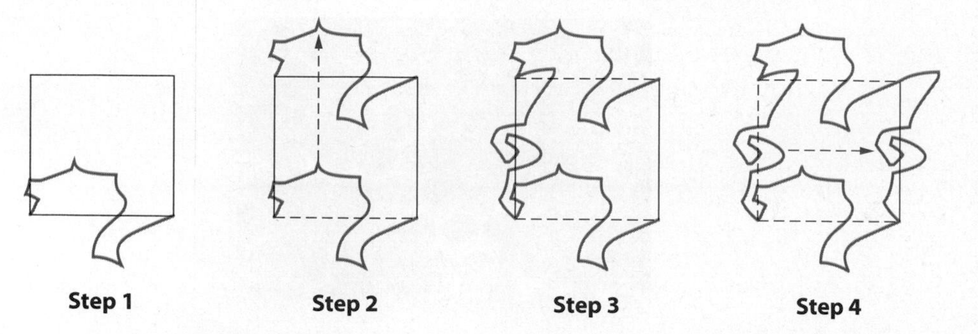

i. Use tracing paper to verify how the modifications of sides are translated to the opposite sides.

ii. How does the area of the Pegasus compare to the area of the initial square?

c. Start with a square, rectangle, or other parallelogram and use a similar process to create your own Escher-type tessellation. Verify that your shape does tile the plane.

9 A beautiful tiling for paving or wallpaper can be made with leaf-like tiles based on a rhombus. Study the process below, which illustrates how this tile may be created.

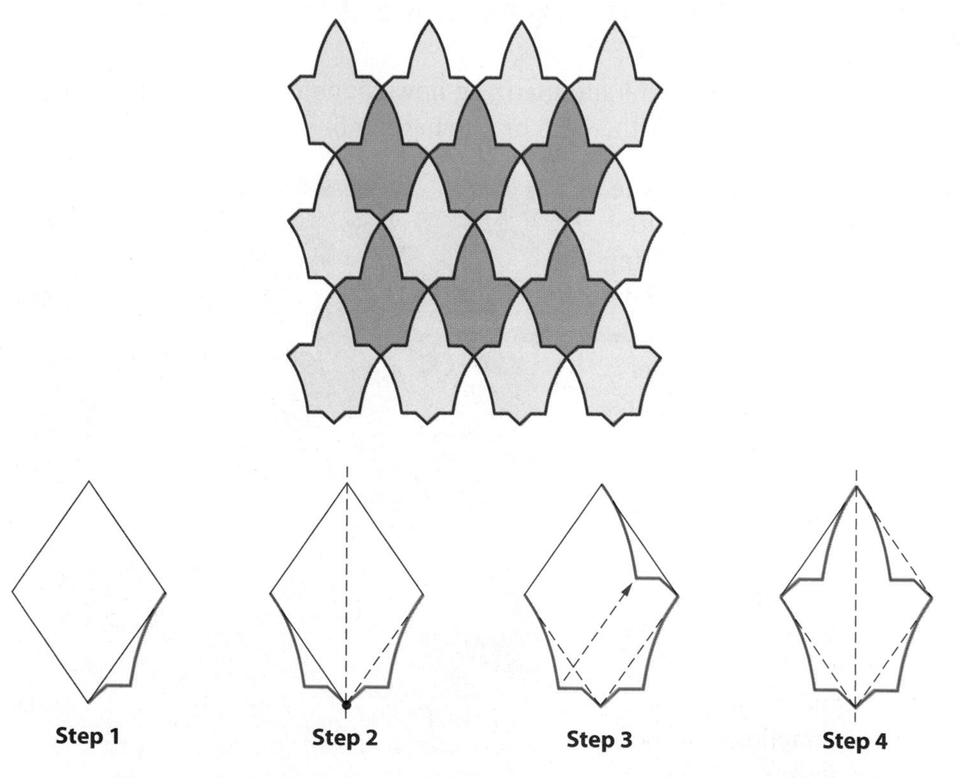

a. Use tracing paper to verify how the modifications of sides of the rhombus are reflected or translated to produce the other sides of the leaf tile.

b. How does the area of the leaf tile compare to the area of the rhombus?

c. Start with a rhombus and use a similar process to create your own paving or wallpaper design.

Connections

10 Psychologists often use figures like that below in their study of human perception.

a. What do you see?

b. Name the colored polygons involved in this task.

c. Which polygons are convex? Nonconvex?

d. Which polygons have symmetry? Describe the symmetries.

11 Graphs of various functions relating variables x and y are shown below. The scale on the axes is 1.

a. For each graph, locate any line(s) of symmetry.

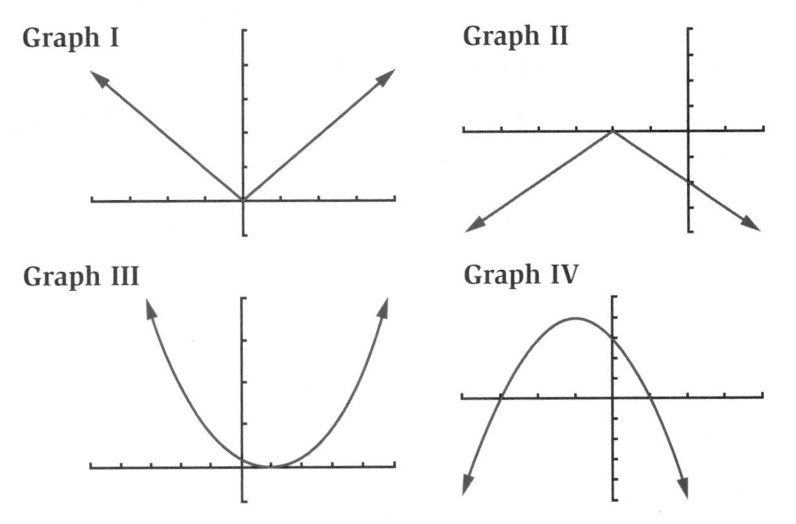

b. What pattern do you see in the coordinates of the points on each line of symmetry? Write the equation of the symmetry line(s).

c. Suppose you have a graph, and its line of symmetry is the y-axis. If one point on the graph has coordinates $(-8, -23)$, what is the y-coordinate of the point on the graph with x-coordinate 8? Explain your reasoning.

12 The initial and first two stages in making a triangular Sierpinski carpet are shown below. In the *Exponential Functions* unit, you explored the area of carpet remaining as a function of the cutout stage. Describe the symmetries that the triangular carpet has at each stage of the process.

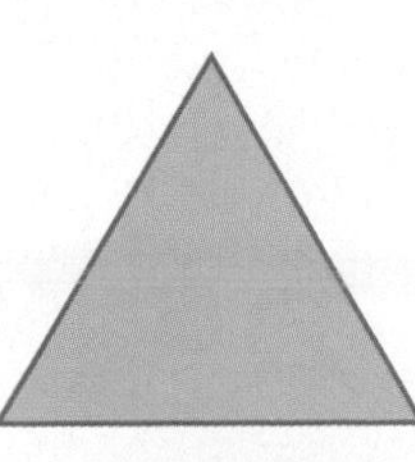
Stage 0

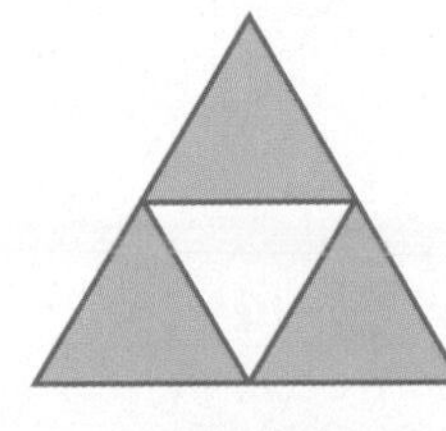
Stage 1

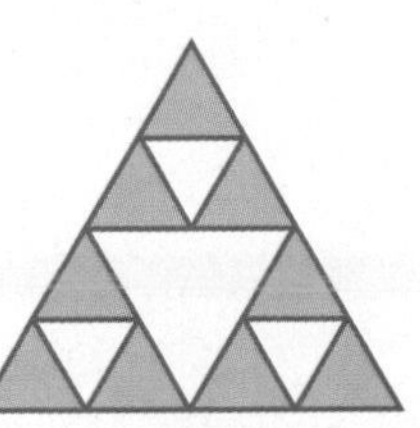
Stage 2

13 Make a sketch of a regular pentagon. Extend one of the sides to form an exterior angle. How could you use your knowledge of exterior angles to calculate the measure of an interior angle of a regular pentagon? Would your method work for other regular polygons? Explain your reasoning.

14 In Investigation 2, you discovered the rule $A = \frac{(n - 2)180°}{n}$, which gives the measure A one interior angle of a regular n-gon.

a. As the number of sides of a regular polygon increases, how does the measure of each of its interior angles change? Is the rate of change constant? Explain.

b. Use your formula to find the measure of one interior angle of a regular 20-gon. Could a tessellation be made of regular 20-gons? Explain your reasoning.

c. When will the measure of each angle of a regular polygon be a whole number?

d. Use your calculator or computer software to produce a table of values for angle measures of various regular polygons. Use your table to help explain why the only regular tiling of the plane is one with regular polygons of 3, 4, or 6 sides.

15 The **dual** of a tessellation by regular polygons is a new tessellation obtained by connecting the centers of polygons that share a common edge. Use equilateral triangular (isometric) dot paper to complete Parts a and b and square dot paper to complete Part c.

a. Draw a portion of a regular hexagon tessellation. Using a different colored pencil, draw and describe the dual of the tessellation.

b. Draw a portion of an equilateral triangle tessellation. Draw and describe the dual of this tessellation.

c. Draw and describe the dual of a tessellation of squares.

16 Strip or *frieze* patterns are used in architecture and interior design. The portions of frieze patterns below came from the artwork on pottery of the San Ildefonso Pueblo, New Mexico.

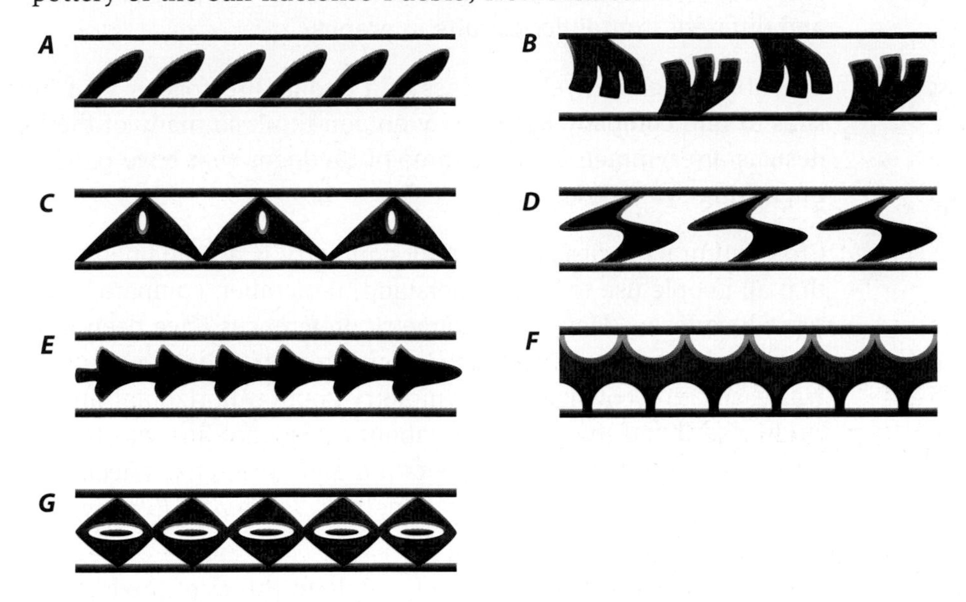

Source: Groups and Geometry in the Ceramic Art of San Ildefonso. *Algebras, Groups and Geometries 2*, no. 3 (September 1985).

Imagine that each frieze pattern extends indefinitely to the right and to the left.

a. Confirm that each pattern has translation symmetry.

b. Examine each frieze pattern for reflection and rotational symmetries. (If a strip has such a symmetry, the strip must appear the same before and after it is reflected across a line or rotated about a point.) Describe your findings for each pattern.

c. Pattern B has a symmetry called *glide-reflection symmetry*. Describe the translation and reflection combination that will move the motif so the pattern coincides with itself.

Reflections

17 Look up the word "polygon" in a dictionary. What is the meaning of its parts "poly" and "gon"? How do these meanings relate to your understanding of "polygon"?

18 A regular polygon is both equilateral and equiangular.

a. Give an example of an equilateral polygon that is not a regular polygon.

b. Give an example of an equiangular polygon that is not a regular polygon.

19 Recall that an Euler circuit is a route through a connected vertex-edge graph such that each edge of the graph is traced exactly once, and the route starts and ends at the same vertex. How are polygons similar to and different from Euler circuits in graphs?

20 Thumb through the yellow pages of a phone directory or visit Internet sites to find company logos. Why do you think so many of the logo designs are symmetric? Draw three of the logos that have particularly interesting symmetries.

21 Cross-cultural studies suggest that symmetry is a fundamental idea that all people use to help understand, remember, compare, and reproduce forms. However, symmetry preferences have been found across cultures. One study found that symmetry about a vertical line was easier to recognize than symmetry about a horizontal line. The study also found that symmetry about a diagonal line was the most difficult to detect. (Source: Orientation and symmetry: effects of multiple, rotational, and near symmetries. *Journal of Experimental Psychology* 4[4]: 1978.)

a. Would the findings of the study apply to the way in which you perceive line symmetry?

b. Describe a simple experiment that you could conduct to test these findings.

22 Look back at your work on Applications Task 1. Do you think your finding applies to other regular polygons? Carefully draw a regular hexagon and label its vertices *PQRSTU*. Name the diagonals that are congruent to each other and give reasons to support your conclusions.

Extensions

23 Commercial artists sometimes use a device similar to the one shown for drawing mirror images. The linkage is assembled so that $APBP'$ is a rhombus and the points at P and P' pivot freely as A and B slide along ℓ. As P traces out a figure, P' traces out its reflection image. Explain why this device works as it does.

24 Use interactive geometry software or other tools to further investigate properties of regular polygons.

a. Is there a relationship between the measure of a central angle and the measure of an interior angle? If so, can you explain why that must *always* be the case?

b. Is there a relationship between the measure of a central angle and the measure of an exterior angle? If so, can you explain why?

25 In Investigation 3, you explored semiregular tessellations by examining various arrangements of regular polygon shapes around a vertex. You can also examine the possibilities using algebra. The diagram below is a start. It shows the case of three regular polygons of m, n, and p sides completely surrounding a vertex with no overlapping.

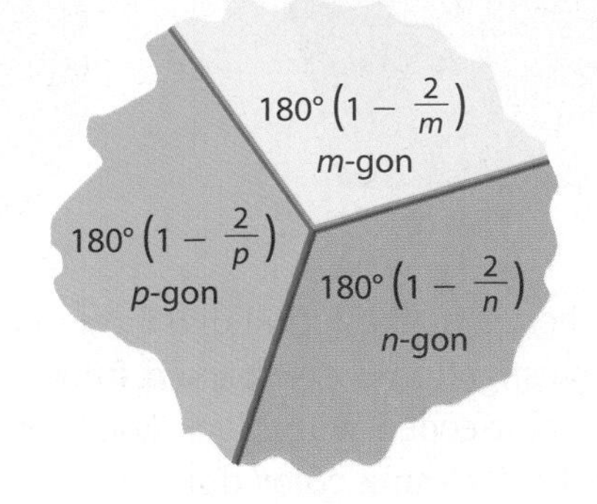

a. Why must the numbers m, n, and p all be integers greater than 2?

b. Explain why the measure of each interior angle is as shown.

c. Write an equation that must be satisfied if the polygons are to form a tessellation.

d. Show that your equation is equivalent to $\frac{1}{m} + \frac{1}{n} + \frac{1}{p} = \frac{1}{2}$. What are the whole number solutions to this equation?

e. Relate one of your solutions in Part d to a semiregular tessellation.

26 In Investigation 1, you explored how a two-mirror kaleidoscope could be used to create regular polygons. For this task, make a three-mirror kaleidoscope by fastening three congruent mirrors together to form an equilateral triangle at the base.

a. Explore what can be created by placing various patterns in the base like those shown below.

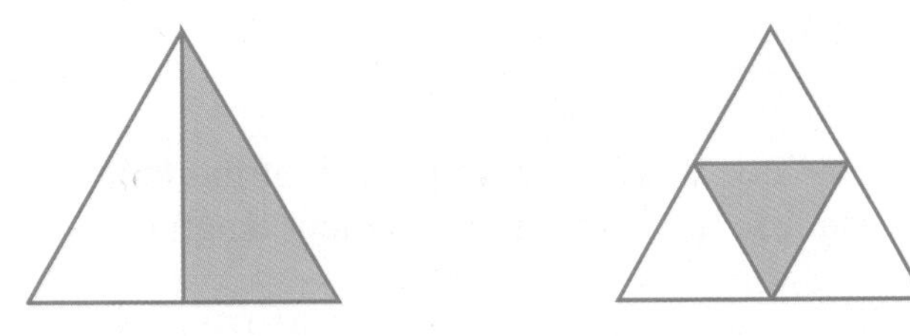

b. Can you create each regular tessellation by placing an appropriate pattern in the base of your kaleidoscope? If so, draw diagrams of the patterns.

c. Can you create a semiregular tessellation by placing an appropriate pattern in the base? If so, draw a diagram of the pattern.

d. How are ideas of symmetry related to your pattern-building explorations?

27 The first two tessellations on page 422 use the same isosceles triangle. Think of each tiling extended indefinitely to cover the plane. A tessellation that fits exactly on itself when translated is called **periodic**; it has translation symmetry. One that does not have translation symmetry is called **nonperiodic**. Many polygonal shapes (like the triangles below) will tile both periodically and nonperiodically. Many shapes (such as Escher's *Pegasus*) will tile *only* periodically.

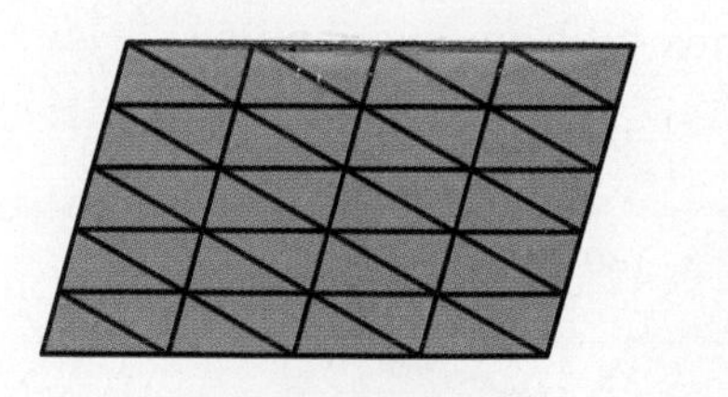

The diagram below at the left shows how the Penrose tiles at the beginning of Investigation 3 are formed from a particular rhombus. If the vertices are color-coded with black dots and gray dots as shown and only vertices with the same color dots are allowed to meet, then the only way these kite and dart shapes can tessellate is nonperiodically.

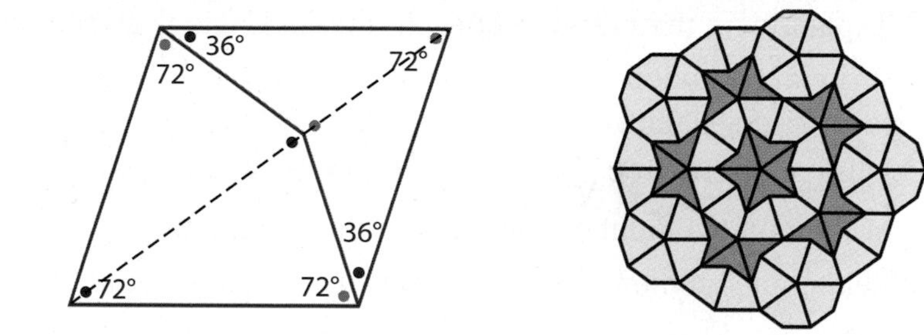

Use the "Tilings with Penrose Tiles" custom tool or tracing paper to make a copy of the kite and dart shown above. Then create two different nonperiodic tilings using the matching rule and explain why they are different.

Review

28 There are three right triangles in each of the following diagrams. In each case, you are given three segment lengths.

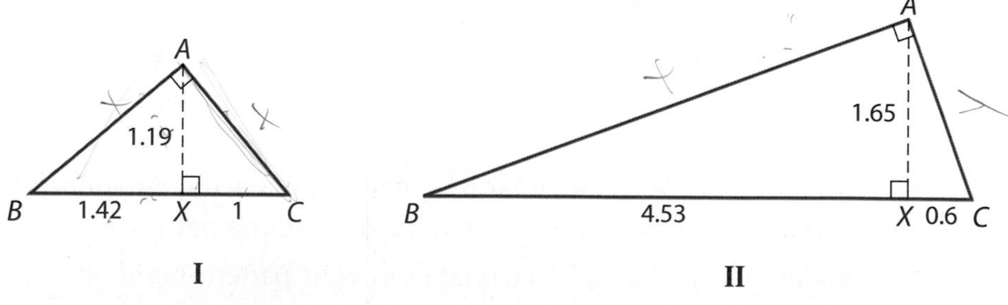

a. Using the marked segment lengths, find AB and AC in each case.

b. Find the ratio AB:AC and AX:XC in each case.

29 Russell's department store is having a 25%-off sale. There is 6% sales tax on all items bought.

a. Liz found a dress she likes that usually costs $65. She has $50. Does she have enough money to buy the dress during this sale?

b. Jerry bought a shirt on sale and paid a total of \$22.26. What is the price of the shirt when it is not on sale?

c. Write a formula that describes the relationship between the original price of an item p and the amount of money that is needed to buy the item during this sale C.

d. Is the relationship described in Part c linear, exponential, or neither? Explain your reasoning.

30 Consider the rule $2x - 4 = y$.

a. How do the y values change as the x values increase by 1?

b. How do the y values change as the x values increase by 5?

c. Where does the graph of this rule intersect the y-axis?

31 Janelle creates and sells small ceramic figurines at arts and crafts fairs. How many figurine boxes, which are 3 inches long, 2 inches wide, and 3 inches tall, could she fit into the shipping box shown at the right?

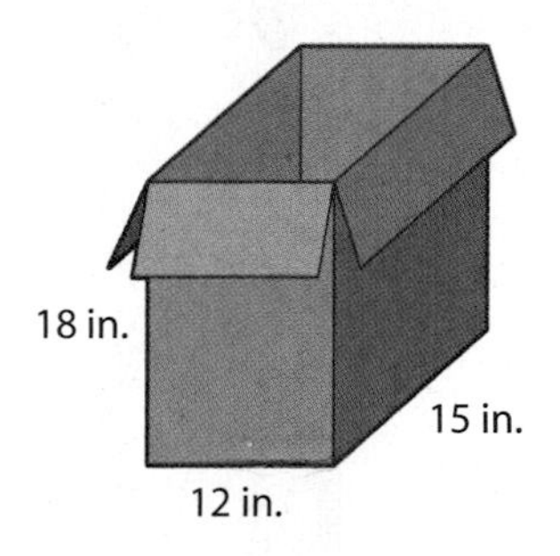

32 Write each of the following in a simpler equivalent exponential form that uses only positive exponents.

a. $(4x^3)^2$

b. $(5x^2y)(-3x^5y)$

c. $\dfrac{24t^6r^2}{8t^3r^5}$

d. $\left(\dfrac{-6d^4}{2}\right)^3$

e. $\left(\dfrac{5x^2y}{7x^5}\right)\left(\dfrac{-7x}{3y}\right)^2$

33 Write each expression in simplest equivalent form. Show your work.

a. $7(4 - x) + 15$

b. $36 - (50 - 15x) + 10x$

c. $0.5x(4 - 20) + \dfrac{6x + 4}{2}$

d. $7 + 4(5x - 7) - 2(15 - 9x)$

34 Complete each table so that its entries show an exponential pattern of change. Then write a rule that expresses y as a function of x.

a.

x	0	1	2	3	4
y	800	200	50		

b.

x	1	2	3	4	5
y	30	45	67.5		

c.

x	0	2	4	6	8
y	1	4	16		

35 Solve each equation.

a. $48 = 3(2^x)$

b. $129 = 6 + 3x$

c. $900(0.4^x) = 90$

d. $5(2x + 4) = 300$

36 Mariah wants to join a gym. In her community she has two choices. The First Street Gym charges a membership fee of \$120 and a monthly fee of \$35. Fitness Center has a \$20 membership fee and charges \$55 per month.

a. For what number of months will the total cost be the same at the two gyms?

b. Under what conditions will First Street Gym be less expensive?

LESSON 3

Three-Dimensional Shapes

In the previous two lessons, you studied two-dimensional shapes and some of their important properties and applications in building, design, and art. In many cases, two-dimensional shapes and their properties have corresponding ideas in three dimensions. In three dimensions, as in two dimensions, the shape of an object helps to determine its possible uses.

In your history classes, you may have noticed that the Greeks and people in other ancient cultures often used columns in the design of their buildings. The Greek Parthenon shown above is made of marble. Thus, the columns had to support great weights. An important design consideration is the shape of the column.

Think About This Situation

Suppose you made three columns by folding and taping sheets of 8.5 × 11-inch paper so that each column was 8.5 inches high with bases shaped, respectively, as equilateral triangles (pictured at the right), squares, and regular octagons. Imagine that you placed a small rectangular piece of cardboard (about 6 by 8 inches) on top of a column. Then you carefully placed a sequence of objects (like notebooks or textbooks) on the platform until the column collapsed.

8.5"

a Which column shape do you think would collapse under the least weight?

b Which column shape do you think would hold the most weight before collapsing?

c Test your conjectures by conducting the experiment with your class. What happens to the maximum weight supported as the number of sides of the column base increases?

d Suppose another column is made with a regular hexagon base. Predict how the number of objects it would support before collapsing would compare to those of the three columns above. Explain your reasoning.

e Why do you think the ancient Greeks chose to use cylindrical columns?

In this lesson, you will use properties of two-dimensional shapes to aid in examining three-dimensional shapes. You will learn how to identify and describe common three-dimensional shapes and how to construct three-dimensional models of them. You will develop skill in visualizing and sketching them in two dimensions and in identifying their symmetries and other important properties. You will also explore some of the many connections between two- and three-dimensional shapes.

Investigation 1 Recognizing and Constructing Three-Dimensional Shapes

Like the columns of the Parthenon, most everyday three-dimensional shapes are designed with special characteristics in mind. In this investigation, you will consider the following questions:

What are important characteristics of common three-dimensional shapes?

How can three-dimensional models of these shapes be constructed?

1 As a class, examine the objects depicted below.

Popcorn Box

Candy Package

Block "L"

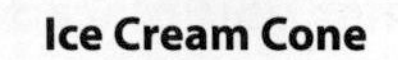

Ice Cream Cone

a. Which of these objects have similar geometric characteristics? What are those characteristics?

b. How would you describe the shapes of the above objects?

c. A **polyhedron** (plural: *polyhedra*) is the three-dimensional counterpart of a polygon. It is made up of a set of polygons that encloses a single region of space. Exactly two polygons (*faces*) meet at each *edge* and three or more edges meet at a *vertex*. Furthermore, the vertices and edges of the polyhedron are vertices and edges of the polygon faces.

- **i.** Which of the above shapes are polyhedra? For each polyhedron, name the polygons that are its faces.
- **ii.** If a shape is not a polyhedron, explain why not.

d. A **convex polyhedron** is a polyhedron in which no segment connecting any two vertices goes outside the polyhedron. Which of the above shapes are convex polyhedra?

e. Name at least two other common objects with different polyhedron shapes.

2 Polyhedron-shaped boxes for packaging products like cereal and candy are often manufactured using a two-dimensional pattern, called a **net**, that can be folded along its edges to form the polyhedron.

a. A **cube** is a polyhedron with six congruent square faces. Examine enlarged copies of the three nets shown below. Which of these nets can be folded to make a cube? For nets that can be folded to make a cube, make matching tick marks on edges that match when the net is folded into a cube.

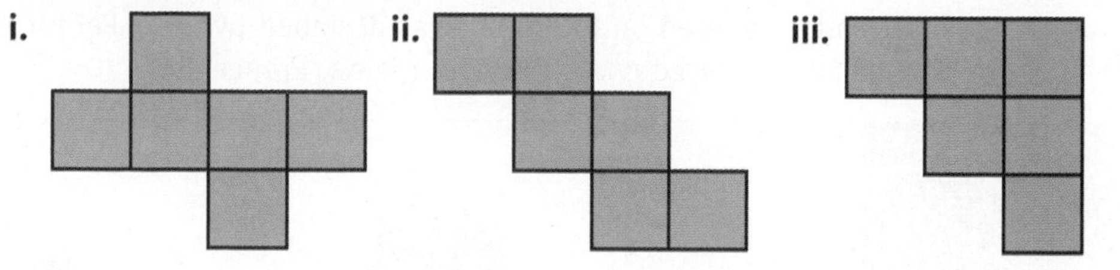

b. Draw a possible net for the candy package shown in Problem 1.

c. Using enlarged copies of the three nets pictured below, fold each net into a polyhedron. Divide the work among your classmates so that each student makes one polyhedron.

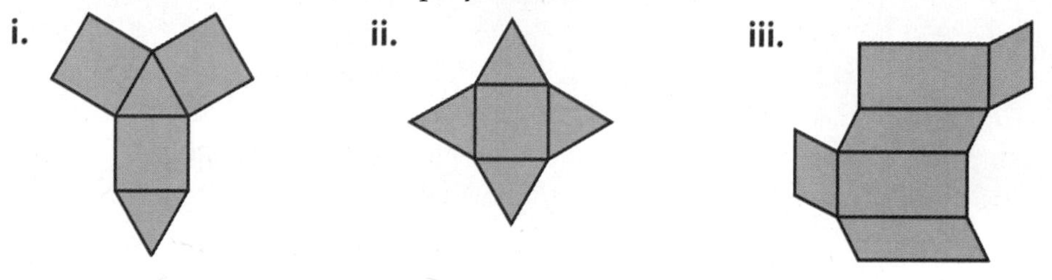

d. Examine each vertex of the three polyhedra you made in Part c. What is the least number of faces that meet at a vertex? What is the least number of edges that meet at a vertex? Would it be possible for fewer faces or edges to meet at a vertex of any polyhedron? Explain your reasoning.

e. When polygons tile the plane, the sum of the measures of angles at a vertex is 360°.

- **i.** What is the sum of the measures of angles that meet at each vertex of the first two polyhedra in Part c?
- **ii.** What is the sum of the measures of the angles that meet at each vertex of the cube in Part a? Of the block "L" in Problem 1?
- **iii.** Make a conjecture about the sum of the measures of angles that meet at each vertex of a convex polyhedron.

 Two important types of polyhedra are *prisms* and *pyramids*. Architects frequently use these shapes and an understanding of their characteristics in the design of buildings. For example, the Flat Iron Building, designed by David Burnham and the oldest surviving skyscraper in New York City, has an unusual prism shape. The glass pyramid, pictured on the right, was designed by I. M. Pei for the entrance to the Louvre museum in Paris, France.

a. A **prism** is a polyhedron with two parallel congruent faces with corresponding edges that are connected by parallelograms (called *lateral faces*). Which of the four polyhedra in Problem 2 are prisms?

b. Either one of a pair of congruent, parallel faces of a prism may be called a *base of the prism*. How many different faces of each of the prisms that you identified in Part a could be considered to be a base of the prism? Explain your answers.

c. Compare the prisms in Problem 2 Part c. One prism is a *right prism* and the other is an *oblique prism*. From what you know about other uses of the term "right" in mathematics, which prism is a right prism? Write a definition of a right prism.

d. A **pyramid** is a polyhedron in which all but one of the faces must be triangular, and the triangular faces share a common vertex called the *apex of the pyramid*. The triangular faces are called *lateral faces*. The face that does not contain the apex may have any polygonal shape, and this face is called the *base of the pyramid*. Identify the base and apex of the pyramid in Problem 2 Part c.

e. Prisms and pyramids are often named by the shapes of their bases. For example, a pyramid with a five-sided base is called a *pentagonal pyramid*. Use this naming method to name a cube as a prism. Name the other prisms and pyramids in Problem 2.

4 Now examine each of the shapes below.

a. Which shapes appear to be right prisms? Oblique prisms? Pyramids? None of these? Explain your choices based on the definitions in Problem 3.

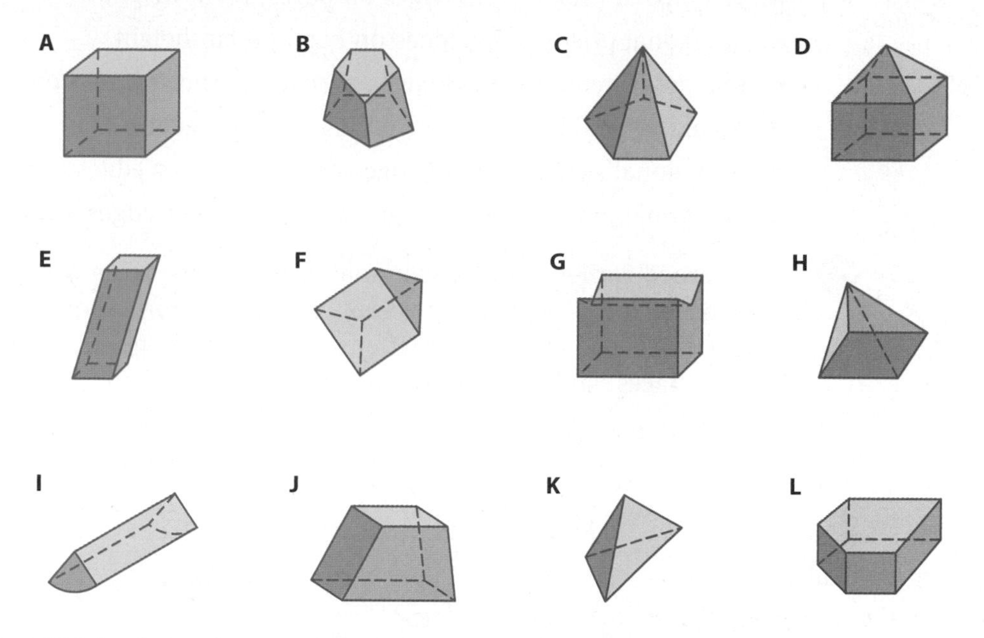

b. Which of the above polyhedra appear to be nonconvex? Explain.

Unless otherwise stated, in the remainder of this unit and in future units, prisms will be assumed to be convex right prisms.

5 Models for polyhedra can be hollow like a folded-up net or an empty box, or they can be solid like a candy bar or brick. Polyhedra can also be modeled as a skeleton that includes only the edges, as in a jungle gym.

Working in a group, make the following models from coffee stirrers or straws and pipe cleaners. Cut the pipe cleaners into short lengths that can be used to connect two straw edges at a vertex, as shown. Divide the work. Each student should build at least one prism and one pyramid. Save the models to explore properties of polyhedra later in this lesson.

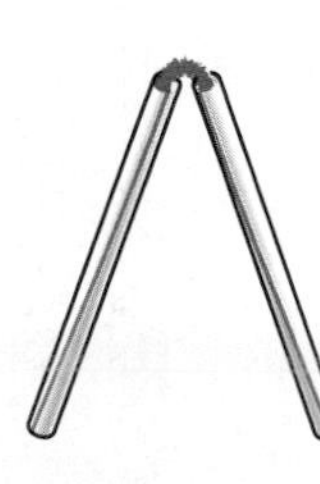

a. cube: 5-cm edges

b. triangular prism: 5-cm edges on bases, 8-cm height

c. square prism: 5-cm edges on bases, 8-cm height

d. pentagonal prism: 5-cm edges on bases, 8-cm height

e. hexagonal prism: 5-cm edges on bases, 8-cm height

f. triangular pyramid: 5-cm edges on bases, other edges 5 cm

g. square pyramid: 5-cm edges on bases, other edges 8 cm

h. pentagonal pyramid: 5-cm edges on bases, other edges 8 cm

i. hexagonal pyramid: 5-cm edges on bases, other edges 8 cm

6 You learned earlier in this lesson that at least three faces and at least three edges must meet at any vertex of a polyhedron. There is a deeper and more surprising relationship among the numbers of vertices, faces, and edges in any polyhedron.

a. Complete a table, like the one below, that includes some of the polyhedra for which your class constructed models.

	Number of Vertices	Number of Faces	Number of Edges
Cube			
Triangular Prism			
Hexagonal Prism			
Triangular Pyramid			
Square Pyramid			

b. Examine the table you made. Describe one or two patterns that you see.

c. Make a conjecture about the relationship among the numbers of faces F, vertices V, and edges E for each of these polyhedra. Compare your conjecture with others and resolve any differences.

d. Test the relationship you discovered in Part c using one of your other polyhedron models. Does it work for that polyhedron, too?

The formula relating the numbers of vertices, faces, and edges of convex polyhedra was first discovered by Swiss mathematician Leonhard Euler (1707–1783) and is called **Euler's Formula for Polyhedra**. The formula applies to all convex polyhedra.

7 Use Euler's Formula for Polyhedra to answer the following questions.

a. If a convex polyhedron has 10 vertices and 8 faces, how many edges does it have?

b. Can a convex polyhedron that has 8 faces, 12 edges, and 7 vertices be constructed? Give reasons for your answer.

8 *Cones* and *cylinders* are two other common three-dimensional shapes.

a. Are cones and cylinders polyhedra? Why or why not?

b. How do cones and cylinders compare to one another? How are they similar to and different from pyramids and prisms?

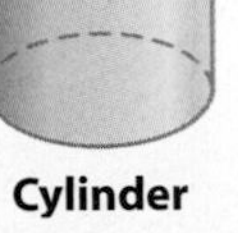

Cone **Cylinder**

Summarize the Mathematics

In this investigation, you explored characteristics of commonly occurring three-dimensional shapes—prisms, pyramids, cylinders, and cones.

a In what ways are polyhedra like polygons? In what ways are they different?

b What is the least number of faces that can meet at a vertex of any polyhedron? What is the least number of edges that can meet at a vertex of any polyhedron?

c How are pyramids like prisms? How are they different?

d Consider a sequence of prisms in which each base is a regular polygon. The base of the first prism has 3 sides, the base of the second has 4 sides, the base of the third has 5 sides, and so on. As the number of sides in the base increases, what shape does the prism begin to resemble?

e Consider a sequence of pyramids with bases like those described in Part d. As the number of sides in the base of a pyramid increases, what shape does the pyramid begin to resemble?

f What formula relates the number of vertices, faces, and edges of the polyhedra that you explored in this investigation?

Be prepared to share your ideas and formula with the class.

Check Your Understanding

Examine the photo of a cereal box shown at the right.

a. Explain why the box is an example of a polyhedron. Then name the polyhedron as precisely as you can.

b. Draw a net for a model of the box.

c. Betsy claimed that this three-dimensional shape is a prism and that any of its faces could be considered to be its base. Do you agree with Betsy? Explain why or why not.

d. If you were to make a model of this three-dimensional shape from straws and pipe cleaners:

  i. how many different length straws would you need?

  ii. how many straws of each length would be needed?

e. Verify that Euler's Formula is satisfied for the polyhedron represented by the cereal box.

Investigation 2 Visualizing and Sketching Three-Dimensional Shapes

It is not always practical to construct models of three-dimensional shapes. For example, you cannot fax a scale model of an off-shore oil rig to an engineer in another country. Rather, the three-dimensional shape needs to be represented in two dimensions in a way that conveys the important information about the shape. In this investigation, you will explore these two main questions:

What are some effective ways to sketch three-dimensional shapes?
What information does each kind of sketch provide about the shape?

There are several methods for representing a three-dimensional shape in a sketch, but since the sketch has only two dimensions, some information about the three-dimensional shape will necessarily be missing. One way to depict three-dimensional shapes is to sketch two-dimensional *face-views* such as a top view, a front view, and a right-side view. Architects commonly use this method, called an **orthographic drawing**. For the house below, a *top view*, a *front view*, and a *right-side view* are shown. Together, these views display the length, depth, and height of the building to scale. You'll notice the top view is different from the other two. Floor plans such as this are frequently used instead of an exterior top view.

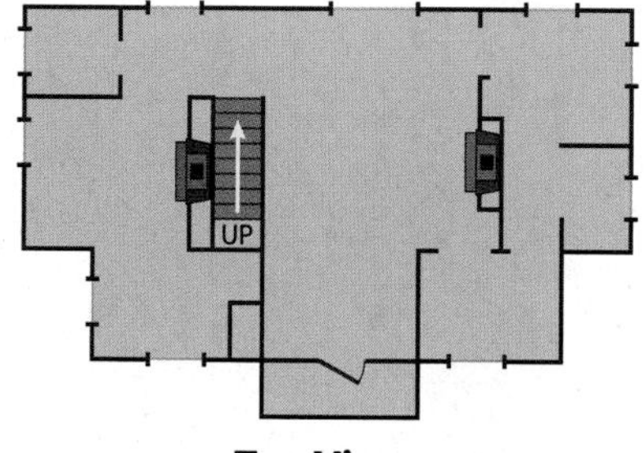

Top View

Front View

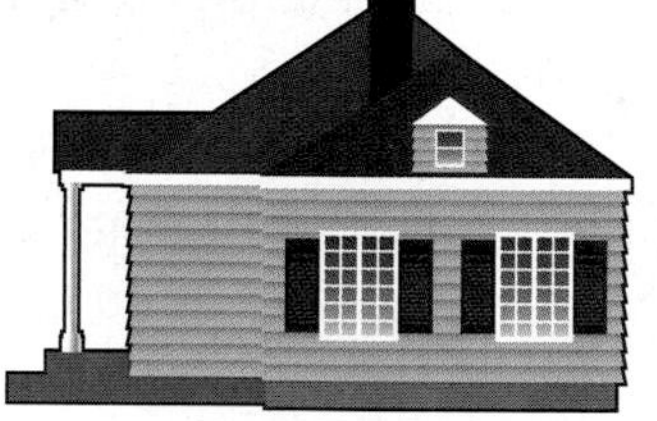

Right-Side View

1 An orthographic drawing of a model of a hotel made from cubes is shown below.

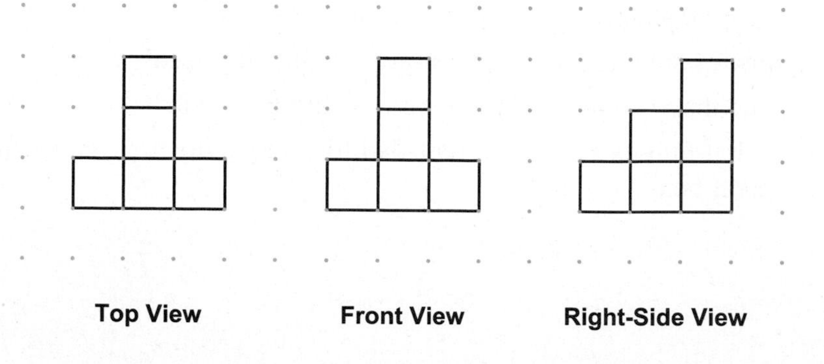

Top View **Front View** **Right-Side View**

a. How many cubes make up the model?

b. Use blocks or sugar cubes to make a model of this hotel. Build your model on a sheet of paper or poster board that can be rotated.

c. Could you make the model using information from only two of these views? Explain.

2 Examine this model of a building built from cubes. Assume any cube above the bottom layer rests on another cube and that there are no other hidden cubes.

a. Make an orthographic sketch of this model.

b. How many cubes are in this model?

c. Would it be possible to make a model with fewer cubes that has the same top, front, and right-side views as this one? Explain.

3 Another way to represent a three-dimensional shape such as a popcorn box is shown below. The sketch on the right, called an **oblique drawing**, is a *top-front-right corner* view of the box as a geometer would draw it. The front face was translated in the direction of the arrow to produce the back face, then edges were drawn to connect vertices. The sketch gives a sense of depth even though it is not drawn in true perspective. The three edges of the box blocked from view are shown as dashed lines.

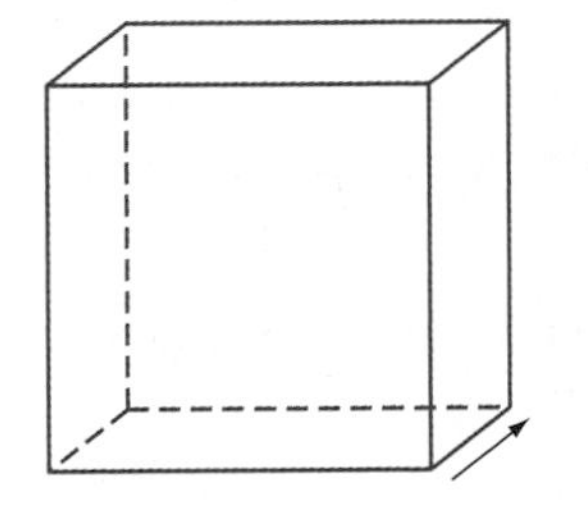

a. What three-dimensional shape is the popcorn box?

b. The actual box is 10 inches high. Three face-views of the box drawn to scale are given below. Find the actual length and width by making appropriate measurements.

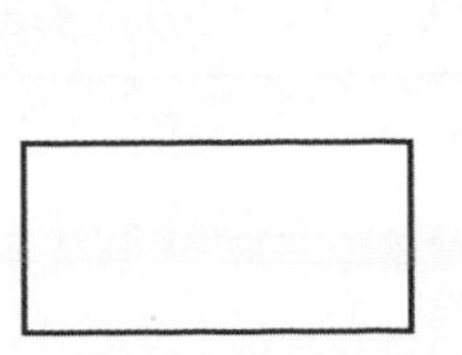

Top View

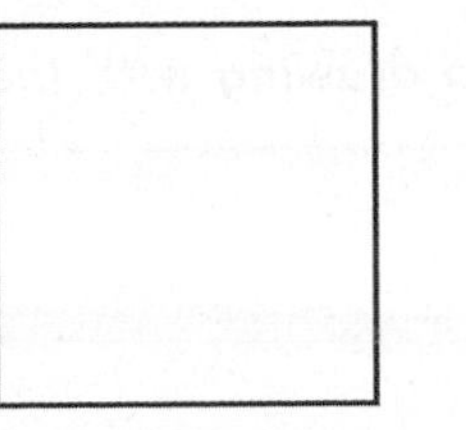

Front View

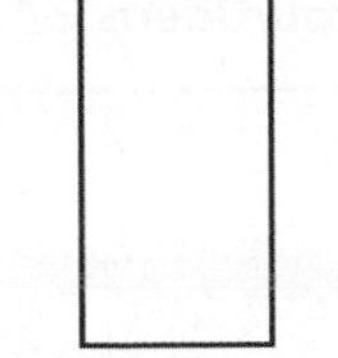

Right-Side View

c. Now examine more carefully the sketch of the box from a top-front-right corner view.

i. What appear to be the shapes of the faces as shown in the drawing? What are the shapes of the faces in the real box?

ii. What edges are parallel in the real box? Are the corresponding edges in the sketch drawn parallel?

d. Sketch the box from a bottom-front-left corner view. Use dashed lines to show "hidden" edges.

For this problem, refer to the straw and pipe cleaner models you previously made. For each of the following models, place the model on your desk so that an edge of a base is parallel to an edge of your desk. Make an oblique sketch of the model. Compare your sketches and strategies for drawing these with those of your classmates.

a. cube

b. square prism

c. square pyramid

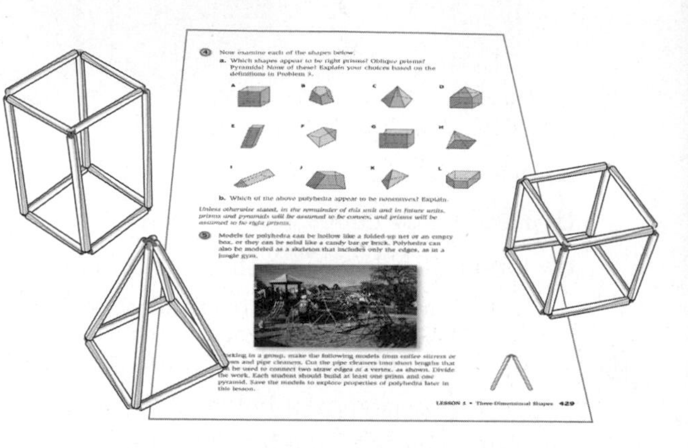

Summarize the Mathematics

A three-dimensional shape can be represented in two dimensions in various ways, including an orthographic (face-views) drawing or an oblique sketch from a particular point of view.

a. When is it helpful to represent a three-dimensional shape by an orthographic drawing? By an oblique sketch?

b. Discuss the similarities and differences between a top-front-right corner sketch of a right rectangular prism and the rectangular prism itself.

c. Consider a convex polyhedron that is made up of two square pyramids sharing a common base. Make an orthographic drawing of this polyhedron. Assume an edge of the common base is parallel to an edge of your desk.

Be prepared to share your ideas and drawing with the class.

Check Your Understanding

Consider the three-dimensional shape formed when a pentagonal pyramid is placed on top of a pentagonal prism. Assume the bases of the two shapes are congruent.

a. Make an orthographic drawing of this shape.

b. Is the shape a convex polyhedron? Explain.

c. Describe a possible real-world application of a shape with this design.

Investigation 3 Patterns in Polyhedra

Like polygons, polyhedra are most useful and interesting when they have certain regularities or symmetries. For example, an "A-frame" is a style of architecture sometimes used in building houses. This attractive shape has a balance, or symmetry, about a vertical plane that contains the top roof line. The symmetry plane splits the basic shape of the house into two parts that are mirror images of one another. In this investigation, you will explore symmetry and other properties of polyhedra. As you work on the following problems, look for clues to this general question:

How are properties of polyhedra such as symmetry and rigidity related to corresponding properties of polygons?

1 The basic shape of an A-frame house is a prism with isosceles triangle bases. In the diagram below, note how the vertical plane cuts the three-dimensional shape into two parts that are mirror images of each other. This plane is called a **symmetry plane** or **mirror plane**. The shape is said to have **reflection symmetry**.

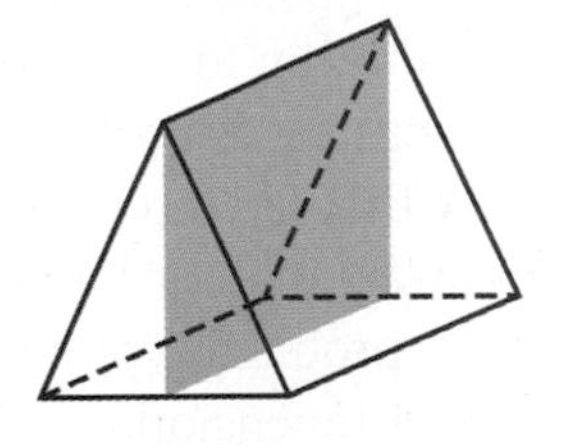

a. Does this isosceles triangular prism have any other symmetry planes? If so, describe or sketch them. If not, explain why not.

b. Next examine the cube and equilateral triangular pyramid models that your class made with straws and pipe cleaners. How many symmetry planes does each shape have? Describe or sketch them.

c. How are the symmetry planes for these three polyhedra related to the symmetry of their faces?

2. Place a model of a square pyramid on the top of a desk or table. Rotate it about the line through its apex and the center of its base.

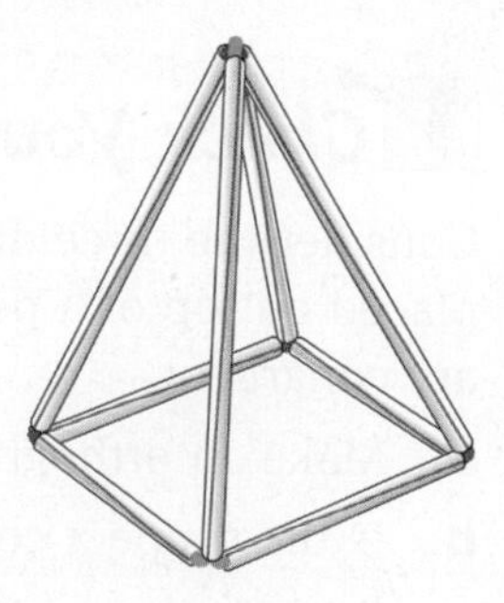

   a. What is the smallest angle that you can rotate the pyramid so that it appears to be in the same position as it was originally?

   b. As you rotate the pyramid through a complete 360° turn, at what other *angles of rotation* does it appear as it did in its original location?

   c. Can you rotate the pyramid through angles less than 360° about other lines so that it appears to be in the same position as originally? If so, describe them. If not, explain why not.

If there is a line about which a three-dimensional shape can be turned less than 360° in such a way that the rotated shape appears in exactly the same position as the original shape, the shape is said to have **rotational symmetry**. The line about which the three-dimensional shape is rotated is called an **axis of symmetry** or **rotation axis**.

3. Next examine the rotational symmetry of the equilateral triangular pyramid you constructed.

   a. How many axes of symmetry are there? Where are the axes of symmetry located?

   b. What are the angles of rotation for each axis of symmetry?

   c. How is the rotational symmetry of the triangular pyramid similar to that of the square pyramid, and how is it different?

4. Now examine the rotational symmetry of the cube model you constructed.

   a. How many axes of symmetry does the cube have? Where are the axes of symmetry located?

   b. What are the angles of rotation for each axis of symmetry?

   c. How is the symmetry of the faces related to the symmetry of the cube?

In addition to symmetry, another important consideration in the design of structures is *rigidity*. In Lesson 1, you discovered that any triangle is rigid; and that polygons with more than three sides can be made rigid by triangulating or bracing them with diagonals. In the next problem, you will investigate the rigidity of three-dimensional shapes.

5 Consider the models that you made in Problem 5 on page 429.

a. Which of the models represent rigid shapes?

b. Add bracing straws to your model of a triangular prism to make it rigid.

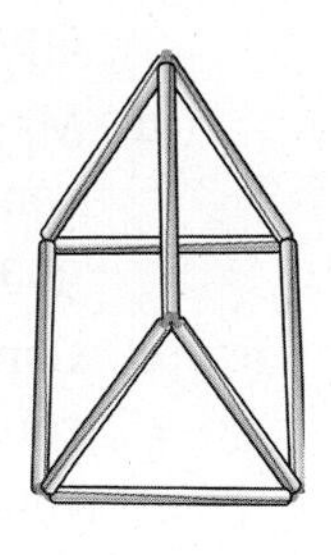

i. Describe where you placed the bracing straws and why you placed them there.

ii. Could you have placed the braces in different positions and still made the triangular prism rigid? Could you have used fewer bracing straws? Explain and illustrate.

c. Add braces to your model of a cube so that it becomes a rigid structure. Note the number of bracing straws that you used and describe the position of each straw. Could you have used fewer bracing straws?

d. Think of a different way to reinforce the cube so that it becomes a rigid structure. Describe the pattern of reinforcing straws.

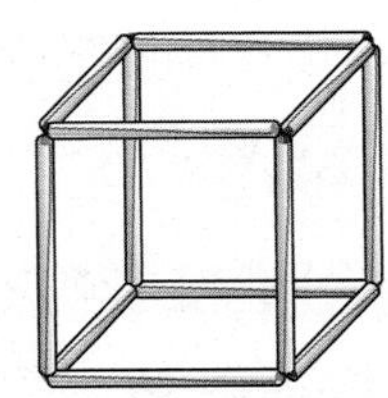

e. Of the methods you used to reinforce the cube, which could best be used to make a rectangular prism-shaped building stand rigidly?

6 Another interesting property of convex polyhedra has to do with the face angles that meet at each vertex. You saw in Investigation 1 that the sum of the measures of the face angles that meet at any vertex of any convex polyhedron must be less than 360°. The positive difference between the angle sum at a vertex and 360° is called the **vertex angle defect**. It is a measure of how close that corner is to being flat.

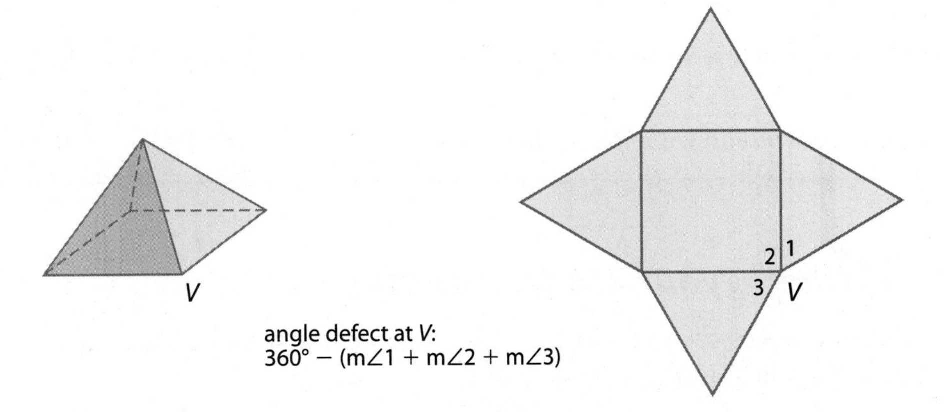

a. What is the angle defect at each vertex of a cube? What is the sum of the angle defects at all vertices of a cube?

b. In a similar way, find the sum of the angle defects for an equiangular triangular pyramid.

c. Find the sum of the angle defects for an equiangular triangular prism and for a regular hexagonal prism. Share the work with a partner.

d. Make a conjecture about the sum of the angle defects in any convex polyhedron. Test your conjecture using a pyramid with a quadrilateral base. Does your conjecture hold true in this case, too?

e. Compare your investigation of the face angles and angle defects in convex polyhedra to what you learned in Lesson 2 about angles of convex polygons. How are the ideas and the results alike? How are they different?

French philosopher and mathematician René Descartes (1596–1650) first discovered that the sum of the angle defects of any convex polyhedron is a constant. The result is called **Descartes' Theorem**.

Summarize the Mathematics

In this investigation, you explored two types of symmetry in three dimensions: reflection symmetry and rotational symmetry. You examined the rigidity of different polyhedra. You also discovered a property about the sum of the angle defects of a polyhedron.

a. Describe how to identify reflection symmetry and rotational symmetry in a polyhedron.

b. Name the rigid polyhedron with the fewest faces and edges.

c. What methods can be used to make a polyhedron rigid?

d. What is true of the sum of the vertex angle defects for the polyhedra that you studied in this investigation?

Be prepared to share your ideas with the entire class.

✓ *Check Your Understanding*

Consider a polyhedron with all edges congruent that looks like a square pyramid joined to the top of a cube.

a. Is this a convex polyhedron? Explain.

b. Describe the reflection symmetry and rotational symmetry of the shape.

c. Describe two ways to add bracing to make this polyhedron rigid. What is the least number of braces needed? Explain.

d. Calculate the sum of the angle defects for this polyhedron and verify that Descartes' Theorem is satisfied.

Investigation 4 Regular Polyhedra

In Lesson 2, you studied regular polygons and their properties. The three-dimensional counterpart of a regular polygon is a regular polyhedron. A **regular polyhedron**, also called a **Platonic solid**, is a convex polyhedron in which all faces are congruent, regular polygons. Furthermore, the arrangement of faces and edges is the same at each vertex. The regular polyhedron "globe" of the Earth shown here was created by R. Buckminster Fuller, inventor of the geodesic dome.

Recall that there are infinitely many different regular polygons, named by the number of sides—regular (or equilateral) triangle, regular quadrilateral (or square), regular pentagon, regular hexagon, and so on. You might think that there would also be infinitely many regular polyhedra. However, that is not the case! That fact is one of the more famous results in the history of geometry. In this investigation, you will explore these two related questions:

How many differently shaped regular polyhedra are possible and why?

What are some of the properties of these polyhedra?

1. Refer to the 9 polyhedra models you constructed in Investigation 1 using straws and pipe cleaners.
   - **a.** Which two models represent *regular* polyhedra? Explain why.
   - **b.** For each of these regular polyhedra, how many faces meet at each vertex?

2. To see whether other regular polyhedra can be constructed, begin by exploring how three or more congruent regular polygons can be arranged at a vertex of a polyhedron.
   - **a.** What is the sum of measures of the face angles that meet at any vertex of a regular triangular pyramid? Draw a partial net to illustrate your answer.
   - **b.** Next suppose 4 equilateral triangles meet at a vertex. If a regular polyhedron with vertices like this could be constructed, what would be the sum of the face angles at each vertex?
   - **c.** Repeat Part b for the case when 5 equilateral triangles meet at a vertex.
   - **d.** Explain why it is impossible for more than 5 equilateral triangles to meet at a vertex of a regular polyhedron.

3 Next, suppose the faces of a regular polyhedron have more than three edges. They might be squares, regular pentagons, regular hexagons, and so on.

a. What is the sum of the measures of the face angles that meet at any vertex of a cube? Illustrate with a partial net.

b. Why is it not possible for 4 or more squares to meet at a vertex of a regular polyhedron?

c. Discuss whether 3 regular pentagons, 3 regular hexagons, or 3 regular septagons could meet at a vertex of a regular polyhedron.

d. Explain why it is impossible for a regular polyhedron to have faces with more than 5 sides.

4 Putting together the steps of your reasoning in Problems 2 and 3, you have shown that there are at most 5 differently shaped regular polyhedra. The partial nets below show the different ways that regular polygon faces could meet at a single vertex of a regular polyhedron.

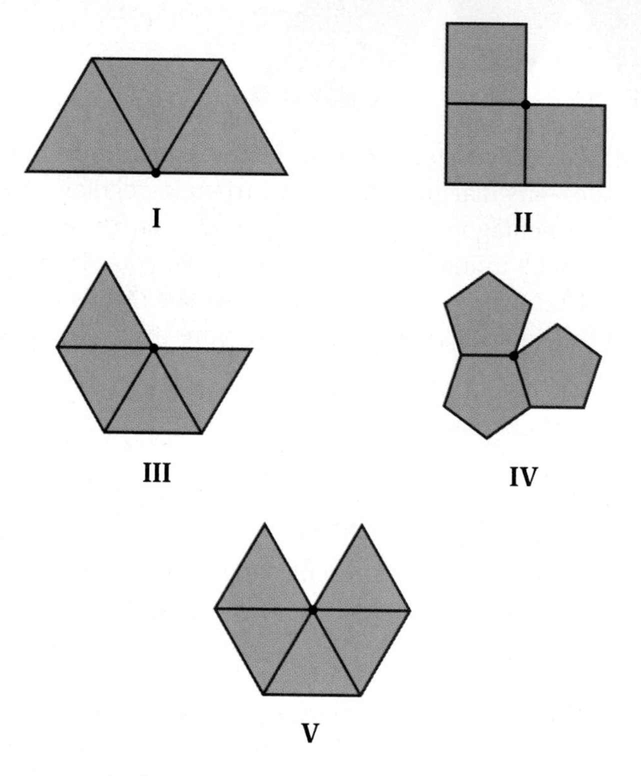

a. Explain as precisely as you can why the above arrangements are the only ones possible.

b. Your reasoning likely has much in common with the reasoning in Lesson 2 that showed that the only regular polygons that tile the plane are the regular quadrilateral, regular triangle, and regular hexagon. How are the two arguments similar? How are they different?

c. For each of the partial nets above, assume that a regular polyhedron can be constructed for which the faces meet at each vertex as illustrated. What is the angle defect at each pictured vertex? How many vertices must each of the regular polyhedra have?

5 You have already constructed models of two of the five regular polyhedra, namely, an equilateral triangular pyramid, also called a **regular tetrahedron** (named from the Greek "tetra" for its 4 faces), and the cube, also called a **regular hexahedron** (named for its 6 faces).

a. Collaborate with your classmates to construct models for a single "corner" of a regular polyhedron using copies of the third, fourth, and fifth partial nets shown in Problem 4. Each student should cut out one partial net, fold, and close the corner by joining two edges with tape. Compare your model to those of classmates who used the same net. Are the models based on the same net identical? If not, resolve the differences.

b. Working in pairs, select either partial net III or IV in Problem 4 to complete the following tasks.

  i. Tape together as many partial nets as needed to form a model for a regular polyhedron. How many copies did you use?

  ii. Describe the polyhedron you formed. How many faces does it have? What would be a good name for the polyhedron? Why?

c. Use the same procedure as in Part b to construct a model for a regular polyhedron using multiple copies of the fifth partial net in Problem 4. Discuss what happens.

d. Although it cannot be constructed only from copies of its "corners" where 5 faces meet, there is a fifth regular polyhedron called a **regular icosahedron** (from the Greek "eikosi" meaning 20). Nets for the last three regular polyhedra are shown below. Construct a model of a regular icosahedron by cutting out a copy of its net and folding and taping.

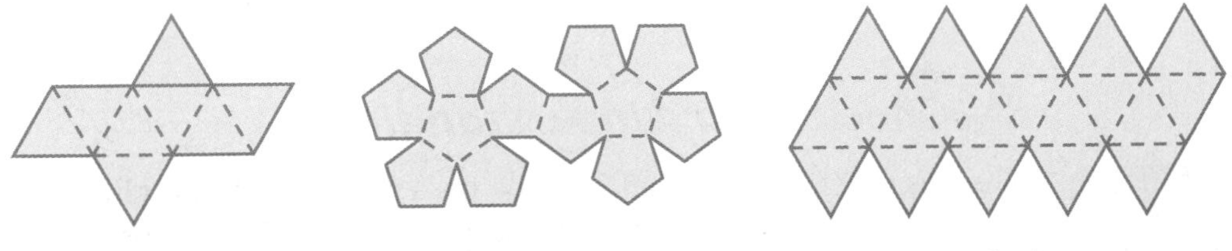

Octahedron **Dodecahedron** **Icosahedron**

6 As you can see from examination of the models, regular polyhedra have many symmetries. You previously explored symmetries of a regular tetrahedron and a regular hexahedron. Study your model of a regular octahedron.

a. How many planes of symmetry does it have? Describe their locations.

b. As for rotational symmetry, a regular octahedron has 6 axes of 180° symmetry, 4 axes of 120° symmetry, and 3 axes of 90° symmetry. Describe the locations of the axes for each of these three types of rotational symmetry.

c. How do the types and number of axes of symmetry for the regular octahedron compare with those of the cube?

Summarize the Mathematics

In this investigation, you demonstrated that there are exactly 5 differently shaped regular polyhedra and examined some of their symmetries.

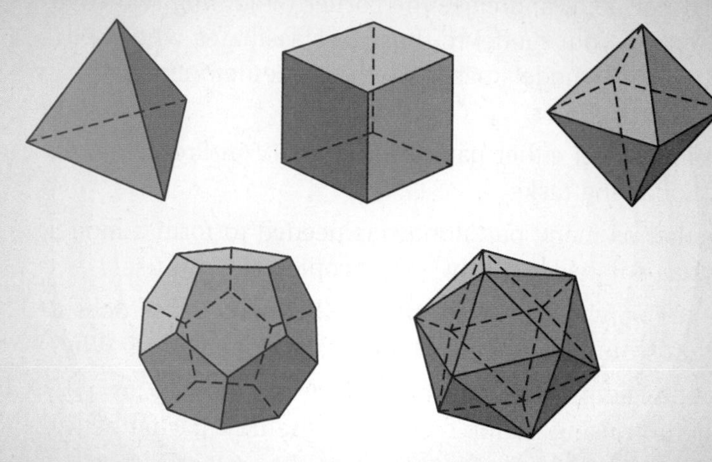

a Name all the regular polyhedra according to the number of faces of each. For each regular polyhedron, describe a face and give the number of faces that meet at each vertex.

b Explain why there cannot be more than 5 differently shaped regular polyhedra.

Be prepared to share your descriptions and explanation with the entire class.

Check Your Understanding

The design of a soccer ball is based on a **semiregular polyhedron**; that is, a polyhedron with faces that are congruent copies of two or more different regular polygons. As in a regular polyhedron, the arrangement of faces and edges is the same at each vertex.

a. Describe the three faces that meet at each vertex of the polyhedron that is used for the soccer ball.

b. What is the sum of the three face angles that meet at each vertex of the polyhedron? What is the angle defect at each vertex?

c. Find the number of vertices of the polyhedron.

d. The polyhedron consists of 20 hexagons and 12 pentagons. How many edges does it have?

On Your Own

Applications

1. Three-dimensional shapes are the basis of atomic structures as well as of common structures for work, living, and play. Often three-dimensional shapes are assembled from a combination of simpler shapes.

   a. Study this photograph of Big Ben in London, England. What three-dimensional shapes appear to be used in this tower? Which are prisms? Which are pyramids?

   b. Scientists use three-dimensional structures to model molecules of compounds, such as the model of a methane molecule shown below. Describe and name the polyhedron with the skeleton that would be formed by joining the outermost points of the four hydrogen atoms that are equally spaced around the central carbon atom.

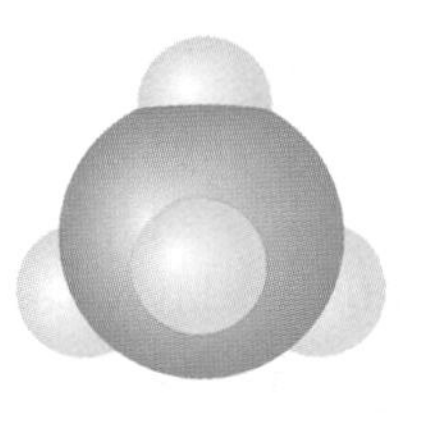

Big Ben clock tower in London, England

2. Make a conjecture about the relationship between the circumference of a circular column and the weight it can support.

   a. Conduct an experiment to test your conjecture about the weight-bearing capability of circular columns. Use columns of the same height but with different circumferences.

   b. Organize your data in a table and display them in a graph.

   c. What appears to be true about the relationship between the circumference of a column and the weight it can support? Why do you think this happens?

3. A net for a square pyramid is shown at the right. Lateral faces are equilateral triangles.

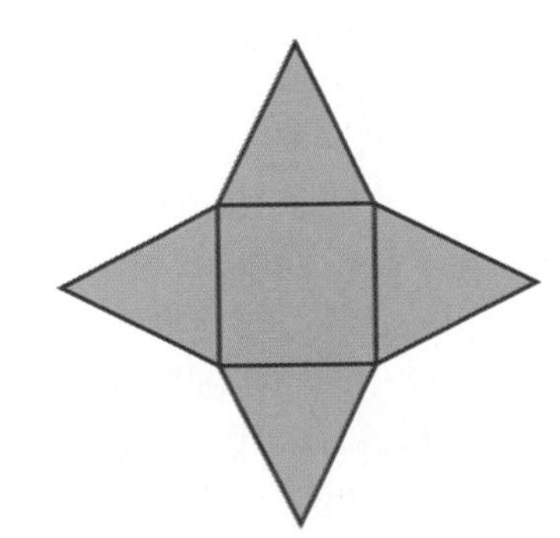

   a. Sketch two other nets that would fold into this pyramid.

   b. How many straight cuts are needed to cut out the net at the right? Each of your nets?

   c. Is there a net that requires even fewer straight cuts than the ones you have examined so far? Explain.

On Your Own

4 Building designers can test their designs by using identical cubes to represent rooms. They can use the cubes to try various arrangements of rooms. Study this drawing of a cube hotel.

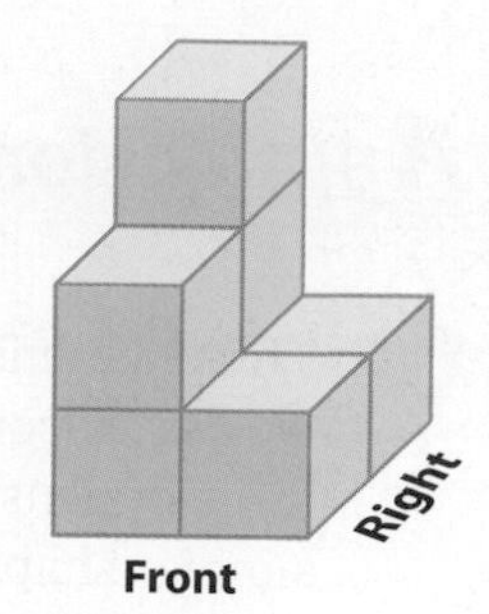

a. How many cubes are there in the model? Assume any cube above the bottom layer rests on another cube.

b. Draw the top, front, and right-side orthographic views of this shape.

c. Is the model pictured on the right a polyhedron? If not, explain why not. If so, is it convex or nonconvex?

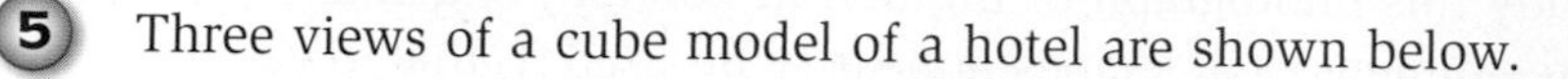

5 Three views of a cube model of a hotel are shown below.

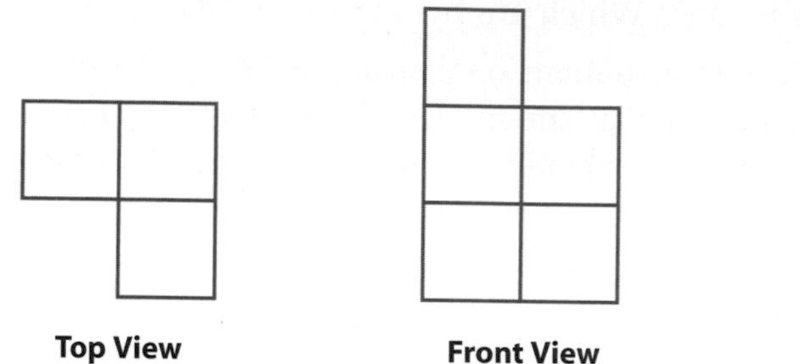

a. Make an oblique drawing of this hotel model from a vantage point that shows clearly all the characteristics of the model. Assume any cube above the bottom layer rests on another cube.

b. Is there more than one model with these three views? If so, make a drawing of a second one from a vantage point that illustrates how this model differs from the one in Part a.

c. How many cubes are there in each hotel model?

6 Both portability and rigidity are design features of a folding "director's chair."

a. How are these features designed into the chair shown at the right?

b. The pair of legs at the front and back are attached at their midpoints. Draw and label a diagram of the front pair of legs and edge of the seat. Using congruent triangles and Connections Task 18, page 391, explain as carefully as you can why these conditions guarantee that the outstretched seat will be parallel to the ground surface.

c. Identify two other commonly used items that can collapse but must remain rigid when "unfolded." Analyze their designs.

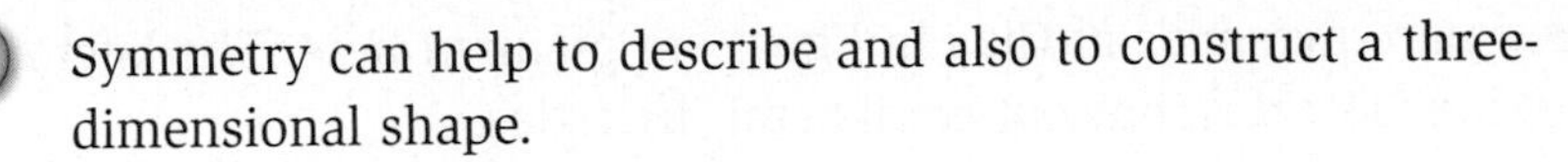

7 Symmetry can help to describe and also to construct a three-dimensional shape.

a. Describe the reflection symmetry and the rotational symmetry of these three-dimensional shapes.

i. a right circular cylinder

ii. a right circular cone

iii. a prism with a parallelogram base

iv. a regular pentagonal pyramid

v. a sphere

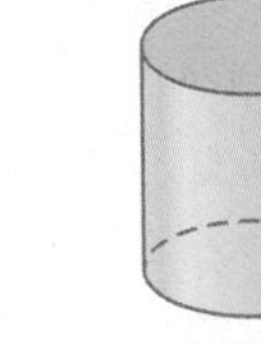

Cone

Cylinder

b. Half of a polyhedron is shown at the right.

i. Sketch the entire polyhedron if it is symmetric about the plane containing the right face.

ii. Sketch the entire polyhedron if it is symmetric about the plane containing the left face.

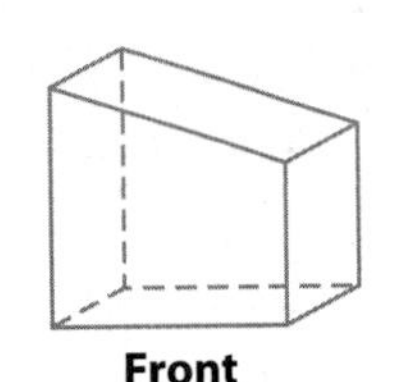

Front

c. Are the polyhedra you sketched in Part b convex? Explain your answers.

d. For each polyhedron in Part b, count the vertices, faces, and edges. Verify that Euler's Formula holds for each polyhedron.

e. Using a polyhedron from Part b above as an example, explain why Descartes' Theorem does not make sense for nonconvex polyhedra.

8 A model of a square prism could be made from a potato or modeling clay. Consider the $5 \times 5 \times 6$ square prism pictured below. The three lines (skewers) intersecting the prism contain the centers of opposite faces.

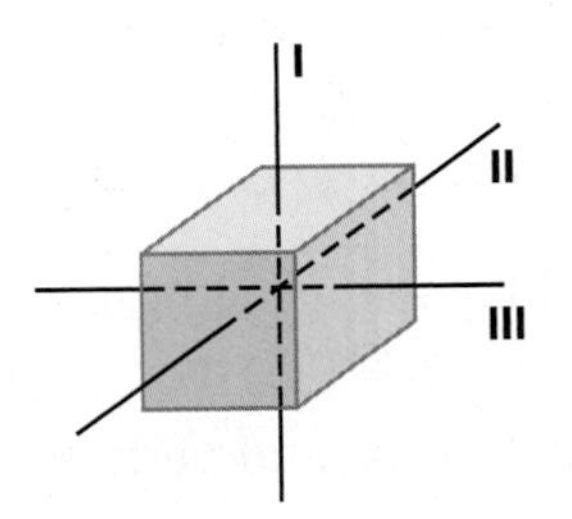

a. Explain why these lines are axes of symmetry for the prism.

b. What angles of rotation are associated with each axis of symmetry? Explain your reasoning.

c. Does this prism have other axes of symmetry? If so, describe their locations and give the angle of rotation associated with each.

d. How is the rotational symmetry of a cube similar to that of the square prism above? How is it different?

On Your Own

9 The square pyramids at Giza in Egypt are pictured here. The lateral faces are isosceles, but not equilateral, triangles.

a. Describe the planes of symmetry and the rotational symmetry of one of the Giza pyramids.

b. In the steepest of these pyramids, the face angles at the apex are each about 40°. Find the angle defect at each of the 5 vertices of this pyramid. Verify that Descartes' Theorem holds for this pyramid.

Connections

10 A pyramid has a square base that is 10 units on a side, and the other faces of the pyramid are congruent isosceles triangles.

a. Suppose you were going to make a model with straws and pipe cleaners of such a pyramid. Could the lateral edges to the apex be 5 units long? Could they be 10 units long? Could they be 20 units long?

b. Is there a minimum length for the lateral edges? Is there a maximum length for the lateral edges? If so, what are they? What property or properties from earlier lessons in this unit would justify your answers?

11 Imagine a model of a cube made of clay or cut from a potato. Make such a model if possible.

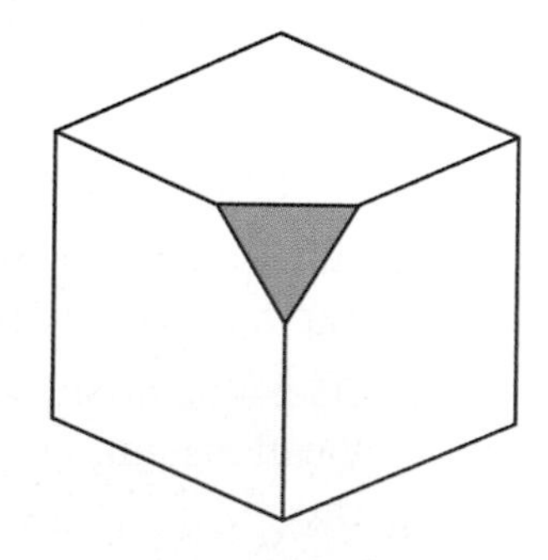

a. How many faces, edges, and vertices does a cube have?

b. Slice a corner off (as shown), making a small triangular face. How many faces, edges, and vertices does the new polyhedron have?

c. Repeat at each corner so that the slices do not overlap. Make a table showing the number of faces, edges, and vertices of the modified cube after each "corner slice."

d. Using *NOW* and *NEXT*, write a rule describing the pattern of change in the number of faces after a slice. Write similar *NOW-NEXT* rules for the number of edges and for the number of vertices after each slice.

e. Do you think Euler's Formula will hold at each stage? Justify your answer using the *NOW-NEXT* rules in Part d.

f. How many faces, edges, and vertices does the new polyhedron have when all the corners are sliced off?

12 When filling three-dimensional containers like boxes or cylindrical cans, a measure of *volume* is needed. The volume of a three-dimensional shape of height 1 unit is numerically equal to the area of the base. For prisms and cylinders, imagine the first layer of 1-unit cubes as shown below.

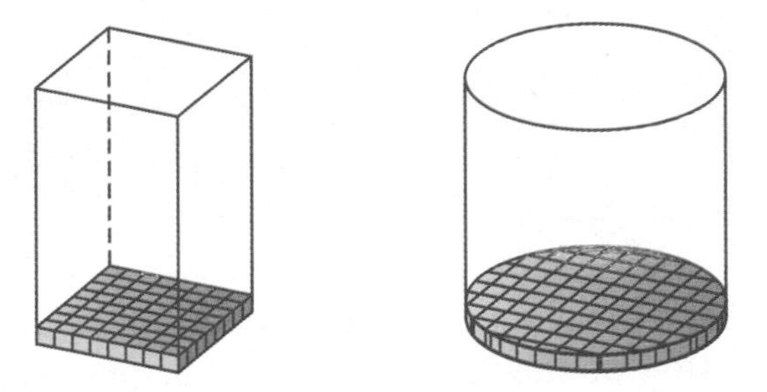

Then add additional layers until the prism or cylinder is filled. The number of layers is the shape's height. This suggests that the volume V of either a prism or a circular cylinder is the product of the *area of the base* B and the *height* h, a formula usually written symbolically as $V = Bh$.

a. Find the volume of a square prism with base edges of 7 cm and height of 10 cm. The volume is in what units?

b. Find the volume of a cylinder in which the radius of the base is 5 in. and the height is 9 in. Indicate the units.

c. Find the volume of a regular triangular prism if all edges are 4 ft long.

d. A right circular cylindrical can is packed snugly into a box as shown here. The base of the box is a square 8 cm on a side and the height of the box is 16 cm.

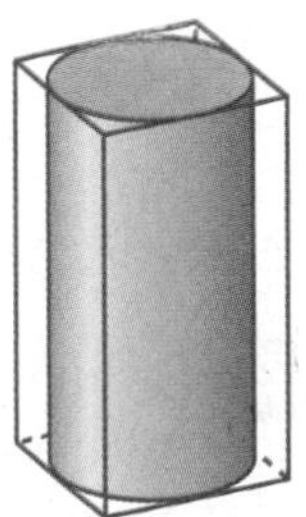

i. Find the volume of the space between the can and the box.

ii. Find the ratio of the volume of the cylinder to the volume of the box. What does the ratio tell you?

e. Suppose the length of the side of a square box as in Part d is s cm and the height is h cm.

i. What is the volume of the box in terms of s and h?

ii. What are the radius and height of the cylinder in terms of s and h?

iii. What is the volume of the cylinder in terms of s and h?

iv. What is the ratio of the volume of the cylinder to the volume of the box? What does the ratio tell you?

13 As seen in Connections Task 12, an important volume formula that holds for both prisms and cylinders is $V = Bh$ where B is the area of the base and h is the height, that is, the perpendicular distance between the bases. A second related formula, $V = \frac{1}{3}Bh$, holds for pyramids and cones. The illustration below shows a triangular prism dissected into three triangular pyramids of equal volume. This explains the use of the fraction $\frac{1}{3}$ in the formula $V = \frac{1}{3}Bh$.

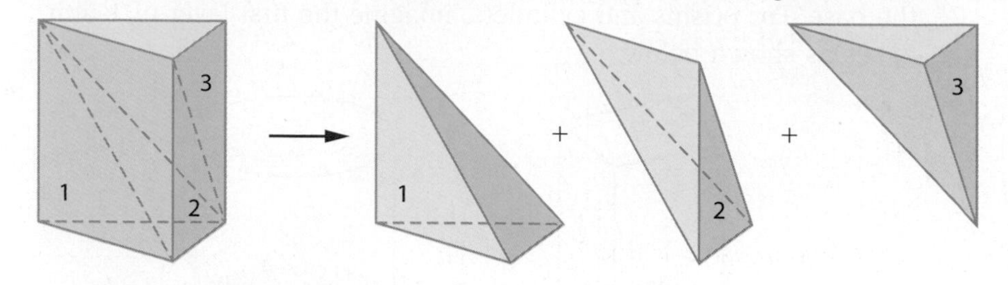

a. Find the volume of the cube, pyramid, cylinder, and cone shown below. How do the volumes of these shapes compare?

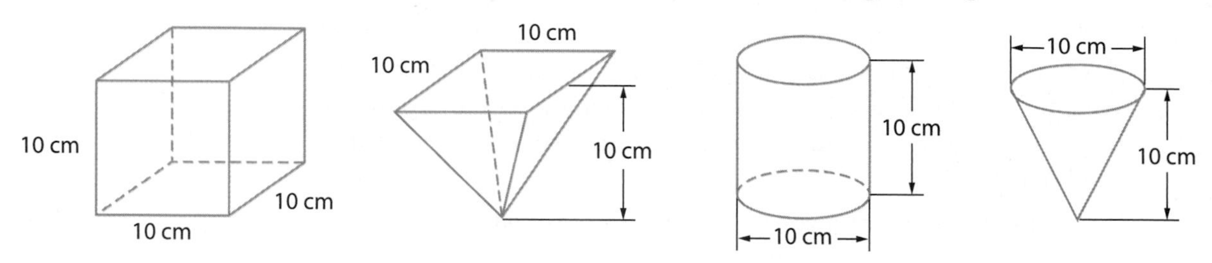

b. A movie theater sells different sizes of popcorn in different shaped containers. One is a cylinder with a height of 17 cm and radius of the base of 8 cm. It sells for $4.50. Another is a rectangular box (prism) with width 10 cm, length 16 cm, and height 18 cm. It sells for $3.75. Which is the best buy? Explain your answer.

c. A third popcorn container is a cone with a height of 24 cm and radius of the base of 10 cm. What is the most its price could be if it is the best buy of the three containers?

14 In three dimensions, the counterpart of a circle is a *sphere*. To discover a formula for the volume of a sphere, consider the following experiment. Imagine or obtain a sphere of radius r that fits snugly inside a cylinder of diameter $2r$ and height $2r$ as shown. Remove the hollow sphere and fill it with water. Pour the water into the cylinder.

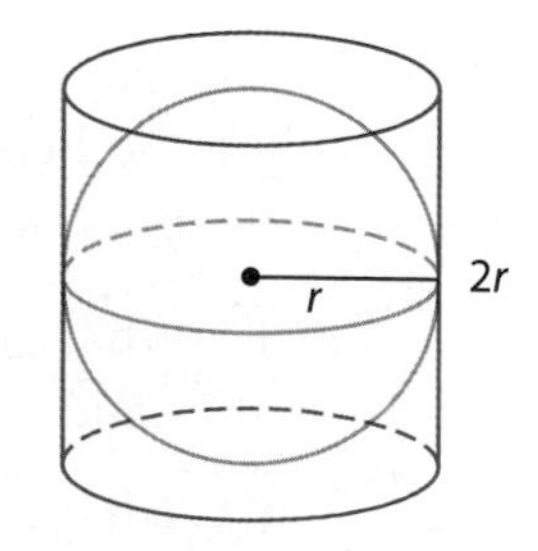

a. Write an expression for the volume of the cylinder in terms of π and r.

b. A class at the Battle Creek Mathematics and Science Center conducted the experiment with different-sized spheres and cylinders. In each case, the height of the water in the cylinder was about $\frac{4}{3}r$. Use the results of their experiment to write a formula for the volume V of a sphere in terms of its radius r.

15 Use your regular polyhedron models and the nets to complete a table like the one below.

	Edges per Face	Faces per Vertex	Number of Faces	Number of Vertices	Number of Edges
Tetrahedron		3	4		
Hexahedron		3		8	
Octahedron		4			12
Dodecahedron		3	12		
Icosahedron		5	20		

a. Describe at least two interesting patterns that you see in the table.

b. Check Euler's Formula for these polyhedra.

c. Eric conjectured that the number of edges in a regular polyhedron is the product of the number of edges per face and the number of faces. Do you agree with Eric? If so, explain why. If not, explain how to correct Eric's statement.

16 In this lesson, you learned how to represent three-dimensional shapes in two dimensions with orthographic (face-view) drawings and with oblique sketches. For the case of a rectangular prism, another way to represent the shape is illustrated below.

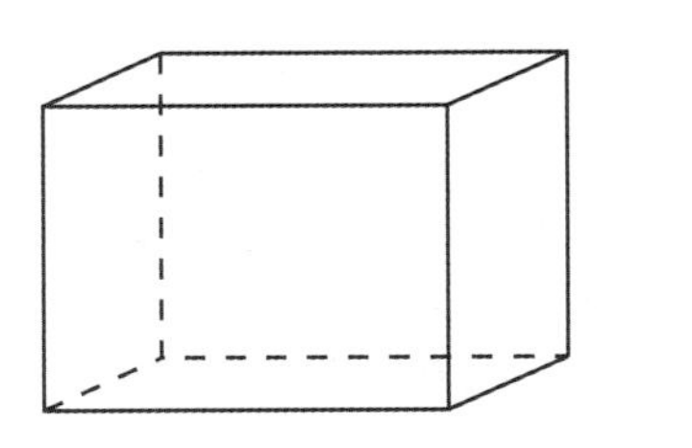

Rectangular Prism

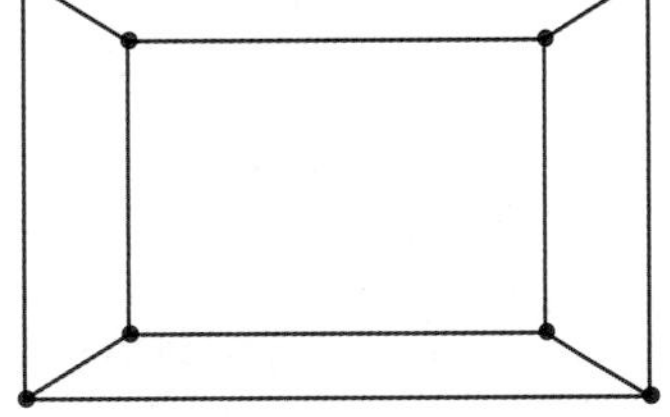

Vertex-Edge Graph

a. The figure on the right is a representation of a rectangular prism as a vertex-edge graph with no edge crossings. (When there are no edge crossings, the graph is called a *planar graph*.) You can think of this graph, called a **Schlegel diagram**, as resulting from "compressing" a rectangular prism with elastic edges down into two dimensions. (You can also think of the graph as resulting from a one-point perspective projection, where the vanishing point is in the center of the back face. Think of looking at a clear box with your nose very close to the front face.) The graph has lost most of the shape of the three-dimensional prism, but it shares many of the prism's properties. Name as many shared properties as you can.

b. How many faces does the rectangular prism have? Explain how all the faces are accounted for in the regions formed by the vertex-edge graph.

c. Use the idea of *compressing* illustrated in Part a to draw Schlegel diagrams for the following polyhedra. After compressing (or projecting) the octahedron, nonintersecting edges in the octahedron should not intersect in the graph. Imagine projecting the model while looking directly at the center of one face.

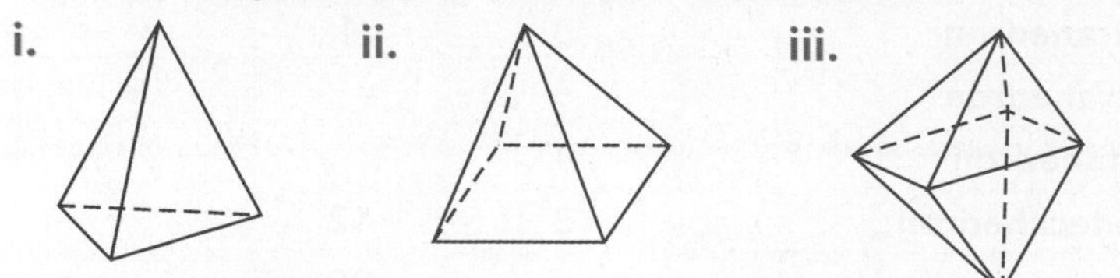

d. Here are Schlegel diagrams for the regular dodecahedron and the regular icosahedron. Which is which? Explain your reasoning.

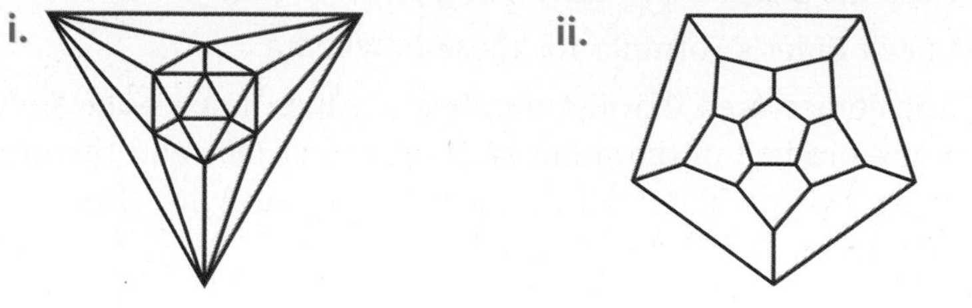

17 There are interesting and useful connections between pairs of regular polyhedra.

a. Refer to the regular tetrahedron pictured on the right. The center of each face is marked.

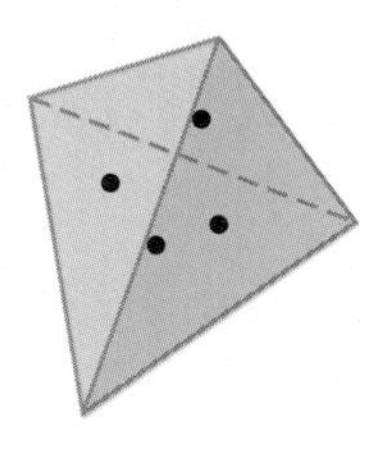

i. How many such centers are there?

ii. Imagine connecting the centers with segments. How many such segments are there?

iii. Visualize the polyhedron having these segments as its edges. What are the shapes of its faces?

iv. What polyhedron is formed by connecting the centers of the faces of a tetrahedron?

b. Now using a model of a cube, imagine the center of its faces.

i. How many such centers are there?

ii. Imagine connecting with segments, each center to the center of the four *adjacent faces*. Adjacent faces are faces that have a common edge. How many such segments are there?

iii. Visualize the polyhedron having these segments as its edges. What are the shapes of its faces? How many faces are there?

iv. What polyhedron is formed by connecting the centers of adjacent faces of a cube?

c. What happens if you start with the new polyhedron in Part b and repeat the process of connecting centers of adjacent faces with segments?

d. Make a conjecture about the polyhedron formed by connecting the center of adjacent faces of a regular dodecahedron. Explain the basis for your conjecture.

Reflections

18 Compare the definitions of convex polygon (page 404) and convex polyhedron (page 426). If all the faces of a polyhedron are convex polygons, must the polyhedron be convex? Explain.

19 Isaiah made the following two conjectures about the symmetry of a right prism. Indicate whether you agree or disagree with each, and write arguments in support of your positions. *Hint:* If you think a statement is false, one counterexample (that is, an example of a right prism that does not have the named symmetry) is a sufficient argument. A more general argument based on the properties of a right prism is required if you think a statement is true.

a. Every right prism has at least one symmetry plane.

b. Every right prism has at least one axis of rotational symmetry.

20 In this unit, as in previous units, you have engaged in important kinds of mathematical thinking. Look back over the four investigations in this lesson and consider some of the mathematical thinking you have done. Describe an example where you did each of the following:

a. experiment

b. search for and explain patterns

c. formulate or find a mathematical model

d. visualize

e. make and check conjectures

f. make connections between geometry and algebra

Extensions

21 The faces of a cube are congruent squares. You know how to use the length of each edge of such a cube to find the lengths of the diagonals on the faces. Cubes also have "body" diagonals such as $\overline{AC}$ in the diagram at the right.

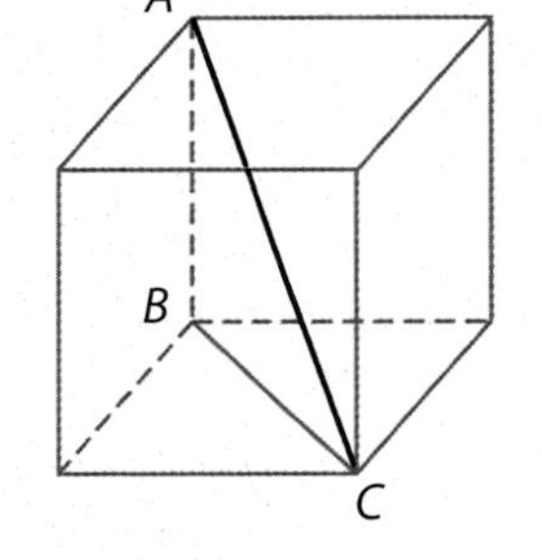

a. Find the length of the body diagonal $\overline{AC}$ of a cube when the cube edges are as given below. Express your answers in simplest radical form.

i. 1 inch long **ii.** 2 inches long **iii.** 3 inches long

b. How many body diagonals does a cube have? Explain as precisely as you can why all body diagonals are the same length.

c. Find a formula for calculating the length of each body diagonal in a cube with edge length x.

22 Analysis of the formulas for the volume of a prism and for the volume of a cylinder suggests that multiplying the dimensions of the shape by a positive constant changes the volume in a predictable way.

a. One large juice can has dimensions twice those of a smaller can. How do the volumes of the two cans compare?

b. One cereal box has dimensions 3 times those of another. How do the volumes of the two boxes compare?

c. If the dimensions of one prism are 5 times those of another, how do the volumes compare?

d. If the dimensions of one prism are k times those of another, how do the volumes compare?

23 In Lesson 2, you identified polygons that tile the plane. In three-dimensional geometry, the related question is: What three-dimensional shapes will fill space? An obvious example of a space-filling, three-dimensional shape is a rectangular prism. In fact, the efficiency with which rectangular prisms can be stacked is what makes their shape so useful as boxes and other containers.

a. What right prisms with regular polygonal bases will fill space? Explain how your answer is related to the regular polygons that will tile the plane.

b. The cells of a honeycomb are approximated by regular hexagonal prisms, which form a three-dimensional tiling for storing honey. Suppose the perimeter of the base of one cell of a honeycomb is 24 mm and the height is 20 mm. What is the *lateral surface area* (surface area not including top and bottom) of a single cell? Explain.

c. Which of the three right prisms with regular polygonal bases that fill space (see Part a) produces the cell with the greatest volume when the perimeter of the base is 24 mm and the height is 20 mm? As the number of sides of the base of a prism with a regular polygonal base of fixed perimeter increases, how does the corresponding volume change?

d. Write a statement summarizing how three-dimensional shape is an important factor in the building of the cells of a honeycomb.

24 A rectangular swimming pool is 28 feet long and 18 feet wide. The shallow end is 3 feet deep and extends for 6 feet. Then for 16 feet horizontally, there is a constant decline toward the 9-foot-deep end.

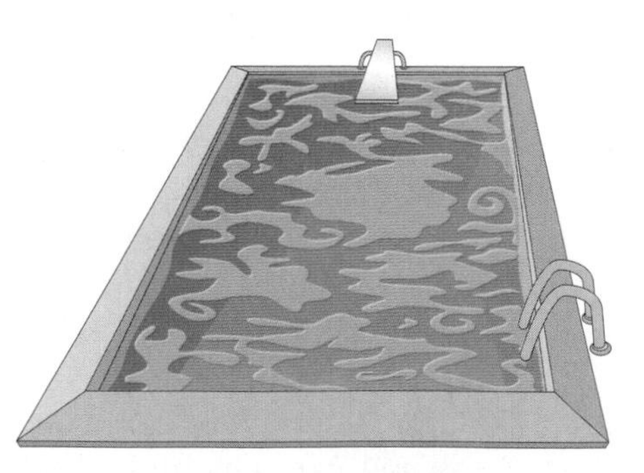

a. Sketch the pool and indicate the measures on the sketch.

b. How much water is needed to fill the pool within 6 inches of the top?

c. One gallon of paint covers approximately 75 square feet of surface. How many gallons of paint are needed to paint the inside of the pool? If the pool paint comes in 5-gallon cans, how many cans should be purchased?

d. How much material is needed to make a rectangular pool cover that extends 2 feet beyond the pool on all sides?

e. About how many 6-inch square ceramic tiles are needed to tile the top 18 inches of the inside faces of the pool?

25 For each figure below, describe how a plane and a cube could intersect so that the intersection (or cross-section) is the figure described. If the figure is not possible, explain your reasoning.

a. a point

b. a segment

c. a triangle

d. an equilateral triangle

e. a square

f. a rectangle

g. a pentagon

h. a hexagon

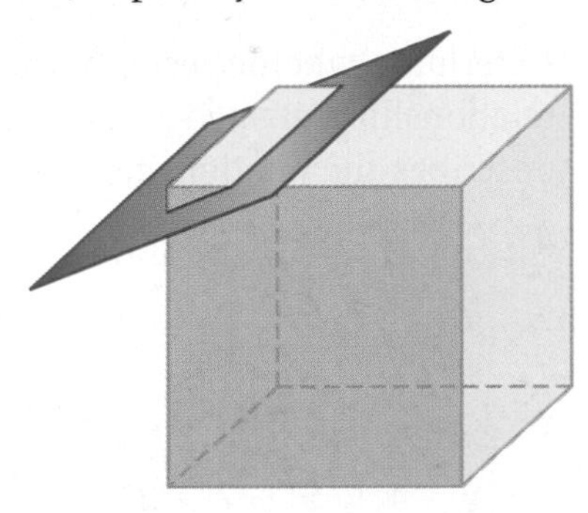

26 The volume formula for a sphere (see Connections Task 14) can be used to derive a formula for the *surface area* of a sphere. Consider a solid sphere with center at point O and radius r. Imagine a tiny polygon-like portion of the sphere's surface. This "polygon" is the base of a pyramid with vertex O and height approximately r.

a. If the area of the base of the polygon is B_1, what is the volume of the pyramid?

b. Next, imagine dividing the entire surface of the sphere into a large number n of tiny, nonoverlapping "polygons," with respective areas of $B_1, B_2, B_3, \ldots, B_n$. Write a formula for the surface area S in terms of $B_1, B_2, B_3, \ldots, B_n$.

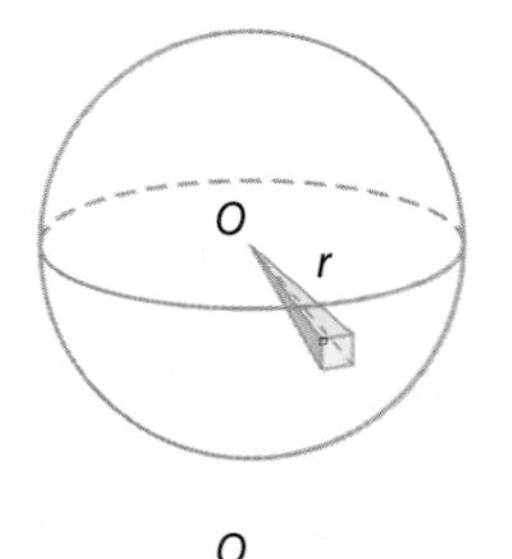

c. Explain why a formula for the volume V of the sphere is

$V = \frac{1}{3}(B_1)(r) + \frac{1}{3}(B_2)(r) + \cdots + \frac{1}{3}(B_n)(r)$.

d. Rewrite the formula in Part c by (i) substituting $\frac{4}{3}\pi r^3$ for V and (ii) factoring the largest common factor from the right-hand side.

e. Use your result from Part d to write a formula for the surface area S of a sphere in terms of its radius r.

Review

27 Examine the information given in the diagram below. Write and solve an equation to find the measure of $\angle ABD$.

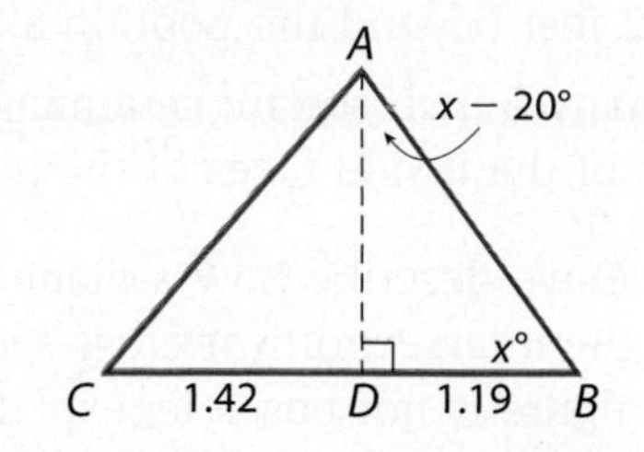

28 Determine whether each of the following tables indicates a linear relationship between x and y. For those that are linear, find a rule that describes the relationship.

a.

x	0	1	2	5
y	1	5	9	21

b.

x	0	1	2	3
y	1	5	7	37

c.

x	−1	0	1	2
y	1	0	−1	−8

29 Use algebraic reasoning to solve each equation or inequality.

a. $150 = 20 - 6x$

b. $150 > 20 - 6x$

c. $6(x + 8) = -72$

d. $6(x + 8) \geq -72$

30 Complete each sentence with the name of a polygon.

a. Every __________ has exactly two lines of symmetry.

b. Every rhombus is also a __________.

c. The diagonals of every _________ are lines of symmetry.

31 Andy kept track of the number of minutes he exercised each day for the last 30 days. His data are shown on the dot plot below.

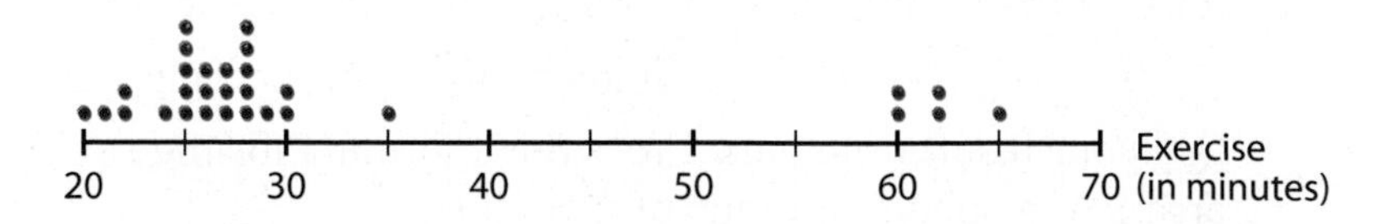

a. What was the greatest amount of time that Andy exercised in one day?

b. On how many days did Andy exercise for 25 minutes?

c. What percentage of the days did Andy exercise for at least one hour?

d. Find the median number of minutes that Andy exercised for these 30 days.

e. Would you expect the mean number of minutes exercised to be greater than or less than the median number of minutes exercised? Explain your reasoning.

32 Produce a graph of the function $y = x^{10}$, for $0 < x < 2$. On the same set of axes, produce a graph of the function $y = 10^x$. About where do the two graphs intersect?

33 Does the vertex-edge graph shown below have an Euler circuit or path? If so, find one. If not, describe how you know. (Remember that an Euler path traces all edges of the graph exactly once but is not a circuit.)

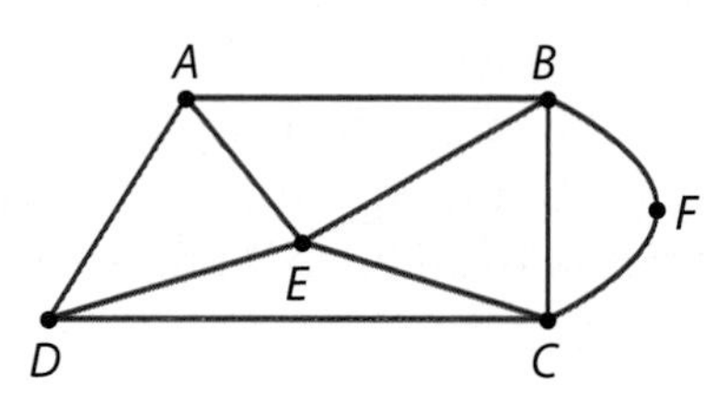

34 In almost all countries, temperature is measured in degrees Celsius. Temperature C in degrees Celsius is related to temperature F in degrees Fahrenheit by the formula

$$C = \frac{5}{9}(F - 32).$$

a. Describe, in words, how to convert from Fahrenheit to Celsius.

b. Write an equivalent formula that relates temperature C in degrees Celsius to temperature F in degrees Fahrenheit.

c. The temperature in New York City's Central Park on July 6, 1999, reached a high of 92° Fahrenheit. Express this temperature in degrees Celsius.

35 Make a copy of the diagram below on a piece of grid paper or square dot paper. The legs of the right triangle each have a length of 1 unit.

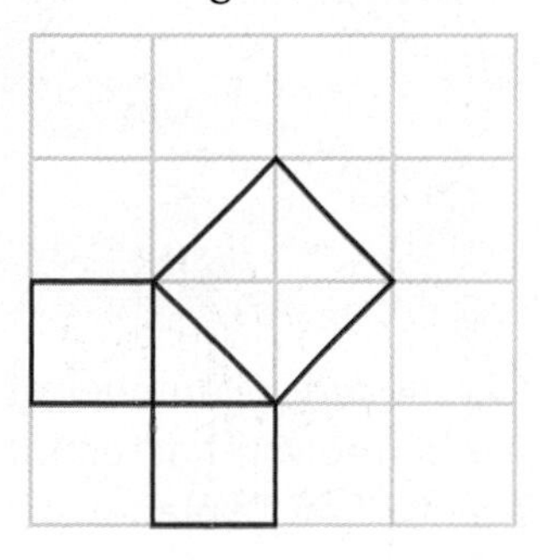

a. What are the areas of the squares shown on the sides of the triangle?

b. On your paper, draw a similar diagram to show two squares, each with area 4 square units and a square with area 8 square units.

c. Use a similar diagram to draw a square with area 5 square units.

d. On your diagrams indicate line segments with lengths $\sqrt{2}$, $\sqrt{8}$, and $\sqrt{5}$.

e. From your diagrams, estimate the lengths $\sqrt{2}$, $\sqrt{8}$, and $\sqrt{5}$.

Looking Back

In this unit, you studied two- and three-dimensional shapes and how they are related. You learned how segment lengths and angle measures determine the shape of special polygons and how polygons determine the shape of polyhedra. You discovered ways to test triangles for congruence and how to use triangles to study useful properties of parallelograms and other polygons. You explored how to visualize three-dimensional shapes and how to represent them in two dimensions. You also learned about symmetry and other properties that make certain shapes useful in design, engineering, and construction. Finally, you learned how knowledge of a few basic properties of shapes could be used to reason to additional properties of those shapes.

The following tasks will help you review, pull together, and apply what you have learned in the process of solving several new problems.

1. Suppose you are given an envelope containing information on separate slips of paper about each of the three side lengths (AB, BC, AC) and each of the three angle measures ($m\angle A$, $m\angle B$, $m\angle C$) of the triangular truss being carried by Diane and Jessie. You randomly draw one slip at a time from the envelope.

   a. What is the *largest* number of slips you would ever need to draw before you had enough information to build a congruent truss? Explain your reasoning.

   b. What is the *smallest* number of slips you could draw and still build the truss? Explain.

2. Some quadrilateral linkages can change rotary motion into "back-and-forth" motion and vice versa. In addition to being used in mechanical devices, the parallelogram linkage serves as the basis for a linkage called a *pantograph*. Pantographs are used for copying drawings and maps to a different scale.

   The pantograph shown has been assembled so that $ABCE$ is a rhombus; $AD = CF$ and these lengths are the same as that of a side of the rhombus. The pantograph is held firm at point D. No matter how the linkage is moved, give reasons why:

   a. $ABCE$ will always be a rhombus.

   b. $DE = EF$

   c. Points D, E, and F will always be on a straight line.

3. Earlier in the unit you saw that the base angles of any isosceles triangle are congruent. This property can be restated as:

   If two sides of a triangle are congruent, then the angles opposite those sides are congruent.

   a. Write the converse of this statement. Do you think the converse is a true statement?

   b. Describe an experiment you could conduct to test whether the converse *might* be true.

   c. The first diagram below shows $\triangle ABC$ with two congruent angles. The second diagram shows the same triangle with the bisector of $\angle A$ drawn.

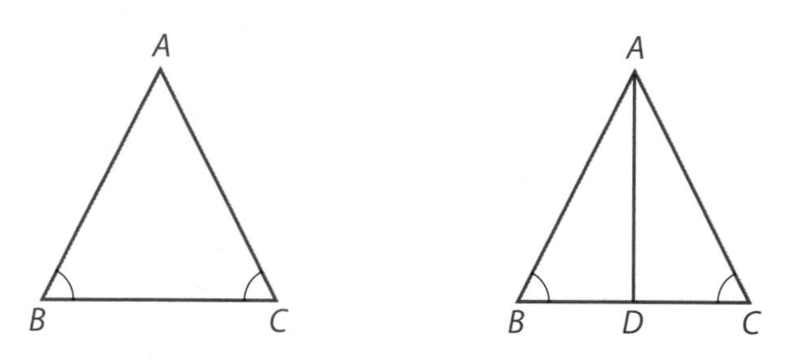

   How could you use careful reasoning and the given information for the second diagram to show that the converse statement is true?

   d. What would the experiment in Part b tell you about this situation? How does that differ from what the reasoning in Part c tells you?

4. Two farms, located at points A and B, are to be connected by separate wires to a transformer on a main power line ℓ.

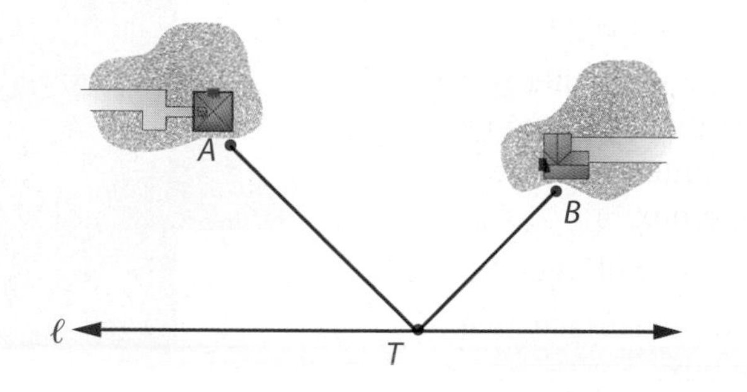

Study the diagram below which shows a method for locating the position of the transformer.

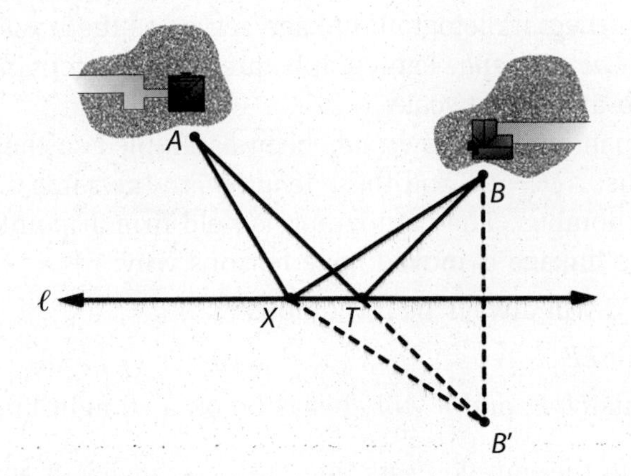

Point B' is located so that the power line ℓ is the perpendicular bisector of $\overline{BB'}$. Then the location T of the transformer is determined by sighting the line AB'. By locating the transformer at point T rather than at any other point X on the line, the power company uses the minimum amount of wire to bring electricity to the two farms.

a. Why is the length of the required wire, $AT + TB$, the same as the length AB'?

b. Explain as carefully and precisely as you can why if any other location X is chosen, then more wire would be required.

5 Make an accurate drawing of a regular decagon.

a. What must be the measure of an interior angle?

b. What must be the measure of an exterior angle?

c. Describe its lines of symmetry and its rotational symmetries.

d. Will a regular decagon tessellate a plane? Explain why or why not.

e. Will copies of an equilateral triangle, a regular decagon, and a regular 15-gon form a semiregular tessellation? Explain your reasoning.

6 Midland Packaging manufactures boxes for many different companies. The net for one type of box manufactured for a candy company is shown below.

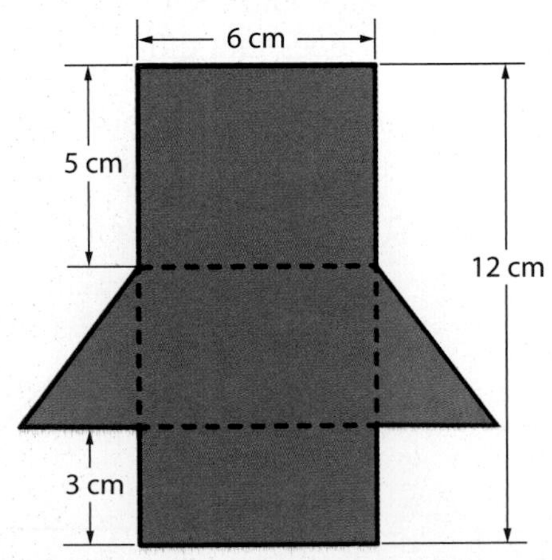

a. Name the three-dimensional shape for which this is a net.

b. Sketch the box showing its hidden edges.

c. Sketch two other possible nets that could be used to manufacture the same box.

d. Find the volume of the box.

e. Find the surface area of the box.

f. Does the box have any symmetries? If so, explain how the symmetries are related to the symmetries of its faces.

7 A common basic shape for a house is a polyhedron formed by placing a triangular prism on top of a rectangular prism. For a particular house, the length, width, and height of the rectangular prism are 70 feet, 50 feet, and 20 feet, respectively. The base of the triangular prism is an isosceles triangle, and the angle at the peak of the roof is 140°.

a. The resulting polyhedron can be viewed as a single prism. Make a scale drawing of the base of the prism formed in this way.

b. Make an orthographic drawing of this prism.

c. Describe the reflection symmetry and the rotational symmetry of the prism.

d. Determine the number of vertices, edges, and faces for this prism. Verify that Euler's Formula holds in this case.

e. Find the angle defect at each vertex. Verify that Descartes' Theorem holds in this case.

8 Two-dimensional concepts often have corresponding, though not usually identical, concepts in three dimensions. Answer the following questions about some of these connections.

a. What property of polygons is related to Descartes' Theorem in three dimensions?

b. What is the three-dimensional counterpart of line symmetry in two dimensions?

c. How are rotational symmetry in two dimensions and rotational symmetry in three dimensions alike? How are they different?

d. What is the three-dimensional counterpart of tiling the plane in two dimensions?

e. What are the three-dimensional counterparts of a triangle, square, circle, rectangle, parallelogram, and regular polygon?

f. How is rigidity of two-dimensional shapes like rigidity of three-dimensional shapes? How is it different?

g. Identify at least one more two-dimensional concept or shape and describe its counterpart in three dimensions.

9 Consider the following two statements, one about shape in two dimensions, the other about shape in three dimensions.

- A regular hexagon tiles the plane.
- It is not possible for a regular polyhedron to have only regular hexagonal faces.

These two statements are true for very similar reasons. Explain why each statement is true, and explain their connections.

Summarize the Mathematics

Shape is a fundamental feature of the world in which you live. Understanding shape involves being able to identify and describe shapes, visualize and represent shapes with drawings, and analyze and apply properties of shapes.

a Triangles and quadrilaterals are special classes of shapes called polygons.

  i. What properties are true of every polygon?

  ii. What properties are true of every quadrilateral? What property of some quadrilaterals makes the shape widely useful as a linkage?

  iii. What properties are true of every triangle?

b What does it mean for two polygons to be congruent?

  i. What information is sufficient to test whether two triangles are congruent?

  ii. Which test in part i could be used to test whether two parallelograms are congruent?

  iii. How can you use the idea of triangle congruence to reason about properties of polygons and parallelograms in particular? What are some of these properties?

c If a statement is true, its converse may or may not be true. What is the converse of the Pythagorean Theorem? Explain why it is true. How is it used in applications?

d Polyhedra are three-dimensional counterparts of polygons.

  i. Compare and contrast polygons and polyhedra.

  ii. Describe a variety of ways that you can represent polyhedra.

  iii. In some cases, congruent copies of a polygon can be used to tile a plane. In other cases, they can be used to form a polyhedron. What must be true about the angle measures at a common vertex in each case?

  iv. Compare tests for symmetries of polygons and other two-dimensional shapes with tests for symmetries of polyhedra and other three-dimensional shapes.

  v. Rigidity is often an important consideration in the design of both two-dimensional and three-dimensional shapes. What is the key idea to bracing shapes for rigidity? Why does this work?

Be prepared to share your ideas and reasoning with the class.

✓ *Check Your Understanding*

Write, in outline form, a summary of the important mathematical concepts and methods developed in this unit. Organize your summary so that it can be used as a quick reference in future units and courses.

When sport balls are kicked, thrown, or hit into the air, the flight paths are parabolas that can be described by quadratic functions like $y = -16x^2 + 40x + 5$. Quadratic functions also provide models for the shape of suspension bridge cables, television dish antennas, and the graphs of revenue and profit functions in business.

The understanding and skill you need to solve problems involving quadratic functions will develop from your work on problems in three lessons of this unit.

QUADRATIC FUNCTIONS

Lessons

1 *Quadratic Patterns*

Explore typical quadratic relations and expressions to discover and explain connections among problem conditions, data tables, graphs, and function rules for quadratic patterns of change.

2 *Equivalent Quadratic Expressions*

Use algebraic properties of number systems to write quadratic expressions in convenient equivalent forms.

3 *Solving Quadratic Equations*

Solve quadratic equations by algebraic methods, including factoring and the quadratic formula.

LESSON 1

Quadratic Patterns

The town of Rehoboth Beach, Delaware, is a popular summer vacation spot along the Atlantic Ocean coast. When Labor Day passes, most beach people leave until the following summer. However, many return in the Fall for the annual *Punkin' Chunkin'* festival. The main attraction of this weekend is a "World Championship" contest to see which team of amateur engineers devised the best machine for launching pumpkins a long distance.

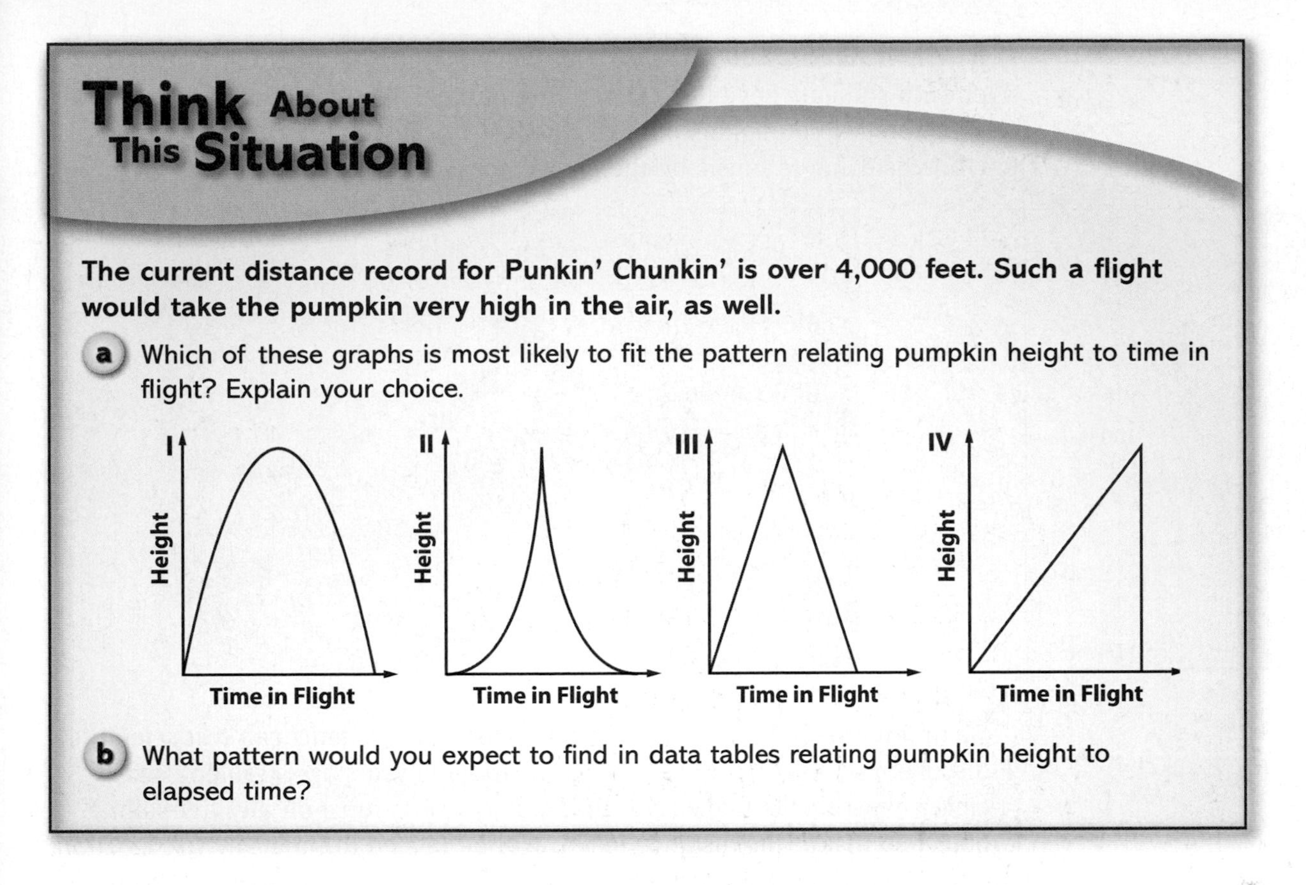

Think About This Situation

The current distance record for Punkin' Chunkin' is over 4,000 feet. Such a flight would take the pumpkin very high in the air, as well.

a Which of these graphs is most likely to fit the pattern relating pumpkin height to time in flight? Explain your choice.

b What pattern would you expect to find in data tables relating pumpkin height to elapsed time?

In work on investigations of this lesson, you will explore several strategies for recognizing, modeling, and analyzing patterns like those involved in the motion of a flying pumpkin.

Pumpkins in Flight

It turns out that the height of a flying pumpkin can be modeled well by a *quadratic* function of elapsed time. You can develop rules for such functions by reasoning from basic principles of science. Then you can use a variety of strategies to answer questions about the relationships. As you work on the problems of this investigation, look for answers to these questions:

What patterns of change appear in tables and graphs of (time, height) values for flying pumpkins and other projectiles?

What functions model those patterns of change?

Punkin' Droppin' At Old Dominion University in Norfolk, Virginia, physics students have their own flying pumpkin contest. Each year they see who can drop pumpkins on a target from 10 stories up in a tall building while listening to music by the group Smashing Pumpkins.

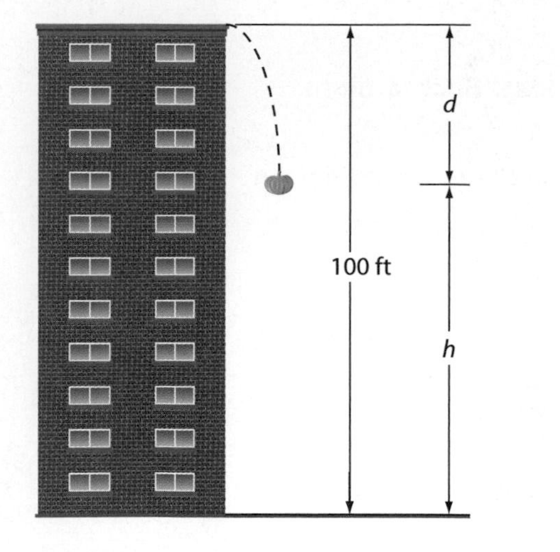

By timing the flight of the falling pumpkins, the students can test scientific discoveries made by Galileo Galilei, nearly 400 years ago. Galileo used clever experiments to discover that gravity exerts force on any free-falling object so that d, the distance fallen, will be related to time t by the function

$$d = 16t^2 \text{ (time in seconds and distance in feet).}$$

For example, suppose that the students dropped a pumpkin from a point that is 100 feet above the ground. At a time 0.7 seconds after being dropped, the pumpkin will have fallen $16(0.7)^2 \approx 7.84$ feet, leaving it $100 - 7.84 = 92.16$ feet above the ground.

This model ignores the resisting effects of the air as the pumpkin falls. But, for fairly compact and heavy objects, the function $d = 16t^2$ describes motion of falling bodies quite well.

Create a table like the one below to show estimates for the pumpkin's distance fallen and height above ground in feet at various times between 0 and 3 seconds.

Time t	Distance Fallen d	Height Above Ground h
0	0	100
0.5	4	$100 - 4 = 96$
1		
1.5		
2		
2.5		
3		

2 Use data relating height and time to answer the following questions about flight of a pumpkin dropped from a position 100 feet above the ground.

a. What function rule shows how the pumpkin's height h is related to time t?

b. What equation can be solved to find the time when the pumpkin is 10 feet from the ground? What is your best estimate for the solution of that equation?

c. What equation can be solved to find the time when the pumpkin hits the ground? What is your best estimate for the solution of that equation?

d. How would your answers to Parts a, b, and c change if the pumpkin were to be dropped from a spot 75 feet above the ground?

High Punkin' Chunkin' Compressed-air cannons, medieval catapults, and whirling slings are used for the punkin' chunkin' competitions.

Imagine pointing a punkin' chunkin' cannon straight upward. The pumpkin height at any time t will depend on its speed and height when it leaves the cannon.

3 Suppose a pumpkin is fired straight upward from the barrel of a compressed-air cannon at a point 20 feet above the ground, at a speed of 90 feet per second (about 60 miles per hour).

a. If there were no gravitational force pulling the pumpkin back toward the ground, how would the pumpkin's height above the ground change as time passes?

b. What function rule would relate height above the ground h in feet to time in the air t in seconds?

c. How would you change the function rule in Part b if the punkin' chunker used a stronger cannon that fired the pumpkin straight up into the air with a velocity of 120 feet per second?

d. How would you change the function rule in Part b if the end of the cannon barrel was only 15 feet above the ground, instead of 20 feet?

Now think about how the flight of a launched pumpkin results from the combination of three factors:

- initial height of the pumpkin's release,
- initial upward velocity produced by the pumpkin-launching device, and
- gravity pulling the pumpkin down toward the ground.

a. Suppose a compressed-air cannon fires a pumpkin straight up into the air from a height of 20 feet and provides an initial upward velocity of 90 feet per second. What function rule would combine these conditions and the effect of gravity to give a relation between the pumpkin's height h in feet and its flight time t in seconds?

b. How would you change your function rule in Part a if the pumpkin is launched at a height of 15 feet with an initial upward velocity of 120 feet per second?

By now you may have recognized that the height of a pumpkin shot straight up into the air at any time in its flight will be given by a function that can be expressed with a rule in the general form

$$h = h_0 + v_0t - 16t^2.$$

In those functions, h is measured in feet and t in seconds.

a. What does the value of h_0 represent? What units are used to measure h_0?

b. What does the value of v_0 represent? What units are used to measure v_0?

When a pumpkin is not launched straight up into the air, we can break its velocity into a vertical component and a horizontal component. The vertical component, the *upward velocity*, can be used to find a function that predicts change over time in the pumpkin's height. The horizontal component can be used to find a function that predicts change over time in the horizontal distance traveled.

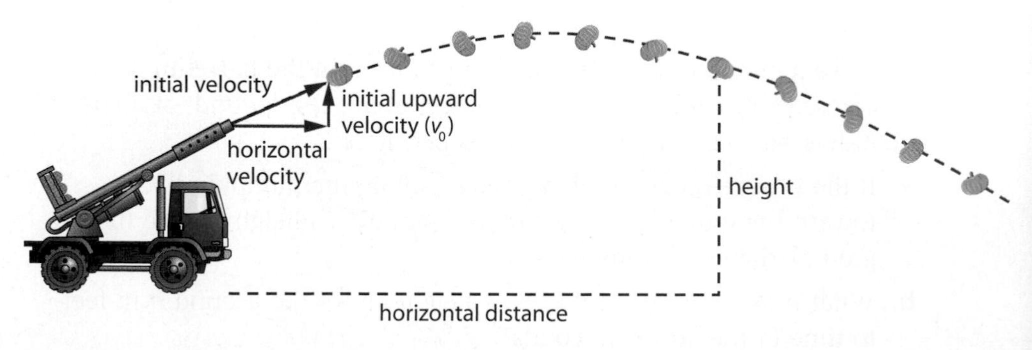

6. The pumpkin's height in feet t seconds after it is launched will still be given by $h = h_0 + v_0t - 16t^2$. It is fairly easy to measure the initial height (h_0) from which the pumpkin is launched, but it is not so easy to measure the initial upward velocity (v_0).

   a. Suppose that a pumpkin leaves a cannon at a point 24 feet above the ground when $t = 0$. What does that fact tell about the rule giving height h as a function of time in flight t?

   b. Suppose you were able to use a stopwatch to discover that the pumpkin shot described in Part a returned to the ground after 6 seconds. Use that information to find the value of v_0.

7. Suppose that you were able to use a ranging tool that records the height of a flying pumpkin every half-second from the time it left a cannon. A sample of the data for one pumpkin launch appears in the following table.

Time (in seconds)	0	0.5	1.0	1.5	2.0	2.5	3.0	3.5	4.0
Height (in feet)	15	40	60	70	70	65	50	30	0

   a. Plot the data on a graph and experiment with several values of v_0 and h_0 in search of a function that models the data pattern well.

   b. Use a calculator or computer tool that offers quadratic curve-fitting to find a quadratic model for the sample data pattern. Compare that automatic curve-fit to what you found with your own experimentation.

   c. Use the rule that you found in Part b to write and solve equations and inequalities matching these questions about the pumpkin shot.

      i. When was the pumpkin 60 feet above the ground?

      ii. For which time(s) was the pumpkin at least 60 feet above the ground?

   d. Use the rule you found in Part b to answer the following questions.

      i. What is your best estimate for the maximum height of the pumpkin?

      ii. How do you know if you have a good estimate? When does the pumpkin reach that height?

Summarize the Mathematics

In this investigation, you used several strategies to find rules for quadratic functions that relate the position of flying objects to time in flight. You used those function rules and resulting tables and graphs to answer questions about the problem situations.

a How can the height from which an object is dropped or launched be seen:

- **i.** In a table of (*time*, *height*) values?
- **ii.** On a graph of height over time?
- **iii.** In a rule of the form $h = h_0 + v_0t - 16t^2$ giving height as a function of time?

b How could you determine the initial upward velocity of a flying object from a rule in the form $h = h_0 + v_0t - 16t^2$ giving height as a function of time?

c What strategies can you use to answer questions about the height of a flying object over time?

Be prepared to share your ideas and strategies with others in your class.

Check Your Understanding

In Game 3 of the 1970 NBA championship series, the L.A. Lakers were down by two points with three seconds left in the game. The ball was inbounded to Jerry West, whose image is silhouetted in today's NBA logo. He launched and made a miraculous shot from beyond midcourt, a distance of 60 feet, to send the game into overtime (there was no 3-point line at that time).

Through careful analysis of the game tape, one could determine the height at which Jerry West released the ball, as well as the amount of time that elapsed between the time the ball left his hands and the time the ball reached the basket.

This information could then be used to write a rule for the ball's height h in feet as a function of time in flight t in seconds.

a. Suppose the basketball left West's hands at a point 8 feet above the ground. What does that information tell about the rule giving h as a function of t?

b. Suppose also that the basketball reached the basket (at a height of 10 feet) 2.5 seconds after it left West's hands. Use this information to determine the initial upward velocity of the basketball.

c. Write a rule giving h as a function of t.

d. Use the function you developed in Part c to write and solve equations and inequalities to answer these questions about the basketball shot.

i. At what other time(s) was the ball at the height of the rim (10 feet)?

ii. For how long was the ball higher than 30 feet above the floor?

iii. If the ball had missed the rim and backboard, when would it have hit the floor?

e. What was the maximum height of the shot, and when did the ball reach that point?

Investigation 2 Golden Gate Quadratics

The quadratic functions that describe the rise and fall of flying pumpkins are examples of a larger family of relationships described by rules in the general form $y = ax^2 + bx + c$. The particular numerical values of the coefficients a and b and the constant c depend on problem conditions. As you work through this investigation, look for answers to these questions:

How can tables, graphs, and rules for quadratic functions be used to answer questions about the situations they represent?

What patterns of change appear in tables and graphs of quadratic functions?

Suspension Bridges Some of the longest bridges in the world are suspended from cables that hang in parabolic arcs between towers. One of the most famous suspension bridges is the Golden Gate Bridge in San Francisco, California.

If you think of one bridge tower as the y-axis of a coordinate system and the bridge surface as the x-axis, the shape of the main suspension cable is like the graph of a quadratic function. For example, if the function defining the curve of one suspension cable is $y = 0.002x^2 - x + 150$, where x and y are measured in feet, the graph will look like that on the top of the next page.

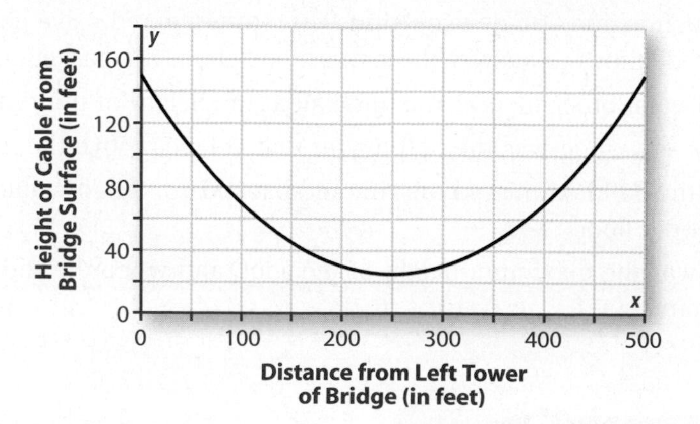

1. Use the function $y = 0.002x^2 - x + 150$ to answer the following questions.

   a. What is the approximate height (from the bridge surface) of the towers from which the cable is suspended?

   b. What is the shortest distance from the cable to the bridge surface, and where does it occur?

   c. For what interval(s) is the suspension cable at least 75 feet above the bridge surface?

   d. Recall that the height function for dropping pumpkins was $h = 100 - 16t^2$ and for a basketball long shot was $h = 8 + 40.8t - 16t^2$.

      i. How is the graph of the suspension cable function similar to and different from the graphs of these two functions?

      ii. How is the rule of the suspension cable function similar to and how is it different from the rules of these two functions?

Fundraising In 1996, the first Tibetan Freedom Concert, regarded by many as the single greatest cultural event in modern rock history, took place in Golden Gate Park in San Francisco. This was the first in a series of benefit concerts organized by the Milarepa Fund to raise awareness about nonviolence and the Tibetan struggle for freedom, as well as to encourage youth activism.

The primary goal for the Tibetan Freedom Concerts was to raise awareness, not money. However, careful planning was needed to ensure that the event would reach a large audience and that it would not *lose* money. The profit from any event will be the difference between income and operating expenses.

2. As organizers planned for the event, they had many variables to consider.

   a. What factors will affect the number of tickets sold for the event?

   b. What kinds of expenses will reduce profit from tickets sales, and how will those expenses depend on the number of people who buy tickets and attend?

3. Suppose that a group of students decided to organize a local concert to raise awareness and funds for the Tibetan struggle, and that planning for the concert led to this information:

   - The relationship between number of tickets sold s and ticket price x in dollars can be approximated by the linear function $s = 4{,}000 - 250x$.
   - Expenses for promoting and operating the concert will include $1,000 for advertising, $3,000 for pavilion rental, $1,500 for security, and $2,000 for catering and event T-shirts for volunteer staff and band members.

   a. Find a function that can be used to predict income I for any ticket price x.

   b. Find a function that can be used to predict profit P for any ticket price x.

   c. How do predicted income and profit change as the concert organizers consider ticket prices ranging from $1 to $20? How are those patterns of change shown in graphs of the income and profit functions?

   d. What ticket price(s) seem likely to give maximum income and maximum profit for the concert? What are those maximum income and profit values? How many tickets will be sold at the price(s) that maximize income and profit?

   e. If event planners are more interested in attracting a large audience without *losing* money on the event than in maximizing profit, what range of ticket prices should they consider? Explain your reasoning.

4. The **break-even point** is the ticket price for which the event's income will equal expenses. Another way to think of the break-even point is the ticket price when profit is $0.

   a. Write and solve an equation that can be used to find the break-even ticket price for this particular planned concert.

   b. Write and solve an inequality that can be used to find ticket prices for which the planned concert will make a positive profit.

   c. Write and solve an inequality that can be used to find ticket prices for which the planned concert will lose money.

5. What similarities and differences do you see in tables, graphs, and rules of the functions relating number of tickets sold, income, and profit to the proposed ticket price?

Summarize the Mathematics

In this investigation, you used several strategies to find and use rules for quadratic functions in different problem situations.

a How are patterns in the tables of values, graphs, and rules of the quadratic function examples in Investigations 1 and 2 similar to each other, and how do they differ from each other?

b How are tables, graphs, and rules of quadratic functions similar to and different from those of other types of functions you have worked with in earlier studies?

c What strategies can you use to:

i. Produce functions that model problem conditions?

ii. Solve quadratic equations or inequalities?

iii. Find maximum or minimum values of quadratic functions?

Be prepared to share your ideas and strategies with others in your class.

Check Your Understanding

The physical forces that determine the shape of a suspension bridge and business factors that determine the graph of profit prospects for a concert apply to other situations as well. For example, the parabolic reflectors that are used to send and receive microwaves and sounds have shapes determined by quadratic functions.

Suppose that the profile of one such parabolic dish is given by the graph of $y = 0.05x^2 - 1.2x$, where dish width x and depth y are in feet.

a. Sketch a graph of the function $y = 0.05x^2 - 1.2x$ for $0 \leq x \leq 25$. Then write calculations, equations, and inequalities that would provide answers for parts i–iv. Use algebraic, numeric, or graphic reasoning strategies to find the answers, and label (with coordinates) the points on the graph corresponding to your answers.

i. If the edge of the dish is represented by the points where $y = 0$, how wide is the dish?

ii. What is the depth of the dish at points 6 feet in from the edge?

iii. How far in from the edge will the depth of the dish be 2 feet?

iv. How far in from the edge will the depth of the dish be at least 3 feet?

b. What is the maximum depth of the dish and at what distance from the edge will that occur? Label the point (with coordinates) on your graph of $y = 0.05x^2 - 1.2x$.

Investigation 3 Patterns in Tables, Graphs, and Rules

In your work on the problems of Investigations 1 and 2, you examined a variety of rules, tables, and graphs for quadratic functions. For example,

A pumpkin's height above ground (in feet) is given by $h = 100 - 16t^2$.

A suspension cable's height above a bridge surface (in feet) is given by $y = 0.002x^2 - x + 150$.

Income for a concert (in dollars) is given by $I = x(4{,}000 - 250x)$.

Profit for a concert (in dollars) is given by $P = x(4{,}000 - 250x) - 7{,}500$.

It turns out that the expressions used in all of these functions are equivalent to expressions in the general form $ax^2 + bx + c$. To solve problems involving quadratic functions, it helps to know how patterns in the expressions $ax^2 + bx + c$ are related to patterns in tables and graphs of the related quadratic functions $y = ax^2 + bx + c$. As you work on the following problems, look for answers to this question:

How are the values of a, b, and c related to patterns in the graphs and tables of values for quadratic functions $y = ax^2 + bx + c$?

To answer a question like this, it helps to use your calculator or CAS software to produce tables and graphs of many examples in which the coefficients a and b and the constant c are varied systematically. The problems of this investigation suggest ways you could do such explorations and some questions that might help in summarizing patterns you notice. Make informal notes of what you observe in the experiments and then share your ideas with your teacher and other students to formulate some general conclusions.

The Basic Quadratic Function When distance traveled by a falling pumpkin is measured in feet, the rule giving distance as a function of time is $d = 16t^2$. When such gravitational effects are studied on the Moon, the rule becomes $d = 2.6t^2$. When distance is measured in meters, the rule is $d = 4.9t^2$ on the Earth and $d = 0.8t^2$ on the Moon.

These are all examples of the simplest quadratic functions—those defined by rules in the form $y = ax^2$.

How can you predict the shape and location of graphs of quadratic functions with rules in the form $y = ax^2$?

1. Study the tables and graphs produced by such functions for several *positive* values of a. For example, you might start by comparing tables and graphs of $y = x^2$, $y = 2x^2$, and $y = 0.5x^2$ for $-10 \leq x \leq 10$.

   a. What do all the graphs have in common? How about all the tables?

   b. How is the pattern in a table or graph of $y = ax^2$ related to the value of the coefficient a when $a > 0$?

2. Next study the tables and graphs produced by such functions for several *negative* values of a. For example, you might start by comparing tables and graphs of $y = -x^2$, $y = -2x^2$, and $y = -0.5x^2$ for $-10 \leq x \leq 10$.

   a. What do all these tables have in common? How about all the graphs?

   b. How is the pattern in a table or graph of $y = ax^2$ related to the value of the coefficient a, when $a < 0$?

3. Now think about *why* the patterns in tables and graphs of functions $y = ax^2$ occur and *why* the coefficient a is helpful in predicting behavior of any particular quadratic in this form.

   a. Consider first the functions $y = ax^2$ when $a > 0$.

      i. Why are the values of y always greater than or equal to zero?

      ii. Why are the graphs always symmetric curves with a minimum point (0, 0)?

   b. Consider next the functions $y = ax^2$ when $a < 0$.

      i. Why are the values of y always less than or equal to zero?

      ii. Why are the graphs always symmetric curves with a maximum point (0, 0)?

Adding a Constant When you designed a quadratic function to model position of a pumpkin at various times after it is dropped from a 100-foot tall building, the rules that made sense were $y = 100 - 16t^2$ or $y = -16t^2 + 100$.

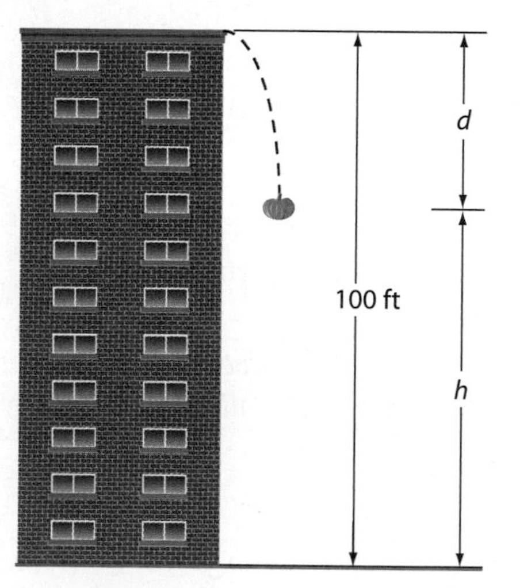

If the building from which pumpkins are to be dropped were taller or shorter, the rules might be $y = -16t^2 + 150$ or $y = -16t^2 + 60$.

These are all examples of another family of quadratic functions—those defined by rules in the form $y = ax^2 + c$.

How can you predict the shape and location of graphs of quadratic functions with rules in the form $y = ax^2 + c$?

4 Study tables and graphs produced by such functions for several combinations of positive and negative values of a and c. You might start by comparing these sets of functions:

Set 1	Set 2	Set 3
$y = x^2$	$y = -x^2$	$y = 2x^2$
$y = x^2 + 3$	$y = -x^2 + 5$	$y = 2x^2 + 1$
$y = x^2 - 4$	$y = -x^2 - 1$	$y = 2x^2 - 3$

a. How is the graph of $y = ax^2 + c$ related to the graph of $y = ax^2$?

b. How is the relationship between $y = ax^2 + c$ and $y = ax^2$ shown in tables of (x, y) values for the functions?

c. What are the values of $y = ax^2 + c$ and $y = ax^2$ when $x = 0$? How do these results help to explain the patterns relating the types of quadratics that you described in Parts a and b?

Factored and Expanded Forms When you studied problems about income from an amusement park bungee jump and promotion of a concert, you looked at functions relating income to ticket price. The resulting income rules had similar forms:

Bungee Jump: $I = p(50 - p)$

Concert Promotion: $I = x(4{,}000 - 250x)$

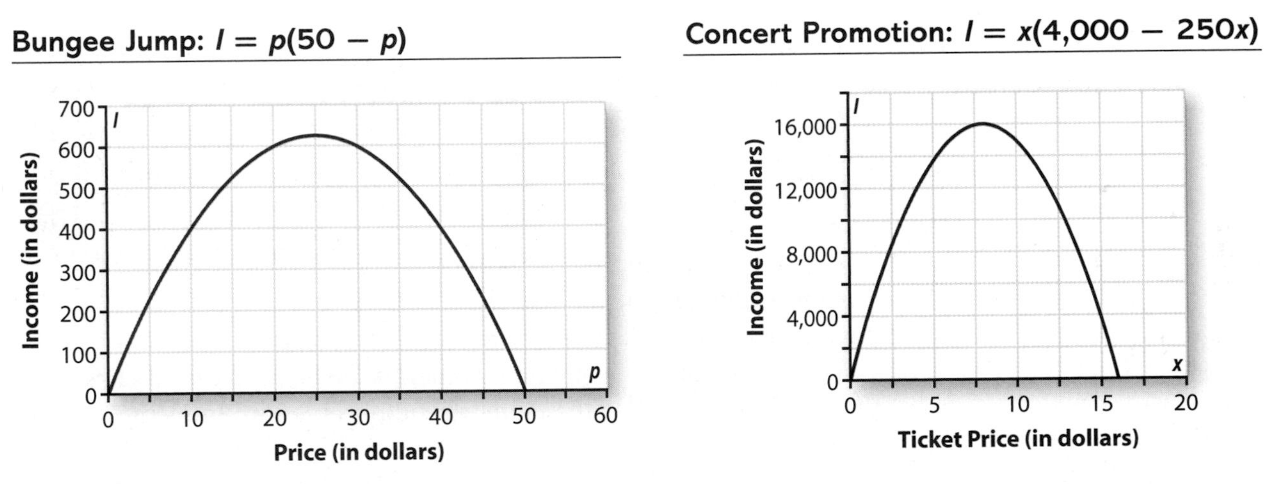

Just as you did with linear expressions in Unit 3, you can apply properties of numbers and operations to rewrite these rules in equivalent *expanded* form.

 To rewrite the rule $I = p(50 - p)$, a student at Sauk Prairie High School reasoned like this:

> Applying the distributive property, $p(50 - p) = 50p - p^2$.
> Rearranging terms, $50p - p^2 = -p^2 + 50p$.
> So $I = -p^2 + 50p$, showing that income is a quadratic function of ticket price.

a. Use similar ideas to rewrite $x(4{,}000 - 250x)$ in an equivalent expanded form.

b. Study graphs of the two income functions: $I = p(50 - p)$ and $I = x(4{,}000 - 250x)$. In each case, find coordinates of:

i. the y-intercept,

ii. the x-intercepts, and

iii. the maximum point.

c. How could you find these special points in Part b by analyzing the symbolic function rules in factored and/or expanded forms?

d. The Sauk Prairie student made the following observations. How do you think the student arrived at those ideas? Do you agree with them? If not, explain why not.

i. It is easiest to find the y-intercept from the *expanded* form $-p^2 + 50p$.

ii. It is easiest to find x-intercepts of the income function graph from the *factored* form $p(50 - p)$.

iii. It is easiest to find the maximum point on the income graph from the x-intercepts.

 The planning committee for Lake Aid, an annual benefit talent show at Wilde Lake High School, surveyed students to see how much they would be willing to pay for tickets. Suppose the committee developed the function $I = -75p^2 + 950p$ to estimate income I in dollars for various ticket prices p in dollars. Use the patterns you observed in Problem 5 to help answer the following questions.

a. Write the function for income using an equivalent factored form of the expression given. What information is shown well in the factored form that is not shown as well in the expanded form?

b. For what ticket prices does the committee expect an income of $0?

c. What ticket price will generate the greatest income? How much income is expected at that ticket price?

d. Use your answers to Parts b and c to sketch a graph of $I = -75p^2 + 950p$.

Adding a Linear Term The income functions you studied in Problems 5 and 6 are examples of another family of quadratics, those with rules in the form $y = ax^2 + bx$.

> *How can you predict the shape and location of graphs of quadratic functions with rules in the form $y = ax^2 + bx$?*

7 Study tables and graphs produced by such functions for several combinations of positive and negative values of a and b. You might start by comparing graphs of the following sets of functions:

Set 1	Set 2	Set 3
$y = x^2$	$y = -x^2$	$y = 2x^2$
$y = x^2 + 4x$	$y = -x^2 + 5x$	$y = 2x^2 + 6x$
$y = x^2 - 4x$	$y = -x^2 - 5x$	$y = 2x^2 - 6x$

Look at graphs of the functions given above to see if you can find patterns that relate the values of a and b in the rules $y = ax^2 + bx$ to location of the features below. It might help to think about the functions using the equivalent factored form, $x(ax + b)$.

a. y-intercepts

b. x-intercepts

c. maximum or minimum point

Putting Things Together The graphs of all quadratic functions are curves called *parabolas*. In work on Problems 1–7, you have learned how to predict the patterns in graphs for three special types of quadratic functions: $y = ax^2$, $y = ax^2 + c$, and $y = ax^2 + bx$. You can use what you know about these quadratic functions to reason about graphs produced by functions when the coefficients a and b and the constant c are not zero.

8 Explore the following examples and look for explanations of the patterns observed.

a. The diagram at the right gives graphs for three of the four quadratic functions below.

$$y = x^2 - 4x$$
$$y = x^2 - 4x + 6$$
$$y = -x^2 - 4x$$
$$y = x^2 - 4x - 5$$

Without using graphing technology:

i. Determine the function with the graph that is missing on the diagram.

ii. Match the remaining functions to their graphs, and be prepared to explain your reasoning.

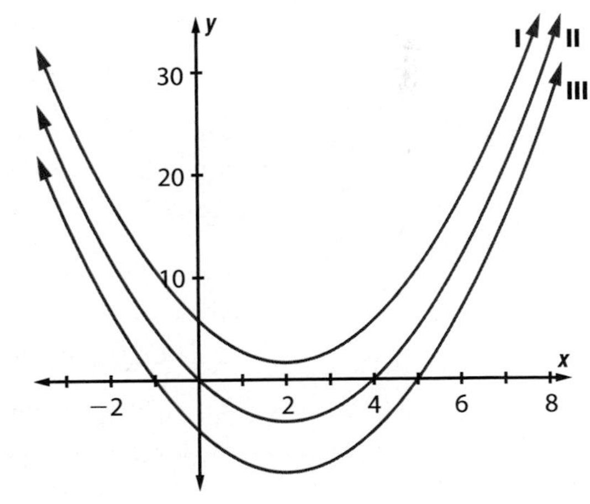

b. Without using graphing technology, sketch the pattern of graphs you would expect for the next set of quadratic functions. Explain your reasoning in making the sketch. Then check your ideas with the help of technology.

$$y = x^2 + 4x$$
$$y = x^2 + 4x - 6$$
$$y = x^2 + 4x + 5$$

c. How would a sketch showing graphs of the following functions be similar to and different from those in Parts a and b? Explain your reasoning. Then check your ideas with the help of technology.

$$y = -x^2 + 4x$$
$$y = -x^2 + 4x - 6$$
$$y = -x^2 + 4x + 5$$

d. How can properties of the special quadratic functions $y = ax^2$, $y = ax^2 + c$, and $y = ax^2 + bx$ help in reasoning about shape and location of graphs for functions in the form $y = ax^2 + bx + c$?

Summarize the Mathematics

In this investigation, you discovered some facts about the ways that patterns in tables and graphs of quadratic functions $y = ax^2 + bx + c$ ($a \neq 0$) are determined by the values of a, b, and c.

a What does the sign of a tell about the patterns of change and graphs of quadratic functions given by rules in the form $y = ax^2$? What does the absolute value of a tell you?

b How are the patterns of change and graphs of quadratic functions given by rules like $y = ax^2 + c$ related to those of the basic quadratic function $y = ax^2$? What does the value of c tell about the graph?

c How are the graphs of functions defined by rules like $y = ax^2 + bx$ ($b \neq 0$) different from those of functions with rules like $y = ax^2$? What does the value of b tell about the graph?

d How can you use what you know about quadratic functions with rules $y = ax^2$, $y = ax^2 + c$, and $y = ax^2 + bx$ to predict the shape and location of graphs for quadratic functions with rules $y = ax^2 + bx + c$ in which none of a, b, or c is 0?

Be prepared to share your ideas with the class.

Check Your Understanding

Use what you know about the relationship between rules and graphs for quadratic functions to match the functions with their graphs. Graphs were all produced with the same windows.

Rule I $y = x^2 + 2$

Rule II $y = x^2 - 5x + 2$

Rule III $y = -x^2 + 2$

Rule IV $y = -0.5x^2 + 2$

Rule V $y = x^2 + 5x + 2$

A

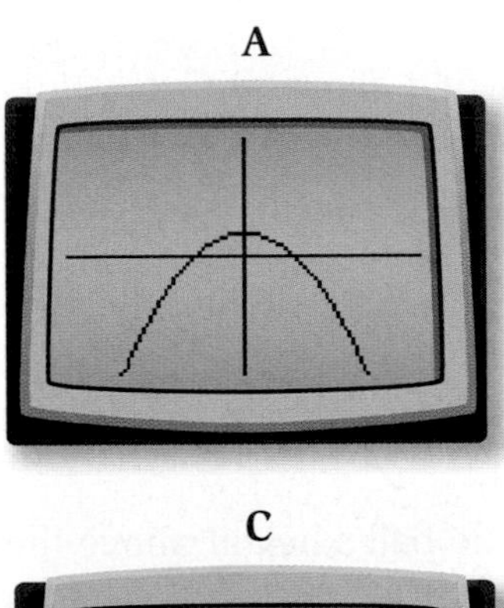

B

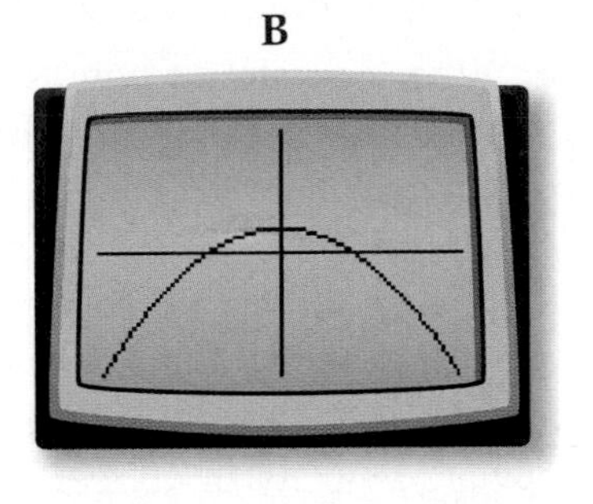

C

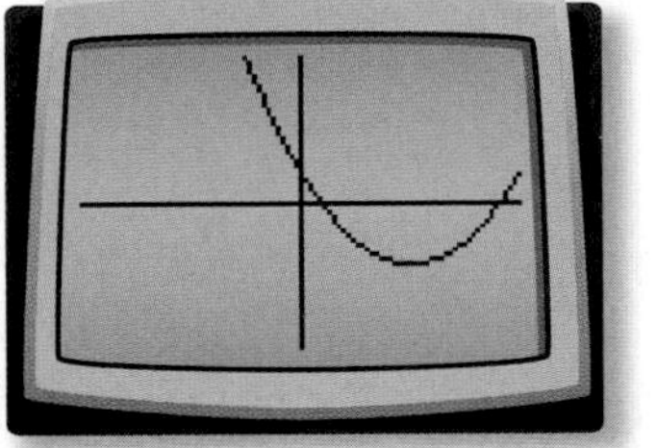

D

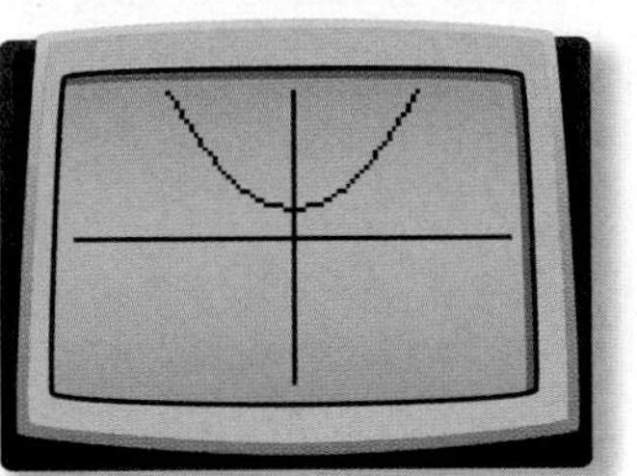

E

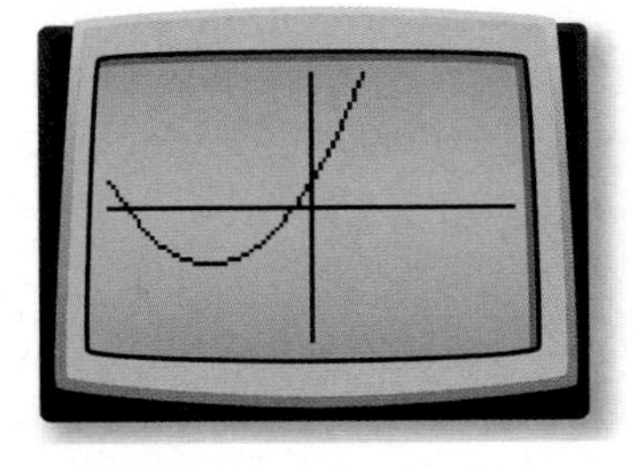

On Your Own

Applications

1. A first-time diver was a bit nervous about his first dive at a swimming pool. To ease his worries about hitting the water after a fall of 15 feet, he decided to push a tennis ball off the edge of the platform to see the effect of landing in the water.

   a. What rule shows how the ball's height above the water h is related to elapsed time in the dive t?

   b. Estimate the time it will take the ball to hit the water.

2. Katie, a goalie for Riverside High School's soccer team, needs to get the ball downfield to her teammates on the offensive end of the field. She punts the ball from a point 2 feet above the ground with an initial upward velocity of 40 feet per second.

   a. Write a function rule that relates the ball's height above the field h to its time in the air t.

   b. Use this function rule to estimate the time when the ball will hit the ground.

   c. Suppose Katie were to kick the ball right off the ground with the same initial upward velocity. Do you think the ball would be in the air the same amount of time, for more time, or for less time? Check your thinking.

3. The opening of the cannon pictured at the left is 16 feet above the ground. The daredevil, who is shot out of the cannon, reaches a maximum height of 55 feet after about 1.56 seconds and hits a net that is 9.5 feet off the ground after 3.25 seconds. Use this information to answer the following questions.

   a. Write a rule that relates the daredevil's height above the ground h at a time t seconds after the cannon is fired.

   b. At what upward velocity is the daredevil shot from the cannon?

   c. If, for some unfortunate reason, the net slipped to the ground at the firing of the cannon, when would the daredevil hit the ground?

4. When a punkin' chunker launches a pumpkin, the goal is long distance, not height. Suppose the relationship between horizontal distance d (in feet) and time t (in seconds) is given by the function rule $d = 70t$, when the height is given by $h = 20 + 50t - 16t^2$.

   a. How long will the pumpkin be in the air?

   b. How far will the pumpkin travel from the chunker by the time it hits the ground?

c. When will the pumpkin reach its maximum height, and what will that height be?

d. How far from the chunker will the pumpkin be (horizontally) when it reaches its maximum height?

5 Imagine you are in charge of constructing a two-tower suspension bridge over the Potlatch River. You have planned that the curve of the main suspension cables can be modeled by the function $y = 0.004x^2 - x + 80$, where y represents height of the cable above the bridge surface and x represents distance along the bridge surface from one tower toward the other. The values of x and y are measured in feet.

a. What is the approximate height (from the bridge surface) of each tower from which the cable is suspended?

b. What is the shortest distance from the cable to the bridge surface and where does it occur?

c. At which points is the suspension cable at least 50 feet above the bridge surface? Write an inequality that represents this question and express the solution as an inequality.

6 One formula used by highway safety engineers relates minimum stopping distance d in feet to vehicle speed s in miles per hour with the rule $d = 0.05s^2 + 1.1s$.

a. Create a table of sample (*speed, stopping distance*) values for a reasonable range of speeds. Plot the sample (*speed, stopping distance*) values on a coordinate graph. Then describe how stopping distance changes as speed increases.

b. Use the stopping distance function to answer the following questions.

i. What is the approximate stopping distance for a car traveling 60 miles per hour?

ii. If a car stopped in 120 feet, what is the fastest it could have been traveling when the driver first noticed the need to stop?

c. Estimate solutions for the following quadratic equations and explain what each solution tells about stopping distance and speed.

i. $180 = 0.05s^2 + 1.1s$ ii. $95 = 0.05s^2 + 1.1s$

7. Use what you know about the connection between rules and graphs for quadratic functions to match the given functions with their graphs that appear below. Each graph is shown in the standard viewing window ($-10 \leq x \leq 10$ and $-10 \leq y \leq 10$).

Rule I $y = x^2 - 4$ **Rule II** $y = 2x^2 + 4$

Rule III $y = -x^2 + 2x + 4$ **Rule IV** $y = -0.5x^2 + 4$

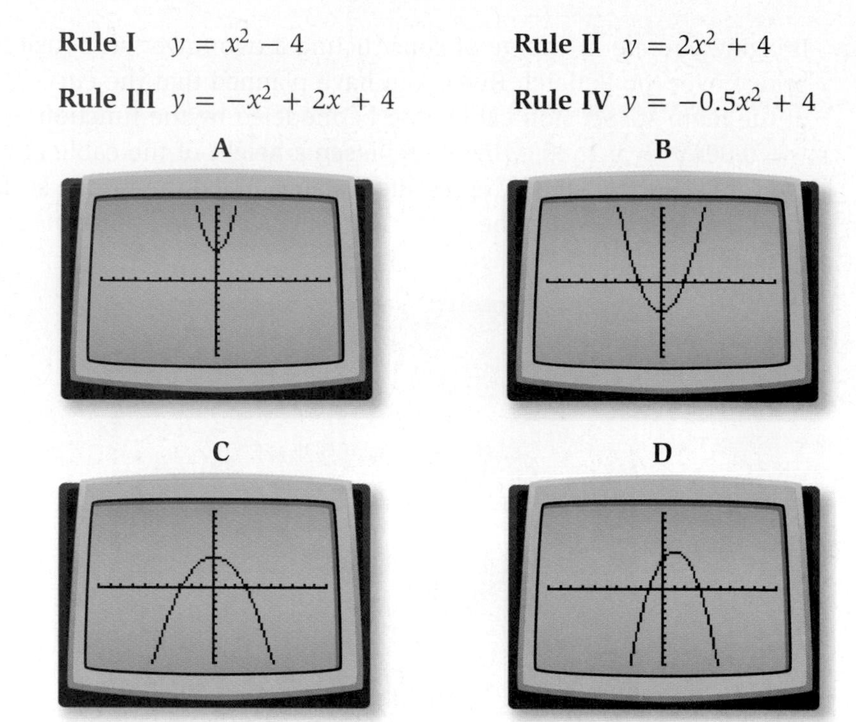

8. In Applications Tasks 1, 2, 5, and 6, you worked with several different quadratic functions. The function rules are restated in Parts a–d below. For each function, explain what you can learn about the shape and location of its graph by looking at the coefficients and constant term in the rule.

a. $h = 15 - 16t^2$

b. $h = 2 + 40t - 16t^2$

c. $y = 0.004x^2 - x + 80$

d. $d = 0.05s^2 + 1.1s$

Connections

9. The following experiment can be used to measure a person's *reaction time,* the amount of time it takes a person to react to something he or she sees.

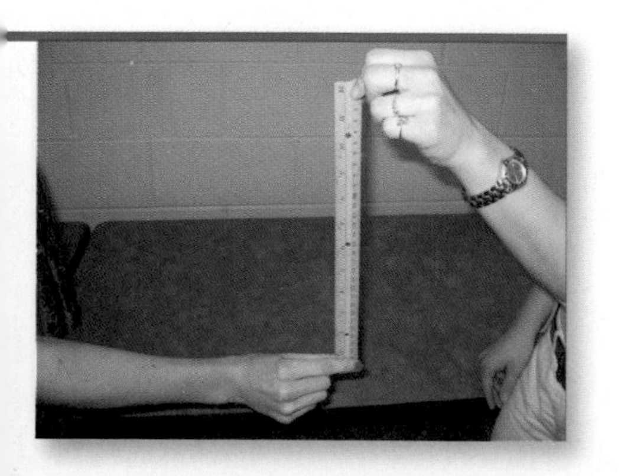

> Hold a ruler at the end that reads 12 inches and let it hang down. Have the subject hold his or her thumb and forefinger opposite the 0-inch mark without touching the ruler. Tell your subject that you will drop the ruler within the next 10 seconds and that he or she is supposed to grasp the ruler as quickly as possible after it is dropped.

The spot on the ruler where it is caught indicates the distance that the ruler dropped.

a. What function describes the distance d in feet that the ruler has fallen after t seconds?

b. Use what you know about the relationship between feet and inches and your function from Part a to estimate the reaction time of a person who grasps the ruler at the 4-inch mark.

c. Conduct this experiment several times and estimate the reaction times of your subjects.

10 Consider some other familiar measurement formulas.

a. Match the formulas A–D to the measurement calculations they express:

I volume of a cube	**A** $y = s^2$
II surface area of a cube	**B** $y = s^3$
III area of a square	**C** $y = 4s$
IV perimeter of a square	**D** $y = 6s^2$

b. Which of the formulas from Part a are those of quadratic functions?

11 The formula $A = \pi r^2$ shows how to calculate the area of a circle from its radius. You can also think about this formula as a quadratic function.

a. With respect to the general form of a quadratic function $y = ax^2 + bx + c$, what are the a, b, and c values for the area-of-a-circle function?

b. For $r > 0$, how does the shape of the graph $A = \pi r^2$ compare to that of $A = r^2$?

c. If the radius of a circle is 6 cm, what is its area?

d. If the area of a circle is about 154 cm^2, what is the approximate radius of the circle?

12 Sketch graphs of the functions $y = 2x$, $y = x^2$, and $y = 2^x$ for $0 \leq x \leq 5$.

a. In what ways are the graphs similar to each other?

b. In what ways do the graphs differ from each other?

c. The values of y for the three functions when x is between 0 and 1 show that $x^2 < 2x < 2^x$. Compare the values of y for the three graphs for the intervals for x below.

i. between 1 and 2

ii. between 2 and 3

iii. between 3 and 4

iv. greater than 4

On Your Own

13. Suppose that a pumpkin is dropped from an airplane flying about 5,280 feet above the ground (one mile up in the air). The function $h = 5{,}280 - 16t^2$ can be used to predict the height of that pumpkin at a point t seconds after it is dropped. But this mathematical model ignores the effects of air resistance.

    a. How would you expect a height function that does account for air resistance to be different from the function $h = 5{,}280 - 16t^2$ that ignores those effects?

    b. The speed of the falling pumpkin at a time t seconds after it is dropped can be predicted by the function $s_1 = 32t$, if you ignore air resistance. If air resistance is considered, the function $s_2 = 120(1 - 0.74^t)$ will better represent the relationship between speed and time.

    i. Make tables of (*time, speed*) values for each function s_1 and s_2 with values of t from 0 to 10 seconds.

    ii. Sketch graphs showing the patterns of change in speed implied by the two functions.

    iii. Describe similarities and differences in patterns of change predicted by the two (*time, speed*) functions.

    c. Air resistance on the falling pumpkin causes the speed of descent to approach a limit called *terminal velocity.* Explore the pattern of (*time, speed*) values for the function $s_2 = 120(1 - 0.74^t)$ for larger and larger values of t to see if you can discover the terminal velocity implied by that speed function.

14. Graphs of quadratic functions are curves called parabolas. Parabolas and other curves can also be viewed as cross sections of a cone—called *conic sections*.

    a. Describe how you could position a plane intersecting a cone so that the cross section is a parabola.

    b. How could you position a plane intersecting a cone so that the cross section is a circle?

    c. What other curve(s) are formed by a plane intersecting a cone? Illustrate your answer.

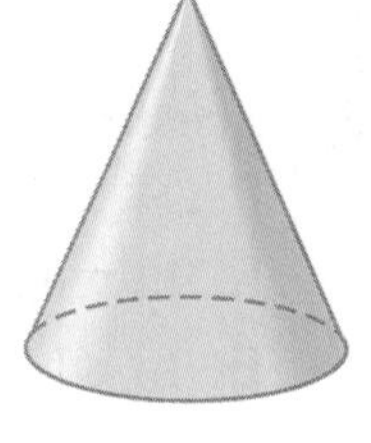

Reflections

15 Suppose that a skateboard rider travels from the top of one side to the top of the other side on a half-pipe ramp.

Which of the following graphs is the best model for the relationship between the rider's speed and distance traveled? Explain your choice.

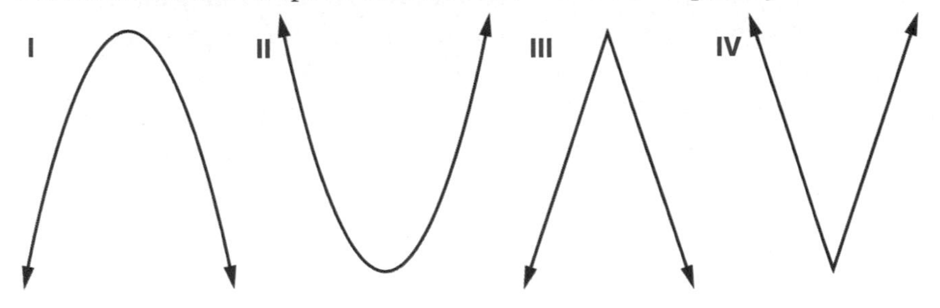

16 In several problems about the relation between income and price for a business venture, you worked with quadratic functions that have graphs like the one shown to the right.

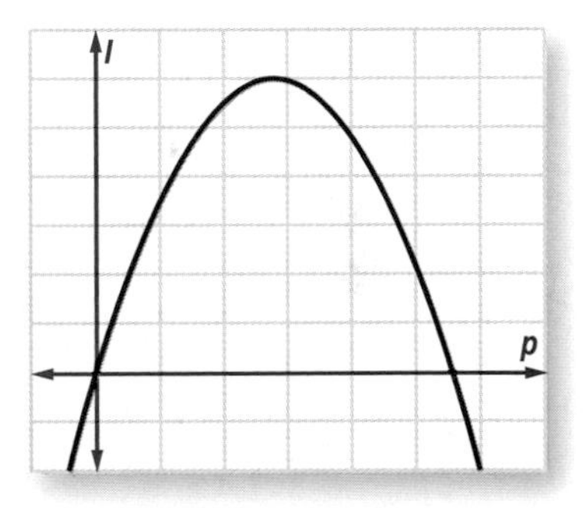

a. How would you describe the pattern of change in predicted income as ticket price increases?

b. Why is that general pattern reasonable in a wide variety of business situations?

17 A student first studying quadratic functions had the idea that in the rule $y = ax^2 + bx + c$, the value of b should tell the slope of the graph and c should tell the y-intercept. Do you agree? How could you use a graph of the function $y = x^2 + 2x + 3$ and other reasoning to support or dispute the student's idea?

18 All linear functions can be described by rules in the form $y = a + bx$. All exponential functions can be described by rules in the form $y = a(b^x)$. All quadratic functions can be described by rules in the form $y = ax^2 + bx + c$. The letters a, b, and c take on specific values in each of the three function forms. What information about each type of function can be learned from the values taken on by the letters a, b, and c?

Extensions

19 For anything that moves, *average speed* can be calculated by dividing the total distance traveled by the total time taken to travel that distance.

For example, a diver who falls from a 35-foot platform in about 1.5 seconds has an average speed of $\frac{35}{1.5}$, or about 23.3 feet per second. That diver will not be falling at that average speed throughout the dive.

a. If a diver falls from 35 feet to approximately 31 feet in the first 0.5 seconds of a dive, what estimate of speed would seem reasonable for the diver midway through that time interval—that is, how fast might the diver be moving at 0.25 seconds?

b. The relation between height above the water and the diver's time in flight can be described by the function $h = 35 - 16t^2$, if time is measured in seconds and distance in feet. Use that function rule to make a table of (*time, height*) data and then estimate the diver's speed at 6 points using your data. Make a table and a graph of the (*time, speed*) estimates.

c. What do the patterns in (*time, speed*) data and the graph tell you about the diver's speed on the way to the water?

d. About how fast is the diver traveling when he hits the water?

e. Write a rule for speed s as a function of time t that seems to fit the data in your table and graph. Use your calculator or computer software to check the function against the data in Part b.

20 When a pumpkin is shot from an air cannon chunker, its motion has two components—vertical and horizontal. Suppose that a pumpkin is shot at an angle of 40° with initial velocity of 150 feet per second and initial height 30 feet. The vertical component of its velocity will be about 96 feet per second; the horizontal component of its velocity will be about 115 feet per second.

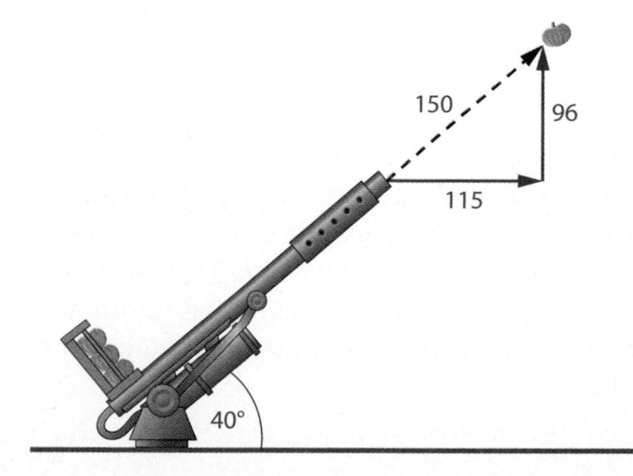

a. What function gives the height h of the pumpkin shot at any time t seconds after it leaves the chunker?

b. What function gives the horizontal distance d traveled by the pumpkin at any time t seconds after it leaves the chunker?

c. Use the functions in Parts a and b to find the horizontal distance traveled by the pumpkin by the time it hits the ground.

d. Rewrite the relation between time and distance in Part b to give time as a function of distance.

e. Combine the rule giving time as a function of horizontal distance and the rule giving height as a function of time to write a function rule giving height as a function of horizontal distance. (*Hint:* Replace each occurrence of t by an equivalent expression involving d.)

f. Use the function developed in Part e to estimate the distance traveled by the pumpkin when it hits the ground. Then compare the result obtained in this way to your answer to Part c.

21 You may have heard of *terminal velocity* in connection with skydiving. Scientific principles predict that a function like $h_2 = 5{,}680 - 120t - 400(0.74^t)$ will predict the height of a pumpkin (in feet) at any time t seconds after it is dropped from an airplane flying at an altitude of one mile. This function (in contrast to the more familiar $h_1 = 5{,}280 - 16t^2$) accounts for the slowing effect of air resistance on the falling pumpkin.

a. Use your calculator to produce a table showing predictions for height of the pumpkin using the functions h_1 and h_2 for times from 0 to 30 seconds. Record the data for times $t = 0, 5, 10, 15, 20, 25,$ and 30 seconds. Describe the patterns of change in height over time that are shown in the (*time, height*) values of the two functions.

b. Extend your table of (*time, height*) values to a point that gives estimates of the time it takes for the pumpkin to hit the ground.

c. Study the patterns of change in height for the last 10 seconds before the pumpkin hits the ground. Explain how the pattern of change in height for function h_2 illustrates the notion of terminal velocity that you explored in Connections Task 13 Part c.

22 Consider the two functions $y = 2^x$ and $y = x^2$.

a. How are the graphs of these two functions alike and how are they different?

b. How many solutions do you expect for the equation $2^x = x^2$? Explain your reasoning.

c. Estimate the solution(s) of the equation $2^x = x^2$ as accurately as possible. Explain or show how you estimated the solution(s).

23 Compare the quadratic function $y = x^2$ and the *absolute value function*, $y = |x|$.

a. Sketch graphs of these two functions and describe ways that they are similar and ways that they are different.

b. Find solutions for $x^2 = |x|$ by reasoning with the symbols themselves and then label the graph points representing the solutions with their coordinates.

c. Find the value(s) of x for which $x^2 > |x|$ and for which $x^2 < |x|$. Then indicate points on the graph representing the coordinates of those solutions.

24 Consider the quadratic functions defined by these rules:

Rule I $y = 3x^2 - 5x + 9$ **Rule II** $y = 1.5x^2 - 5x + 9$

Rule III $y = 3x^2 + 4x - 23$ **Rule IV** $y = x^2 - 5x - 23$

a. Examine graphs and/or tables of the functions to determine which of the functions have values that are relatively close to each other for large values of x.

b. Based on your findings in Part a, which coefficient (a or b) in the quadratic standard form determines **right end behavior**—the values of y for large positive values of x?

c. Can the same be said for **left end behavior** (which you may have guessed is for negative values of x with large absolute values)?

25 A computer spreadsheet can be a useful tool for exploring the effect of each coefficient and the constant term on the pattern of change of a quadratic function with rule $y = ax^2 + bx + c$. For example, the table below was produced with a spreadsheet to study $y = 2x^2 - 5x + 7$.

Quadratic Patterns.xls

◇	A	B	C	D	E	F	G
1	x	ax^2	bx	c	$ax^2 + bx + c$	$a =$	2
2	−5	50	25	7	82	$b =$	−5
3	−4	32	20	7	59	$c =$	7
4	−3	18	15	7	40		
5	−2	8	10	7	25		
6	−1	2	5	7	14		
7	0	0	0	7	7		
8	1	2	−5	7	4		
9	2	8	−10	7	5		
10	3	18	−15	7	10		
11	4	32	−20	7	19		
12	5	50	−25	7	32		

a. What numerical and formula entries and other spreadsheet techniques are needed to produce the x values in cells A2–A12?

b. What spreadsheet formulas can be used to produce entries in cells B2, C2, D2, and E2?

c. What formulas will appear in cells B3, C3, D3, and E3?

d. How could the spreadsheet be modified to study the function $y = x^2 + 3x - 5$?

26 Important questions about quadratic functions sometimes require solving inequalities like $10 > x^2 + 2x - 5$ or $-2x^2 + 6x \geq -8$.

a. What is the goal of the process in each case?

b. How can you use graphs to solve the inequalities?

c. How can you use tables of values to solve the inequalities?

d. How many solutions would you expect for a quadratic inequality?

Review

27 At many basketball games, there is a popular half-time contest to see if a fan can make a half-court shot. Some of these contests offer prizes of up to \$1,000,000! You may wonder how schools and other organizations could afford such payouts. In many cases, the organization offering the contest has purchased insurance to cover the costs, in the rare event that someone happens to make the shot. Imagine you've decided to start *Notgonnahappen Insurance* and your company will specialize in insuring \$1 million prizes.

a. If you charge organizations \$2,000 per contest for insurance, how many contests would you need to insure to cover the cost of a single event in which a contestant makes a million dollar half-court shot?

b. Suppose that for the first \$1,000,000 you collect in insurance fees, there are no payouts. You decide to invest this money in a savings account for future contests.

i. If the account earns 4% interest compounded annually, how much would the account be worth in one year if no deposits or withdrawals were made?

ii. Write two rules for calculating the account balance b at the end of t years—one using the *NOW-NEXT* approach and the other "$b = \ldots$."

iii. How much would the account be worth in 10 years if no deposits or withdrawals were made?

28 Write an equation for the line that:

a. Contains the points (0, 4) and (5, −3).

b. Contains the points (−2, 3) and (−5, 6).

c. Contains the point (7, 5) and has slope $\frac{2}{3}$.

d. Contains the point (−2, 5) and is parallel to the line $y = 1.5x + 6$.

29 In the diagram at the right, $\overline{AC} \cong \overline{BD}$ and $\angle BAC \cong \angle ABD$. Using only this information:

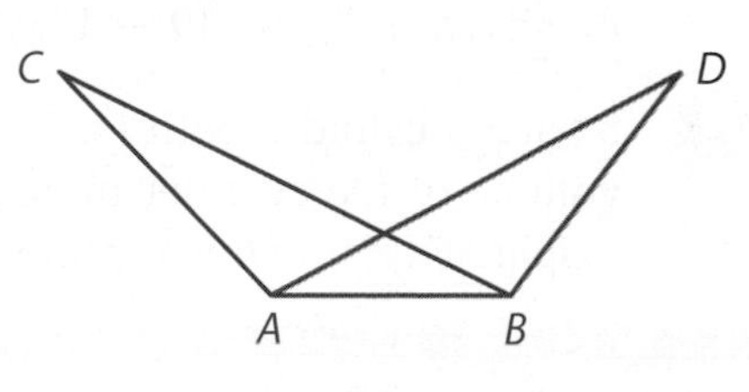

a. Explain why $\triangle ABC \cong \triangle BAD$.

b. Explain why $\overline{AD} \cong \overline{BC}$.

30 Write each of the following exponential expressions in the form 5^x for some integer x.

a. $(5^3)(5^4)$

b. $5^7 \div 5^3$

c. $(5^2)^3$

d. $4(5^3) + 5^3$

31 Suppose that the 15 numbers below are a sample of fares (in dollars) collected by drivers for *Fast Eddy's Taxi* company from trips on one typical day.

13, 23, 20, 22, 27, 21, 29, 31, 12, 10, 11, 21, 5, 19, 36

a. What are the mean and median of this sample of fares collected?

b. What are the range and the standard deviation of the sample of fares?

c. If there is a \$2 local government tax on each taxi fare, what are the mean, median, range, and standard deviation of the sample of fares after taxes have been deducted?

Fast Eddy's Taxi allows each driver to keep 70% of each after-tax fare as her or his pay.

d. What are the mean and median driver earnings from the sample of trips?

e. What are the range and standard deviation of driver earnings from the sample of trips?

32 In the *Patterns in Shape* unit, you revisited how to calculate areas and volumes of shapes.

a. What is the area of an equilateral triangle with side lengths of 10 units?

b. What is the area of a regular hexagon with side lengths of 10 units?

c. What additional information do you need to find the volume of a prism with a regular hexagonal base that has side lengths of 10 units?

d. How would the volumes of a prism and a pyramid compare if they had the same hexagonal base and same height?

33 Find the prime factorization of each number.

a. 28

b. 105

c. 72

d. 297

34 Rewrite each expression in a simpler equivalent form by first using the distributive property and then combining like terms.

a. $6x(3x - 5) + 12x$

b. $22 - 2(15 - 4x)$

c. $\frac{1}{2}(12x + 7) + \frac{2}{3}(9 - 15x)$

d. $15 - (3x - 8) + 5x(6 + 3x)$

35 Sketch a cylinder with radius r and height h. What will change the volume of the cylinder more, doubling the radius or doubling the height? Explain your reasoning.

LESSON 2

Equivalent Quadratic Expressions

When the freshman class officers at Sturgis High School were making plans for the annual end-of-year class party, they had a number of variables to consider:

- the number of students purchasing tickets and attending the party
- the price charged for tickets
- expenses, including food, a DJ, security, and clean-up

A survey of class members showed that the number of students attending the party would depend on the price charged for tickets. Survey data suggested a linear model relating price x and number of tickets sold n: $n = 200 - 10x$.

The students responsible for pricing food estimated an expense of $5 per student.

The students responsible for getting a DJ reported that the person they wanted would charge $150 for the event.

The school principal said that costs of security and clean-up by school crews would add another $100 to the cost of the party.

Think About This Situation

Think about how you could use the above information to set a price that would guarantee the freshman class would not lose money on the party.

a What party profit could be expected if ticket price is set at $5 per person? What if the ticket price is set at $10 per person?

b How could the given information be combined to figure out a price that would allow the class to break even or maybe even make some profit for the class treasury?

In this lesson, you will explore ways to develop expressions for functions that model quadratic patterns of change and to write and reason with equivalent forms of those expressions.

Investigation 1 Finding Expressions for Quadratic Patterns

When the freshman class officers at Sturgis High School listed all the variables to be considered in planning their party, they disagreed about how to set a price that would guarantee a profitable operation. To help settle those arguments, they tried to find a single function showing how profit would depend on the ticket price. They came up with two different profit functions and wondered whether they were algebraically equivalent. As you work on the problems in this investigation, look for answers to these questions:

What strategies are useful in finding rules for quadratic functions?

In deciding whether two quadratic expressions are equivalent?

In deciding when one form of quadratic expression is more useful than another?

1 One way to discover functions that model problem conditions is to consider a variety of specific pairs of (x, y) values and look for a pattern relating those values.

a. Use the information on page 492 to find ticket sales, income, costs, and profits for a sample of possible ticket prices. Record results in a table like this:

	Ticket Price (in $)				
	0	5	10	15	20
Number of Tickets Sold	200	150			
Income (in $)	0	750			
Food Cost (in $)	1,000				
DJ Cost (in $)	150				
Security/Cleanup Cost (in $)	100				
Profit (in $)	−1,250				

b. Plot the sample (*ticket price, profit*) values and describe the kind of function that you would expect to model the data pattern well.

2 Ms. Parkhurst, one of the Sturgis mathematics teachers, suggested another way to find a profit function that considers all factors. Check each step of her reasoning and explain why it is correct.

(1) Since the number of tickets sold n is related to the ticket price x by the linear function $n = 200 - 10x$, income from ticket sales will be related to ticket price by the function $I = x(200 - 10x)$.

(2) The cost c for food is related to the number of tickets sold n by the function $c = 5n$, so the cost for food will be related to the ticket price by the function $c = 5(200 - 10x)$.

(3) The costs for a DJ, security, and cleanup total $250, regardless of the number of students who attend the party.

(4) So, the profit of the party can be predicted from the ticket price x using the function $P = x(200 - 10x) - 5(200 - 10x) - 250$.

3 Some students followed Ms. Parkhurst's reasoning up to the point where she said the profit function would be $P = x(200 - 10x) - 5(200 - 10x) - 250$. But they expected a quadratic function like $P = ax^2 + bx + c$.

a. Is the expression $x(200 - 10x) - 5(200 - 10x) - 250$ equivalent to an expression in the $ax^2 + bx + c$ form? If so, what is the expression in that form? If not, how do you know?

b. What are the advantages of expanded and simplified expressions in reasoning about the function relating party profit to ticket price?

4 Ms. Parkhurst's rule for the function relating party profit to ticket price involves products and sums of linear functions, but the result is a quadratic function. For each of the following pairs of linear functions:

i. Graph the sum and describe the type of function that results from that operation. For example, for the functions in Part a, the sum is $y = (x + 2) + 0.5x$.

ii. Graph the product and describe the type of function that results from that operation. For example, for the functions in Part a, the product is $y = (x + 2)(0.5x)$.

iii. Write each sum and product in simpler equivalent form.

Be prepared to explain the reasoning you used to produce the equivalent function expressions.

a. $y_1 = x + 2$ and $y_2 = 0.5x$

b. $y_1 = 2x - 3$ and $y_2 = -1.5x$

c. $y_1 = -3x$ and $y_2 = 5 - 0.5x$

d. $y_1 = x + 2$ and $y_2 = 2x + 1$

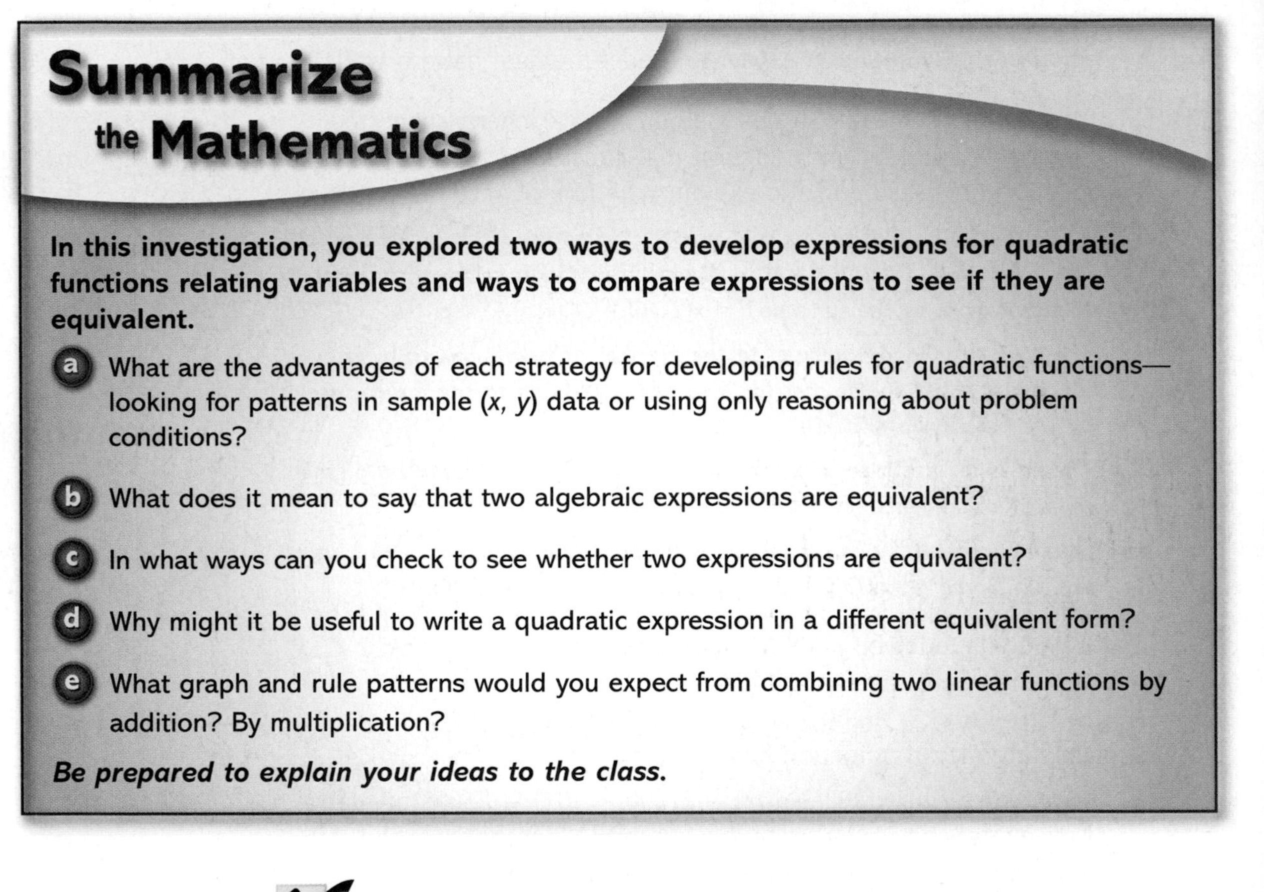

Summarize the Mathematics

In this investigation, you explored two ways to develop expressions for quadratic functions relating variables and ways to compare expressions to see if they are equivalent.

a What are the advantages of each strategy for developing rules for quadratic functions—looking for patterns in sample (*x*, *y*) data or using only reasoning about problem conditions?

b What does it mean to say that two algebraic expressions are equivalent?

c In what ways can you check to see whether two expressions are equivalent?

d Why might it be useful to write a quadratic expression in a different equivalent form?

e What graph and rule patterns would you expect from combining two linear functions by addition? By multiplication?

Be prepared to explain your ideas to the class.

Check Your Understanding

Use your understanding of equivalent expressions to help complete the following tasks.

a. Which of the following pairs of algebraic expressions are equivalent, and how do you know?

i. $x^2 + 5x$ and $x(x + 5)$

ii. $m(100 - m) + 25$ and $-m^2 + 100m + 25$

iii. $43 - 5(x - 10)$ and $33 - 5x$

b. Which of these functions are linear, which are quadratic, which are neither, and how can you justify your conclusions algebraically?

i. $J = (4p)(7p - 3)$

ii. $y = (3x + 2) - (2x - 4)$

iii. $d = (6t - 4) \div (2t)$

Investigation 2 Reasoning to Equivalent Expressions

In your work with linear functions, you learned that the form $y = mx + b$ was very useful for finding the slope and intercepts of graphs. However, you also found that linear functions sometimes arise in ways that make other equivalent expressions natural and informative. The work on party planning in Investigation 1 showed that the same thing can happen with quadratic relations.

It is relatively easy to do some informal checking to see if two given quadratic expressions might be equivalent—comparing graphs or entries in tables of values. But there are also some ways that properties of numbers and operations can be used to prove equivalence of quadratic expressions and to write any given expression in useful equivalent forms. As you complete the following problems, look for answers to this question:

What strategies can be used to transform quadratic expressions into useful equivalent forms?

One basic principle used again and again to produce equivalent expressions is the *Distributive Property of Multiplication over Addition (and Subtraction)*. It states that for any numbers a, b, and c, $a(b + c) = (ab) + (ac)$ and $a(b - c) = (ab) - (ac)$. For example, $5(x + 7) = 5x + 35$ and $5(x - 7) = 5x - 35$.

1 Use the distributive property to expand and combine like terms to write each of the following expressions in equivalent standard form $ax^2 + bx + c$. Be prepared to explain your reasoning in each case.

a. $(3 + x)x$

b. $5x(4x - 11)$

c. $7x(11 - 4x)$

d. $7x(x + 2) - 19$

e. $-9(5 - 3x) + 7x(x + 4)$

f. $mx(x + n) + p$

 Use the distributive property to write each of these quadratic expressions in equivalent form as a product of two linear factors. Be prepared to explain your reasoning in each case.

a. $7x^2 - 11x$

b. $12x + 4x^2$

c. $-3x^2 - 9x$

d. $ax^2 + bx$

3 Sometimes you need to combine expanding, factoring, and rearrangement of terms in a quadratic expression in order to produce a simpler form that gives useful information. For example, the following work shows how to write a complex expression in simpler expanded and factored forms.

$$\begin{aligned} 5x(6x - 8) + 4x(2 - 3x) &= 30x^2 - 40x + 8x - 12x^2 \\ &= 18x^2 - 32x \\ &= 2x(9x - 16) \end{aligned}$$

Use what you know about ways of writing algebraic expressions in equivalent forms to produce simplest possible expanded and (where possible) factored forms of these expressions.

a. $(14x^2 + 3x) - 7x(4 + x)$

b. $-x + 4x(9 - 2x) + 3x^2$

c. $5x(2x - 1) + 4x^2 - 2x$

d. $(5x^2 - 4) - 3(4x + 8x^2) - 25x$

The distributive property is used many places in algebra to write expressions in equivalent forms. In fact, the operations of *expanding* and *factoring* expressions like those for quadratic functions are now built into computer algebra systems. The following screen shows several results that should agree with answers you got in Problems 1, 2, and 3.

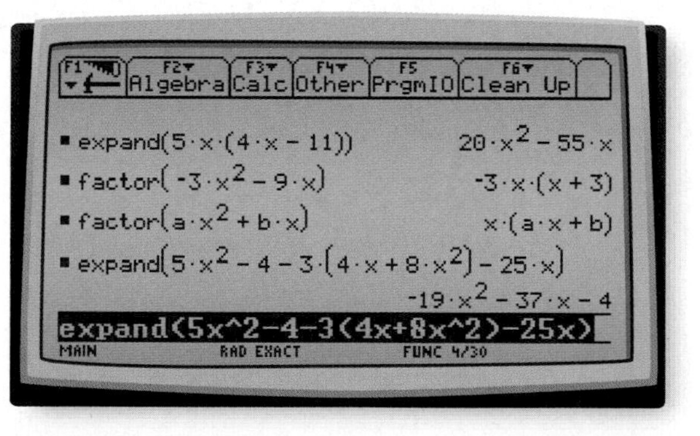

You can use such a computer algebra tool to check your answers as you learn how expanding and factoring work and when you meet problems that require complicated symbol manipulation.

4 In some situations, a quadratic expression arises as the product of two linear expressions. In those cases, you can use the distributive property twice to expand the factored quadratic to standard form. Study the steps in these examples, and then apply similar reasoning to expand the expressions in Parts a–e.

Strategy 1

$$\begin{aligned}(x + 5)(x - 7) &= (x + 5)x - (x + 5)7 \\ &= x^2 + 5x - 7x - 35 \\ &= x^2 - 2x - 35\end{aligned}$$

Strategy 2

$$\begin{aligned}(x + 5)(x - 7) &= x(x - 7) + 5(x - 7) \\ &= x^2 - 7x + 5x - 35 \\ &= x^2 - 2x - 35\end{aligned}$$

a. $(x + 5)(x + 6)$

b. $(x - 3)(x + 9)$

c. $(x + 10)(x - 10)$

d. $(x - 5)(x + 1)$

e. $(x + a)(x + b)$

5 The next five expressions have a special form $(x + a)^2$ in which both linear factors are the same. Use the distributive property to find equivalent expanded forms for each given expression and look for a consistent pattern in the calculations. Remember $(x + a)^2 = (x + a)(x + a)$.

a. $(x + 5)^2$

b. $(x - 3)^2$

c. $(x + 7)^2$

d. $(x - 4)^2$

e. $(x + a)^2$

6 The next four expressions also have a special form in which the product can be expanded to a standard-form quadratic. Use the distributive property to find expanded forms for each expression. Then look for a pattern and an explanation of why that pattern works.

a. $(x + 4)(x - 4)$

b. $(x + 5)(x - 5)$

c. $(3 - x)(3 + x)$

d. $(x + a)(x - a)$

7 When algebra students see the pattern $(x + a)(x - a) = x^2 - a^2$, they are often tempted to take some other "shortcuts" that lead to errors. How would you help another student to find and correct the mistakes in these calculations?

a. $(x + 5)(x - 3) = x^2 - 15$

b. $(m + 7)^2 = m^2 + 49$

 The next screen shows how a computer algebra system would deal with the task of expanding products of linear expressions like those in Problems 4–6. Compare these results to your own work and resolve any differences.

```
F1     F2▾     F3▾  F4▾   F5     F6▾
     Algebra Calc Other PrgmIO Clean Up

■ expand((x + m)·(x + p))
                       x^2 + m·x + p·x + m·p
■ expand((x + 5)^2)           x^2 + 10·x + 25
■ expand((x + 4)·(x - 4))             x^2 - 16
expand((x+4)*(x-4))
MAIN          RAD EXACT          FUNC 3/30
```

Summarize the Mathematics

In this investigation, you explored several of the most common ways that quadratic expressions can be written in equivalent factored and expanded forms.

a What is the standard expanded form equivalent to the product $2x(5x + 3)$?

b What is a factored form equivalent to $ax^2 + bx$?

c What is the standard expanded form equivalent to the product $(x + a)(x + b)$?

d What are the standard expanded forms equivalent to the products $(x + a)(x - a)$ and $(x + a)^2$?

Be prepared to compare your answers with those of your classmates.

Check Your Understanding

Write each of the following quadratic expressions in equivalent factored or expanded form.

a. $9x(4x - 5)$

b. $9x^2 + 72x$

c. $3(x^2 + 5x) - 7x$

d. $(x + 3)(x + 7)$

e. $(x + 2)^2$

f. $(x + 6)(x - 6)$

In Parts g and h, find values for the missing numbers that will make the given expressions equivalent.

g. $x^2 + 12x + \_\_ = (x + 4)(x + \_\_)$

h. $x^2 + \_\_x - 8 = (x + 4)(x + \_\_)$

On Your Own

Applications

1 Planners of a school fund-raising carnival considered the following factors affecting profit prospects for a rental bungee jump attraction:

- The number of customers n will depend on the price per jump x (in dollars) according to the linear function $n = 100 - x$.
- Insurance will cost $4 per jumper.
- Costs include $250 for delivery and setup and $100 to pay a trained operator to supervise use of the jumping equipment.

a. Complete a table like that begun here, showing number of customers, income, costs, and profit expected for various possible prices.

	Price per Jump (in $)						
	0	15	30	45	60	75	90
Number of Customers							
Income (in $)							
Insurance Cost (in $)							
Delivery/Setup Cost (in $)							
Operator Pay (in $)							
Profit (in $)							

b. Plot the (*price per jump, profit*) data. Then find a function that models the pattern relating those variables.

c. Write a rule showing how profit p depends on price per jump x by replacing each variable name in the following verbal rule with an expression using numbers and symbols:

profit = *income* − *insurance cost* − *delivery/setup cost* − *operator pay*

d. Check to see if the expressions for profit derived in Parts b and c are equivalent and explain how you reached your conclusion.

On Your Own

2 Students in a child development class at Caledonia High School were assigned the task of designing and building a fenced playground attached to their school as shown in the following sketch. They had a total of 150 feet of fencing to work with.

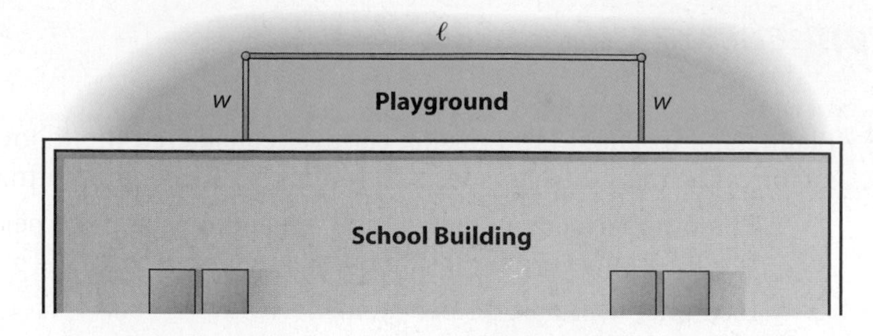

a. Complete a table like that begun here, showing how the length and area of the playground depend on choice of the width w.

Width (in feet)	10	20	30	40	50	60	70
Length (in feet)	130						
Area (in ft^2)	1,300						

b. Plot the data relating area to width and find the function that models the pattern in that relationship.

c. Write a function rule showing how length ℓ depends on width w and then another showing how area A depends on width.

d. Compare the two functions from Parts b and c relating area to width and decide whether they are equivalent. Explain evidence or reasoning that supports your answer.

e. Graph the area function and estimate the value of w that will produce the playground with largest possible area. Find the corresponding value of ℓ.

3 In many mountainous places, rope bridges provide the only way for people to get across fast rivers and deep valleys. A civil engineering class at a Colorado university got interested in one such rope bridge located in the mountains near their campus.

They came up with a function that they believed would give the distance in feet from the bridge to the river at any point. The function proposed was $d = 0.02x(x - 100) + 110$, where x measures horizontal distance (0 to 80 feet) from one side of the river to the other.

a. Use the given function to calculate the distance from the bridge to the river below at points 0, 10, 20, 30, 40, 50, 60, 70, and 80 feet from one end of the bridge. Sketch a graph showing the bridge shape in relation to the mountain sides and to the river below.

b. Estimate the low point of the bridge and its height above the water.

c. One brave student decided to check the proposed model of the distance from the bridge to the river below. She walked across the bridge and used a range-finding device to get data relating bridge height to horizontal distance. Her data are shown in the following table.

Horizontal Distance *x* (in feet)	15	25	35	45	55	65	75
Distance to the River *d* (in feet)	85	70	65	60	60	65	75

Find a function that models the pattern in these data well.

d. Compare the function proposed by the civil engineering students (who used only a few data points to derive their model) to that based on the range-finder data and decide whether you think the two models are equivalent or nearly so.

e. Write the first function $d = 0.02x(x - 100) + 110$ in standard quadratic form and explain how that form either supports or undermines your decision in Part d.

4 Write each of the following quadratic expressions in equivalent standard form.

a. $(3x + 4)x$

b. $m(3m - 15)$

c. $2p(3p - 1)$

d. $3d(5d + 2) + 29$

5 Write each of these quadratic expressions in equivalent form as the product of two linear factors.

a. $3x^2 + 9x$

b. $2x - 5x^2$

c. $-7d^2 - 9d$

d. $cx + dx^2$

6 Write each of these quadratic expressions in two equivalent forms—one expanded and one factored—so that both are as short as possible.

a. $2x(5 - 3x) + 4x$

b. $-3(2s^2 + 4s) - (3s + 5)7s$

c. $(9m + 18)m - 3m^2 - 5m$

d. $6x(8x + 3) + 4(2x - 7) - 2x$

7 Expand each of the following products to equivalent expressions in standard quadratic form.

a. $(x + 2)(x + 7)$

b. $(p + 2)(p - 2)$

c. $(x + 6)(x - 6)$

d. $(x + 6)(x + 6)$

e. $(R + 1)(R - 4)$

f. $(m - 7)(7 + m)$

8 Expand each of the following products to equivalent expressions in standard quadratic form.

a. $(t + 9)(t - 5)$

b. $(m + 1)^2$

c. $(x + 9)(9 - x)$

d. $(3x + 6)(3x - 6)$

Connections

9 The diagrams below are vertex-edge graphs that you can think of as maps that show cities and roads connecting them. In the first two "maps," every city can be reached from every other city by a direct road. In the third "map," every city can be reached from every other city, but some trips would require passing through another city along the way.

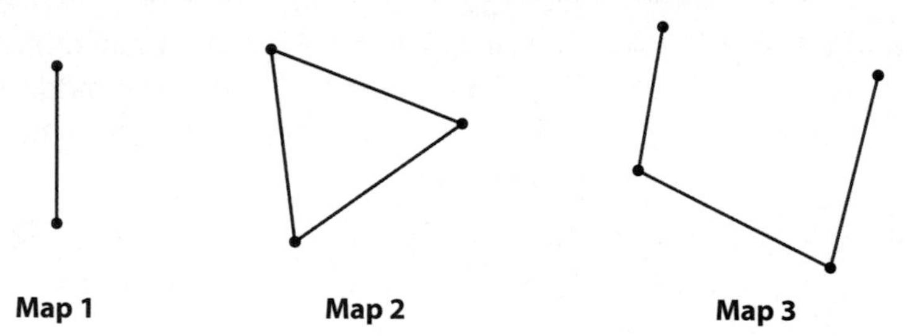

a. Sketch maps with 5, 6, and 7 cities and the smallest number of connecting roads to enable travel from any city on the map to any other city. Record the (*number of cities, number of roads*) data in a table like this:

Number of Cities *c*	2	3	4	5	6	7	8	9
Number of Roads *r*	1	2	3					

i. Use the pattern of results from the sketches to find a rule for calculating the number of roads r for any number of cities c if the number of connecting roads is to be a minimum in each case.

ii. Describe the type of function relating r and c. Explain how the rule could be justified.

b. Next sketch maps with 4, 5, 6, and 7 cities and direct roads connecting each pair of cities. Record the (*number of cities, number of roads*) data in a table like this:

Number of Cities *c*	2	3	4	5	6	7	8
Number of Roads *r*	1	3					

i. Use the pattern of results from the sketches to find a rule for calculating the number of direct roads r for any number of cities c.

ii. Study the following argument:

Each of the c cities must be connected by a road to the $c - 1$ other cities, so it looks like there must be $c(c - 1)$ direct roads. But that counting will include each road twice, so the actual number of direct roads r in a map with c cities is given by $r = \frac{c(c - 1)}{2}$.

Do the reasoning and the resulting function seem right?

iii. Explain how you know that the results from data analysis in part i and from the reasoning approach in part ii are or are not equivalent.

10 Your work with the Pythagorean Theorem often involved expressions with radicals like $\sqrt{49}$ or $\sqrt{2x^2}$, where $x > 0$. Write each of the following radical expressions in equivalent form with the simplest possible whole number or expression under the radical sign.

a. $\sqrt{18}$ b. $\sqrt{\frac{9}{4}}$ c. $\sqrt{2x^2}$ d. $\sqrt{\frac{3x^2}{4}}$

11 You can think of expanding products of linear expressions in terms of geometric models based on the area formula for rectangles.

a. What is the expanded form of $(x + 2)(x + 4)$? How is that result shown in the following diagram? (*Hint:* How can the area of the whole rectangle be calculated in two ways?)

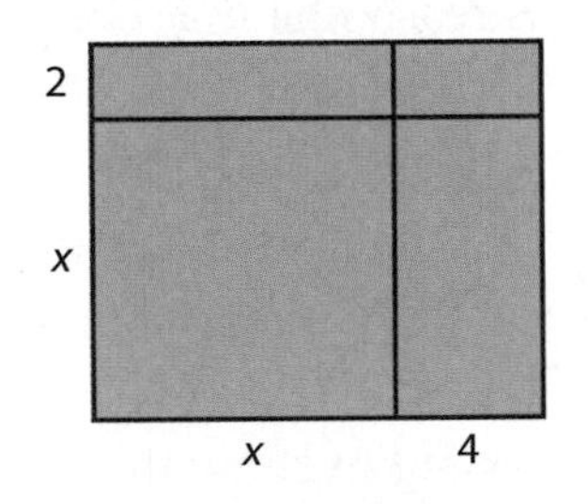

b. Find the expanded form of $(x + 3)(x + 7)$ and illustrate the result with a sketch similar to that given in Part a.

c. Make an area sketch like that in Part a to illustrate the general rule for expanding an expression in the form $(x + k)^2$.

d. What general result does the next sketch show?

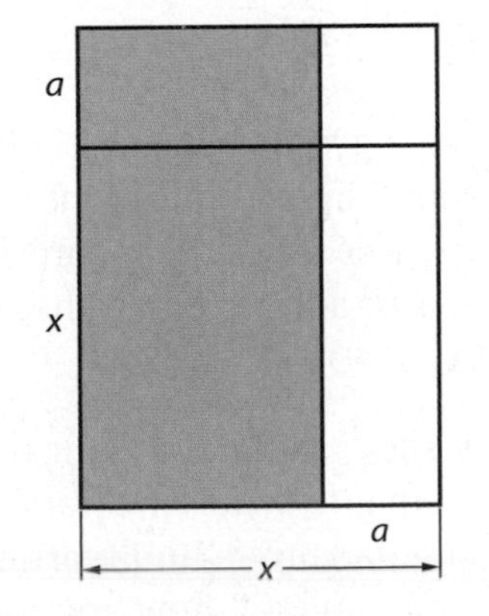

On Your Own

12 What kind of function do you think will result when two exponential functions are added or when the same two functions are multiplied? Use the examples in Parts a–d to develop conjectures from exploration of tables and graphs. Then summarize your ideas by answering Parts e and f, and use what you know about combining exponential expressions to confirm your ideas.

a. $y_1 = 1.5^x$ and $y_2 = 2^x$

b. $y_1 = 0.8^x$ and $y_2 = 2^x$

c. $y_1 = 0.5^x$ and $y_2 = 0.9^x$

d. $y_1 = 3^x$ and $y_2 = 2(3^x)$

Based on these examples (and others you might choose to test), how would you answer the questions in Parts e and f?

e. Is the sum of two exponential functions (always, sometimes, never) an exponential function?

f. Is the product of two exponential functions (always, sometimes, never) an exponential function?

13 When working with exponential growth functions it is often important to compare the value of the function at one time to the value at some future time. You can make the comparison by division or by subtraction.

a. What properties of numbers, operations, and exponents justify each step in this reasoning that claims $3^{x+1} - 3^x = 2(3^x)$?

$$\begin{aligned} 3^{x+1} - 3^x &= (3^x)(3^1) - (3^x)(1) && (1) \\ &= 3^x(3 - 1) && (2) \\ &= 2(3^x) && (3) \end{aligned}$$

b. Use similar reasoning to find an expression of the form $k(3^x)$ equivalent to $3^{x+2} - 3^x$.

c. Use similar reasoning to show that for any number n, $3^{x+n} - 3^x = 3^x(3^n - 1)$.

d. What property of exponents guarantees that for any x, $\frac{3^{x+1}}{3^x} = 3$?

e. What property of exponents guarantees that for any x, $\frac{3^{x+n}}{3^x} = 3^n$?

Reflections

14 When you are working with a quadratic function with rule like $y = 5x^2 + 15x$, what kinds of questions would be most easily answered using the rule in that standard form and what kinds of questions are easier to answer when the rule uses the equivalent expression $y = 5x(x + 3)$?

15 The claims below show some of the most common errors that people make when attempting to write quadratic expressions in equivalent forms by expanding, factoring, and rearranging terms. Spot the error(s) in each claim and tell how you would help the person who made the error correct his or her understanding.

a. Claim: $5x(4 + 3x)$ is equivalent to $23x$.

b. Claim: $7x - 5(2x + 4)$ is equivalent to $-3x + 20$.

c. Claim: $5x^2 + 50x$ is equivalent to $5x(x + 50)$.

d. Claim: $5x + 7x^2$ is equivalent to $12x^3$.

16 You know at least four different strategies for checking to see if two algebraic expressions are equivalent or not—comparing tables or graphs of (x, y) values, comparing the reasoning that led from the problem conditions to the expressions, or using algebraic reasoning based on number system properties like the distributive property.

a. What do you see as the advantages and disadvantages of each strategy?

b. How do you decide which strategy to use in a given situation?

c. Which strategy gives you most confidence in your judgment about whether the given expressions are or are not equivalent?

17 A rectangle is divided into 4 regions by equal-length segments as shown.

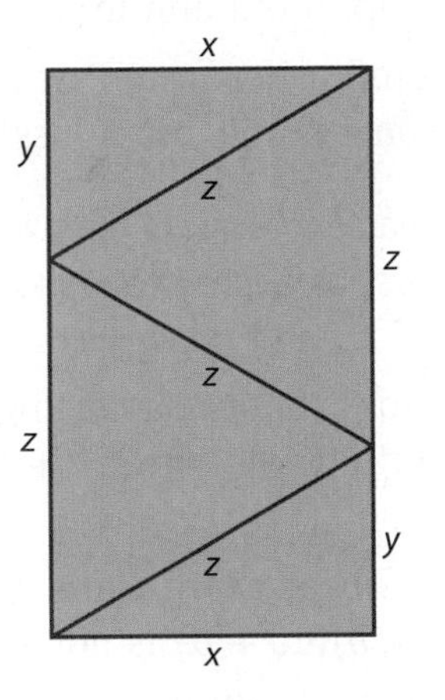

a. Bushra says the area of the rectangle is $x(y + z)$. Is she correct? If so, explain the reasoning she likely used.

b. Kareem says that the area of the rectangle is $\frac{1}{2}zx + \frac{1}{2}zx + \frac{1}{2}xy + \frac{1}{2}xy$. Is he correct? If so, explain his possible reasoning.

c. Fatmeh says that the area is $2x + 2y + 2z$. Is she correct? If so, explain her reasoning.

Extensions

18 Suppose that the conditions for operation of a rented bungee jump at the school carnival are as follows:

- The number of customers n will depend on the price per jump x (in dollars) according to the linear function $n = 80 - 0.75x$.
- Insurance will cost \$500 plus \$2 per jumper.
- Costs include \$250 for delivery and setup and \$3 per jumper to pay a trained operator to supervise use of the jumping equipment.

a. Write two rules for calculating projected profit for this attraction as a function of price per jump x. Write one rule in a form that shows how each income and cost factor contributes and another that is more efficient for calculation. Explain how you are sure that the two rules are equivalent.

b. Use one of the profit rules from Part a to estimate the price that will yield maximum profit and to find what that profit is.

c. Use one of the profit rules from Part a to estimate price(s) that will assure at least some profit (not a loss) for the attraction.

19 The sum of two linear expressions is always a linear expression, but the product of two linear expressions is *not always* a quadratic expression.

a. Use what you know about rearranging and combining terms in a linear expression to prove that $(ax + b) + (cx + d)$ is always a linear expression.

b. Find examples of linear expressions $ax + b$ and $cx + d$ for which the product $(ax + b)(cx + d)$ is not a quadratic expression.

c. Under what conditions will the product of two linear expressions *not be* a quadratic expression?

20 Consider the following mathematical question.

If a 5-ft tall person stands in one spot on the equator of Earth for 24 hours, how much farther will that person's head travel than his or her feet as Earth rotates about its axis?

a. Before doing any calculations, which of the following would you guess as an answer to the question?

1 foot	30 feet	300 feet	3,000 feet
1 mile	30 miles	300 miles	3,000 miles

b. Use the facts that the radius of Earth is about 4,000 miles at the equator and there are 5,280 feet in one mile to find the answer.

c. Use algebraic reasoning to answer this more general question:

If a tower that is k feet tall stands upright on the equator of a sphere of radius r feet, how much farther will the top of the tower travel than the base of the tower as the sphere makes one complete revolution about its axis?

Show that your answer works for the case of the person standing on Earth's equator.

d. The next sketch shows a disc inside a shaded ring. If the disc has radius 1 inch and the ring adds 0.25 inches to the radius of the figure, what is the area of the shaded ring?

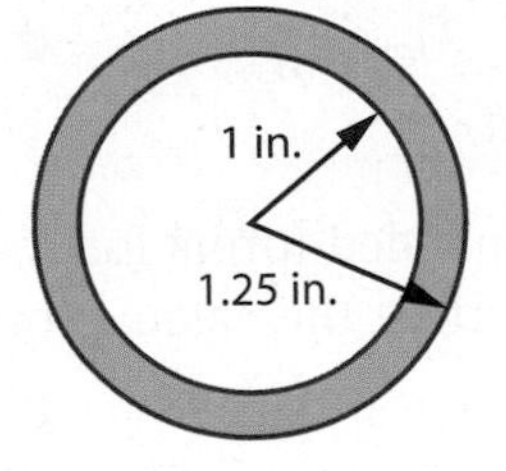

e. If a disc of radius r is inside a shaded ring that adds k to the radius, what rule gives the area of the shaded ring? Express that rule using the simplest possible expression involving r and k.

f. Natasha thought you might be able to calculate the area of the shaded ring by thinking about unwrapping the ring from around the disc. She said it would be pretty close to a rectangle, so its area could be estimated by multiplying the circumference of the ring by the width of the ring. She thought of three possible calculations to estimate this area.

$$2\pi rk \qquad 2\pi(r + k)k \qquad 2\pi\left(r + \frac{k}{2}\right)k$$

i. What thinking would have led Natasha to each of these expressions for area of the shaded ring?

ii. Which, if any, of the expressions gives a correct way of estimating the area of the shaded ring?

21 Study the geometric design in the figure below—a square with sides of length s surrounded by four congruent rectangles.

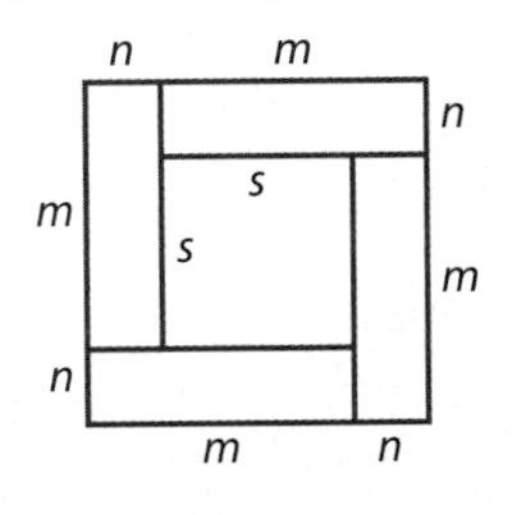

a. Express s in terms of m and n.

b. Express the area of the large square in two equivalent ways—each a function of m and n.

c. Equate the two expressions for area you found in Part b. Then explain how this equation implies that $(m + n)^2 \geq 4mn$ for any positive numbers m and n.

d. Use the result in Part c to explain why $\frac{m + n}{2} \geq \sqrt{mn}$ for any positive numbers m and n. This is known as the *arithmetic mean-geometric mean inequality*.

22 Expand each expression and look for a pattern to shortcut the calculations.

a. $(3x + 5)(2x + 1)$ **b.** $(5x - 3)(x + 4)$
c. $(-2x + 7)(4x - 3)$ **d.** $(7x - 4)(x + 2)$
e. $(ax + b)(cx + d)$

23 Find equivalent expanded forms for each given expression and look for a pattern to shortcut the calculations.

a. $(3x + 5)^2$ **b.** $(5x - 3)^2$
c. $(-2x + 7)^2$ **d.** $(7x - 4)^2$
e. $(ax + b)^2$

24 Find equivalent expanded forms for each given expression and look for a pattern to shortcut the calculations.

a. $(3x + 5)(3x - 5)$ **b.** $(2x - 3)(2x + 3)$
c. $(-2x + 7)(-2x - 7)$ **d.** $(8 - 4x)(8 + 4x)$
e. $(ax + b)(ax - b)$

25 Consider all quadratic functions with rules of the form $y = ax^2 + bx$.

a. How can any expression $ax^2 + bx$ be written as a product of linear factors?

b. Why does the factored form in Part a imply that $ax^2 + bx = 0$ when $x = 0$ and when $x = \frac{-b}{a}$?

c. How does the information from Part b imply that the maximum or minimum point on the graph of any function $y = ax^2 + bx$ occurs where $x = \frac{-b}{2a}$?

Review

26 Solve these linear equations.

a. $3.5x + 5 = 26$ **b.** $7(2x - 9) = 42$
c. $\frac{5x}{2} + 12 = 99$ **d.** $3.5x - 8 = 12 + 5x$
e. $7x + 4 = 4(x + 7)$ **f.** $3(x + 2) = 1.5(4 + 2x)$

27 Amy's last 10 scores on 50-point quizzes are 30, 32, 34, 34, 35, 35, 35, 36, 40, 42.

a. She says she can find her mean score by adding 3.0, 3.2, 3.4, 3.4, 3.5, 3.5, 3.5, 3.6, 4.0, and 4.2. Her friend Bart says she has to add and divide the total by 10, $\frac{30 + 32 + 34 + 34 + 35 + 35 + 35 + 36 + 40 + 42}{10}$. Who is correct? Explain why.

b. Amy is hoping for a 44 on the next 50-point quiz. If so, she says she will add 4.4 to her current mean to get the new mean. Is she correct?

c. Amy says that if she gets a 44 on the next quiz, her median score will increase. Bart disagrees. Who is correct? Explain.

28 In the diagram at the right, $\overline{AB} \cong \overline{BC}$ and $m\angle 1 = 100°$. Use geometric reasoning to determine:

a. $m\angle 2$ **b.** $m\angle 4 + m\angle 5$

c. $m\angle 4$ **d.** $m\angle 3$

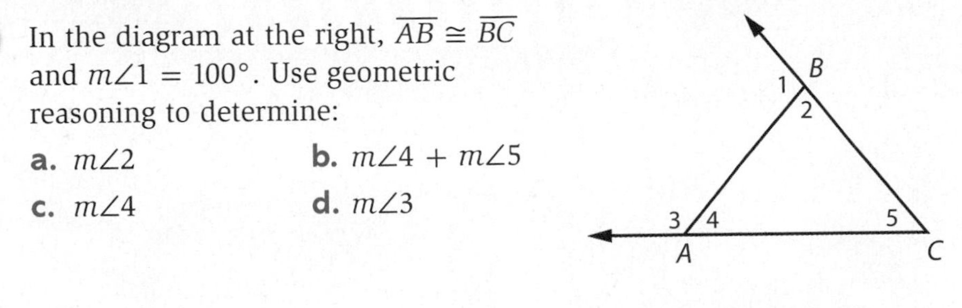

29 Try using various values of a, b, and c to help you determine if each pair of expressions are equivalent. If you think they are not equivalent, show this with an example using values for a, b, and c. If you think they are equivalent, use algebraic reasoning to transform one into the other.

a. $(a + b)^2$ and $a^2 + b^2$

b. $a(b - c)$ and $ab - ac$

c. $\sqrt{a^2 + b^2}$ and $a + b$

d. $\frac{\sqrt{9a^2}}{3}$ and a $(a \geq 0)$

e. $(ab)^c$ and $a^c b^c$

f. $\sqrt[3]{a^3 b^6}$ and ab^2

30 Alex takes two servings of fruit with him every day to school. Suppose that there are always apples, oranges, grapes, and watermelon in his house.

a. List all the possibilities for his daily fruit if he wants two different types of fruit.

b. How much longer would your list be if he didn't care if the types of fruit were different? Explain your reasoning.

31 Draw a triangle that has a perimeter of 12 cm.

a. How long are the sides of your triangle?

b. Suppose that another student in your class drew a triangle with the same side lengths as yours. Will the two triangles be guaranteed to be congruent? Explain your reasoning.

c. Heather wants to draw a right triangle that has perimeter 12 cm. Give possible lengths of the sides that she could use. Explain how you know they will give her a right triangle.

d. Christine indicates that the triangle she drew has sides of length 7 cm, 3 cm, and 2 cm. Is that possible? Explain why or why not.

32 Explain, as precisely as you can, why a rectangle with a pair of consecutive sides the same length is a square.

LESSON 3

Solving Quadratic Equations

Many key questions about quadratic functions require solving equations. For example, you used the function $h = 8 + 40.8t - 16t^2$ to answer questions about the flight of a long basketball shot.

To find the time when the shot would reach the 10-foot height of the basket, you solved the equation $8 + 40.8t - 16t^2 = 10$.

To find the time when an "air ball" would hit the floor, you solved the equation $8 + 40.8t - 16t^2 = 0$.

The values of t that satisfy the equations are called the **solutions** of the equations.

In each problem, you could get good estimates of the required solutions by searching in tables of values or by tracing coordinates of points on the graphs of the height function.

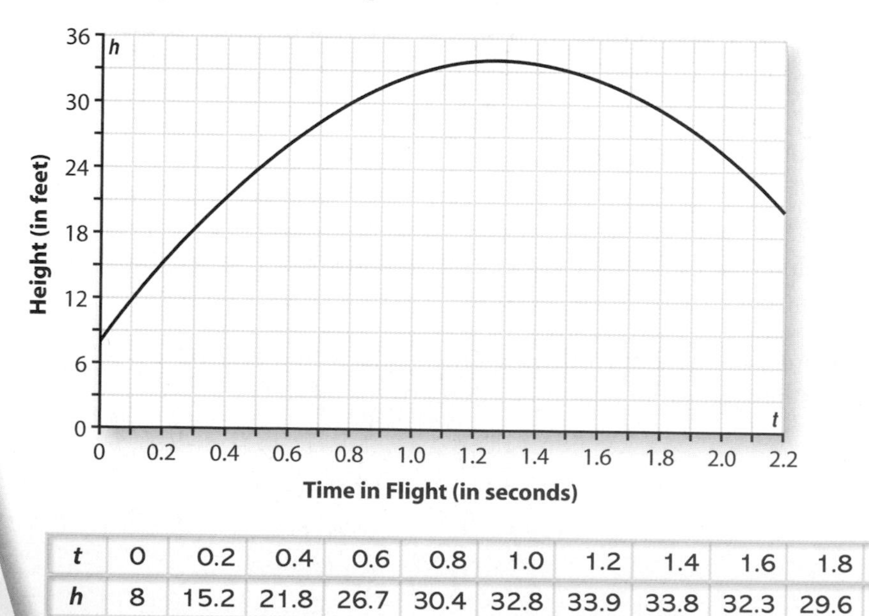

t	0	0.2	0.4	0.6	0.8	1.0	1.2	1.4	1.6	1.8	2.0	2.2
h	8	15.2	21.8	26.7	30.4	32.8	33.9	33.8	32.3	29.6	25.6	20.3

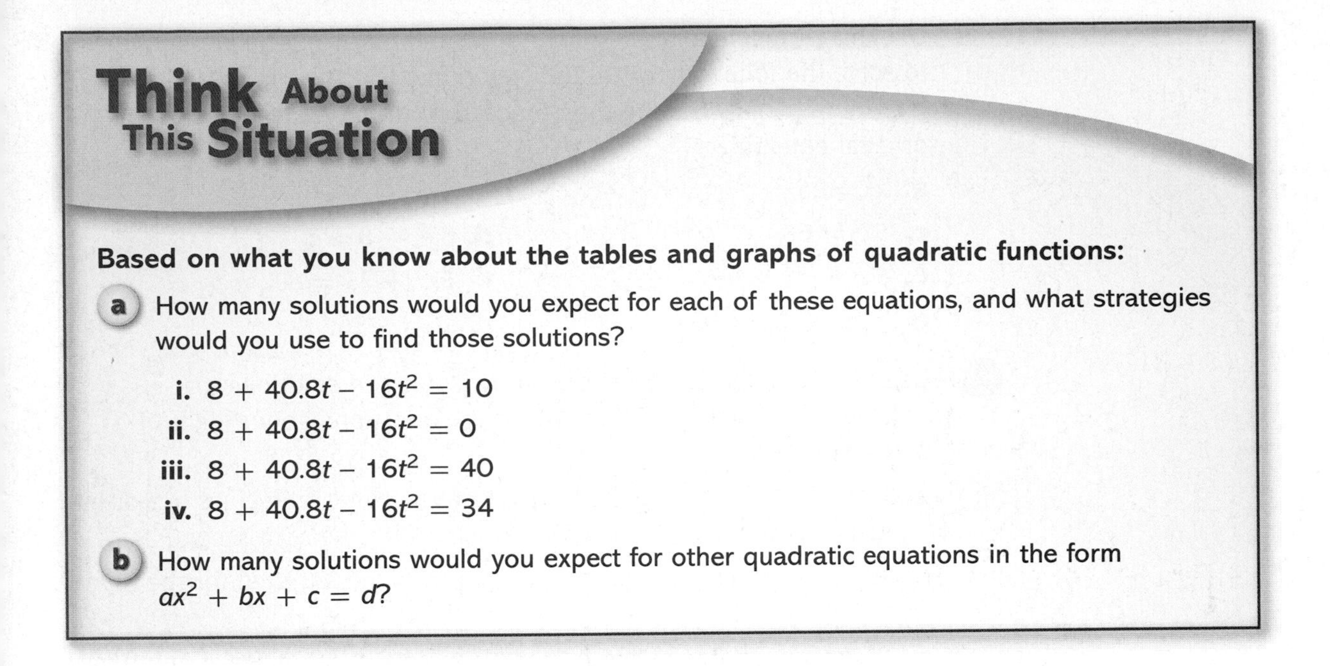

Think About This Situation

Based on what you know about the tables and graphs of quadratic functions:

a How many solutions would you expect for each of these equations, and what strategies would you use to find those solutions?

i. $8 + 40.8t - 16t^2 = 10$

ii. $8 + 40.8t - 16t^2 = 0$

iii. $8 + 40.8t - 16t^2 = 40$

iv. $8 + 40.8t - 16t^2 = 34$

b How many solutions would you expect for other quadratic equations in the form $ax^2 + bx + c = d$?

In this lesson, you will explore methods for finding exact solutions of quadratic equations.

Investigation 1 Solving $ax^2 + c = d$ and $ax^2 + bx = 0$

Calculator or computer tables, graphs, and symbol manipulation tools are helpful in finding approximate or exact solutions for quadratic equations and inequalities. But there are times when it is easier to use algebraic reasoning alone. As you work on the problems of this investigation, look for answers to this question:

What are some effective methods for solving quadratic equations algebraically?

1 Some quadratic equations can be solved by use of the fact that for any positive number n, the equation $x^2 = n$ is satisfied by two numbers: $\sqrt{n}$ and $-\sqrt{n}$.

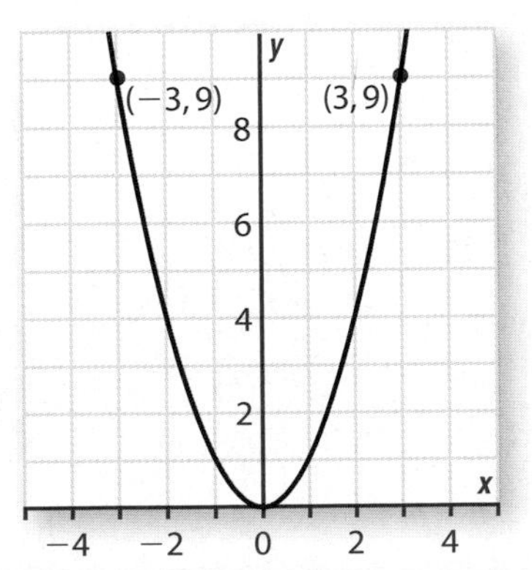

Use this principle and what you know about solving linear equations to solve the following quadratic equations. In each case, check your reasoning by substituting the proposed solution values for x in the original equation.

a. $x^2 = 25$

b. $x^2 = 12$

c. $5x^2 = 60$

d. $5x^2 + 8 = 8$

e. $5x^2 + 15 = 60$

f. $5x^2 + 75 = 60$

g. $-5x^2 + 75 = 60$

h. $x^2 = -16$

2 Use the methods you developed in reasoning to solutions for equations in Problem 1 to answer these questions about flight of a platform diver.

a. If the diver jumps off a 50-foot platform, what rule gives her or his distance fallen d (in feet) as a function of time t (in seconds)?

b. Write and solve an equation to find the time required for the diver to fall 20 feet.

c. What function gives the height h (in feet) of the diver at any time t (in seconds) after she or he jumps from the platform?

d. Write and solve an equation to find the time when the diver hits the water.

3 If a soccer player kicks the ball from a spot on the ground with initial upward velocity of 24 feet per second, the height of the ball h (in feet) at any time t seconds after the kick will be approximated by the quadratic function $h = 24t - 16t^2$. Finding the time when the ball hits the ground again requires solving the equation $24t - 16t^2 = 0$.

a. Check the reasoning in this proposed solution of the equation.

i. The expression $24t - 16t^2$ is equivalent to $8t(3 - 2t)$. Why?

ii. The expression $8t(3 - 2t)$ will equal 0 when $t = 0$ and when $3 - 2t = 0$. Why?

iii. So, the solutions of the equation $24t - 16t^2 = 0$ will be 0 and 1.5. Why?

b. Adapt the reasoning in Part a to solve these quadratic equations.

i. $0 = x^2 + 4x$

ii. $0 = 3x^2 + 10x$

iii. $0 = x^2 - 4x$

iv. $-x^2 - 5x = 0$

v. $-2x^2 - 6x = 0$

vi. $x^2 + 5x = 6$

4 Solving quadratic equations like $3x^2 - 15 = 0$ and $3x^2 - 15x = 0$ locates x-intercepts on the graphs of the quadratic functions $y = 3x^2 - 15$ and $y = 3x^2 - 15x$.

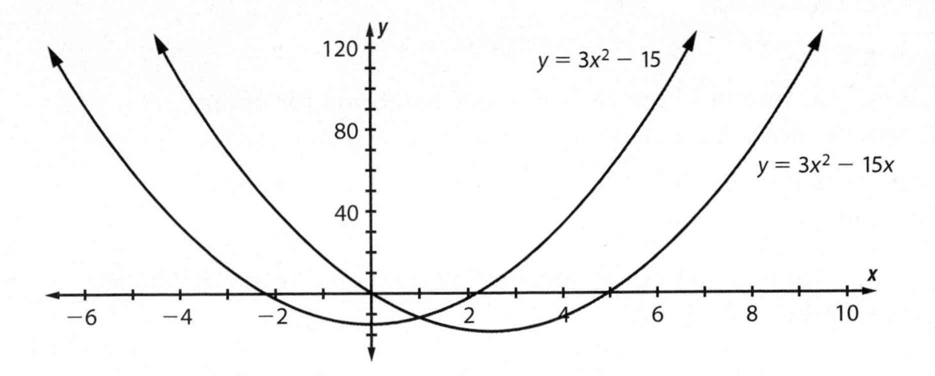

a. Using the graphs above, explain how the symmetry of these parabolas can be used to relate the location of the minimum (or maximum) point on the graph of a quadratic function to the x-intercepts.

b. Use the results of your work in Problem 3 to find coordinates of the maximum or minimum points on the graphs of these quadratic functions.

i. $y = x^2 + 4x$ **ii.** $y = 3x^2 + 10x$

iii. $y = x^2 - 4x$ **iv.** $y = -x^2 - 5x$

v. $y = -2x^2 - 6x$ **vi.** $y = ax^2 + bx$

5 Use what you know about solving quadratic equations and the graphs of quadratic functions to answer these questions.

a. What choices of values for a and d will give equations in the form $ax^2 = d$ that have two solutions? Only one solution? No solutions? Explain the reasoning behind your answers and illustrate that reasoning with sketches of graphs for the related function $y = ax^2$.

b. What choices of values for a, c, and d will give equations in the form $ax^2 + c = d$ that have two solutions? Only one solution? No solutions? Explain the reasoning behind your answers and illustrate that reasoning with sketches of graphs for the related function $y = ax^2 + c$.

c. Why must every equation in the form $ax^2 + bx = 0$ (neither a nor b zero) have exactly two solutions? Explain the reasoning behind your answer and illustrate that reasoning with sketches of graphs for the related function $y = ax^2 + bx$.

Summarize the Mathematics

In this investigation, you learned how to find exact solutions for some forms of quadratic equations by algebraic reasoning.

a Describe a process that uses rules of algebra to find solutions for any quadratic equation in the form $ax^2 + c = d$.

b Describe a process that uses rules of algebra to find solutions for any quadratic equation in the form $ax^2 + bx = 0$.

c What are the possible numbers of solutions for equations in the form $ax^2 = d$? For equations in the form $ax^2 + c = d$? For equations in the form $ax^2 + bx = 0$?

d How can graphs of quadratic functions in the form $y = ax^2$, $y = ax^2 + c$, and $y = ax^2 + bx$ be used to illustrate your answers to Part c?

e How can you locate the maximum or minimum point on the graph of a quadratic function with rule in the form:

i. $y = ax^2$ **ii.** $y = ax^2 + c$ **iii.** $y = ax^2 + bx$

Be prepared to share your ideas and reasoning strategies with the class.

Check Your Understanding

Quadratic equations, like linear equations, can often be solved more easily by algebraic reasoning than by estimation using a graph or table of values for the related quadratic function.

a. Solve each of these quadratic equations algebraically.

i. $3x^2 = 36$

ii. $-7x^2 = 28$

iii. $5x^2 + 23 = 83$

iv. $-2x^2 + 4 = 4$

v. $7x^2 - 12x = 0$

vi. $x^2 + 2x = 0$

b. Use algebraic methods to find coordinates of the maximum and minimum points on graphs of these quadratic functions.

i. $y = -5x^2 - 2$

ii. $y = 5x^2 + 3$

iii. $y = x^2 - 8x$

iv. $y = -x^2 + 2x$

Investigation 2 The Quadratic Formula

Many problems that require solving quadratic equations involve *trinomial* expressions like $15 + 90t - 16t^2$ that are not easily expressed in equivalent factored forms. So, solving equations like

$$15 + 90t - 16t^2 = 100$$

(When is a flying pumpkin 100 feet above the ground?)

is not as easy as solving equations like those in Investigation 1.

Fortunately, there is a **quadratic formula** that shows how to find all solutions of any quadratic equation in the form $ax^2 + bx + c = 0$. For any such equation, the solutions are

$$x = \frac{-b}{2a} + \frac{\sqrt{b^2 - 4ac}}{2a} \text{ and } x = \frac{-b}{2a} - \frac{\sqrt{b^2 - 4ac}}{2a}.$$

In Course 3 of *Core-Plus Mathematics*, you will prove that the quadratic formula gives the solutions to any quadratic equation. For now, to use the quadratic formula in any particular case, all you have to do is

- be sure that the quadratic expression is set equal to 0 as is prescribed by the formula;
- identify the values of a, b, and c; and
- substitute those values where they occur in the formula.

As you work on the problems in this investigation, make notes of answers to these questions:

What calculations in the quadratic formula give information on the number of solutions of the related quadratic equation?

What calculations provide information on the x-intercepts and maximum or minimum point of the graph of the related quadratic function?

1 Solve each quadratic equation by following the procedure for applying the quadratic formula.

- Give the values of a, b, and c that must be used to solve the equations by use of the quadratic formula.
- Evaluate $\frac{-b}{2a}$ and $\frac{\sqrt{b^2 - 4ac}}{2a}$.
- Evaluate $x = \frac{-b}{2a} + \frac{\sqrt{b^2 - 4ac}}{2a}$ and $x = \frac{-b}{2a} - \frac{\sqrt{b^2 - 4ac}}{2a}$.
- Check that the solutions produced by the formula actually satisfy the equation.

a. $2x^2 - 2x - 12 = 0$

b. $15 + 90t - 16t^2 = 100$

2 Test your understanding and skill with the quadratic formula by using it to find solutions for the following equations. In each case, check your work by substituting proposed solutions in the original equation and by sketching a graph of the related quadratic function to show how the solutions appear as points on the graph.

a. $x^2 - 7x + 10 = 0$

b. $x^2 - x - 8 = 0$

c. $-x^2 - 3x + 10 = 0$

d. $2x^2 - 12x + 18 = 0$

e. $13 - 6x + x^2 = 0$

f. $-x^2 - 4x - 2 = 2$

The formula for calculating solutions of quadratic equations is a complex set of directions. You can begin to make sense of the formula, by connecting it to patterns in the graphs of quadratic functions.

Consider the related quadratic functions:

$y = 2x^2 - 12x$
$y = 2x^2 - 12x + 10$
$y = 2x^2 - 12x - 14$
$y = 2x^2 - 12x + 24$

The graphs of these functions are shown in the following diagram.

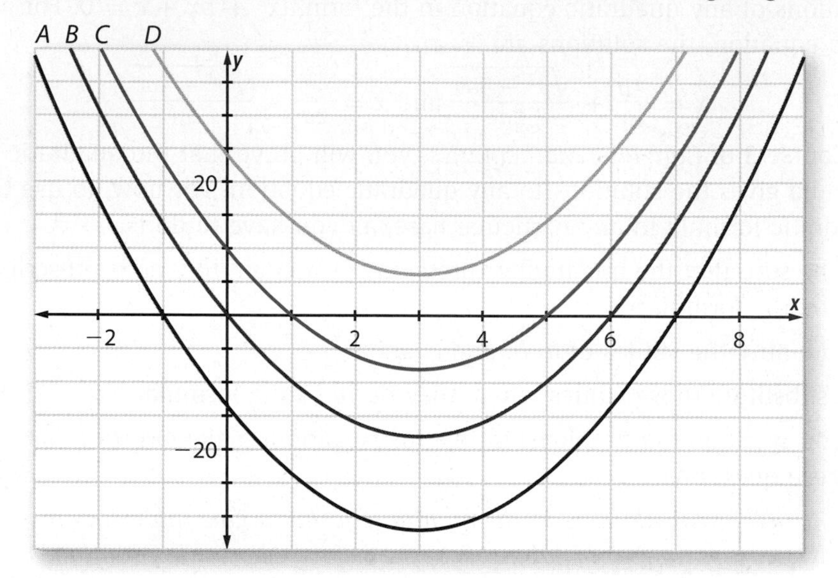

a. Match each function with its graph.

b. What are the x-coordinates of the minimum points on each graph? How (if at all) are those x-coordinates related to the x-intercepts of the graphs?

c. What are the coordinates of the x-intercepts on the graph of a quadratic function with rule $y = ax^2 + bx$? What is the x-coordinate of the minimum or maximum point for such a graph?

d. How will the x-coordinate of the maximum or minimum point of $y = ax^2 + bx + c$ be related to that of $y = ax^2 + bx$?

Now look back at the quadratic formula and think about how the results of Problem 3 help to explain the connections between these parts of the formula and the graph of $y = ax^2 + bx + c$.

a. $\dfrac{-b}{2a}$

b. $\dfrac{\sqrt{b^2 - 4ac}}{2a}$

c. $\dfrac{-b}{2a} + \dfrac{\sqrt{b^2 - 4ac}}{2a}$

d. $\dfrac{-b}{2a} - \dfrac{\sqrt{b^2 - 4ac}}{2a}$

Summarize the Mathematics

This investigation focused on a method for finding exact solutions for any quadratic equation.

- **a** What are the key steps in using the quadratic formula to solve a quadratic equation?
- **b** How can you tell from the calculation $\sqrt{b^2 - 4ac}$ that is part of the quadratic formula when a given quadratic equation has 2, 1, or 0 solutions? Give specific examples that illustrate the three possibilities.
- **c** How are the locations of the maximum or minimum point and x-intercepts of a quadratic function graph shown by calculations in the quadratic formula?
- **d** How do you choose among solution strategies—estimation using a table or graph, using arithmetic operations and square roots, factoring, or the quadratic formula—for the different quadratic equations that arise in solving problems?

Be prepared to share your thinking with the entire class.

Check Your Understanding

Use what you have learned about the quadratic formula to complete the following tasks.

a. Solve each of these equations. Show or explain the steps in each solution process.

i. $x^2 - 6x + 8 = 0$

ii. $3x^2 - 8 = -2$

iii. $x^2 - 2x + 8 = 2$

iv. $-7x + 8 + x^2 = 0$

v. $15 = 7 + 4x^2$

vi. $x^2 = 36$

b. Suppose that a group of students made these statements in a summary of what they had learned about quadratic functions and equations. With which would you agree? For those that you don't believe to be true, what example or argument would you offer to correct the proposer's thinking?

i. "Every quadratic equation has two solutions."

ii. "The quadratic formula cannot be applied to an equation like $-7x + 8 + x^2 = 0$."

iii. "To use the quadratic formula to solve $-7x + 8 + x^2 = 0$, you let $a = -7$, $b = 8$, and $c = 1$."

iv. "To use the quadratic formula to solve $x^2 - 6x + 8 = 0$, you let $a = 0$, $b = 6$, and $c = 8$."

v. "To use the quadratic formula to solve $3x^2 + 5x - 7 = 8$, you let $a = 3$, $b = 5$, and $c = -7$."

On Your Own

Applications

1. Solve each of these quadratic equations by using only arithmetic operations and square roots. Show the steps of your solution process and a check of the solutions.

   a. $x^2 = 20$

   b. $s^2 + 9 = 25$

   c. $x^2 - 11 = -4$

   d. $3m^2 + 9 = 5$

   e. $-2x^2 + 24 = 2$

   f. $29 - 3n^2 = 5$

2. An engineer designed a suspension bridge so that the main cables would lie along parabolas defined by the function $h = 0.04x^2 + 15$ where h is the distance from the cable to the bridge surface at a point x feet from the *center* of the bridge. The bridge is to be 50 feet long.

   a. Sketch a graph of the function for $-25 \leq x \leq 25$ and indicate on that sketch the height of each tower and the shortest distance from the cable to the bridge surface.

   b. Write and solve an equation to answer the question, "At what location(s) on the bridge surface is the suspension cable exactly 20 feet above the surface?"

3. The exterior of the St. Louis Abbey Church in Missouri shows a collection of parabolic faces.

Suppose that an arch in the lower ring of the parabolic roof line is defined by the function $y = -0.075x^2 + 30$, with y giving the distance in feet from the roof to the ground at any point x feet from the line of symmetry for the arch.

   a. What is the maximum height of the arch and how can you find that without any calculation?

b. Write and solve an equation that determines the distance from the line of symmetry to the points where the arch would meet the ground.

c. Write and solve an equation that determines the points where the roof is exactly 15 feet above the ground.

4 Solve each of these quadratic equations algebraically. Show your work.

a. $5x^2 + 60x = 0$

b. $-5x^2 + 23x = 0$

c. $-12x + 7x^2 = 0$

d. $2x - x^2 = 0$

5 Without graphing, find coordinates of the maximum and minimum points on graphs of these quadratic functions.

a. $y = 5x^2 + 60x$

b. $y = -5x^2 + 23x$

c. $y = -12x + 7x^2$

d. $y = 2x - x^2$

6 In football, when a field goal attempt is kicked, it leaves the ground on a path for which the height of the ball h in feet at any time t seconds later might be given by a function like $h = 45t - 16t^2$.

a. Write and solve an equation that tells time(s) when the ball hits the ground at the end of its flight.

b. Write and solve an equation that tells time(s) when the ball is at the height of the end zone crossbar (10 feet above the ground).

c. Find the maximum height of the kick and when it occurs.

7 Using the quadratic function $y = x^2 - 3x + 2$, choose values of y to write equations that have the prescribed number of solutions. In each case, show on a graph of the function how the condition is satisfied.

a. Two solutions

b. One solution

c. No solutions

8 Test your understanding and skill with the quadratic formula by using it to find solutions for the following equations. In each case, check your work by substituting proposed solutions in the original equation and by sketching a graph of the related quadratic function.

a. $x^2 + 4x + 5 = 0$

b. $x^2 - 7x + 8 = -2$

c. $x^2 - x - 8 = 4$

d. $5x + x^2 - 3 = 0$

e. $3x^2 - 18x + 27 = 0$

f. $5x^2 - x = 8$

Connections

9 The Pythagorean Theorem says that in any right triangle with legs of length a and b and hypotenuse of length c, $c^2 = a^2 + b^2$. Write and solve quadratic equations that provide answers for these questions:

a. What is the length of the hypotenuse of a right triangle with legs of length 5 and 7?

b. If a right triangle has one leg of length 9 and the hypotenuse of length 15, what is the length of the other leg?

c. If a right triangle has hypotenuse of length 20 and one leg of length 10, what is the length of the other leg?

d. What are the lengths of the sides of a square that has diagonal length 15?

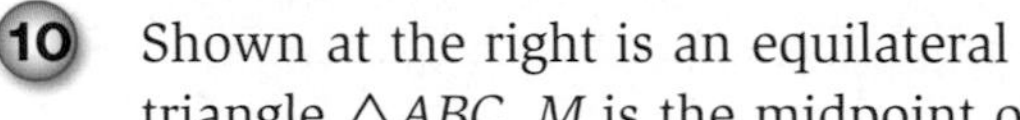

10 Shown at the right is an equilateral triangle $\triangle ABC$. M is the midpoint of $\overline{AC}$.

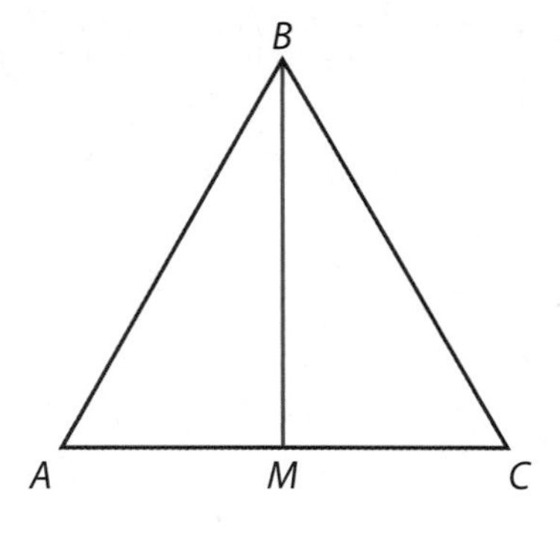

a. What are the measures of the angles in $\triangle ABM$?

b. If side $\overline{AB}$ has length 8, what is the length of side $\overline{AM}$? What is the length of side $\overline{MB}$?

c. If side $\overline{AB}$ has length x, what are the lengths of sides $\overline{AM}$ and $\overline{MB}$ in terms of x?

11 At the right is square $ABCD$ with one diagonal drawn in.

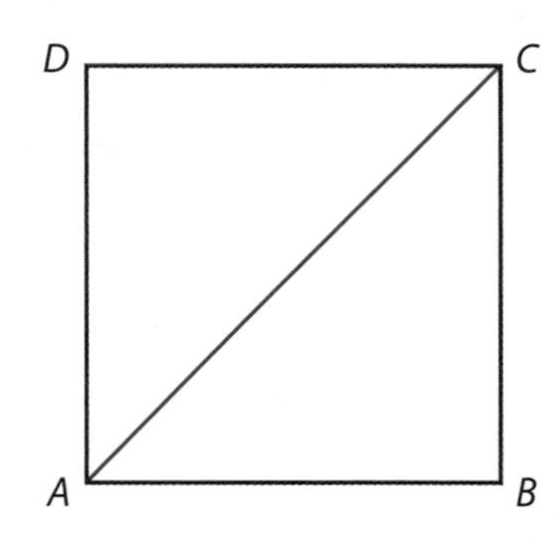

a. If side $\overline{AB}$ has length 6, what is the length of the diagonal $\overline{AC}$?

b. If side $\overline{AB}$ has length x, show that the length of the diagonal $\overline{AC}$ is $x\sqrt{2}$.

12 Consider all linear equations of the type $2x + 1 = k$.

a. What are the greatest and the least numbers of solutions for such equations?

b. How can your answer to Part a be supported with a sketch of the graph for the function $y = 2x + 1$?

13 Consider all equations of the type $5(2^x) = k$. How many solutions can there be for such equations? Give equations with specific values of k that illustrate the possibilities. How can you support your answer with a graph of the function $y = 5(2^x)$?

14 The quadratic formula gives a rule for finding all solutions of equations in the form $ax^2 + bx + c = 0$. Now consider all linear equations of the type $ax + b = k$ (where $a \neq 0$). Write a rule that shows how to calculate all solutions of such linear equations, using arithmetic operations involving the values of a, b, and k.

Reflections

15 What are the possible numbers of solutions for a quadratic equation? How would you use the graph of a quadratic function to explain those possibilities?

16 Mathematicians call "$x^2 + 5x + 6$" an expression and "$x^2 + 5x + 6 = 0$" an equation. How would you explain the difference between *expressions* and *equations*?

17 How are the possible numbers of solutions for a quadratic equation seen by examining the quadratic formula?

18 You now know three different ways to solve quadratic equations using algebraic methods: by reducing the problem to solving $x^2 = n$, by reducing the problem to solving $x(x + m) = 0$, and by using the quadratic formula.

a. What do you see as the advantages and disadvantages of each strategy?

b. Give an example of a quadratic equation that you would prefer to solve by factoring.

c. Give an example of a quadratic equation that you would prefer to solve by use of the quadratic formula.

Extensions

19 Find all solutions of the following equations that involve products of three or four linear terms. Then sketch graphs of the functions involved in these equations for $-5 \leq x \leq 5$ and explain how the solutions to the equations are shown on the graphs. Compare the pattern relating graphs and equations in these examples to the patterns you met in dealing with quadratic functions.

a. $x(x - 4)(x + 2) = 0$

b. $(2x + 3)(x - 1)(x + 3) = 0$

c. $x(2x + 3)(x - 1)(x + 3) = 0$

d. $x(x - 4)(x + 2) = -9$

20 Shown at the right are the dimensions for a multi-use gift box.

a. Will a pen 11 cm long fit in the bottom of the box? Explain why or why not.

b. Find the length of the longest pencil that will fit inside the box. Illustrate and explain how you found your answer.

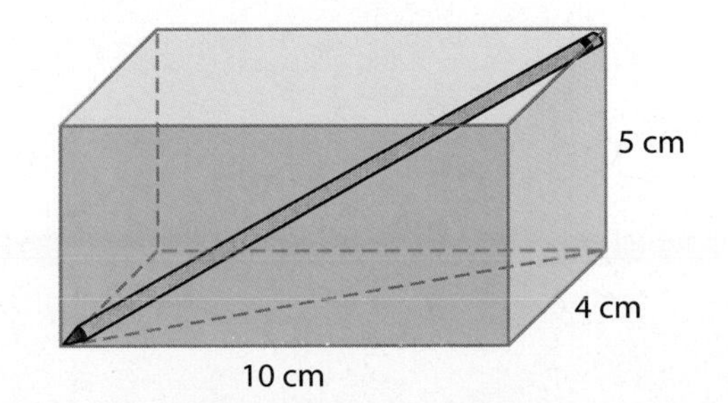

21 What values of x satisfy these quadratic inequalities?

a. $9 - x^2 < 0$

b. $9 - x^2 > 5$

c. $x^2 - 5x < 0$

d. $x^2 - 5x > 0$

22 In Investigation 1, you considered a quadratic model for a soccer ball kicked from a spot on the ground using the model $h = 24t - 16t^2$, where height was measured in feet and time in seconds.

a. What question can be answered by solving the inequality $24t - 16t^2 > 8$?

b. Solve the inequality and answer the question you posed.

c. Write an inequality that can be used to answer the question, "At what times in its flight is the ball within 8 feet of the ground?"

d. Solve the inequality you wrote in Part c.

23 Find rules for the functions with the graphs given below. Be prepared to explain how you developed the rules in each case.

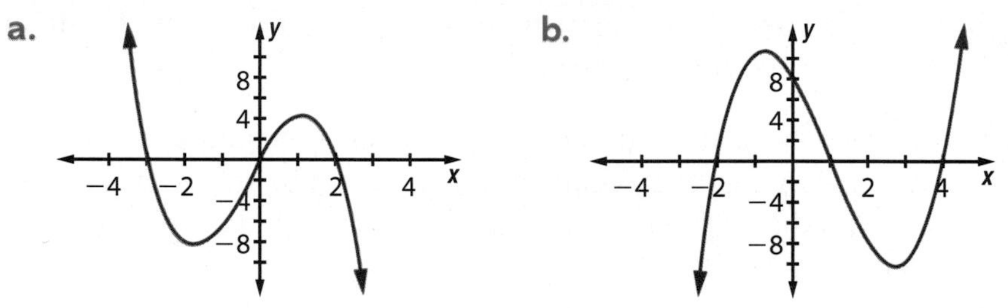

24 What results will you expect from entering these commands in a calculator or computer algebra system?

a. **Solve(a*x^2+c=d,x)**

b. **Solve(a*x^2+b*x=0,x)**

c. **Solve(a*x^2+b*x+c=0,x)**

Execute the commands and resolve any differences between what you expected and what occurred.

25 If you had a job that required solving many quadratic equations in the form $ax^2 + bx + c = 0$, it might be helpful to write a spreadsheet program that would do the calculations after you entered the values of a, b, and c. The table that follows was produced with one such spreadsheet and the equation $3x^2 - 6x - 24$.

Quadratic Equations.xls

◇	A	B	C	D
1	1	3	a =	3
2	Solution 1 =	−2	b =	−6
3	Solution 2 =	4	c =	−24
4				

a. What formulas would you expect to find in cells A1 and B1?

b. What formulas would you expect to find in cells B2 and B3?

c. How could you modify the spreadsheet to solve $-2x^2 + 4x - 7 = 0$?

26 The graphs of $y_1 = x^2 + x - 8$ and $y_2 = 2x + 4$ are shown at the right.

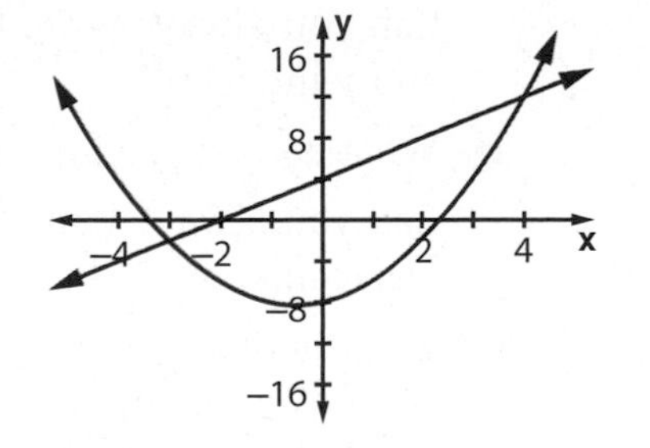

a. Estimate the coordinates of points where the two graphs intersect.

b. Solve the equation $x^2 + x - 8 = 2x + 4$ algebraically and compare the results to your estimate in Part a.

c. What are the possible numbers of solutions to an equation in the form *quadratic expression* = *linear expression*? Give examples and sketches of function graphs to illustrate your ideas.

Review

27 Write each of the following radical expressions in equivalent form with the smallest possible integer inside the radical sign.

a. $\sqrt{24}$

b. $\sqrt{125}$

c. $\sqrt{\frac{25}{16}}$

d. $\sqrt{12}\sqrt{8}$

e. $\sqrt[3]{-1{,}000}$

f. $\sqrt[4]{16}$

28 Examine this portion of a semiregular tessellation made with regular octagons and squares.

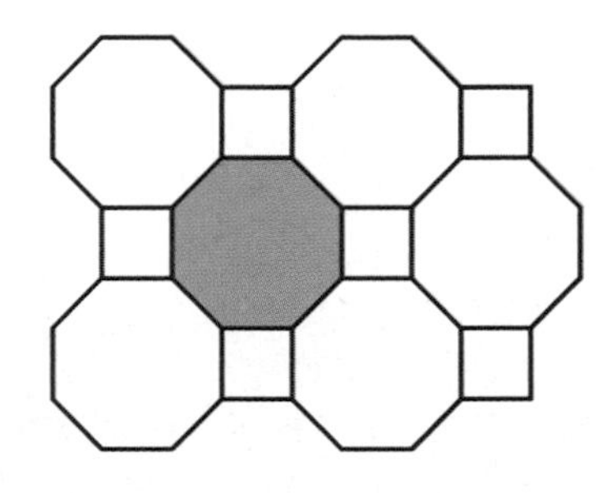

a. What is the measure of each angle in a regular octagon?

b. Explain why the octagons and squares in this semiregular tessellation fit together exactly.

c. If the shaded polygon is removed, in how many ways can the tile be replaced to fit the same outline, using both rotation and reflection?

29 Given that two vertex-edge graphs both have 8 edges and 6 vertices, which of the following statements is true? Give examples to support your answers.

a. The graphs must be identical.

b. Both graphs must have Euler circuits.

c. Both graphs must have Euler paths.

30 You know that the general form of a linear function is $y = mx + b$, and that the general form of an exponential function is $y = a(b^x)$.

a. What is the equation of a line with slope $\frac{2}{3}$ and y-intercept at (0, 3)?

Can you always write the equation of a line if you know its slope and y-intercept?

b. What is the equation of a line with slope $\frac{2}{3}$ and x-intercept at (3, 0)?

Can you always write the equation of a line if you know its slope and x-intercept?

c. What is the rule for an exponential function that has a graph with y-intercept at (0, 5) and growth factor 2?

Can you always write the rule for an exponential function if you know the y-intercept of its graph and the growth factor?

d. Is there an exponential function matching a graph that has x-intercept at (3, 0) and a decay factor of 0.5?

Can you always find an exponential function when given an x-intercept and a decay factor?

31 Determine if the triangles in each pair are congruent. If they are congruent, indicate how you know the triangles are congruent and write the congruence relation. If they are not congruent, explain how you know.

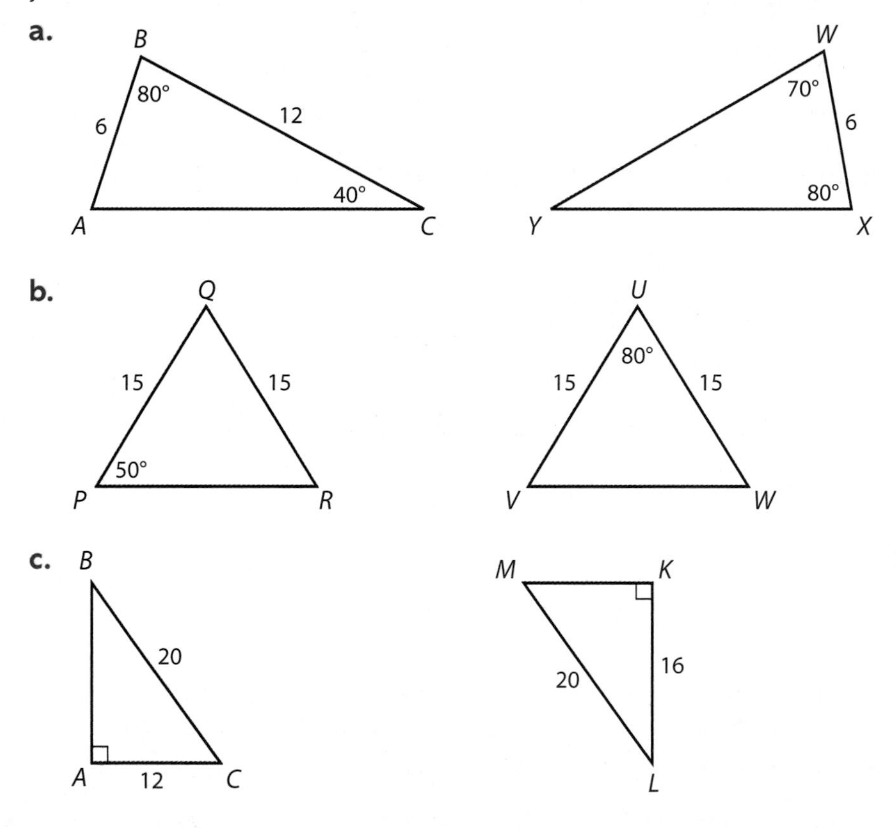

32 Write results of these calculations in standard scientific notation $N \times 10^x$, $1 \leq N < 10$.

a. $(5 \times 10^8)(4 \times 10^{11})$

b. $(5 \times 10^8)(4 \times 10^{-3})$

c. $(5 \times 10^8) \div (4 \times 10^5)$

33 Find the value of each absolute value expression.

a. $|-5|$

b. $-|6|$

c. $|7 - 4|$

d. $|4 - 7|$

e. $|3(-2) + 5|$

f. $|-3 - 10|$

34 In kite $ABCD$, $m\angle BAD = 78°$ and $m\angle BCD = 40°$. Calculate the measures of $m\angle ABC$ and $m\angle ADC$.

Looking Back

In your work on problems and explorations of this unit, you studied patterns of change in variables that can be modeled well by quadratic functions.

The height of a basketball shot increases to a maximum and then falls toward the basket or the floor, as a function of time after the shot is taken.

The projected income from a business venture rises to a maximum and then falls, as a function of price charged for the business product.

The main cables of a suspension bridge dip from each tower to a point where the bridge surface is closest to the water below.

The stopping distance for a car increases in a quadratic pattern as its speed increases.

The functions that model these patterns of change can all be expressed with rules like

$$y = ax^2 + bx + c.$$

Since quadratic expressions play such an important role in reasoning about relations among quantitative variables, you explored a variety of strategies for combining and expanding those expressions to produce useful equivalent forms. You also learned how to solve quadratic equations by algebraic reasoning and use of the quadratic formula.

As a result of your work on Lessons 1–3, you should be better able to recognize situations in which variables are related by quadratic functions, to use data tables and graphs to display patterns in those relationships, to use symbolic expressions to describe and reason about the patterns, and to use graphing calculators and computer algebra systems to answer questions that involve quadratic functions.

In future units of *Core-Plus Mathematics*, you will extend your understanding and skill in use of quadratic functions and expressions to solve problems. The tasks in this final lesson ask you to put your current knowledge to work in solving several new problems.

Mystic Mountain Mike and Tanya grew up in the mountains of Idaho, but they are now mathematics teachers in a large eastern city. They have a dream of going into business with a restaurant and entertainment complex they plan to call Mystic Mountain.

Not surprisingly, several of Mike and Tanya's ideas for Mystic Mountain involve quadratic functions and their graphs.

1. **The Entry Bridge** Their first idea is to have customers enter Mystic Mountain by walking across a rope bridge suspended over an indoor river that will be 40 feet wide. Tanya worked out the function $h = 0.02x^2 - 0.8x + 15$ to describe the arc of the bridge. In her rule, x gives horizontal distance from the entry point toward the other side of the river, and h gives height of the bridge above the water, both measured in feet.

   a. Sketch a graph of this function for $0 \leq x \leq 40$. On the sketch, show the coordinates of the starting and ending points and the point where the bridge surface is closest to the water.

   b. Show how coordinates for those key points on the graph can be calculated exactly from the rule, not simply estimated by tracing the graph or scanning a table.

   c. Write and estimate solutions for an equation that will locate point(s) on the bridge surface that are 10 feet above the water.

2. **Mike's Water Slide** Mike wanted to design a parabolic water slide for customers. The curve he wanted is shown in the following sketch. He wanted to have the entry to the slide at the point $(-10, 35)$, the low point to be $(0, 5)$, and the exit point to be $(5, 12.5)$.

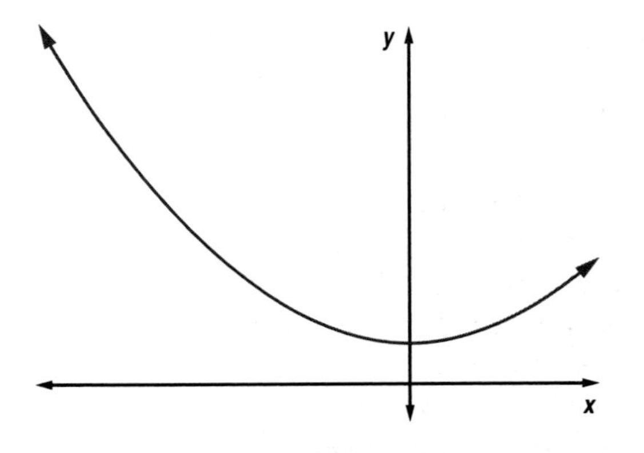

   a. What rule will define a function with a graph in the shape that Mike wants?

   b. Use the rule from Part a to write and solve an equation that gives the x-coordinate(s) of point(s) on the slide that are exactly 20 feet above the ground.

   c. Mike tried to find the rule he wanted by looking first at functions like $y = ax^2$. He started by trying to find a so that $(-10, 30)$ and $(0, 0)$ would be points on the graph.

      i. How do you suppose Mike was thinking about the problem that led him to this approach?

      ii. How could he have figured out the value of a with that start?

      iii. How could he then adjust the rule so the graph would go through the point $(-10, 35)$?

      iv. Verify that the graph of the adjusted rule in part iii goes through the point $(5, 12.5)$.

3 **Flying Off the Slide** Mike figured that people would come off the end of his water slide at the point (5, 12.5) with an upward speed of about 20 feet per second. He wanted to know about the resulting flight into the air and down to a splash in the pool lying at the end of the slide.

a. What rule would give good estimates of the slider's height above the water as a function of time (in seconds) after they leave the end of the slide?

b. What would be the maximum height of the slider?

c. How long would it take the slider to make the trip from the end of the slide, into the air, and down to the water?

4 **Water Slide Business** Tanya liked Mike's water slide idea, but she wondered how much they should charge customers for the experience. She asked a market research company to survey how the number of customers would depend on the admission price.

a. The market research data suggested that daily number of customers n would be related to admission price x by $n = 250 - 10x$. What function shows how predicted income depends on admission price?

b. What price gives maximum daily income? Explain how you can locate that point by estimation and by reasoning that gives an exact answer.

c. If Mike and Tanya expect operation of the water slide attraction to cost $450 per day, what function shows how predicted daily profit depends on admission price?

d. What water slide admission price leads to maximum profit?

e. What water slide admission price leads to a break-even operation of the water slide? Explain how you can locate the point(s) by estimation and by reasoning that gives an exact answer.

5 Use algebraic methods to find exact solutions for these quadratic equations. Show the steps in your reasoning and a check of each solution.

a. $2x^2 + 11x + 12 = 0$

b. $6x^2 + 10x = 0$

c. $2x^2 + 11 = 19$

Summarize the Mathematics

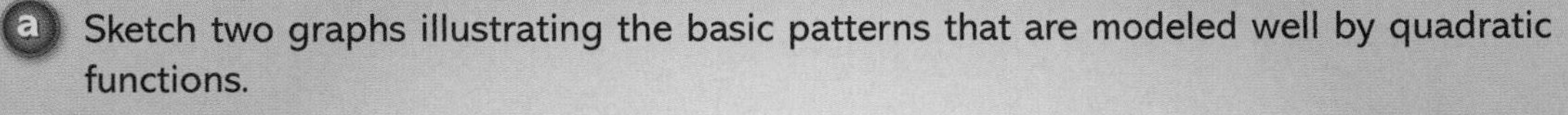

When two variables are related by a quadratic function, that relationship can be recognized from patterns in tables and graphs of (x, y) data, from the rules that show how to calculate values of one variable from given values of the other, and from key features of the problem situations.

a Sketch two graphs illustrating the basic patterns that are modeled well by quadratic functions.

- **i.** For each graph, write a brief explanation of the pattern shown in the graph and describe a problem situation that involves the pattern.
- **ii.** Then give "$y = \ldots$" rules that would produce each given graph pattern and explain how those rules alone could be used to determine the pattern of change in the dependent variable as the independent variable increases.

b Suppose that you develop or discover a rule that shows how a variable y is a quadratic function of another variable x. Describe the different strategies you could use to complete tasks like these:

- **i.** Find the value of y associated with a specific given value of x.
- **ii.** Find the value of x that gives a specific target value of y.
- **iii.** Describe the way that the value of y changes as the value of x increases or decreases.

c What information about the graph of a quadratic function can be obtained easily from each of these types of rules?

- **i.** $y = ax^2 + c$
- **ii.** $y = ax^2 + bx$
- **iii.** $y = ax(x - m)$

d For questions that call for solving quadratic equations,

- **i.** How many solutions would you expect, and how is that shown by the graphs of quadratic functions? By the quadratic formula?
- **ii.** How would you decide on a solution strategy?
- **iii.** How would you find the solution(s)?
- **iv.** How would you check the solution(s)?

Be prepared to share your examples and descriptions with the class.

✓ Check Your Understanding

Write, in outline form, a summary of the important mathematical concepts and methods developed in this unit. Organize your summary so that it can be used as a quick reference in future units and courses.

Patterns in Chance

Many events in life occur by chance. They cannot be predicted with certainty. For example, you cannot predict whether the next baby born will be a boy or a girl. However, in the long run, you can predict that about half of the babies born will be boys and half will be girls. Probability is the study of events that occur (or don't occur) at random from one observation to the next but that occur a fixed proportion of the time in the long run. In this unit, you will study two methods for solving problems in probability: calculating theoretical probabilities using mathematical formulas or using geometric models and estimating probabilities using simulation.

Key ideas for solving problems involving chance will be developed through your work in two lessons.

Lessons

1 *Calculating Probabilities*

Construct probability distributions from sample spaces of equally likely outcomes. Use the Addition Rule to solve problems involving chance.

2 *Modeling Chance Situations*

Design simulations using a table of random digits or random number generators to estimate answers to probability questions. Use geometric models to solve probability problems.

LESSON 1

Calculating Probabilities

Backgammon is the oldest game in recorded history. It originated before 3000 B.C. in the Middle East. Backgammon is a two-person board game that is popular among people of all ages. When it is your turn, you roll a pair of dice and move your checkers ahead on the board according to what the dice show. The object is to be the first to move your checkers around and then off of the board. It's generally a good thing to get doubles because then you get to use what the dice show twice instead of just once.

Think About This Situation

Suppose you and a friend are playing a game of backgammon.

a Which probability should be larger?

- the probability of rolling doubles on your first turn
- the probability of rolling doubles on your first turn or on your second turn

Explain your thinking.

b What do you think is the probability of rolling doubles on your first turn? Explain your reasoning.

c What assumptions are you making about the dice in finding the probability in Part b?

d Suppose you rolled doubles on each of your first three turns. Your friend did not roll doubles on any of his first three turns. Who has the better chance of rolling doubles on their next turn? Explain.

In this lesson, you will learn to construct probability distributions from sample spaces of *equally likely outcomes* and use them to solve problems involving chance.

Investigation 1 Probability Distributions

Because of the symmetry in a fair die—each side is equally likely to end up on top when the die is rolled—it is easy to find the probabilities of various outcomes. As you work on the problems in this investigation, look for answers to this question:

How can you find and organize the probabilities associated with random events like the roll of two dice?

Rolling Two Dice

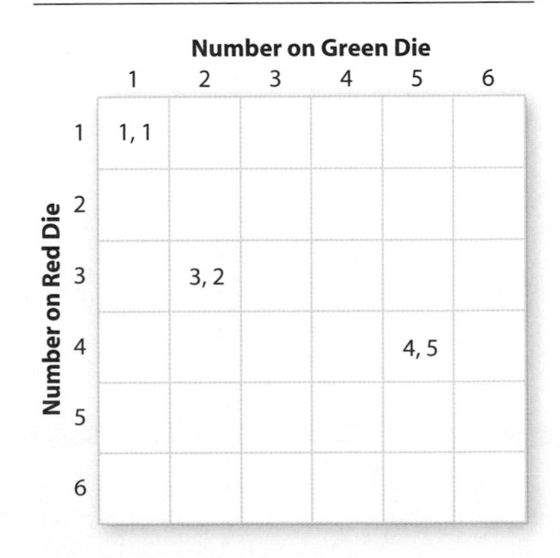

Number on Red Die \ Number on Green Die	1	2	3	4	5	6
1	1, 1					
2						
3		3, 2				
4					4, 5	
5						
6						

1 Suppose a red die and a green die are rolled at the same time.

a. What does the entry "3, 2" in the chart mean?

b. Complete a copy of the chart at the right, showing all possible outcomes of a single roll of two dice.

c. How many possible outcomes are there?

d. What is the probability of rolling a (1, 2), that is, a 1 on the red die and a 2 on the green die? What is the probability of rolling a (2, 1)? A (4, 4)?

e. Would the chart be any different if both dice had been the same color?

 The chart you completed in Problem 1 is called a sample space for the situation of rolling two dice. A **sample space** is the set of all possible outcomes. For fair dice, all 36 outcomes in the sample space are equally likely to occur. **Equally likely** means that each outcome has the same probability. When outcomes are equally likely, the probability of an event is given by

$$P(event) = \frac{\text{number of successful outcomes}}{\text{number of possible outcomes}}.$$

If two dice are rolled, what is the probability of getting

a. doubles?

b. a sum of 7?

c. a sum of 11?

d. a 2 on at least one die or a sum of 2?

e. doubles and a sum of 8?

f. doubles or a sum of 8?

 Suppose two dice are rolled.

a. What is the probability that the sum is no more than 9?

b. What is the probability that the sum is at least 9?

c. What is the probability that the sum is 2 or 3? Is greater than 3? Is at least 3? Is less than 3?

4 A **probability distribution** is a description of all possible numerical outcomes of a random situation, along with the probability that each occurs. A probability distribution differs from a sample space in that all of the outcomes must be a single number and the probabilities must be specified. For example, the probability distribution table below shows all possible sums that you could get from the roll of two dice.

Probability Distribution for the Sum of Two Dice

Sum	Probability
2	
3	
4	
5	
6	
7	
8	
9	
10	
11	
12	

a. Complete a copy of this probability distribution by filling in the probabilities.

b. What is the sum of all of the probabilities?

c. How could you use your probability distribution table to find each of the probabilities in Problem 3?

5. Other probability distributions can be made from the sample space in Problem 1 for the roll of two dice. Suppose that you roll two dice and record the larger of the two numbers. (If the numbers are the same, record that number.)

   a. Use your sample space from Problem 1 to help you complete a probability distribution table for this situation.

   b. What is the probability that the larger of the two numbers is 3? Is 2 or 3? Is 3 or less? Is more than 3?

6. Now suppose you roll two dice and record the absolute value of the difference of the two numbers.

   a. Use your sample space from Problem 1 to help you complete a probability distribution table for this situation.

   b. What is the probability that the absolute value of the difference is 3? Is 2 or 3? Is at least 2? Is no more than 2?

7. If you flip a coin, {heads, tails} is a sample space, but not a probability distribution. However, you can make a probability distribution by recording the number of heads as your outcome, as shown in the table below. Fill in the two missing probabilities.

Probability Distribution for the Number of Heads

Number of Heads	Probability
0	
1	

8. Now suppose that you flip a coin twice.

   a. Complete a chart that shows the sample space of all possible outcomes. It should look like the chart for rolling two dice except that only heads and tails are possible for each coin rather than the six numbers that are possible for each die.

   b. Make a probability distribution table that gives the probability of getting 0, 1, and 2 heads.

   c. What is the probability that you get exactly one head if you flip a coin twice? What is the probability that you get at least one head?

Summarize the Mathematics

In this investigation, you learned how to construct probability distributions from the sample space of equally likely outcomes from the roll of two dice and from the flip(s) of a coin.

a What is the difference between a sample space and a probability distribution?

b How would you make a probability distribution table for the product of the numbers from the roll of two dice?

c Why is it that the outcomes in the sample space for rolling two dice are equally likely?

Be prepared to share your ideas and reasoning with the class.

Check Your Understanding

Suppose that you flip a coin three times.

a. List the sample space of all 8 possible outcomes. For example, one outcome is heads, tails, tails (HTT).

b. Are the outcomes in your sample space equally likely? Explain.

c. Make a probability distribution table for the number of heads.

d. What is the probability that you will get exactly 2 heads? At most 2 heads?

Investigation 2 The Addition Rule

In the previous investigation, you constructed the probability distribution for the sum of two dice. You discovered that to find the probability that the sum is 2 or 3, you could *add* the probability that the sum is 2 to the probability that the sum is 3, $\frac{1}{36} + \frac{2}{36} = \frac{3}{36}$. As you work on the following problems, look for an answer to this question:

Under what conditions can you add individual probabilities to find the probability that a related event happens?

Some people have shoes of many different colors, while others prefer one color and so have all their shoes in just that color. As a class, complete a copy of the following two tables on the color of your shoes.

In the first table, record the number of students in your class that today are wearing each shoe color. (If a pair of shoes is more than one color, select the color that takes up the largest area on the shoes.)

Color of Shoes You Are Wearing Today	Number of Students
Blue	
Black	
White	
Brown, Beige, or Tan	
Red	
Other	

Now complete the second table by recording the number of students in your class who own a pair of shoes of that color. For example, a student who has all shoes in the colors blue or black would identify themselves for only those two colors.

Color of Shoes You Own	Number of Students
Blue	
Black	
White	
Brown, Beige, or Tan	
Red	
Other	

In mathematics, the word "or" means "one or the other or both." So, the event that *a student owns white shoes or owns black shoes* includes all of the following outcomes:

- The student owns white shoes but doesn't own black shoes.
- The student owns black shoes but doesn't own white shoes.
- The student owns both white and black shoes.

a. Which question below can you answer using just the data in your tables? Answer that question.

I What is the probability that a randomly selected student from your class is wearing shoes today that are black or wearing shoes that are white?

II What is the probability that a randomly selected student from your class owns shoes that are black or owns shoes that are white?

b. Why can't the other question in Part a be answered using just the information in the tables?

c. As a class, collect information that can be used to answer the other question.

2 The table below gives the percentage of high school sophomores who say they engage in various activities at least once a week.

Weekly Activities of High School Sophomores

Activity	Percentage of Sophomores
Use personal computer at home	71.2
Drive or ride around	56.7
Work on hobbies	41.8
Take sports lessons	22.6
Take class in music, art, language	19.5
Perform community service	10.6

Source: National Center for Education Statistics. *Digest of Education Statistics 2004, Table 138.* Washington, DC: nces.ed.gov/programs/digest/d04/tables/dt04_138.asp

Use the data in the table to help answer, if possible, each of the following questions. If a question cannot be answered, explain why not.

a. What is the probability that a randomly selected sophomore takes sports lessons at least once a week?

b. What is the probability that a randomly selected sophomore works on hobbies at least once a week?

c. What is the probability that a randomly selected sophomore takes sports lessons at least once a week or works on hobbies at least once a week?

d. What is the probability that a randomly selected sophomore works on hobbies at least once a week or uses a personal computer at home at least once a week?

3 You couldn't answer the "or" questions in Problem 2 by adding the numbers in the table. However, you could answer the "or" questions in Problems 3, 4, 5, and 6 of the previous investigation by adding individual probabilities in the tables. What characteristic of a table makes it possible to add the probabilities to answer an "or" question?

4 The Minnesota Student Survey asks teens questions about school, activities, and health. Ninth-graders were asked, "How many students in your school are friendly?" The numbers of boys and girls who gave each answer are shown in the table below.

How Many Students in Your School Are Friendly?

Answer	Boys	Girls	Total
All	480	303	783
Most	13,199	14,169	27,368
Some	7,920	8,874	16,794
A few	1,920	1,815	3,735
None	480	50	530
Total	23,999	25,211	49,210

Source: 2004 Minnesota Student Survey, Table 4, www.mnschoolhealth.com/resources.html?ac=data

Suppose you pick one of these students at random.

a. Find the probability that the student said that all students are friendly.

b. Find the probability that the student said that most students are friendly.

c. Find the probability that the student is a girl.

d. Find the probability that the student is a girl and said that all students are friendly.

e. Think about how you would find the probability that the student said that all students are friendly or said that most students are friendly. Can you find the answer to this question using your probabilities from just Parts a and b? If so, show how. If not, why not?

f. Think about the probability that the student is a girl or said that all students are friendly.

  i. Can you find the answer to this question using just your probabilities from Parts a and c? If so, show how. If not, why not?

  ii. Can you find the answer if you also can use your probability in Part d? If so, show how. If not, why not?

5 Two events are said to be **mutually exclusive** (or **disjoint**) if it is impossible for both of them to occur on the same outcome. Which of the following pairs of events are mutually exclusive?

a. You roll a sum of 7 with a pair of dice; you get doubles on the same roll.

b. You roll a sum of 8 with a pair of dice; you get doubles on the same roll.

c. Isaac wears white shoes today to class; Isaac wears black shoes today to class.

d. Yen owns white shoes; Yen owns black shoes.

e. Silvia, who was one of the students in the survey described in Problem 2, works on hobbies; Silvia takes sports lessons.

f. Pat, who was one of the students in the survey described in Problem 4, is a boy; Pat said most students in his school are friendly.

g. Bernardo, who was one of the students in the survey described in Problem 4, said all students are friendly; Bernardo said most students are friendly.

6 Suppose two events A and B are mutually exclusive.

a. Which of the *Venn diagrams* below better represents this situation?

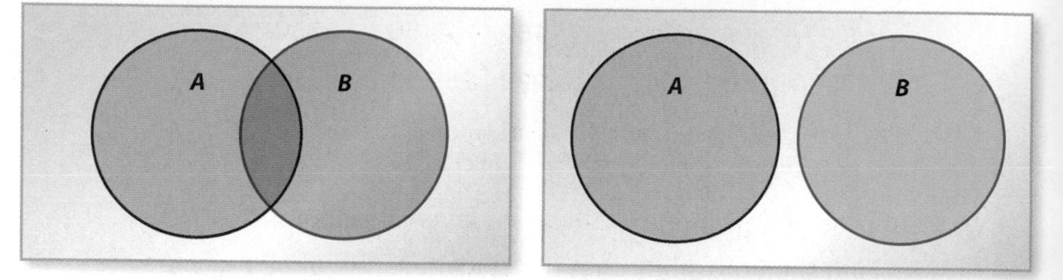

b. What does the fact that A and B are mutually exclusive mean about ***P*(*A* and *B*)**—the probability that A and B both happen on the same outcome?

c. When A and B are mutually exclusive, how can you find the probability that A happens or B happens (or both happen)?

d. Write a symbolic rule for computing the probability that A happens or B happens, denoted ***P*(*A* or *B*)**, when A and B are mutually exclusive. This rule is called the **Addition Rule for Mutually Exclusive Events**.

Suppose two events A and B are *not* mutually exclusive.

a. Which diagram in Problem 6 better represents this situation?

b. What does the fact that A and B are not mutually exclusive mean about $P(A \text{ and } B)$? Where is this probability represented on the Venn diagram you chose?

c. Review your work in Problems 1 and 4 and with the Venn diagram. Describe how you can modify your rule from Problem 6, Part d to compute $P(A \text{ or } B)$ when A and B are not mutually exclusive.

d. Write a symbolic rule for computing $P(A \text{ or } B)$. This rule is called the **Addition Rule**.

Test your rules on the following problems about rolling a pair of dice.

a. Find the probability that you get doubles or a sum of 5.

b. Find the probability that you get doubles or a sum of 2.

c. Find the probability that the absolute value of the difference is 3 or you get a sum of 5.

d. Find the probability that the absolute value of the difference is 2 or you get a sum of 11.

Summarize the Mathematics

In this investigation, you learned how to compute the probability that event *A* happens or event *B* happens on the same outcome.

a Give an example of two mutually exclusive events different from those in the investigation.

b How do you find the probability that event *A* happens or event *B* happens if events *A* and *B* are mutually exclusive?

c How must you modify the rule in Part b if the two events are not mutually exclusive?

Be prepared to share your ideas and reasoning with the class.

Check Your Understanding

Use what you have learned about mutually exclusive events and the Addition Rule to complete the following tasks.

a. Which of the following pairs of events are mutually exclusive? Explain your reasoning.

- **i.** rolling a pair of dice: getting a sum of 6; getting one die with a 6 on it
- **ii.** flipping a coin 7 times: getting exactly 3 heads; getting exactly 5 heads
- **iii.** flipping a coin 7 times: getting at least 3 heads; getting at least 5 heads

b. Use the appropriate form of the Addition Rule to find the probability of rolling a pair of dice and

- **i.** getting a sum of 6 or getting one die with a 6 on it.
- **ii.** getting a sum of 6 or getting doubles.

c. Janet, a 50% free-throw shooter, finds herself in a two-shot foul situation. She needs to make at least one of the shots.

- **i.** List a sample space of all possible outcomes. Are the outcomes equally likely?
- **ii.** Find the probability that she will make the first shot or the second shot.
- **iii.** What assumptions did you make about her shooting?

On Your Own

Applications

 The game of Parcheesi is based on the Indian game pachisi.

In Parcheesi, two dice are rolled on each turn. A player cannot start a pawn until he or she rolls a five. The five may be on one die, or the five may be the sum of both dice. What is the probability a player can start a pawn on the first roll of the dice?

2. Suppose you roll a die and then roll it again. The die has the shape of a regular tetrahedron and the numbers 1, 2, 3, and 4 on it.

   a. Make a chart that shows the sample space of all possible outcomes.

   b. How many possible outcomes are there? Are they equally likely?

   c. Make a probability distribution table for the difference of the two rolls (*first die* − *second die*).

   d. What difference are you most likely to get?

   e. What is the probability that the difference is at most 2?

3. Suppose that you roll a tetrahedral die and a six-sided die at the same time.

   a. Make a chart that shows the sample space of all possible outcomes.

   b. How many possible outcomes are there? Are they equally likely?

   c. Make a table for the probability distribution of the sum of the two dice.

   d. What sum are you most likely to get?

   e. What is the probability that the sum is at most 3?

4 Use your work from Applications Task 2 and the appropriate form of the Addition Rule to answer these questions about a roll of two tetrahedral dice.

a. What is the probability that you get a difference of 3 or you get a 2 on the first die?

b. What is the probability you get a difference of 2 or you get doubles?

c. What is the probability you get a difference of 0 or you get doubles?

d. What is the probability you get a difference of 0 or a sum of 6?

5 The Titanic was a British luxury ship that sank on its first voyage in 1912. It was en route from Southampton, England, to New York City. The table below gives some information about the passengers on the Titanic.

Passengers Aboard the Titanic

	Men	Women and Children	Total
Survived	138	354	492
Died	678	154	832
Total	816	508	1,324

Source: http://www.titanicinquiry.org/USInq/USReport/AmInqRep03.html#a8

a. Suppose a passenger is selected at random. Use the table above to find the probability of each of the following events.

i. The passenger is a man.

ii. The passenger survived.

iii. The passenger is a man and survived.

b. Now use your results from Part a and the appropriate form of the Addition Rule to find the probability that a randomly selected passenger is a man or a survivor. Check your answer by adding the appropriate entries in the table.

c. Suppose a passenger is selected at random. Find the probability of each of the following events.

i. The passenger is a woman/child.

ii. The passenger died.

iii. The passenger is a woman/child and died.

iv. The passenger is a woman/child or died.

6. In almost all states, it is illegal to drive with a blood alcohol concentration (BAC) of 0.08 grams per deciliter (g/dL) or greater. The table below gives information about the drivers involved in a crash in which someone died. The age of the driver is given in the left column. The BAC of the driver is given across the top row.

Drivers Involved in Fatal Crashes

Age of Driver (in years)	Number with BAC Lower than 0.08 g/dL	Number with BAC 0.08 g/dL or Higher	Total Number of Drivers
16–20	6,395	1,314	7,709
21–24	4,313	2,069	6,382
25–34	8,171	3,008	11,179
35–44	8,217	2,465	10,682
45–54	7,418	1,684	9,102
55–64	4,882	691	5,573
65–74	2,835	222	3,057
75+	2,999	143	3,142
Total	25,230	11,596	56,826

Source: *Traffic Safety Facts, 2004 Data*, National Center for Statistics and Analysis, U.S.

Suppose that you select a driver at random from these 56,826 drivers involved in fatal crashes.

a. Find the probability that the driver was age 16–20.

b. Find the probability that the driver was age 21–24.

c. Find the probability that the driver had a BAC of 0.08 or greater.

d. Find the probability that the driver was age 16–20 or was age 21–24.

e. Can you find the answer to Part d using just your probabilities from Parts a and b? Why or why not?

f. Find the probability that the driver was age 16–20 or had a BAC of 0.08 or greater.

g. Can you find the answer to Part f just by adding the two probabilities from Parts a and c? Why or why not?

Connections

7. Make a histogram of the information in the probability distribution table that you created for the sum of two dice in Problem 4 (page 534) of Investigation 1. Probability will go on the y-axis.

a. What is the shape of this distribution?

b. What is its mean?

c. Estimate the standard deviation using the approximation that about two-thirds of the probability should be within one standard deviation of the mean. (While this distribution isn't normal, this approximation works relatively well.)

8 Graph the points that represent the information in the probability distribution table that you created for the sum of two dice in Problem 4 (page 534) of Investigation 1. For example, the first point would be $\left(2, \frac{1}{36}\right)$ and the second point would be $\left(3, \frac{2}{36}\right)$.

a. Write a linear equation with a graph going through the points for $x = 2, 3, 4, 5, 6, 7$.

b. Write a second linear equation with a graph going through the points for $x = 7, 8, 9, 10, 11, 12$.

c. What are the slopes of these two lines?

9 Graph the points that represent the information in the probability distribution table that you created for the difference of two tetrahedral dice in Applications Task 2. For example, the first point would be $\left(-3, \frac{1}{16}\right)$ and the second point would be $\left(-2, \frac{2}{16}\right)$.

a. Write a linear equation with a graph going through the points for $x = -3, -2, -1, 0$.

b. Write a second linear equation with a graph going through the points for $x = 0, 1, 2, 3$.

c. What are the slopes of these two lines?

10 Recall that there are five regular polyhedra: tetrahedron (4 faces), hexahedron or cube (6 faces), octahedron (8 faces), dodecahedron (12 faces), and icosahedron (20 faces).

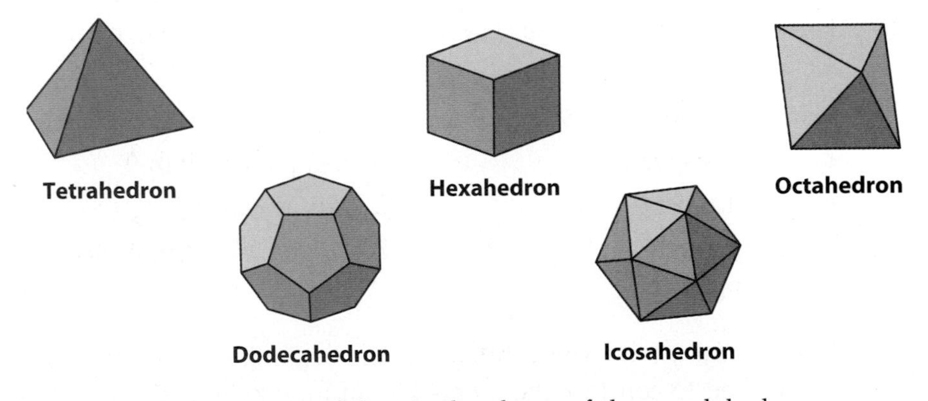

Find or imagine pairs of dice in the shape of these polyhedra. The numbers on the faces of the tetrahedron are 1, 2, 3, and 4. The hexahedron (regular die) has the numbers 1, 2, 3, 4, 5, and 6. The remaining three pairs of dice have the numbers 1 to 8, 1 to 12, and 1 to 20 on their faces, respectively.

a. Make a sample space chart like that in Investigation 1, Problem 1 (page 533) for the octahedral dice.

b. On which of the five pairs of dice is the probability of getting doubles the greatest? Explain why this is the case.

c. If the number of faces on a pair of regular polyhedral dice is n, what is the probability (in terms of n) of rolling doubles with that pair of dice?

d. For each type of dice, what is the mean of the probability distribution of the sum?

11 On Problem 6 on page 540, you saw how to draw Venn diagrams to represent the situation where events A and B are mutually exclusive and the situation where events A and B are *not* mutually exclusive. Now, think about three events, A, B, and C.

a. Draw a Venn diagram that represents the situation where A and B are mutually exclusive, A and C are mutually exclusive, and B and C are mutually exclusive.

b. Draw a Venn diagram that represents the situation where A and B are mutually exclusive, A and C are mutually exclusive, but B and C are not mutually exclusive.

c. Draw a Venn diagram that represents the situation where A and B are not mutually exclusive, A and C are not mutually exclusive, and B and C are not mutually exclusive.

Reflections

12 Some chance situations have exactly two outcomes.

a. Is it true that the two outcomes must be equally likely? Explain.

b. Spin a coin by holding it on edge and flicking it with your finger. Do this 30 times and record the results. Does it appear that heads and tails are equally likely? How could you decide for sure?

13 For each phrase below, select the corresponding symbolic phrase:

$x \geq 9$ $\quad$ $x > 9$ $\quad$ $x \leq 9$ $\quad$ $x < 9$

a. x is no more than 9

b. x is less than 9

c. x is 9 or greater

d. x is no less than 9

e. x is at most 9

f. x is at least 9

g. x is greater than 9

h. x is a maximum of 9

i. x is 9 or more

14 In the Think About This Situation at the beginning of this lesson, you were asked to consider the probability of rolling doubles on your first turn or on your second turn.

a. Is it true that the probability of rolling doubles on your first turn or your second turn is $\frac{1}{6} + \frac{1}{6}$? Give an explanation for your answer.

b. Is it true that the probability of rolling doubles on at least one of your first six turns is $\frac{1}{6} + \frac{1}{6} + \frac{1}{6} + \frac{1}{6} + \frac{1}{6} + \frac{1}{6}$?

15 Refer to Applications Task 6. The age group of 21–24 had only 6,382 drivers involved in fatal accidents. This is less than all age groups until age 55 and up. Does this mean the drivers in this age group are less likely to get in a fatal crash? Explain your answer.

16 Why does it make sense to title Investigation 2 "The Addition Rule" rather than "Addition Rules"?

Extensions

17 Suppose that you flip a coin four times and record head (H) or tail (T) in the order it occurs.

a. Make a list of all 16 possible outcomes.

b. Are these outcomes equally likely?

c. Make a table of the probability distribution of the number of heads.

d. What is the probability that you will get exactly 2 heads? At most 2 heads?

18 Flavia selects one of the special dice with faces shown below, and then Jack selects one of the remaining two.

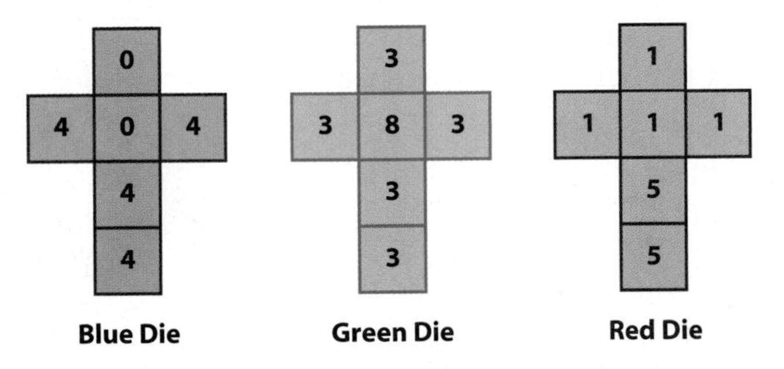

They each roll their die. The person with the larger number wins. To help you decide if it is better to use, for example, the blue die or the green die, you could complete a table like the following.

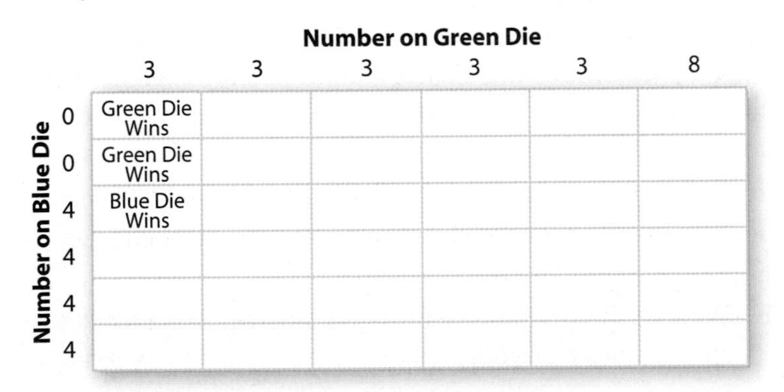

Number on Blue Die \ Number on Green Die	3	3	3	3	3	8
0	Green Die Wins					
0	Green Die Wins					
4	Blue Die Wins					
4						
4						
4						

Suppose Flavia picks the blue die. To have the best chance of winning, which die should Jack choose? If Flavia picks the green die, which die should Jack choose? If Flavia picks the red die, which die should Jack choose? What is the surprise here?

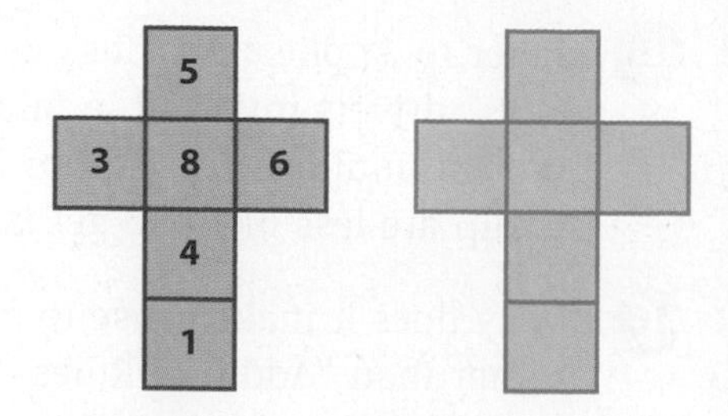

19 Refer to the probability distribution table for the sum of two standard six-sided dice that you constructed in Problem 4 (page 534) of Investigation 1. Two nonstandard six-sided dice have this same probability distribution table. A net for one of those nonstandard dice is shown above. Using positive whole numbers, label a copy of the net for the other nonstandard die. The other die may be different from the one given, and numbers may be repeated on its faces.

20 Refer to the Venn diagrams you made in Connections Task 11. Use the appropriate one to help you write an Addition Rule that you can use to determine $P(A \text{ or } B \text{ or } C)$ when

a. A and B are mutually exclusive, A and C are mutually exclusive, and B and C are mutually exclusive.

b. A and B are mutually exclusive, A and C are mutually exclusive, but B and C are not mutually exclusive.

c. A and B are not mutually exclusive, A and C are not mutually exclusive, and B and C are not mutually exclusive.

21 In the game of backgammon, if you "hit" your opponent's checker exactly, that checker must go back to the beginning and start again. For example, to hit a checker that is three spaces ahead of your checker, you need to move your checker three spaces. You can do this either by getting a 3 on one die or by getting a sum of 3 on the two dice. In addition, if you roll double 1s, you can also hit your opponent's checker because if you roll doubles, you get to move the numbers that show on the die twice each. So you move one space, then one space again, then one more space, hitting your opponent's checker, then move your last space. (If you roll, say, double 4s, you can't hit your opponent's checker—you must skip over it as you move your first four spaces.)

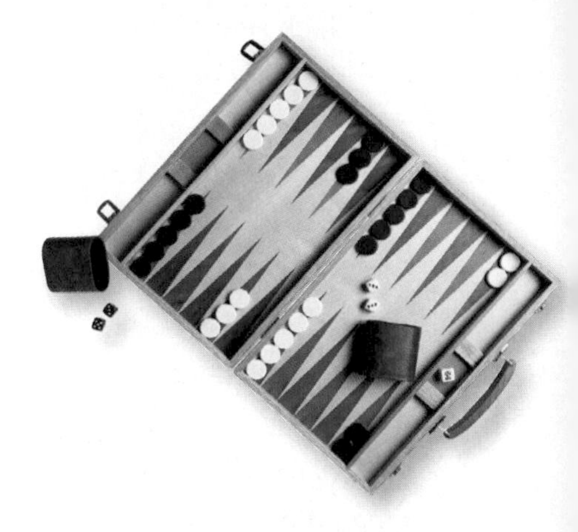

a. What is the probability of being able to hit a checker that is three spaces ahead of you?

b. What is the probability of being able to hit a checker that is five spaces ahead of you?

c. What is the probability of being able to hit a checker that is twelve spaces ahead of you?

d. If you want to have the best chance of hitting a checker ahead of you, how many spaces should it be in front of you?

Review

22 Determine if it is possible to draw zero, one, or more than one triangle that will fit the given description. Explain your reasoning.

a. triangle XYZ with $m\angle X = 120°$, $m\angle Y = 30°$, and $XY = 8$ cm

b. a triangle with two right angles

c. an isosceles triangle with legs of length 5 cm and base of length 4 cm

d. a triangle with sides of length 2 in., 3 in., and 5 in.

e. a triangle with angles measuring 45°, 65°, and 70°

f. a right triangle with sides of length 10 m, 6 m, and 7 m

23 The mean number of people who attended five high school basketball games was 468. The tickets cost $3 per person. What is the total amount received from ticket sales?

24 The height in feet of a punted football t seconds after a punt can be found using the equation $h = 1.8 + 50t - 16t^2$.

a. How long was the football in the air?

b. How high did the football go?

c. When was the football 20 feet above the ground?

25 Solve each inequality and graph the solution on a number line. Substitute at least one number from your solution set back into the inequality to check your work.

a. $3x + 7 \leq -59$

b. $16 \geq 12 - 2x$

c. $25 < 5(2x + 12)$

d. $4x + 18 > 6x - 24$

e. $\frac{1}{6}(5x + 12) - 8 \geq 0$

26 In the rectangle shown, is the sum of the areas of $\triangle APD$ and $\triangle BCP$ greater than, less than, or equal to the area of $\triangle DPC$? Justify your answer.

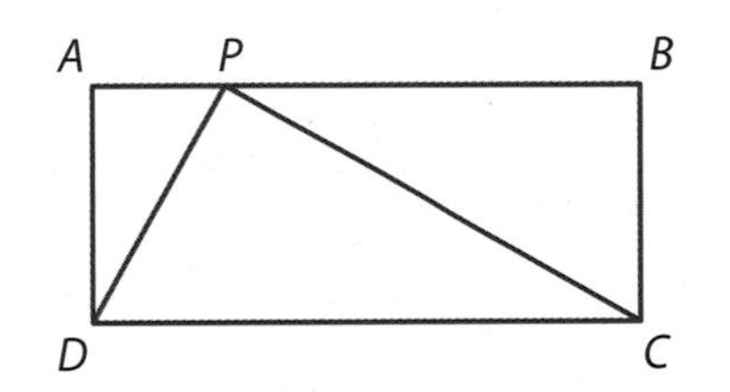

27 Find the following sums without using your calculator.

a. $0.4 + 0.23$

b. $0.05 + 0.24 + 0.15$

c. $-0.62 + 0.82$

d. $\frac{1}{6} + \frac{2}{3}$

e. $\frac{2}{10} + \frac{1}{4} - \frac{3}{5}$

28 Find the value of x in each polygon below.

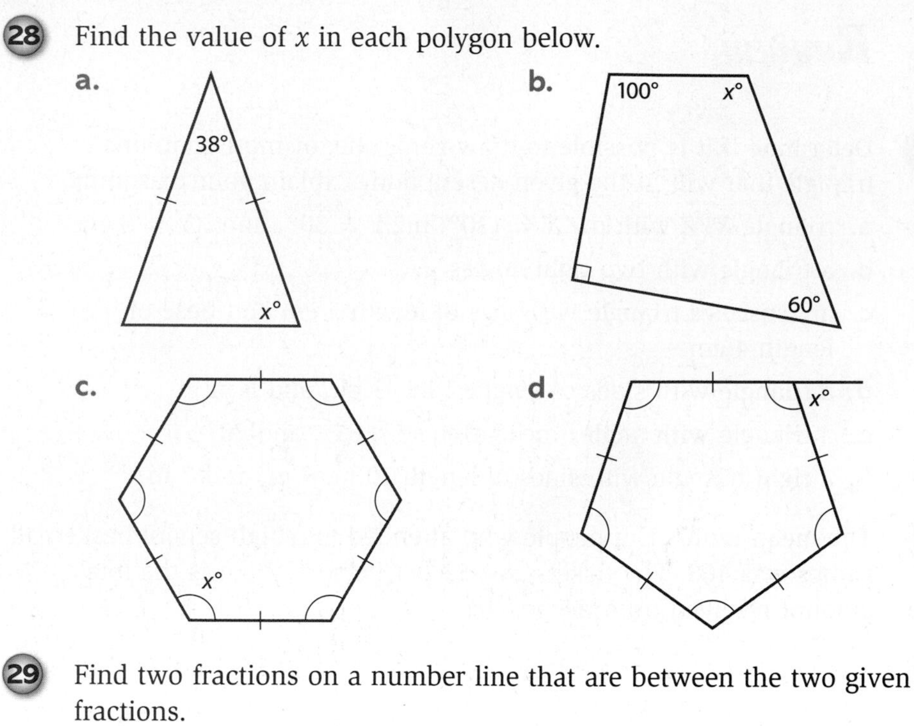

29 Find two fractions on a number line that are between the two given fractions.

a. $\frac{1}{3}$ and $\frac{1}{2}$

b. $\frac{5}{6}$ and $\frac{6}{7}$

c. $\frac{a}{b}$ and $\frac{c}{d}$

LESSON 2

Modeling Chance Situations

In some cultures, it is customary for a bride to live with her husband's family. As a result, couples who have no sons and whose daughters all marry will have no one to care for them in their old age.

In 2000, China had a population of over 1,200,000,000. In an effort to reduce the growth of its population, the government of China had instituted a policy to limit families to one child. The policy has been very unpopular among rural Chinese who depend on sons to carry on the family farming and care for them in their old age.

Think About This Situation

Customs of a culture and the size of its population often lead to issues that are hard to resolve. But probability can help you understand the consequences of various policies.

a In a country where parents are allowed to have only one child, what is the probability that their one child will be a son? What is the probability they will not have a son? What assumption(s) are you making when you answer these questions?

b If each pair of Chinese parents really had only one child, do you think the population would increase, decrease, or stay the same? Explain your reasoning.

c Describe several alternative plans that the government of China might use to control population growth. For each plan, discuss how you might estimate the answers to the following questions.

i. What is the probability that parents will have a son?
ii. Will the total population of China grow, shrink, or stay about the same?
iii. Will China end up with more boys than girls or with more girls than boys?
iv. What is the mean number of children per family?

In this lesson, you will learn to estimate probabilities by designing simulations that use random devices such as coin flipping or random digits. In our complex world, simulation is often the only feasible way to deal with a problem involving chance. Simulation is an indispensable tool to scientists, business analysts, engineers, and mathematicians. In this lesson, you will also explore how geometric models can be used to solve probability problems.

Investigation 1 When It's a 50-50 Chance

Finding the answers to the questions in Part c of the Think About This Situation may be difficult for some of the plans you proposed. If that is the case, you can estimate the effects of the policies by creating a mathematical model that **simulates** the situation by copying its essential characteristics. Although slightly more than half of all births are boys, the percentage is close enough to 50% for you to use a probability of $\frac{1}{2}$ to investigate different plans. As you work on the problems of this investigation, look for answers to the following question:

How can you simulate chance situations that involve two equally likely outcomes?

1 Consider, first, the plan that assumes each family in China has exactly two children.

a. Construct a sample space of the four possible families of two children.

b. Use your sample space to answer these questions from the Think About This Situation:

i. What is the probability that parents will have a son?

ii. Will the total population of China grow, shrink, or stay about the same?

iii. Will China end up with more boys than girls or with more girls than boys?

iv. What is the mean number of children per family?

2 Your class may have discussed the following plan for reducing population growth in rural China.

Allow parents to continue to have children until a boy is born. Then no more children are allowed.

Suppose that all parents continue having children until they get a boy. After the first boy, they have no other children. Write your best prediction of the answer to each of the following questions.

a. In the long run, will the population have more boys or more girls, or will the numbers be approximately the same?

b. What will be the mean number of children per family?

c. If all people pair up and have children, will the population increase, decrease, or stay the same?

d. What percentage of families will have only one child?

e. What percentage of the children will belong to single-child families?

3 To get a good estimate of the answers to the questions in Problem 2, you could simulate the situation. To do this, you can design a **simulation model** that imitates the process of parents having children until they get a boy.

a. Describe how to use a coin to conduct a simulation that models a family having children until they get a boy.

b. When you flip a coin to simulate one family having children until they get the first boy, you have conducted one **run** of your simulation. What is the least number of times you could have to flip the coin on a run? The most? If it takes n flips to get the first "boy", how many "girls" will be in the family?

c. Carry out one run of your simulation of having children until a boy is born. Make a table like the one below. Make a tally mark (/) in the tally column opposite the number of flips it took to get a "boy."

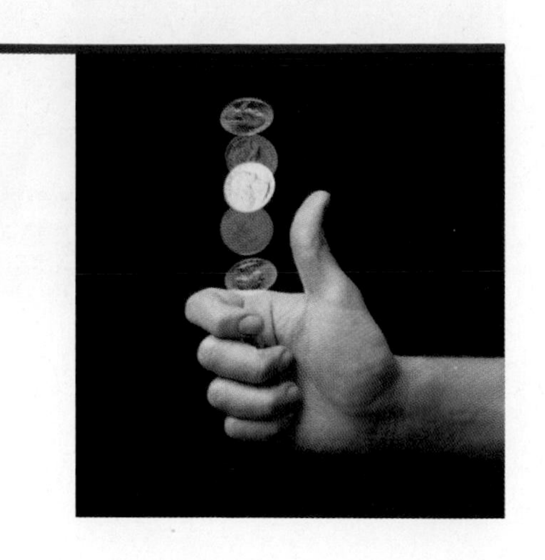

Number of Flips to Get a "Boy"	Tally	Frequency (Number of Tallies)	Relative Frequency
1			
2			
3			
4			
5			
6			
7			
8			
9			
10			
⋮			
Total	200 runs	200 runs	1.00

d. Continue the simulation until your class has a total of 200 "families." Record your results in the frequency table. Add as many additional rows to the table as you need.

e. How many "boys" were born in your 200 "families?" How many "girls?"

f. What assumption(s) are you making in this simulation?

g. Make a relative frequency histogram on a copy of the graph shown below and describe its shape.

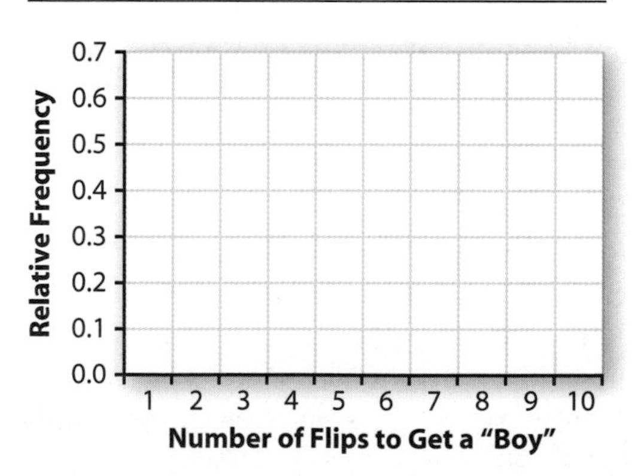

h. Compare the median number of children to the mean number of children in the families.

4 Now reconsider the questions from Problem 2 which are reproduced below. Estimate the answers to these questions using your completed table from Problem 3.

a. In the long run, will the population have more boys or more girls, or will the numbers be approximately the same?

b. What will be the mean number of children per family?

c. If all people pair up and have children, will the population increase, decrease, or stay the same?

d. What percentage of families will have only one child?

e. What percentage of the children will belong to single-child families?

5 Compare your estimates in Problem 4 with your original predictions in Problem 2. For which questions did your initial prediction vary the most from the estimate for the simulation? (If most of your original predictions were not accurate, you have a lot of company. Most people aren't very good at predicting the answers to probability problems. That's why simulation is such a useful tool.)

In the previous simulation, your class was asked to produce 200 runs. There is nothing special about that number except that it is about the most that it is reasonable for a class to do by hand. The **Law of Large Numbers** says that the more runs there are, the better your estimate of the probability tends to be. For example, according to the Law of Large Numbers, when you flip a coin a large number of times, eventually you should get a proportion of heads that is close to 0.5. (This assumes that the flips are independent and the probability of a head on each flip is 0.5.)

6 To illustrate the Law of Large Numbers, some students made the graph below. They flipped a coin 50 times and after each flip recorded the *cumulative* proportion of flips that were heads.

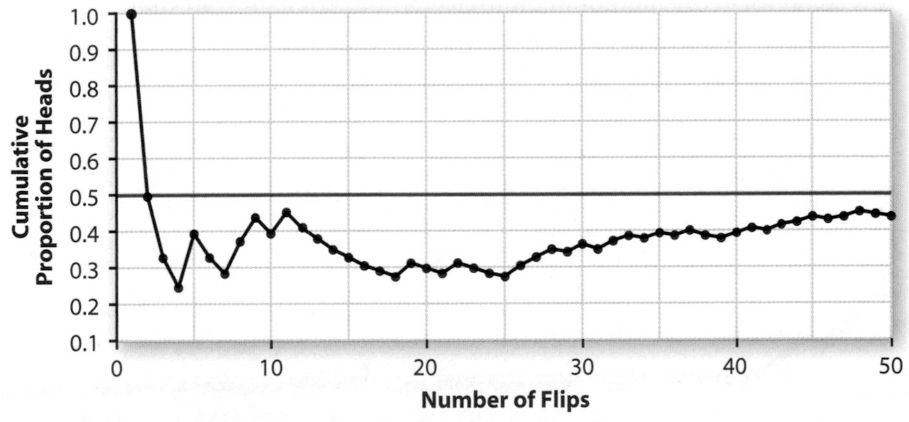

Source: Tim Erickson, *Fifty Fathoms*, Key Curriculum, 2002, pages 48–49.

a. Was the first flip heads or tails? How can you tell from the graph whether the next flip was a head or a tail? What were the results of each of the first 10 coin flips?

b. To three decimal places, what was the proportion of heads at each step in the first 10 coin flips?

c. Could the graph eventually go back above 0.5? Explain.

d. Why do the lengths of the line segments between successive flips tend to get smaller as the number of flips increases?

e. At the end of 50 flips, there were 22 heads and 28 tails. Continue on from there, doing 20 more flips. As you go along, complete a copy of the following table and a copy of the graph on the previous page, extending them to 70 flips.

Number of Flips	Frequency of Heads	Proportion of Heads
10	4	0.400
20	6	0.300
30	11	0.367
40	16	0.400
50	22	0.440
60		
70		

f. Explain how your completed graph and table illustrate the Law of Large Numbers.

More flips of the coin or more runs of a simulation are better when estimating probabilities, but in practice it is helpful to know when to stop. You can stop when the distribution stabilizes; that is, you can stop when it seems like adding more runs isn't changing the proportions very much.

7 Refer to your graph showing the results of your coin flips from Problem 6. Does the proportion of heads appear to be stabilizing, or is there still a lot of fluctuation at the end of 70 flips?

8 The graph below shows the results of 200 runs of a simulation of the plan for reducing population growth in Problem 1. In that plan, each family has exactly two children. The proportion of families that had two girls is plotted. How does this graph illustrate the Law of Large Numbers?

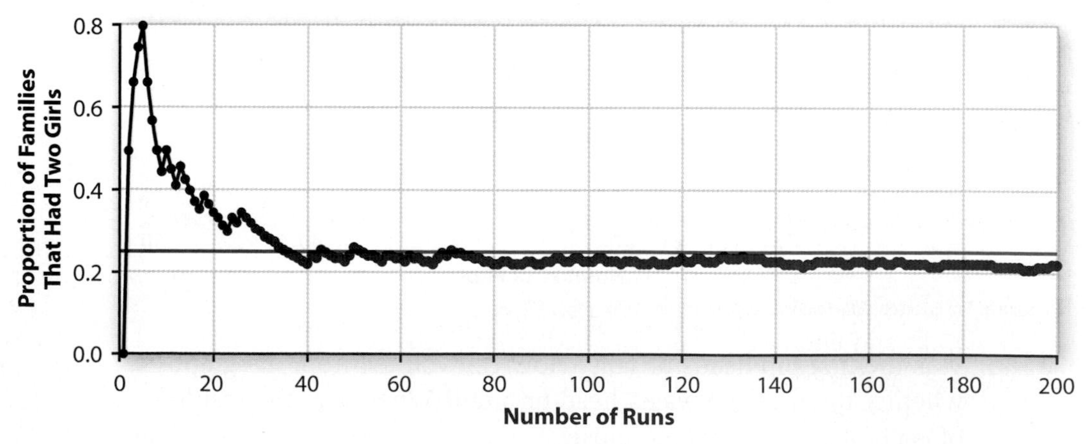

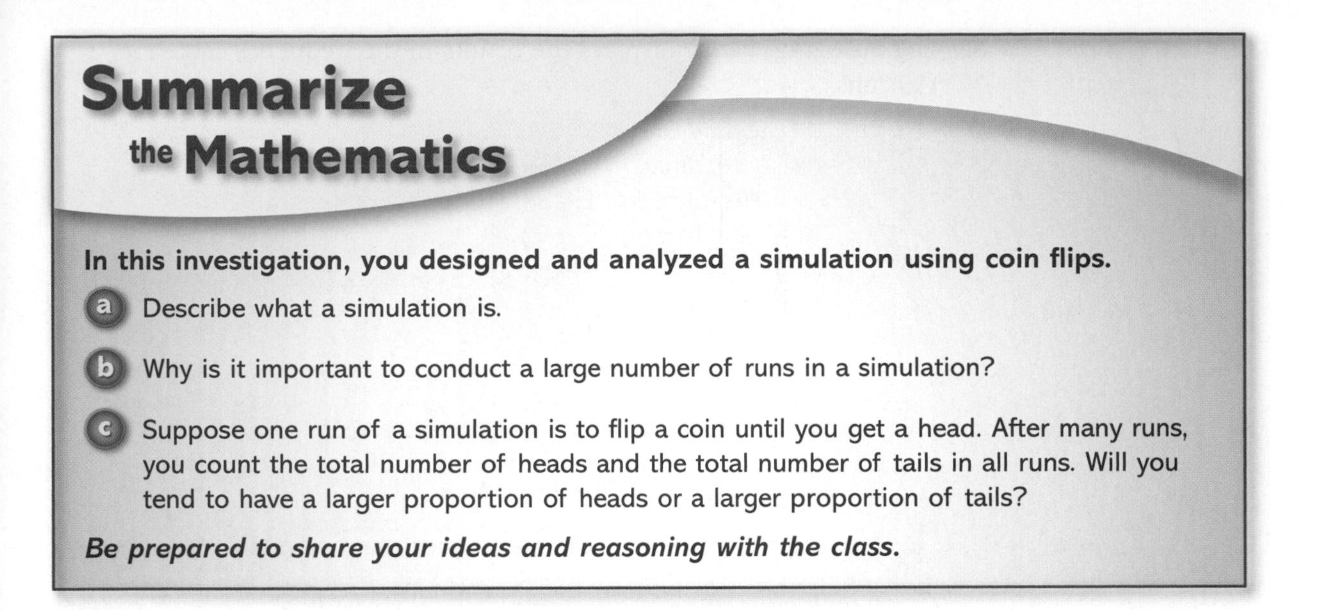

Summarize the Mathematics

In this investigation, you designed and analyzed a simulation using coin flips.

a Describe what a simulation is.

b Why is it important to conduct a large number of runs in a simulation?

c Suppose one run of a simulation is to flip a coin until you get a head. After many runs, you count the total number of heads and the total number of tails in all runs. Will you tend to have a larger proportion of heads or a larger proportion of tails?

Be prepared to share your ideas and reasoning with the class.

Check Your Understanding

When asked in what way chance affected her life, a ninth-grader in a very large Los Angeles high school reported that students are chosen randomly to be checked for weapons. Suppose that when this policy was announced, a reporter for the school newspaper suspected that the students would not be chosen randomly, but that boys would be more likely to be chosen than girls. The reporter then observed the first search and found that all 10 students searched were boys.

a. If there are the same number of boys and girls in the high school and a student is in fact chosen randomly, what is the probability that the student will be a boy?

b. Write instructions for conducting one run of a simulation that models selecting 10 students at random and observing whether each is a boy or a girl.

c. What assumptions did you make in your model?

d. Perform five runs of your simulation and add your results to the frequency table at the right.

Number of Boys	Frequency (Before)	Frequency (After)
0	0	
1	1	
2	9	
3	20	
4	41	
5	45	
6	43	
7	23	
8	11	
9	0	
10	2	
Total	195 runs	200 runs

e. The histogram below displays the results in the frequency table for 195 runs.

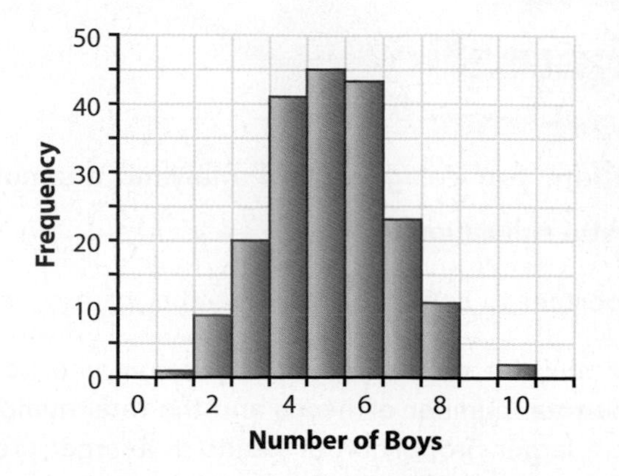

i. Describe its shape and estimate its mean and standard deviation.

ii. Why is this distribution almost symmetric?

iii. If you added your results to the histogram, would the basic shape change?

f. Using results from 200 runs of your simulation, estimate the probability that all 10 students will be boys if students are selected at random to be searched. What conclusion should the reporter make?

g. How much did adding your five runs change the probability of getting 10 boys? What can you conclude from this?

Investigation 2 Simulation Using Random Digits

In the previous investigation, the problem situations were based on two equally likely outcomes that you could simulate with a coin flip. In other situations, it is possible that there are more than two outcomes or the two outcomes aren't equally likely. In these situations, you can use random digits in designing a simulation. You can get strings of random digits from your calculator or from a random digit table produced by a computer. As you work on the following problems, look for answers to this question:

How can you use random digits when designing simulations?

 Examine this table of random digits generated by a computer.

2 4 8 0 3	1 8 6 5 6	4 2 0 3 0	9 1 4 9 6
7 6 8 6 3	0 5 6 8 2	5 0 7 4 5	6 7 3 6 3
0 9 5 8 1	7 3 0 9 9	8 7 7 7 7	1 6 2 7 2
0 2 6 8 6	2 5 5 4 1	5 9 8 1 0	1 5 2 9 7
4 1 2 9 0	8 6 7 0 3	3 8 2 5 1	8 4 1 4 1
1 5 8 0 9	5 7 3 5 6	5 0 2 0 3	6 6 5 0 3
9 7 6 2 5	9 2 6 3 5	0 3 1 9 3	9 7 2 6 3
2 1 0 9 6	0 1 8 5 5	2 2 6 8 6	0 6 6 6 3

a. How many digits are in the table? About how many 6s would you expect to find? How many are there?

b. About what percentage of digits in a large table of random digits will be even?

c. About what percentage of the 1s in a large table of random digits will be followed by a 2?

d. About what percentage of the digits in a large table of random digits will be followed by that same digit?

2 Explain how you can use the table of random digits in Problem 1 to simulate each situation given below. (You may have to disregard certain digits.) Then perform one run of your simulation.

a. Flip a coin and see if you get heads or tails.

b. Observe five coin flips and record how many heads you get.

c. Observe whether it rains or not on one day when the prediction is 80% chance of rain.

d. Select three cars at random from a large lot where 20% of the cars are black, 40% are white, 30% are green, and 10% are silver, and record the color.

e. Spin the spinner shown here four times and record the colors.

f. Roll a die until you get a 6 and record the number of rolls you needed.

g. Select three different students at random from a group of ten students. How is this problem different from the others you have done?

h. Select three different students at random from a group of seven students.

3 Suppose three meteorites are predicted to hit the United States. You are interested in how many will fall on publicly owned land. About 30% of the land in the United States is owned by the public.

a. Describe how to conduct one run that simulates this situation. What assumptions are you making in this simulation?

b. Combine results with your class until you have 100 runs and place the results in a frequency table that shows how many of the three meteorites fell on public land.

c. What is your estimate of the probability that at least two of the meteorites will fall on public land?

4 In trips to a grocery store, you may have noticed that boxes of cereal often include a surprise gift such as one of a set of toy characters from a current movie or one of a collection of stickers. Cheerios, a popular breakfast cereal, once included one of seven magic tricks in each box.

a. What is the least number of boxes you could buy and get all seven magic tricks?

b. If you buy one box of Cheerios, what is the probability that you will get a multiplying coin trick? To get your answer, what assumptions did you make about the tricks?

Collect All 7 and Put On Your Own Magic Show!

c. Suppose you want to estimate the number of boxes of Cheerios you will have to buy before you get all seven magic tricks. Describe a simulation model, including how to conduct one run using a table of random digits. (Recall that you may ignore certain digits.)

d. Compare your simulation model with that of other students. Then as a class, decide on a simulation model that all students will use.

e. Perform five runs using the agreed upon simulation. Keep track of the number of "boxes" you would have to buy. Add your numbers and those for the other students in your class to a copy of this frequency table.

Number of Boxes to Get All 7 Tricks	Frequency (Before)	Frequency (After)
7	1	
8	3	
9	4	
10	11	
11	13	
12	18	
13	22	
14	13	
15	9	
16	8	
17	10	
18	11	
19	13	
20	12	
21	4	
22	8	
23	3	
24	0	
25	0	
⋮		
Total		

f. Add the runs from your class to a copy of the histogram below.

  i. Describe the shape of the distribution.

  ii. Compare the shape of this distribution to others you have constructed in this unit.

g. Based on the simulation, would it be unusual to have to buy 15 or more boxes to get the 7 magic tricks? Explain.

h. What could be some possible explanations why a person would, in fact, end up buying a much larger number of boxes than you would expect from this simulation?

In the previous simulations, you have been able to use single digits from your table of random digits. In other cases, you may need to use groups of two or more consecutive random digits.

5 When playing tennis, Sheila makes 64% of her first serves. Describe how to use pairs of random digits to conduct one run of a simulation of a set where Sheila tries 35 first serves. Then conduct one run of your simulation and count the number of serves that Sheila makes.

6 Twenty-nine percent of the students at Ellett High School reported that they have been to a movie in the previous two weeks. Connie found that only 1 of her 20 (or 5%) closest friends at the school had been to a movie in the previous two weeks. Connie wants to know if this smaller-than-expected number can reasonably be attributed to chance or if she should look for another explanation.

a. Describe how to use pairs of random digits to conduct one run of a simulation to find the number of recent moviegoers in a randomly selected group of 20 students.

b. The table and histogram below show the results of 195 simulations of the number of recent moviegoers in groups of 20 randomly selected students. Conduct 5 runs, adding your results to a copy of the table and histogram.

Number of Students	Frequency (Before)	Frequency (After)
0	1	
1	3	
2	8	
3	18	
4	25	
5	35	
6	44	
7	30	
8	15	
9	7	
10	8	
11	1	
Total	195	200

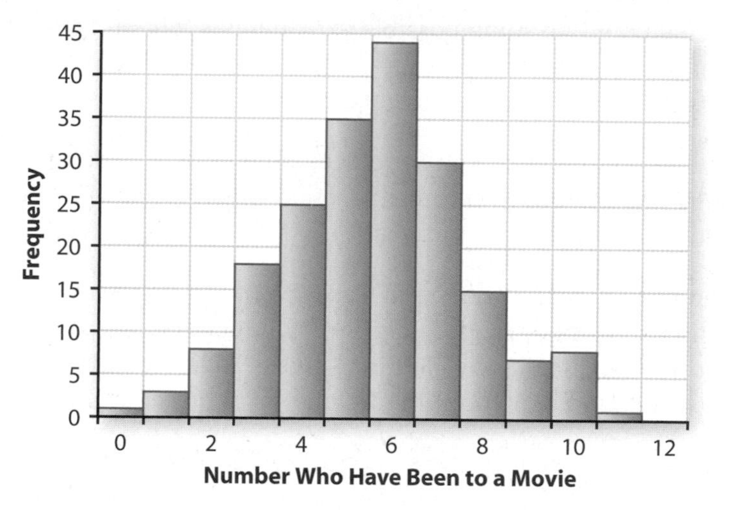

c. Based on your simulation, estimate the probability that no more than 1 of 20 randomly selected students have been to a movie in the previous two weeks.

d. Is the result for Connie's friends about what might be expected for a randomly selected group of 20 students? If not, what are some possible explanations?

A table of random digits is a convenient tool to use in conducting simulations. However, calculators and computer software with random number generators are more versatile tools.

7 Investigate the nature of the integers produced by the random integer generator on your calculator or computer software. You may find the command **randInt** under the probability menu. If, for example, you want six integers randomly selected from $\{1, 2, 3, 4, \ldots, 19, 20\}$, enter **randInt (1,20,6)**.

a. If the same integer can be selected twice, the random integer generator is said to select *with replacement*. If the same integer cannot appear more than once in the same set, the random integer generator is said to select *without replacement*. Does your **randInt** command select with replacement or without replacement?

b. Your calculator probably selects integers at random with replacement. If this is the case, describe how you can get a random selection of six numbers from $\{1, 2, 3, 4, \ldots, 19, 20\}$ without replacement.

8 Explain how to use the **randInt** command to simulate each of the following situations. Then perform one run of the simulation.

a. Roll a die five times and record the number on top.

b. Flip a coin 10 times and record whether it is heads or tails on each flip.

c. Select three different students at random from a group of seven students.

d. Roll a die until you get a 6 and count the number of rolls needed.

e. Check five boxes of Cheerios for which of seven magic tricks they contain.

f. Draw a card from a well-shuffled deck of 52 playing cards and record whether it is an ace.

Summarize the Mathematics

In this investigation, you explored the properties of random digits and learned how to use them in designing a simulation.

a What is a table of random digits? What command do you use to get random digits on your calculator?

b Give an example of when you would want to select random digits with replacement. Without replacement.

c How do you use random digits in a simulation when the probability of the event you want is 0.4? When the probability of the event is 0.394?

d As a tie-in to the opening of a baseball season, each box of Honey Bunches of Oats cereal contained one of six Major League Baseball CD-ROMs (one for each division). Suppose you want to collect all six and wonder how many boxes you can expect to have to buy. How could you simulate this situation efficiently using a table of random digits? Using your calculator? What assumptions are you making?

Be prepared to share your ideas and reasoning with the class.

Check Your Understanding

A teacher notices that of the last 20 single-day absences in his class, 10 were on Friday. He suspects that this did not happen just by random chance.

a. Assuming that absences are equally likely to occur on each day of the school week, describe how to use a table of random digits to simulate the days of the week that 20 single-day absences occur.

i. Conduct 5 runs. Add your results to a copy of the frequency table and histogram at the right that show the number of absences that are on Friday for 295 runs.

Number of the 20 Absences that Are on Friday	Frequency (Before)	Frequency (After)
0	6	
1	13	
2	37	
3	62	
4	65	
5	47	
6	43	
7	13	
8	4	
9	5	
10	0	
Total	295	300

ii. Based on the simulation, what is your estimate of the probability of getting 10 or more absences out of 20 on Friday just by chance? What should the teacher conclude?

iii. From the simulation, what is your estimate of the mean number of absences on Friday, assuming that the 20 absences are equally likely to occur on each day of the week? Does this make sense? Why or why not?

iv. How could you get better estimates for parts ii and iii?

b. Describe how to use your calculator to conduct one run that simulates this situation.

Investigation 3 Using a Random Number Generator

Sometimes a simulation requires that you select numbers from a continuous interval. For example, suppose you are painting a person on the backdrop in a school play. You can't decide how tall to make the person and so decide just to select a height at random from the interval 60 inches to 72 inches. There are infinitely many heights in that interval. For example, possible heights are 60 inches, 60.1 inches, 60.11 inches, 60.111 inches, 60.1111 inches, etc. As you work on problems in this investigation, look for answers to the following question:

How can you design simulations to solve problems when numbers can come from a continuous interval?

1 Investigate the nature of the numbers produced by selecting the command "rand" in the probability menu of your calculator and then pressing ENTER repeatedly.

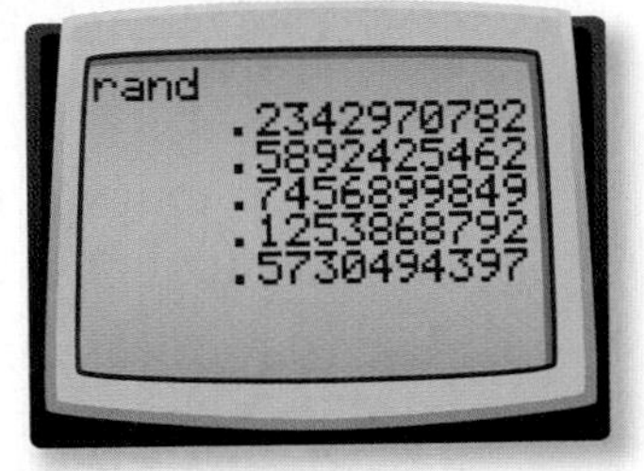

a. How many decimal places do the numbers usually have? (Some calculators leave off the last digit if it is a 0.)

b. Between what two whole numbers do all the random numbers lie?

2 Now explore how to generate random numbers in other continuous intervals.

a. Generate random numbers of the form "6 rand" (or "6 × rand"). Between what two whole numbers do all the random numbers lie?

b. Between what two whole numbers do the random numbers lie when you use **10rand**? **36rand**? **rand+2**? **100rand+2**?

c. Write a **rand** command that selects a number at random from the interval between 1 and 7. Between 4 and 5. Between 60 inches and 72 inches.

d. Suppose you select two random numbers between 0 and 12 and want to estimate the probability that they are both more than 7.

i. Use your calculator to select two numbers from this interval. Record whether both numbers are more than 7.

ii. How can you simulate this situation using a spinner?

3 Julie wakes up at a random time each morning between 6:00 and 7:00. If she wakes up after 6:35, she won't have time for breakfast before school. Conduct one run of a simulation to estimate the probability that Julie won't have time for breakfast the next two school days. Repeat this 10 times. What is your estimate of the probability that Julie won't have time for breakfast the next two school days?

4 Recall the triangle-building experiment in the *Patterns in Shape* unit. In this problem, suppose you have a 10-inch strand of uncooked spaghetti and select two places at random along its length. You break the strand at those places and try to make a triangle out of the three pieces.

a. What do you think the probability is that the three pieces will form a triangle? Make a conjecture.

b. If the breaks are at 2-inch mark and 7.5-inch mark from the left end (see diagram below), how long are each of the three pieces? Can you make a triangle out of the three pieces?

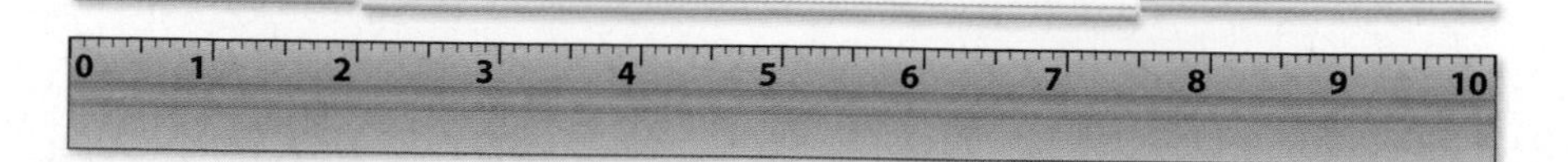

c. If the breaks are at 2.5-inch mark and at 6-inch mark from the left end, how long are each of the three pieces? Can you make a triangle out of the three pieces?

d. Use your calculator to simulate one run of the situation of breaking a 10-inch strand of spaghetti at two randomly selected places. Can you make a triangle out of the three pieces?

e. Carry out a simulation to estimate the probability that a triangle can be formed by the three pieces of a strand of spaghetti broken at two places at random along its length. Compare your estimated probability to your conjecture in Part a.

5 The square region pictured below is 250 feet by 250 feet. It consists of a field and a pond which lies below the graph of $y = 0.004x^2$. Imagine a totally inept skydiver parachuting to the ground and trying to avoid falling into the pond. The skydiver can be sure of landing somewhere inside the region, but the spot within it is random.

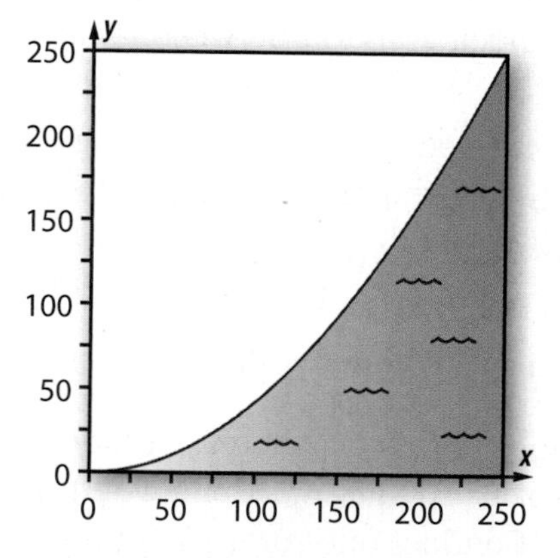

a. Describe how to use the **rand** function of your calculator to simulate the point where the skydiver will land. (You will need an x-coordinate between 0 and 250 and a y-coordinate between 0 and 250.)

b. How can you tell from the coordinates (x, y) of the simulated landing whether the skydiver landed in the pond?

c. Simulate one landing and tell whether the skydiver landed in the pond or not.

d. Simulate a second landing and tell whether the skydiver landed in the pond or not.

6 Al and Alice work at the counter of an ice cream store. Al takes a 10-minute break at a random time between 12:00 and 1:00. Alice does the same thing, independently of Al.

a. Suppose that Al takes his break at 12:27. If Alice goes on her break at 12:35, would there be an overlap in the two breaks? What times for Alice to go on her break would result in an overlap of the two breaks?

b. Use the **rand** command to simulate this situation. Did the two breaks overlap?

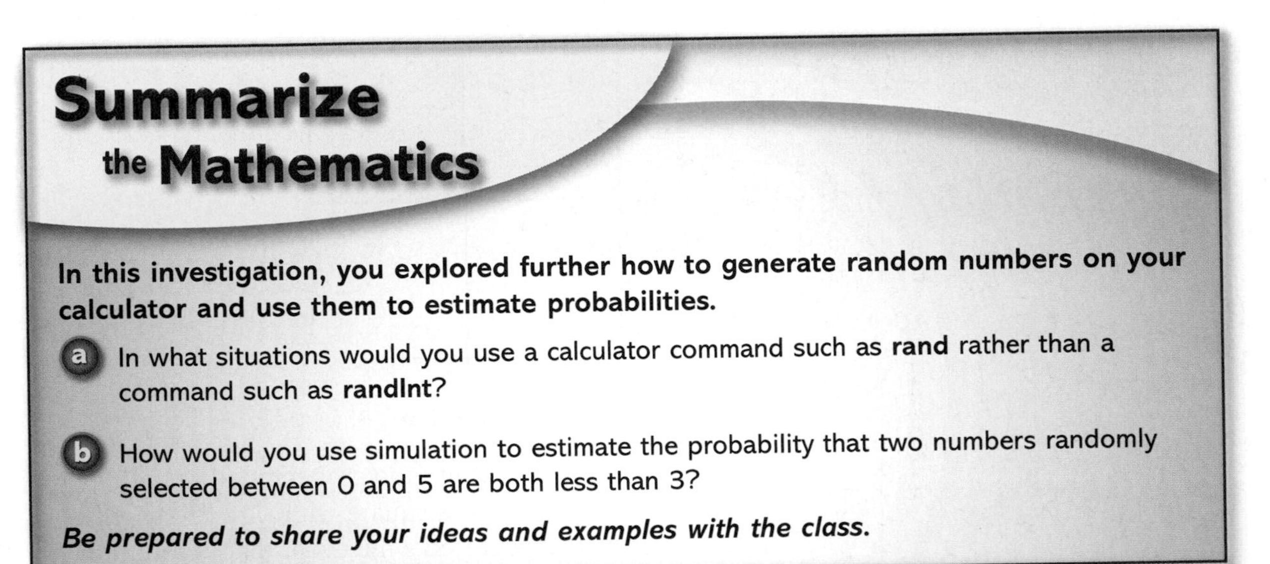

In this investigation, you explored further how to generate random numbers on your calculator and use them to estimate probabilities.

(a) In what situations would you use a calculator command such as **rand** rather than a command such as **randInt**?

(b) How would you use simulation to estimate the probability that two numbers randomly selected between 0 and 5 are both less than 3?

Be prepared to share your ideas and examples with the class.

Check Your Understanding

Jerome arrives at school at a random time between 7:00 and 7:30. Nadie leaves independently of Jerome and arrives at a random time between 6:45 and 7:15. Suppose you want to estimate the probability that Nadie arrives at school before Jerome.

a. Describe how to use a calculator command to simulate the time that Jerome arrives at school. To simulate the time that Nadie arrives at school.

b. Perform one run of a simulation. Did Jerome or Nadie get to school first?

c. Continue until you have a total of 10 runs. What is your estimate of the probability that Nadie gets to school before Jerome?

Investigation 4 Geometric Probability

You can solve some problems similar to those in the previous investigation without a simulation. The method involves drawing a geometric diagram and reasoning with areas. As you work on the problems in this investigation, look for answers to the following question:

How can you use area models to solve probability problems when numbers can come from a continuous interval?

1 Suppose you select two random numbers that are both between 0 and 1.

a. You want to find the probability that both numbers are more than 0.5. Explain how this problem can be solved using the diagram below.

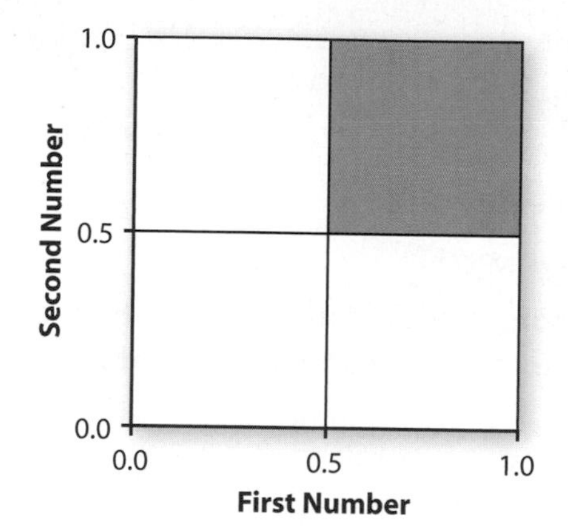

b. Now suppose that you want to find the probability that both numbers are less than 0.2. Draw a square with sides of length 1 and shade in the area that represents the event that both numbers are less than 0.2. What is the probability that they are both less than 0.2?

c. Use an area model to find the probability that both numbers are less than 0.85.

2 Suppose you select two random numbers between 0 and 12 and want to find the probability that they both are greater than 7. Use an area model to represent this situation and find the probability.

3 The National Sleep Association estimates that teens generally require 8.5 to 9.25 hours of sleep each night. However only 15% of teens report that they sleep 8.5 hours or more on school nights. (Source: www.sleepfoundation.org/_content/hottopics/sleep_and_teens_report1.pdf) Suppose you pick two teens at random and ask them if they sleep 8.5 hours or more on school nights.

a. Draw a square with sides of length 1 and shade in the area that represents the probability that at least one teen says that they sleep 8.5 hours or more on school nights.

b. Use your diagram from Part a to find the probability that at least one teen says they sleep 8.5 hours or more on school nights.

4 Suppose you select two random numbers between 0 and 1.

a. Draw a square with sides of length 1 and shade in the area that represents the event that the sum of the two numbers is between 0 and 1.

i. What is the probability that their sum is less than 1?

ii. What is the equation of the line that divides the shaded and unshaded regions in your area model?

b. Draw a square with sides of length 1 and shade in the area that represents the event that the sum of the two numbers is less than 0.6.

i. What is the probability that their sum is less than 0.6?

ii. What is the equation of the line that divides the shaded and unshaded regions in this area model?

5 In Investigation 3 Problem 6 (page 567), you used simulation methods to estimate the probability that the breaks of two employees would overlap. The problem conditions were that Al takes a 10-minute break at a random time between 12:00 and 1:00 and Alice does the same thing, independently of Al. Now consider how you could calculate the exact probability using an area model.

a. On a copy of the diagram below, identify points that correspond to identical start times for breaks by Al and Alice.

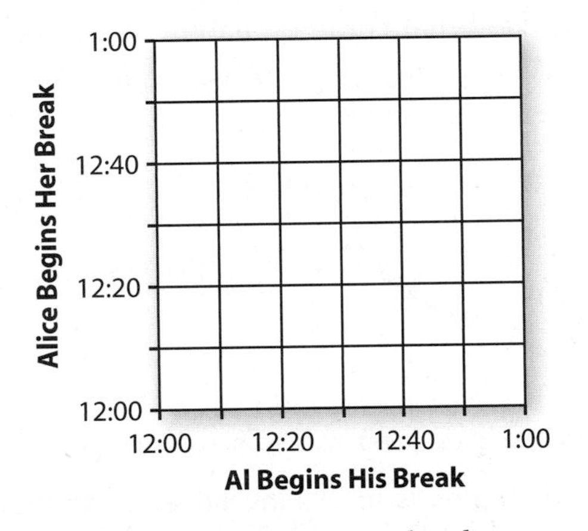

b. For any given start time for a break by Al, what start times for a break by Alice would overlap with Al's break?

c. On your diagram, shade in the beginning times of the two breaks that would result in overlap in their breaks.

d. What is the probability that their two breaks overlap?

6 Imagine an archer shooting at the target shown. The square board has a side length of 6 feet. The archer can always hit the board, but the spot on the board is random. Use geometric reasoning to find the probability that an arrow lands in the circle.

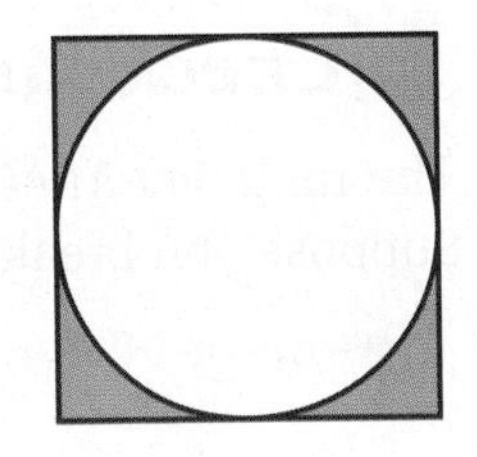

 In an old carnival game, players toss a penny onto the surface of a table that is marked off in 1-inch squares. The table is far enough away that it is random where the penny lands with respect to the grid, but large enough that the penny always lands on the table.

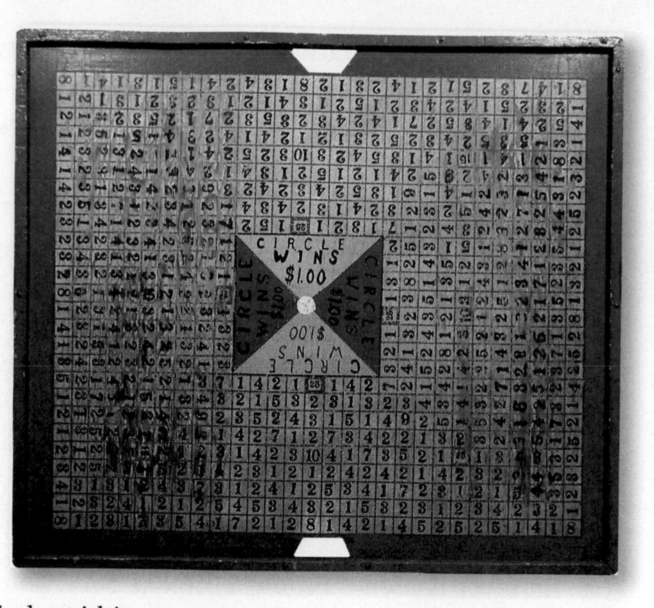

If the penny lies entirely within a square, the player wins a prize. If the penny touches a line, the player loses his or her penny. A penny is about 0.75 inch in diameter.

a. In order to win a prize, how far must the center of the penny be from any side of a 1-inch square?

b. Draw a 1-inch square. Shade the region where the center of the penny must land in order to win a prize. What is the probability that the center of the penny lands in that region and the player wins the prize?

Summarize the Mathematics

In this investigation, you learned to use area models to find probabilities exactly.

a When can you use the geometric methods in this investigation to find probabilities?

b How would you use an area model to find the probability that two numbers randomly selected between 0 and 5 are both less than 3?

Be prepared to share your ideas and examples with the class.

Check Your Understanding

The midpoint M of a 10-inch piece of spaghetti is marked as shown below. Suppose you break the spaghetti at a randomly selected point.

A M B

a. What is the probability that the break point is closer to point M than to point A?

b. What is the probability that the break point is closer to point A or point B than to point M?

On Your Own

Applications

1. Suppose that a new plan to control population growth in China is proposed. Parents will be allowed to have at most three children and must stop having children as soon as they have a boy.

a. Describe how to use a coin to conduct one run that models one family that follows this plan.

b. Conduct 5 runs. Copy the following frequency table, which gives the results of 195 runs. Add your results to the frequency table so that there is a total of 200 runs.

Type of Family	Frequency (Before)	Frequency (After)	Relative Frequency
First Child is a Boy	97		
Second Child is a Boy	50		
Third Child is a Boy	26		
Three Girls and No Boy	22		
Total	195	200	1.0

c. Estimate the percentage of families that would not have a son.

d. A histogram of the 195 results in the frequency table is given below. Describe its shape. If you added your results to the histogram, would the basic shape change? Explain your reasoning.

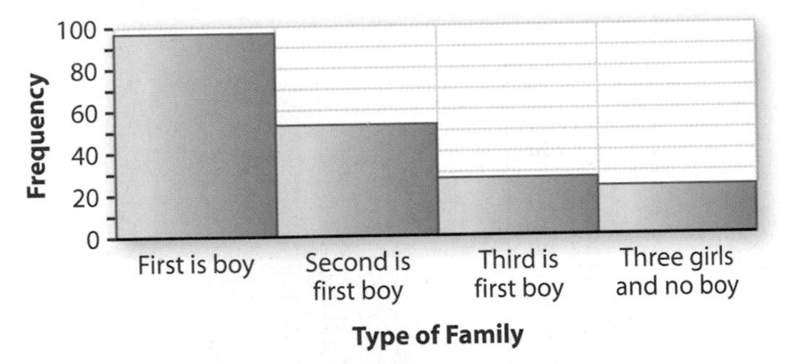

e. How does the shape of the histogram based on 200 runs differ from that of the have-children-until-you-have-a-boy plan from Problem 3 of Investigation 1 on page 553? Explain why that makes sense.

f. What is the mean number of children per family? Will the total population increase or decrease under this plan, or will it stay the same?

g. In the long run, will this population have more boys or more girls, or will the numbers be about equal? Explain your reasoning.

2 Jeffrey is taking a 10-question true-false test. He didn't study and doesn't even have a reasonable guess on any of the questions. He answers "True" or "False" at random.

a. Decide how to use a coin to conduct one run that models the results of this true-false test.

b. Conduct 5 runs. Copy the following frequency table, which gives the results of 495 runs. Add your results to the frequency table so that there is a total of 500 runs.

Number Correct	Frequency (Before)	Frequency (After)
0	1	
1	6	
2	24	
3	57	
4	98	
5	127	
6	100	
7	61	
8	17	
9	3	
10	1	
Total Number of Runs	495	500

c. A histogram of the 495 results in the frequency table is given below. Add your results to a copy of the histogram. Describe its shape and estimate its mean and standard deviation.

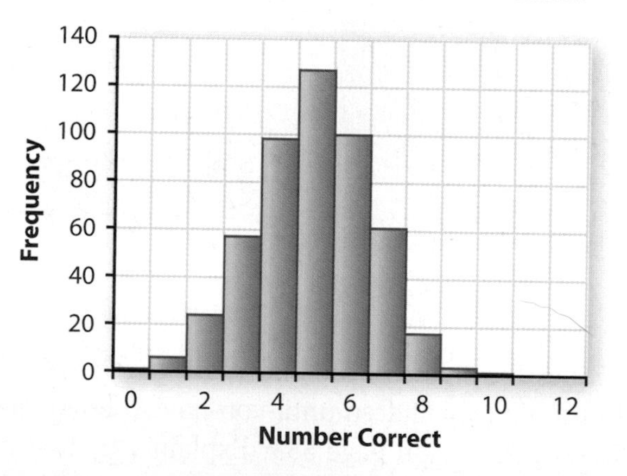

d. On average, how many questions should Jeffrey expect to get correct using his random guessing method? How does this compare to the mean from the simulation?

e. If 70% is required to pass the test, what is your estimate of the probability that Jeffrey will pass the test?

f. Considering the Law of Large Numbers, should Jeffrey prefer a true-false test with many questions or with few questions?

3 The winner of baseball's World Series is the first of the two teams to win four games.

a. What is the fewest number of games that can be played in the World Series? What is the greatest number of games that can be played in the World Series? Explain.

b. Suppose that the two teams in the World Series are evenly matched. Describe how to use a table of random digits to conduct one run simulating the number of games needed in a World Series.

c. Conduct 5 runs. Construct a frequency table similar to the one shown below and add your 5 results so that there is a total of 100 runs.

Number of Games Needed in the Series	Frequency (Before)	Frequency (After)
4	11	
5	21	
6	30	
7	33	
Total Number of Runs	95	100

d. What is your estimate of the probability that the series will go seven games?

e. By how much did your 5 results change the probability that the series will go seven games? What can you conclude from this?

f. A histogram of the 95 results in the frequency table is given below. If you added your results to the histogram, how, if at all, would the basic shape change? Explain your reasoning.

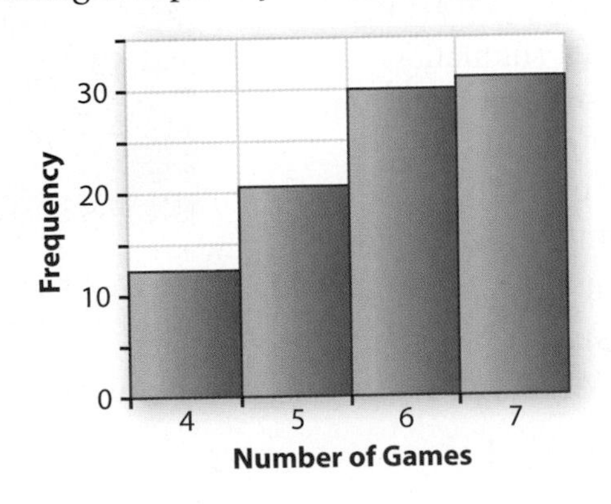

g. The following table and histogram give the actual number of games played in each World Series from 1940 through 2005. (There was no series in 1994.) Compare the results of the simulation with that of the actual World Series. What conclusions can you draw? (Source: www.infoplease.com/ipsa/A0112302.html)

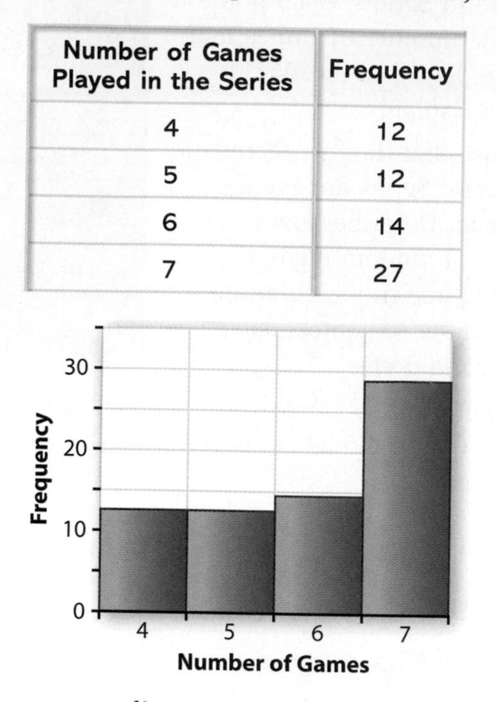

Number of Games Played in the Series	Frequency
4	12
5	12
6	14
7	27

4 For some rock concerts, audience members are chosen at random to have their bags checked for cameras, food, and other restricted items. Suppose that you observe the first 25 males and 25 females entering a concert. Ten of them are chosen to be searched. All 10 are male. You will simulate the probability of getting 10 males if you randomly select 10 people from a group of 25 males and 25 females.

a. When simulating the rock concert situation, why can't you use a flip of a coin like you did in the high school weapons search problem in the Check Your Understanding on page 557 of Investigation 1?

b. Describe a simulation using slips of paper drawn from a bag. How are your assumptions different from those in the Check Your Understanding in Investigation 1 on page 557?

c. How would you conduct this simulation using a table of random digits? Would you select with or without replacement?

d. Conduct 5 runs of your simulation using the method of your choice. Record the results in a copy of the frequency table below, which shows the number of males selected.

Number of Males	Frequency (Before)	Frequency (After)
0	2	
1	7	
2	23	
3	40	
4	47	
5	48	
6	44	
7	20	
8	17	
9	2	
10	0	
Total Number of Runs	250	255

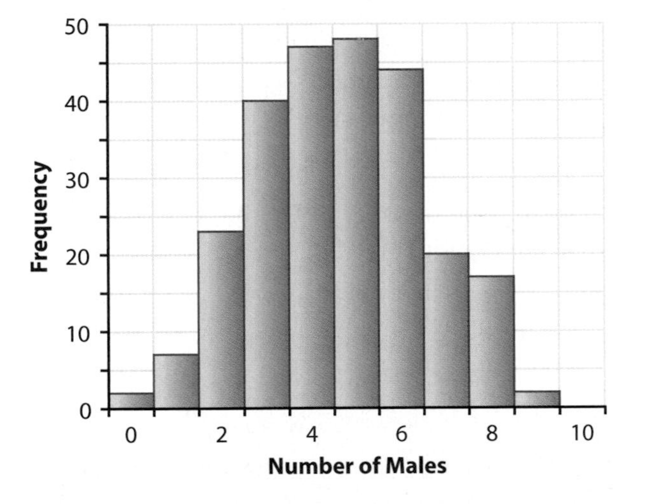

e. What is your conclusion about the probability that all 10 concert attendees chosen to be searched would be male? Is your conclusion different from your conclusion in the high school weapons search problem? Explain.

5 According to the U.S. Department of Education report, *The Condition of Education 2003*, about 62% of high school graduates enroll in college immediately after high school graduation. (Source: nces.ed.gov/pubs2003/2003067_3.pdf)

a. Describe how to use your random digit generator to simulate the situation of picking a randomly selected high school graduate and finding out if he or she enrolls in college immediately after graduation.

b. Describe how to conduct one run of a simulation of the situation of selecting 30 high school graduates at random and counting the number who enroll in college immediately after graduation.

c. Conduct 5 runs of your simulation. Copy the frequency table below that shows the results of 195 runs. Add your results to the table.

Number of Graduates Who Immediately Enroll in College	Frequency (Before)	Frequency (After)
10	1	
11	0	
12	1	
13	1	
14	6	
15	15	
16	13	
17	21	

Number of Graduates Who Immediately Enroll in College	Frequency (Before)	Frequency (After)
18	25	
19	38	
20	25	
21	24	
22	9	
23	11	
24	4	
25	1	
Total Number of Runs	195	200

d. Use your results to estimate answers to these questions.

i. What is the probability that 12 or fewer of a randomly selected group of 30 high school graduates immediately enroll in college?

ii. Would it be unusual to find that 24 or more of 30 randomly selected graduates immediately enroll in college?

e. Suppose you select a classroom of seniors at random in your school. You find that all 30 of the seniors in the class plan to enroll immediately in college. List as many reasons as you can why this could occur.

6 About 44.5% of violent crimes in the United States are committed by someone who is a stranger to the victim. (Source: *Statistical Abstract of the United States,* 2006, page 203, Table 311; www.census.gov/prod/www/statistical-abstract.html) Suppose that you select four violent crimes at random and count the number committed by strangers.

a. Describe how to use the **randInt** function of your calculator to conduct one run that simulates this situation.

b. Describe how to use the **rand** function of your calculator to conduct one run that simulates this situation.

c. Conduct 10 runs using the calculator function of your choice and place the results in a frequency table that shows how many of the 10 violent crimes were committed by strangers.

d. What is your estimate of the probability that at least half of the four violent crimes were committed by strangers?

7. Suppose that you select two numbers at random from between 0 and 2. Draw a geometric diagram to help find the following probabilities.
   a. What is the probability that both are less than 1.2?
   b. What is the probability that their sum is less than 1?

8. Jerome arrives at school at a random time between 7:00 and 7:30. Nadie leaves independently of Jerome and arrives at a random time between 6:45 and 7:15.
   a. Draw a geometric diagram to help find the probability that Nadie arrives at school before Jerome.
   b. Compare your answers to that obtained in the Check Your Understanding on page 567.

Connections

9. You can use random devices other than coins to simulate situations with two equally likely outcomes.
   a. Is rolling the die below equivalent to flipping a coin? Explain.

One View

Another View

   b. How could a regular, six-sided die marked with the numbers 1 through 6 be used to simulate whether a child is a boy or a girl?
   c. How could a tetrahedral die marked with the numbers 1 through 4 be used to simulate whether a child is a boy or a girl?
   d. Identify other geometric shapes of dice that could be used to simulate a birth.

10. Make an accurate drawing of a spinner that simulates rolling a die. Describe the characteristics of this spinner.

11. Refer to your frequency table from Problem 3 from Investigation 1 (page 553).
   a. Make a scatterplot of the ordered pairs (*number of flips to get a "boy," relative frequency*).
   b. What function is a reasonable model of the relationship between *number of flips to get a "boy"* and *relative frequency*?

12. Explain how you can use each of the devices in the situation described.
   a. How could you use an icosahedral die to simulate picking random digits?

b. How could you use a deck of playing cards to conduct one run in a simulation of collecting Cheerios tricks? (See Problem 4 on page 559 of Investigation 2.)

c. How could you use a deck of playing cards to generate a table of random digits?

13 Describe how you could use simulation to estimate the area shaded in the diagram to the right. The area lies between the x-axis and the graph of $y = x^2$ on the interval $0 \le x \le 1$.

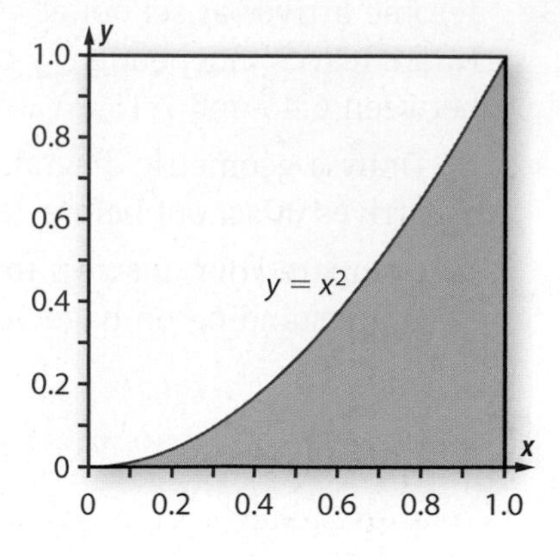

14 In Investigation 4, Problem 3 (page 568), you used geometric probability to find the probability that at least one of the two randomly selected teens says they sleep 8.5 hours or more on school nights. Dana suggested that since the probability for each teen reporting 8.5 or more hours of sleep is 15%, the probability that at least one teen says they sleep 8.5 hours or more is 30%, or 0.3. Use a geometric diagram and what you learned about the Addition Rule in Lesson 1 to explain why Dana's thinking is incorrect.

Reflections

15 A school is selling magazine subscriptions to raise money. A group of students wants to simulate the situation of asking 10 people if they will buy a magazine and counting the number who will say yes. Jason proposes that the group flip a coin 10 times and count the number of heads since a person either says "yes" (heads) or says "no" (tails). Is Jason's simulation model a reasonable one? Explain your position.

16 Suppose you want to estimate the probability that a family of two children will have at least one girl. You will count the number of heads in two flips of a coin. Does it matter if you flip *one* coin twice or flip *two* coins at the same time? Explain your reasoning.

17 The Law of Large Numbers tells you that if you flip a coin repeatedly, the percentage of heads tends to get closer to 50%. This is something that most people intuitively understand: the more "trials," the closer $\frac{\text{number of heads}}{\text{number of flips}}$ should be to $\frac{1}{2}$.

a. Jack flips a coin 300 times and gets 157 heads. What is his estimate of the probability that a coin will land heads?

b. Julie flips a coin 30 times and uses her results to estimate the probability a coin will land heads. Would you expect Jack or Julie to have an estimate closer to the true probability of getting a head?

c. Find the missing numbers in the table below. Do the results illustrate the Law of Large Numbers? Why or why not?

Number of Flips	Number of Heads	Percentage of Heads	Expected Number of Heads	Excess Heads
10	6	60	5	1
100	56	56	50	6
1,000	524			
10,000	5,140			

d. What surprising result do you see in the completed table?

18 When presented with a problem involving chance, how would you decide whether to calculate the probability using a formula, using a geometric probability model, or by conducting a simulation?

Extensions

19 Suppose a cereal company is thinking about how many different prizes to use in its boxes. It wants children to keep buying boxes, but not get too discouraged. The company conducted very large simulations where all prizes are equally likely to be in each box. It could then estimate quite accurately the mean number of boxes that must be purchased to get all of the prizes. The company's estimates are given in the table below.

Number of Possible Prizes	Mean Number of Boxes Purchased
1	1
2	3
3	5.5
4	8.3
5	11.4
6	14.7
7	18.1

a. Examine this scatterplot of (*number of prizes, mean number of boxes purchased*). Find a model that fits these data reasonably well.

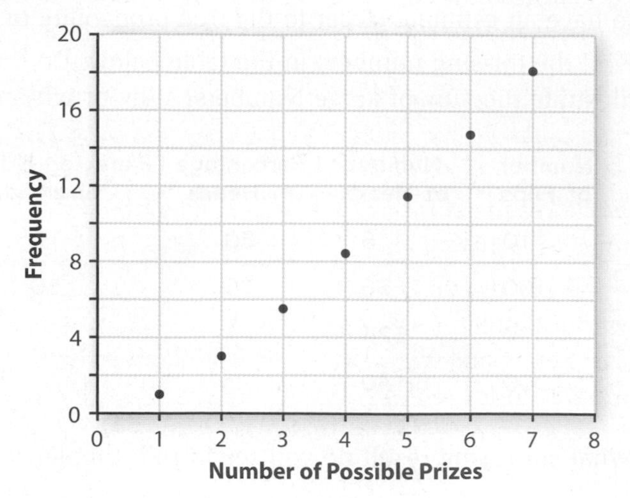

b. Below is an exact formula that can be used to find the mean number of boxes that must be purchased when there are n equally likely prizes:

$$M(n) = n\left(1 + \frac{1}{2} + \frac{1}{3} + \cdots + \frac{1}{n}\right)$$

i. Check the values in the table using this formula.

ii. What is the mean number of boxes you would have to buy to get all of 20 possible prizes?

20 The card below shows a scratch-off game. There are ten asteroids. Two say "Zap," two say "$1," and the other six name six larger cash prize amounts. All of the asteroids were originally covered. The instructions say:

Start anywhere. Rub off silver asteroids one at a time. Match 2 identical prizes BEFORE a "ZAP" appears and win that prize. There is only one winning match per card.

a. Describe how to conduct one run of a simulation to estimate the probability of winning a prize with this card. Conduct 10 runs of your simulation and estimate the probability.

b. Would the estimated probability be different if the prizes with no match weren't on the card?

21 Toni doesn't have a key ring and so just drops her keys into the bottom of her backpack. Her four keys—a house key, a car key, a locker key, and a key to her bicycle lock—are all about the same size.

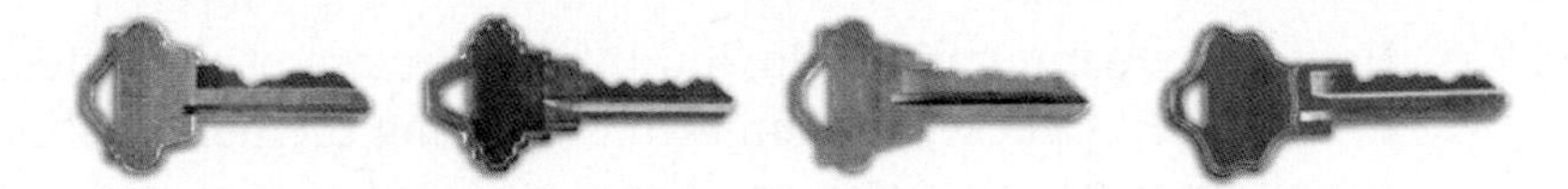

a. If she reaches into her backpack and grabs the first key she touches, what is the probability it is her car key?

b. If the key drawn is not her car key, she holds onto it. Then, without looking, she reaches into her backpack for a second key. If that key is not her car key, she holds on to both keys drawn and reaches in for a third key. Do the chances that Toni will grab her car key increase, decrease, or remain the same as she continues? Explain your reasoning.

c. The frequency table below gives the results of 1,000 simulations of this situation. From the frequency table, it appears that the numbers of keys needed are equally likely. Explain why this is the case.

Number of Grabs Toni Needs to Get Her Car Key	Frequency
1	255
2	252
3	236
4	257
Total Number of Runs	1,000

d. Estimate the probability that all four keys have to be drawn before Toni gets her car key.

22 Toni's key selection problem in Extensions Task 21 is one that depends on *order*—the order in which she chooses the keys. One way to model the problem would be to list all the possible orders in which the keys could be selected. An ordering of a set of objects is called a **permutation** of the objects. For example, the six permutations of the letters A, B, and C are:

ABC ACB BAC BCA CAB CBA

a. List all of the permutations of the letters A and B. How many are there?

b. List all of the permutations of the letters A, B, C, and D. How many are there?

c. Look for a pattern relating the number of permutations to the number of different letters.

d. How many permutations are there of Toni's four keys? What is the probability that all four keys have to be drawn before Toni gets her car key? Compare your answer with that for Extensions Task 21 Part d.

e. How many permutations do you think there are of the letters A, B, C, D, and E? Check your conjecture by using the permutations option on your calculator or computer software. (For most calculators, you need to enter "5 nPr 5". This means the number of permutations of 5 objects taken 5 at a time.)

23 In the history of National Basketball Association finals, the home team has won about 71% of the games. Suppose that the Los Angeles Lakers are playing the Philadelphia 76ers in the NBA finals. The two teams are equally good, except for this home team advantage. The finals are a best-of-seven series. The first two games will be played in Philadelphia, the next three (if needed) in Los Angeles, and the final two (if needed) in Philadelphia.

a. What is the probability that the 76ers will win a game if it is at home? What is the probability that the 76ers will win a game if it is played in Los Angeles? What is the probability the 76ers will win the first game of the series? The second game? The third game? The fourth game? The fifth game? The sixth game? The seventh game?

b. Describe how to conduct one run to simulate this series.

c. Conduct 5 runs and add your results to a copy of the frequency table below to make a total of 200 runs.

Number of Games Won by the 76ers	Frequency (Before)	Frequency (After)
0	9	
1	33	
2	21	
3	20	
4	112	
Total Number of Runs	195	200

d. What is your estimate of the probability that the 76ers win the finals?

e. Suppose that, to cut travel costs, the NBA schedules three games in Los Angeles followed by four in Philadelphia.

i. Design a simulation to estimate the probability that the 76ers win the finals in this situation.

ii. Conduct 5 runs, adding your results to those in a copy of the frequency table below.

Number of Games Won by the 76ers	Frequency (Before)	Frequency (After)
0	21	
1	21	
2	22	
3	19	
4	112	
Total Number of Runs	195	200

iii. What is your estimate of the probability that the 76ers win this series?

f. Compare the estimated probability that the 76ers win the series in Parts d and e. Does it matter which way the series is scheduled?

24 Suppose that you break a 10-inch strand of spaghetti at two random places.

a. On a copy of the diagram below, shade in the region that includes the points where a triangle can be formed from the three pieces.

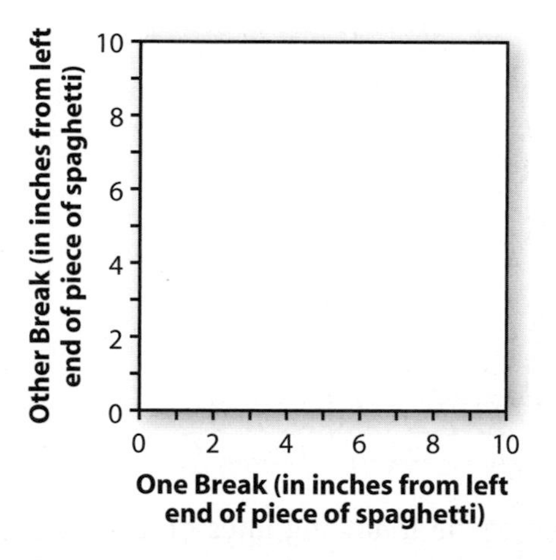

b. Find the probability that a triangle can be formed from the three pieces.

Review

25 Find the number that lies halfway between the two given numbers on a number line.

a. 0.005 and 0.006

b. 0.12345 and 0.123451

c. 0.52 and 0.520001

On Your Own

26 In reporting health statistics like births, deaths, and illnesses, rates are often expressed in phrases like "26 per 1,000" or "266 per 100,000." Express each of the following in equivalent percent language.

a. Infant mortalities occur in about 7 of every 1,000 live births in the United States.

b. Each year about 266 out of every 100,000 Americans die from heart disease.

c. Twins, triplets, or other multiple births occur in about 255 out of every 10,000 births in the United States.

27 Find the area of the shaded portion on each graph below.

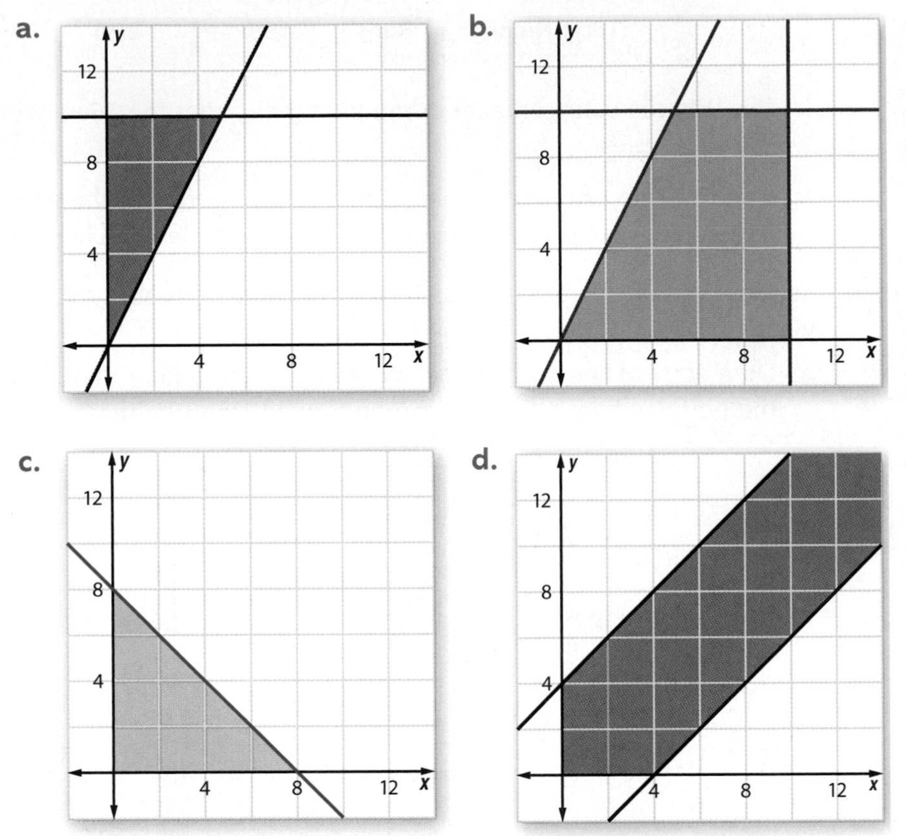

28 Graph each of the following lines on a separate coordinate system.

a. $y = -2$

b. $y = 8 - \frac{5}{3}x$

c. $x + y = 4$

d. $x = 3$

e. $y = 2x - 3$

29 John could take any of four different routes to get from his home to school and then take any one of five different routes to get to after-school soccer practice. How many different ways are there for him to travel from home to school to soccer practice? Illustrate your answer using a geometric diagram.

30 For each of the tables of values below:

i. Decide if the relationship between x and y can be represented by a linear, exponential, or quadratic function.

ii. Find an appropriate function rule for the relationship.

iii. Use your rule to find the y value that corresponds to an x value of 10.

a.

x	−2	−1	0	1	2	3	4	5
y	12	5	0	−3	−4	−3	0	5

b.

x	−2	−1	0	1	2	3	4	5
y	1	2.5	4	5.5	7	8.5	10	11.5

c.

x	−2	−1	0	1	2	3	4	5
y	1	2	4	8	16	32	64	128

31 Draw and label two right triangles PQR and WXY for which: $\angle Q$ and $\angle X$ are right angles, $\overline{PR} \cong \overline{WY}$, and $\angle R \cong \angle Y$. Explain as precisely as you can whether or not $\triangle PQR \cong \triangle WXY$.

32 Rewrite each expression in equivalent standard quadratic form, $ax^2 + bx + c$.

a. $4x(6 - x)$

b. $(x + 5)(x + 10)$

c. $x - 2x(x + 3)$

d. $(7x - 2)^2$

e. $(3x + 6)(3x - 6)$

f. $(3x + 5)(x - 9)$

33 Write each exponential expression in simplest possible equivalent form using only positive exponents.

a. $\dfrac{(2x)^3}{x^5}$

b. $(-4a^3bc^4)^3$

c. $(4x^3y)(-6x^4y)$

d. $3x^{-2}$

e. $(4n)^{-1}$

f. $\left(\dfrac{6xy^2}{2y}\right)^3$

34 Rhombus $ABCD$ has sides of length 11 cm.

a. Could the length of diagonal $\overline{AC}$ be 6 cm? What about 24 cm? Explain your reasoning.

b. If $m\angle A = 50°$, find the measures of the other three angles.

c. If rhombus $ABCD$ is a square, find the length of diagonal $\overline{BD}$.

Looking Back

When outcomes are equally likely, such as those from the roll of a die, you often can calculate probabilities exactly. In this unit, you learned how to write out sample spaces for such situations and to make probability distributions based on them. For more complex situations, you used geometric models to find probabilities or simulation to estimate probabilities. An important feature of your simulation models was the use of coins, dice, spinners, or random numbers to produce random outcomes. In each case, the simulation had the same mathematical characteristics as those of the original problem.

The tasks in this final lesson will help you review and apply key ideas for constructing probability distributions and using simulation and geometric models to solve problems involving chance.

1. Suppose that you roll two octahedral dice, which have the numbers 1 through 8 on each one.

   a. Make a sample space that shows all possible outcomes.

   b. Make a probability distribution table for the sum of the two dice.

   c. What is the probability that the sum is 8? At least 8?

   d. Make a probability distribution table for the absolute value of the difference of the two dice.

   e. What is the probability that the difference is 6? At most 6?

2. Show how to use the appropriate form of the Addition Rule to answer these questions about rolling two octahedral dice.

   a. What is the probability you get doubles or a sum of 7?

   b. What is the probability you get doubles or a sum of 8?

   c. What is the probability you get a sum of 7 or a sum of 8?

3 According to the 2000 census, about 10% of the population of Los Angeles is African American.

a. Juries have 12 members. Design a simulation model to determine the probability that a jury randomly selected from the population of Los Angeles would have no African American members.

b. Conduct 5 runs. Add your results to a copy of the frequency table below so that there is a total of 200 runs.

Number of African Americans on the Jury	Frequency (before)	Frequency (after)
0	56	
1	73	
2	45	
3	16	
4	4	
5	1	
⋮	⋮	⋮
Total Number of Runs	195	200

c. Add your results to a copy of the histogram below. Describe its shape and estimate the mean.

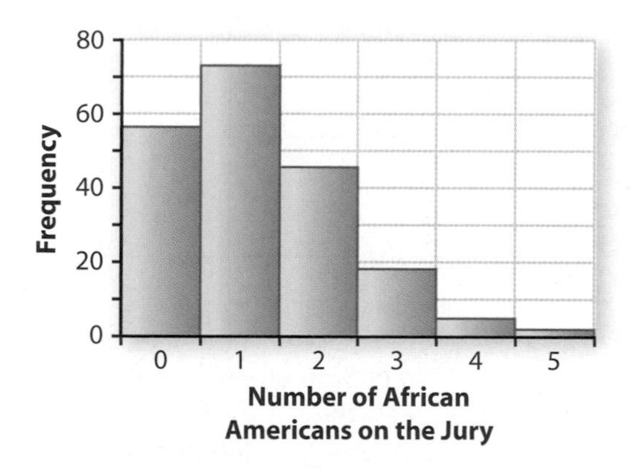

d. What is your estimate of the probability that a randomly selected jury of 12 people would have no African American members?

e. A *grand jury* decides whether there is enough evidence against a person to bring him or her to trial. A grand jury generally consists of 24 people. Do you think the probability that a randomly selected grand jury in Los Angeles would have no African American members is more, less, or the same as your answer to Part d? Why?

4 This roller coaster has 7 cars. Ranjana stands in a long line to get on the ride. When she gets to the front, the attendant directs her to the next empty car. No one has any choice of cars, but must take the next empty one in the coaster. Ranjana goes through the line 10 times, each time hoping she gets to sit in the front car.

a. Each time she goes through the line, what is the probability that Ranjana will sit in the front car? Do you think Ranjana has a good chance of sitting in the front car at least once in her 10 rides? Explain your reasoning.

b. Describe how to use random digits to conduct one run simulating this situation.

c. Perform 15 runs. Place the results in a frequency table that lists the number of times out of the 10 rides that Ranjana sits in the front car.

d. From your simulation, what is your estimate of the probability that Ranjana will sit in the front car at least once?

e. How could you get a better estimate of the probability that Ranjana will sit in the front car at least once?

5 Mark arrives at the gym at a random time between 7:00 and 7:30. Susan arrives at a random time between 7:10 and 7:40.

a. Shade in the region on the following diagram that represents the event that Mark gets to the gym before Susan.

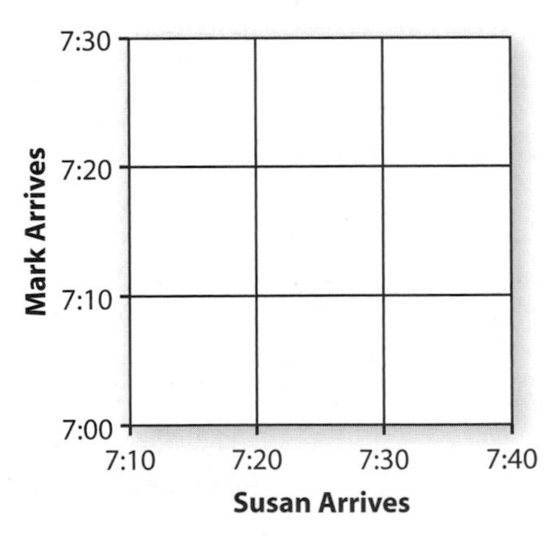

b. What is the probability that Mark gets to the gym before Susan?

c. What is the probability that Susan gets to the gym before Mark?

Summarize the Mathematics

In this unit, you learned how to find exact probabilities using a sample space of equally likely outcomes. You also learned how to use simulation and geometric diagrams to model more complex situations.

a. What is a probability distribution?

b. What are mutually exclusive events? Give an example of two mutually exclusive events. Give an example of two events that aren't mutually exclusive.

c. How can you find $P(A \text{ or } B)$ when A and B are mutually exclusive? When they aren't?

d. Summarize the steps involved in using simulation to model a problem involving chance.

e. Why doesn't simulation give you an "exact" probability? What does the Law of Large Numbers tell you about how to get a more precise answer?

f. Summarize the main ideas involved in using geometric probability models to solve problems involving chance.

Be prepared to share your ideas and reasoning with the class.

✓ *Check Your Understanding*

Write, in outline form, a summary of the important mathematical concepts and methods developed in this unit. Organize your summary so that it can be used as a quick reference in future units and courses.

Glossary/Glosario

Math Online A mathematics multilingual glossary is available at www.math.glencoe.com/multilingual_glossary. The Glossary includes the following languages:

Arabic	English	Korean	Tagalog
Bengali	Hatian Creole	Russian	Urdu
Cantonese	Hmong	Spanish	Vietnamese

English	Español
A	
Acute triangle (p. 68) A triangle with three *acute angles* (angles whose measures are less than 90°).	**Triángulo acutángulo** (pág. 68) Triángulo con tres *ángulos agudos* (ángulos cuyas medidas son menos de 90°).
Adjacency matrix (p. 248) A matrix representation of a vertex-edge graph in which each entry of the matrix is the number of edges between the corresponding pair of vertices.	**Matriz de adyacencia** (pág. 248) Representación matricial de un grafo en que cada entrada de la matriz es el número de aristas entre pares correspondientes de vértices.
Adjacent vertices (p. 248) Two vertices are adjacent if there is an edge between them.	**Vértices adyacentes** (pág. 248) Dos vértices son adyacentes si hay una arista entre ellos.
Algorithm (p. 246) A list of step-by-step instructions or a systematic step-by-step procedure.	**Algoritmo** (pág. 246) Lista de instrucciones detalladas o procedimiento detallado.
Altitude In a triangle, the perpendicular segment from a side to the opposite vertex (p. 389). In a parallelogram, a perpendicular segment from the line containing the base to the opposite side (p. 389). In a prism, a segment that is perpendicular to the planes of both bases (p. 428). In a pyramid, the perpendicular segment from the plane of the base to the apex (p. 428). In a cone, the perpendicular segment from the plane of the base to the vertex. In a cylinder, a perpendicular from the plane of one base to the plane of the other base.	**Altura** En un triángulo, el segmento perpendicular de un lado al vértice opuesto (pág. 389). En un paralelogramo, un segmento perpendicular de la recta que contiene la base al lado opuesto (pág. 389). En un prisma, segmento perpendicular a ambas bases (pág. 428). En una pirámide, el segmento perpendicular del plano de la base al vértice (pág. 428). En un cono, el segmento perpendicular del plano de la base al vértice. En un cilindro, la perpendicular del plano de una de las bases al plano de la otra.
Apex *see* **pyramid**	**Vértice** *véase* **pirámide**
Approximately normal distribution (mound-shaped) (p. 77) A data or probability distribution that has one peak and tapers off on both sides. Normal distributions are **symmetric**—the two halves look like mirror images of each other.	**Distribución aproximadamente normal (campaniforme)** (pág. 77) Datos o distribución probabilística que tiene un pico y que disminuye en ambos lados. Las distribuciones normales son **simétricas**: las dos mitades son imágenes especulares mutuas.
B	
Back-to-back stemplot (p. 98) A stemplot used to compare two sets of data. The center column contains the stem of the data, while the right leaf contains one data set and the left leaf the other.	**Esquema de tallos consecutivo** (pág. 98) Un esquema de tallos que se usa para comparar dos conjuntos de datos. La columna central lleva los tallos de los datos, la hoja derecha lleva uno de los conjuntos de datos y la izquierda el otro.
Base *see* **exponential expression, parallelogram, prism, pyramid,** and **triangle**	**Base** *véase* **expresión exponencial, paralelogramo, prisma, pirámide** y **triángulo**
Base angles of a triangle (p. 372) Angles opposite the congruent sides of an isosceles triangle. These angles are congruent.	**Ángulos basales de un triángulo** (pág. 372) Ángulos opuestos a los lados congruentes de un triángulo isósceles. Dichos ángulos son congruentes.

Glossary/Glosario

English

Bimodal distribution (p. 96) A distribution with two well-defined peaks.

Bisector of an angle (p. 385) A ray that begins at the vertex of an angle and divides the angle into two angles of equal measure.

Box plot (box-and-whiskers plot) (p. 109) A statistical graphic in which only the minimum, lower quartile, median, upper quartile, maximum, and outliers are displayed.

C

Categorical data (p. 94) Data that fall into categories such as male/female or freshman/sophomore/junior/senior.

Central angle of a regular polygon (p. 401) An angle whose vertex is at the center of the polygon and whose sides (rays) extend through the endpoints of a side of the polygon.

Chromatic number (p. 275) The smallest number of colors needed to color all the vertices of a graph. (*see* **color a graph**)

Coefficient (p. 157) A number in front of a variable. For example, in the expression $x^2 - 10x + 25$, the coefficient of x^2 is 1 and the coefficient of x is -10.

Color a graph (p. 269) Assign a color to each vertex of a graph so that adjacent vertices have different colors. This may also be referred to as *graph coloring* or *vertex coloring*.

Coloring the edges (p. 282) Assign a color to each edge of a graph so that edges that share a vertex have different colors.

Computer Algebra System (CAS) (p. 54) Software that directs a calculator or computer to perform numeric, graphic, and symbolic mathematical operations required in arithmetic, algebra, calculus, and beyond.

Cone (p. 430) A figure formed by a circular region (the *base*), a point (the *vertex*) not in the plane of the base, and all of the segments joining the vertex to the base.

Español

Distribución bimodal (pág. 96) Distribución con dos picos bien definidos.

Bisectriz de un ángulo (pág. 385) Rayo que parte del vértice de un ángulo y que lo divide en dos ángulos congruentes.

Diagrama de caja (diagrama de caja y patillas) (pág. 109) Gráfica estadística en que sólo se muestran el mínimo, el cuartil inferior, la mediana, el cuartil superior, el máximo y los valores atípicos.

Datos categóricos (pág. 94) Datos que caen en categorías como masculino/femenino o primer año/segundo año/tercer año/cuarto año.

Ángulo central de un polígono regular (pág. 401) Ángulo cuyo vértice está en el centro del polígono y cuyos lados (rayos) se extienden por los extremos de un lado del polígono.

Número cromático (pág. 275) Mínimo de colores que se requieren para pintar todos los vértices de un grafo. (*véase* **coloración de un grafo**)

Coeficiente (pág. 157) Número delante de una variable. Por ejemplo, en la expresión $x^2 - 10x + 25$, el coeficiente de x^2 es 1 y el de x es -10.

Colorear un grafo (pág. 269) Asignación de un color a cada vértice de un grafo de modo que vértices adyacentes tengan colores distintos. Esto se llama también *coloración de un grafo* o *coloración de vértices*.

Coloración de aristas (pág. 282) Asignación de un color a cada arista de un grafo de modo que las aristas con vértices comunes tengan colores distintos.

Sistema algebraico computacional (SAC) (pág. 54) Programas que hacen que una computadora ejecute operaciones matemáticas numéricas, gráficas o simbólicas que se requieren en aritmética, álgebra, cálculo y otros.

Cono (pág. 430) Figura formada por una región circular (*la base*), un punto (*el vértice*) fuera del plano de la base y todos los segmentos que unen el vértice a la base.

Glossary/Glosario

English

Congruent figures (p. 370) Figures that have the same shape and size, regardless of position or orientation.

Connected graph (p. 243) A graph that is all in one piece. That is, from each vertex there is at least one path to every other vertex.

Consecutive angles (p. 387) In a polygon, two angles whose vertices are adjacent.

Constant term (p. 157) The term in an algebraic expression in which a variable does not appear. For example, in the expression $x^2 - 10x + 25$, the 25 is the constant term.

Counterexample (p. 451) A statement or diagram that shows that a given statement is not always true.

Converse (p. 380) A statement formed by interchanging the "if"-clause and the "then"-clause of a given "if-then" statement.

Convex polygon (p. 404) A polygon in which no segment connecting any two vertices of the polygon contains points in the exterior of the polygon. Otherwise the polygon is called *nonconvex*.

Convex polyhedron (p. 426) A polyhedron in which no segment connecting any two vertices of the polyhedron contains points in the exterior of the polyhedron. Otherwise the polyhedron is called *nonconvex*.

Cube (regular hexahedron) (p. 427) A regular polyhedron with six congruent, square faces.

Cycle graph (p. 279) A vertex-edge graph consisting of a single cycle—a route that uses each edge and vertex exactly once and ends where it started.

Cylinder (p. 430) A figure formed by two congruent circular regions (the *bases*) contained in parallel planes along with all segments having an endpoint on each base and parallel to the line joining the centers of the bases.

D

Dart (p. 422) A nonconvex quadrilateral with two distinct pairs of consecutive sides the same length.

Data distribution (pp. 75–82) The collection of data values, typically summarized in a table or plotted so that the number or proportion of times that each value occurs can be observed.

Español

Figuras congruentes (pág. 370) Figuras de la misma forma y tamaño, sea cual sea su posición u orientación.

Grafo conexo (pág. 243) Un grafo que es de una sola pieza, o sea, de cada vértice hay por lo menos un camino a cada uno de los otros vértices.

Ángulos consecutivos (pág. 387) En un polígono, dos ángulos cuyos vértices son adyacentes.

Término constante (pág. 157) Término de una expresión algebraica sin variables. Por ejemplo, en la expresión $x^2 - 10x + 25$, 25 es un término constante.

Contraejemplo (pág. 451) Enunciado o diagrama que prueba que un enunciado dado no es verdadero.

Recíproco (pág. 380) Enunciado que resulta al intercambiar las cláusulas "si" y "entonces" de un enunciado "si-entonces."

Polígono convexo (pág. 404) Polígono en que no hay segmento que una dos de sus vértices y que contenga puntos fuera del mismo. De lo contrario, el polígono se llama *no convexo*.

Poliedro convexo (pág. 426) Poliedro en que no hay segmento que una dos de sus vértices y que contenga puntos fuera del mismo. De lo contrario, el poliedro se llama *no convexo*.

Cubo (hexaedro regular) (pág. 427) Poliedro regular con seis caras cuadradas congruentes.

Grafo cíclico (pág. 279) Grafo que consta de un solo ciclo: un camino que pasa por cada arista y cada vértice solo una vez y que termina donde empezó.

Cilindro (pág. 430) Figura formada por dos regiones circulares congruentes (las *bases*) contenidas en planos paralelos, junto con todos los segmentos que tienen un extremo en cada base y que son paralelos a la recta que une los centros de las bases.

Dardo (pág. 422) Cuadrilátero no convexo con dos pares de lados consecutivos que tienen la misma longitud.

Distribución de datos (págs. 75–82) Conjunto de datos típicamente tabulados o graficados de modo que pueda observarse el número o proporción de veces que ocurre cada valor.

Glossary/Glosario

English

Data transformation (p. 124) A change in each value in a set of data such as adding the same constant to each value, taking the square root of each value, or dividing each value by the same constant; often used to change units of measure.

Degree of a vertex (p. 245) The number of edges touching a vertex. If an edge loops back to the same vertex, that counts as two edge-touchings.

Dependent variable (p. 6) A dependent variable is one whose value changes in response to change in one or more related independent variables.

Deviation from the mean (p. 118) The difference between a data value and the mean of its distribution.

Directed edge (p. 262) An edge in a vertex-edge graph with a direction indicated.

Directed graph (digraph) (p. 262) A vertex-edge graph in which all the edges are directed.

Dot plot (number line plot) (p. 76) A statistical graphic where dots that represent data values are plotted above a number line.

Doubling time (p. 314) For a quantity growing exponentially, the time it takes for the quantity to double.

Dual of a tessellation (p. 418) A tessellation obtained by connecting the centers of regular polygons that share a common edge in a given tessellation of regular polygons.

E

Edge (of a vertex-edge graph) (p. 241) Segment or arc joining two vertices in a vertex-edge graph.

Equally-likely outcomes (p. 534) Outcomes that all have the same probability of occurring.

Equation A statement using symbols indicating that two expressions are equivalent.

Equilateral polygon (p. 419) A polygon in which all sides have equal length.

Equivalent expressions (p. 215) Expressions that produce equal output values from all possible equal input values.

Español

Transformación de datos (pág. 124) Cambio de cada valor en un conjunto de datos, como la adición de la misma constante a cada valor, la extracción de la raíz cuadrada de cada valor o la división de cada valor entre la misma constante; se usan a menudo para convertir unidades de medida.

Grado de un vértice (pág. 245) Número de aristas que concurren en un vértice. Los lazos se cuentan dos veces.

Variable dependiente (pág. 6) Variable cuyos valores cambian en respuesta a cambios en una o más variables independientes relacionadas.

Desviación de la media (pág. 118) Diferencia entre un dato y la media de su distribución.

Arista dirigida (pág. 262) Arista de un grafo en que se indica la dirección de la misma.

Grafo dirigido (digrafo) (pág. 262) Grafo en que todas las aristas poseen dirección.

Esquema de puntos (esquema lineal numérico) (pág. 76) Gráfica estadística en que puntos que corresponden a los datos se grafican encima de una recta numérica.

Tiempo de duplicación (pág. 314) Tiempo que tarda en duplicarse una cantidad que crece exponencialmente.

Dual de un teselado (pág. 418) En un teselado dado de polígonos regulares, teselado que se obtiene al unir los centros de los polígonos regulares que tienen una arista común.

E

Arista (de un grafo) (pág. 241) Segmento o arco que une dos vértices de un grafo.

Resultados equiprobables (pág. 534) Resultados que tienen la misma probabilidad de ocurrir.

Ecuación Enunciado que usa símbolos y que indica que dos expresiones son iguales.

Polígono equilátero (pág. 419) Polígono cuyos lados tienen todos la misma longitud.

Expresiones equivalentes (pág. 215) Expresiones que producen los mismos valores de salida para todos los valores de entrada posibles iguales.

Glossary/Glosario

English	Español
Euler circuit (p. 243) A route through a connected graph such that (1) each edge is used exactly once, and (2) the route starts and ends at the same vertex.	**Circuito de Euler** (pág. 243) Camino en un grafo conexo de modo que (1) cada arista se recorre sólo una vez y (2) el camino empieza y termina en el mismo vértice.
Euler path (p. 256) A route through a connected graph that traces each edge of the graph exactly once.	**Camino de Euler** (pág. 256) Camino en un grafo conexo que traza cada arista sólo una vez.
Expanding (p. 221) Rewriting an algebraic expression with parentheses as an equivalent expression that does not contain parentheses.	**Desarrollo** (pág. 221) Volver a escribir una expresión algebraica con paréntesis como una expresión equivalente sin paréntesis.
Exponent (power) *see* **exponential expression**	**Exponente (potencia)** *véase* **expresión exponencial**
Exponential decay (p. 323) Process in which change of the dependent variable can be modeled by an exponential function with rule in the form $y = a(b^x)$ where $a > 0$ and $0 < b < 1$.	**Desintegración exponencial** (pág. 323) Proceso en que el cambio de la variable dependiente viene dado por una función exponencial de la forma $y = a(b^x)$, con $a > 0$ y $0 < b < 1$.
Exponential expression (p. 304) An algebraic expression in the form b^n, where b and n are real numbers or variables. The number b is called the *base* of the exponential expression, and n is called the **exponent** or the **power**.	**Expresión exponencial** (pág. 304) Expresión algebraica de la forma b^n, con b y n números reales o variables. El número b es la base de la expresión exponencial y n es su **exponente** o **potencia**.
Exponential function (pp. 296; 328) A function of the form $y = a(b^x)$ where $a \neq 0$ and $0 < b < 1$ or $b > 1$.	**Función exponencial** (págs. 296; 328) Función de la forma $y = a(b^x)$, con $a \neq 0$ y $0 < b < 1$ ó $b > 1$.
Exponential growth (p. 291) Process in which change of the dependent variable can be modeled by an exponential function with rule in the form $y = a(b^x)$ where $a > 0$ and $b > 1$.	**Crecimiento exponencial** (pág. 291) Proceso en que el cambio de la variable dependiente viene dado por una función exponencial de la forma $y = a(b^x)$, con $a > 0$ y $b > 1$.
Expression A symbolic representation of a calculation procedure.	**Expresión** Representación simbólica de un procedimiento de cálculo.
Exterior angle of a convex polygon (p. 405) An angle formed at a vertex of the polygon by one side and the extension of the adjacent side.	**Ángulo exterior de un polígono convexo** (pág. 405) Ángulo formado en un vértice del polígono por un lado y la extensión del lado adyacente.

F

English	Español
Factoring (p. 221) Rewriting an algebraic expression in an equivalent form as a product of several expressions.	**Factorización** (pág. 221) Replanteamiento de una expresión algebraica en una equivalente que sea un producto de varias expresiones.
Fibonacci sequence (p. 35) The sequence of numbers 1, 1, 2, 3, 5, 8, 13, 21, 34, … .	**Sucesión de Fibonacci** (pág. 35) La sucesión de números 1, 1, 2, 3, 5, 8, 13, 21, 34, … .
Five-number summary (p. 108) The minimum, lower quartile (Q_1), median (Q_2), upper quartile (Q_3), and maximum of a data set.	**Resumen de cinco números** (pág. 108) El mínimo, el cuartil inferior (Q_1), la mediana (Q_2), el cuartil superior (Q_3) y el máximo de un conjunto de datos.

Glossary/Glosario

English	Español

Frequency table (p. 87) A summary table for numerical data, where typically the column on the left gives the different data values and the column on the right gives the number of times each value occurs.

Tabla de frecuencias (pág. 87) Una tabla sumarial de datos numéricos, donde la columna de la izquierda lleva típicamente los diversos valores de los datos y la columna de la derecha lleva el número de veces que aparece cada valor.

Function (in one variable) (p. 69) A relationship between two variables in which each value of the independent variable corresponds to exactly one value of the dependent variable.

Función (de una variable) (pág. 69) Relación entre dos variables en que a cada valor de la variable independiente corresponde un solo valor de la variable dependiente.

Function graph The set of points (x, y) on a coordinate grid whose coordinates are related by a function.

Gráfica de una función El conjunto de puntos (x, y) en un cuadriculado cuyas coordenadas están relacionadas por una función.

G

Graph *see* **vertex-edge graph, function graph**

Gráfica *véase* **grafo, gráfica de función**

Graph coloring *see* **color a graph**

Coloración de un grafo *véase* **colorear un grafo**

H

Half-life (p. 327) For a quantity decaying exponentially, the amount of time it takes for the quantity to diminish by half.

Media vida (pág. 327) Tiempo que tarda en reducirse a la mitad una cantidad que decrece exponencialmente.

Height of a figure (p. 389) The length of the figure's altitude. (*see* **altitude**)

Altura de una figura (pág. 389) Longitud de la altura de una figura. (*véase* **altura**)

Histogram (p. 78) A statistical graphic for numerical data, where the height of a bar shows the *frequency* or count of the values that lie within the interval covered by the bar.

Histograma (pág. 78) Gráfica estadística para datos numéricos en que la altura de una barra muestra la *frecuencia* o cuenta de los valores que yacen en el intervalo cubierto por la barra.

I

Independent variable (p. 6) Variables whose values are free to be changed in ways that are restricted by the context of the problem or by mathematical restrictions on allowed values. These variables influence the values of other variables called *dependent variables.*

Variable independiente (pág. 6) Variable cuyos valores cambian libremente según las restricciones de un problema o por restricciones matemáticas sobre los valores permisibles. Estas variables influyen en los valores de otras variables, las llamadas *variables dependientes.*

Interquartile range (IQR) (p. 108) A measure of spread; the distance between the first and third quartiles.

Rango intercuartílico (RI) (pág. 108) Medida de dispersión; distancia entre los cuartiles primero y tercero.

Isosceles triangle (p. 68) A triangle with at least two sides of equal length. The noncongruent side, if any, is called the *base*, and the angles that lie opposite the congruent sides are called the *base angles.*

Triángulo isósceles (pág. 68) Triángulo con por lo menos dos lados congruentes. El lado no congruente, si es que existe, se llama *base* y los ángulos opuestos a los lados congruentes se llaman *ángulos basales.*

Glossary/Glosario

English	Español

K

Kite (p. 366) A convex quadrilateral with two distinct pairs of consecutive sides the same length.

Deltoide (pág. 366) Cuadrilátero convexo con exactamente dos pares distintos de lados consecutivos de la misma longitud.

L

Lateral face *see* **prism, pyramid**

Cara lateral *véase* **prisma, pirámide**

Linear equation (p. 189) An equation in which expressions on both sides of the equal sign are either numbers or linear expressions.

Ecuación lineal (pág. 189) Ecuación en que las expresiones en ambos lados del signo de igualdad son números o expresiones lineales.

Linear expression (p. 216) An expression that defines a linear function.

Expresión lineal (pág. 216) Expresión que define una función lineal.

Linear function (p. 150) A function of the form $y = a + bx$ where a and b are real numbers.

Función lineal (pág. 150) Función de la forma $y = a + bx$, donde a y b números reales.

Linear inequality (p. 189) An inequality in which expressions on both sides of the inequality sign are either numbers or linear expressions.

Desigualdad lineal (pág. 189) Desigualdad en que las expresiones en ambos lados del signo de desigualdad son o números o expresiones lineales.

Linear regression (p. 166) A systematic method of finding linear mathematical models for patterns in (x, y) data sets.

Regresión lineal (pág. 166) Método sistemático para hallar modelos matemáticos lineales de patrones en conjuntos de datos de la forma (x, y).

Loop (p. 262) An edge in a graph connecting a vertex to itself.

Lazo (pág. 262) Arista de un grafo que une un vértice a sí mismo.

M

Main diagonal of a matrix (p. 248) The entries in a square matrix running from the top-left corner of the matrix to the bottom-right corner.

Diagonal principal de una matriz (pág. 248) Las entradas de una matriz que van de la esquina superior izquierda de la matriz a la esquina inferior derecha.

Mathematical model (pp. 162; 240) A symbolic or pictorial representation including only the essential features of a problem situation.

Modelo matemático (págs. 162; 240) Representación simbólica o pictórica de un problema que sólo incluye sus características esenciales.

Matrix (plural: matrices) (p. 248) A rectangular array of numbers.

Matriz (plural: Matrices) (pág. 248) Arreglo rectangular de números.

Mean (arithmetic average) (p. 84) The sum of the values in a data set divided by how many values there are; the balance point of the distribution.

Media (promedio aritmético) (pág. 84) Suma de los valores de un conjunto de datos dividida entre el número de valores; punto de equilibrio de la distribución.

Mean absolute deviation (MAD) (p. 140) A measure of variability in a data set found by computing the mean of the absolute values of the deviations from the mean of the distribution.

Desviación absoluta media (DAM) (pág. 140) Medida de variabilidad de un conjunto de datos que se halla calculando la media de los valores absolutos de las desviaciones de la media de la distribución.

Glossary/Glosario

English	Español
Measure of center (p. 83) Numerical summary of the center of a distribution, such as the mean or median.	**Medida central** (pág. 83) Resumen numérico del centro de una distribución, como la media o la mediana.
Measure of position (p. 104) A number that tells the position of a data value in its distribution, such as a percentile or a deviation from the mean.	**Medida de posición** (pág. 104) Número que indica la posición de un dato en su distribución, como el percentil o la desviación de la media.
Measure of spread (measure of variability) Numerical summary of the variability of the values in a distribution, such as the range, interquartile range, or standard deviation.	**Medida de dispersión (medida de variabilidad)** Resumen numérico de la variabilidad de los valores de una distribución, como el rango, el rango intercuartílico o la desviación estándar.
Median (second quartile, Q_2) (p. 84) The value in the middle of an ordered list of data; the 50th percentile. If there are an even number of values, the mean of the two values in the middle.	**Mediana (segundo cuartil, Q_2)** (pág. 84) Valor central de una lista ordenada de datos; percentil quincuagésimo. Si hay un número par de valores, la mediana es la media de los dos valores centrales.
Median of a triangle (p. 95) The line segment joining a vertex to the midpoint of the opposite side.	**Mediana de un triángulo** (pág. 95) Segmento de recta que une un vértice al punto medio del lado opuesto.
Mode (p. 94) The most frequent value in a set of numerical data; the category with the highest frequency in a set of categorical data is called the modal category.	**Moda** (pág. 94) Valor más frecuente de un conjunto de datos numéricos; en un conjunto de datos categóricos, la categoría de mayor frecuencia, la llamada categoría modal.
Mutually-exclusive events (disjoint) (p. 539) Events that cannot occur on the same outcome of a probability experiment.	**Eventos mutuamente excluyentes (disjuntos)** (pág. 539) Eventos que no pueden ocurrir en el mismo resultado de un experimento probabilístico.

N

English	Español
Net (p. 426) A two-dimensional pattern consisting of polygons that can be folded along edges to form a polyhedron.	**Red** (pág. 426) Patrón bidimensional que consta de polígonos que forman un poliedro al plegarse a lo largo de sus aristas.
Nonconvex *see* **convex**	**No convexo** *véase* **convexo**
Nonperiodic tessellation *see* **periodic tessellation**	**Teselado aperiódico** *véase* **teselado periódico**
Normal distribution *see* **approximately normal**	**Distribución normal** *véase* **aproximadamente normal**
***NOW-NEXT* rule** (p. 29) An equation that shows how to calculate the value of the next term in a sequence from the value of the current term.	**Regla de recurrencia** (pág. 29) Ecuación que muestra cómo hallar el valor del término siguiente a partir del valor del término actual.

O

English	Español
Oblique drawing (p. 433) A way to depict three-dimensional objects that maintains parallelism of lines.	**Proyección oblicua** (pág. 433) Forma de presentar sólidos, que mantiene el paralelismo de rectas.

Glossary/Glosario

English

Oblique prism (p. 428) A prism in which some lateral faces are parallelograms that are not rectangles.

Obtuse triangle (p. 68) A triangle with an *obtuse angle* (an angle with measure greater than 90°).

Opposite angles (p. 375) In a triangle $\triangle ABC$, $\angle A$ is opposite $\overline{BC}$, $\angle B$ is opposite $\overline{AC}$, and $\angle C$ is opposite $\overline{AB}$. In a quadrilateral $ABCD$ $\angle A$ is opposite $\angle C$ and $\angle B$ is opposite $\angle D$.

Orthographic drawing (p. 432) A way to depict three-dimensional objects by sketching several two-dimensional face-views such as a top view, a front view, and a right-side view.

Outlier (p. 77) A data value that lies far away from the bulk of the other values; for single-variable data, an unusually large or an unusually small value.

P

***P*(*A* and *B*)** (p. 540) The probability that *A* and *B* both happen on the same outcome.

***P*(*A* or *B*)** (p. 540) The probability that *A* or *B* occurs.

Parabola (p. 470) The shape of the graph of a quadratic function.

Parallel lines (p. 177) Lines that are coplanar and do not intersect.

Parallel planes Planes that do not intersect.

Parallelogram (pp. 365; 389) A quadrilateral with opposite sides of equal length. Any side may be designated the *base*, and an *altitude* to that base is a perpendicular segment from the line containing the base to the opposite side.

Percentile (p. 104) A way of describing the position of a value in a distribution. The 60th percentile, for example, is the value that separates the smallest 60% of the data values from the largest 40%.

Periodic tessellation (p. 421) A tessellation that fits exactly on itself when translated in different directions. Such a tessellation has translation symmetry. A tessellation that does not have any translation symmetry is called *nonperiodic*.

Español

Prisma oblicuo (pág. 428) Prisma en que algunas caras laterales son paralelogramos que no son rectángulos.

Triángulo obtusángulo (pág. 68) Triángulo con un ángulo obtuso (uno que mide más de 90°).

Ángulos opuestos (pág. 375) En un triángulo $\triangle ABC$, $\angle A$ se opone a $\overline{BC}$, $\angle B$ se opone a $\overline{AC}$, y $\angle C$ se opone a $\overline{AB}$. En un cuadrilátero $ABCD$ $\angle A$ se opone a $\angle C$ y $\angle B$ se opone a $\angle D$.

Proyección ortogonal (pág. 432) Forma de presentar sólidos mediante el bosquejo de varias vistas fisonómicas bidimensionales, como las vistas superior, frontal o derecha.

Valor atípico (pág. 77) Dato que está muy alejado del grueso de los otros valores; para datos de una variable, un valor inusualmente grande o inusualmente pequeño.

P

***P*(*A* y *B*)** (pág. 540) Probabilidad que *A* y *B* ocurran ambos en el mismo resultado.

***P*(*A* o *B*)** (pág. 540) Probabilidad de que ocurra *A* o *B*.

Parábola (pág. 470) La forma de la gráfica de una función cuadrática.

Rectas paralelas (pág. 177) Rectas coplanarias que no se intersecan.

Planos paralelos Planos que no se intersecan.

Paralelogramo (págs. 365; 389) Cuadrilátero de lados opuestos de la misma longitud. Cualquier lado es la *base* y la *altura* correspondiente es el segmento perpendicular trazado de la recta que contiene la base al lado opuesto.

Percentil (pág. 104) Forma de describir la posición de un valor en una distribución. El percentil sexagésimo, por ejemplo, es el valor que separa el 60% inferior de los datos del 40% superior de los mismos.

Teselado periódico (pág. 421) Teselado que encaja perfectamente en sí mismo cuando se traslada en diversas direcciones. Tal teselado posee simetría de traslación. Un teselado que carece de tal simetría se llama *aperiódico*.

Glossary/Glosario

Permutation ↔ English/Español two-column glossary merged in reading order.

English

Permutation (p. 581) A rearrangement of a finite set of objects.

Perpendicular bisector of a segment (p. 374) A line that is perpendicular to a segment and contains its midpoint.

Perpendicular lines (p. 396) Lines that intersect to form a right angle (an angle with measure of 90°).

Planar graph (p. 278) A vertex-edge graph that can be drawn in the plane so that edges intersect only at the vertices.

Polygon (p. 398) A closed figure in a plane, formed by connecting line segments (sides) endpoint-to-endpoint (vertices) with each segment meeting exactly two other segments. Polygons with four, five, six, seven, and eights sides are called quadrilaterals, pentagons, hexagons, septagons, and octagons respectively. An n-gon is a polygon with n sides.

Polyhedron (*plural:* polyhedra) (p. 426) A three-dimensional counterpart of a polygon, made up of a set of polygons that encloses a single region of space. Exactly two polygons (faces) meet at each edge and three or more edges meet at each vertex.

Prism (p. 428) A polyhedron with two congruent polygonal faces, called *bases*, contained in parallel planes, and joined by parallelogram faces called *lateral faces*.

Probability distribution (p. 534) A description of all possible numerical outcomes of a random situation, along with the probability that each occurs; may be in table, formula, or graphical form.

Pyramid (p. 428) A polyhedron in which all but one of the faces must be triangular and share a common vertex. The triangular faces are called *lateral faces*, and the *apex* is the vertex that is common to the lateral faces. The *base* is the face that does not contain the apex.

Q

Quadratic equation (p. 511) An equation in which expressions on both sides of the equal sign are either numbers, linear expressions, or quadratic expressions and at least one of those expressions is quadratic.

Español

Permutación (pág. 581) Una reordenación de un conjunto finito de objetos.

Mediatriz de un segmento (pág. 374) Recta perpendicular a un segmento y que contiene por su punto medio.

Rectas perpendiculares (pág. 396) Rectas que se intersecan en ángulo recto (uno que mide 90°).

Grafo planar (pág. 278) Grafo que puede trazarse en el plano de modo que sus aristas se intersequen sólo en los vértices.

Polígono (pág. 398) Figura cerrada planar, que consta de segmentos de recta (los lados), unidos extremo a extremo (los vértices) y cada segmento sólo interseca a otros dos segmentos. Los polígonos de cuatro, cinco, seis, siete y ocho lados se llaman cuadriláteros, pentágonos, hexágonos, heptágonos y octágonos, respectivamente. Un e$n$ágono es un polígono de n lados.

Poliedro (pág. 426) Homólogo tridimensional de un polígono, compuesto por un conjunto de polígonos que encierran una sola región del espacio. Sólo se intersecan dos polígonos (caras) en cada arista y tres o más aristas concurren en cada vértice.

Prisma (pág. 428) Poliedro con dos polígonos congruentes y paralelos (las bases), unidas por paralelogramos (las caras laterales).

Distribución probabilística (pág. 534) Descripción de todos los resultados numéricos posibles de una situación aleatoria, junto con la probabilidad de cada uno; puede darse en una tabla, fórmula o gráfica.

Pirámide (pág. 428) Poliedro en que todas las caras, salvo una, son triangulares y tienen un vértice común. Las caras triangulares se llaman *caras laterales* y el *vértice* es común a todas ellas. La *base* es la cara que no contiene el vértice.

Ecuación cuadrática (pág. 511) Ecuación con por lo menos una expresión cuadrática y en la cual las expresiones en ambos lados del signo de igualdad son números, expresiones lineales o expresiones cuadráticas.

Glossary/Glosario

English	Español

Quadratic expression (p. 494) An expression that defines a quadratic function.

Expresión cuadrática (pág. 494) Expresión que define una función cuadrática.

Quadratic function (p. 470) A function of the form $y = ax^2 + bx + c$ where a, b, and c are real numbers and $a \neq 0$.

Función cuadrática (pág. 470) Función de la forma $y = ax^2 + bx + c$, donde a, b, c son números reales y $a \neq 0$.

Quartile, lower (first quartile, Q_1) (p. 108) The value that divides the ordered list of data into the smallest one-fourth and the largest three-fourths; the median of the smaller half of the values; the 25th percentile.

Cuartil inferior (primer cuartil, Q_1) (pág. 108) Valor que divide una lista ordenada de datos en el cuarto inferior y los tres cuartos superiores; mediana de la mitad inferior de los valores; percentil vigésimo quinto.

Quartile, upper (third quartile, Q_3) (p. 108) The value that divides the ordered list of data into the smallest three-fourths and the largest one-fourth; the median of the larger half of the values; the 75th percentile.

Cuartil superior (tercer cuartil, Q_3) (pág. 108) Valor que divide una lista ordenada de datos en los tres cuartos inferiores y el cuarto superior; mediana de la mitad superior de los valores; percentil septuagésimo quinto.

R

Random digit (p. 558) A digit selected from 0, 1, 2, 3, 4, 5, 6, 7, 8, 9 in a way that makes each of the digits equally likely to be chosen (has probability $\frac{1}{10}$); successive random digits should be independent, which means that if you know what random digits have already been selected, each digit from 0 through 9 still has probability $\frac{1}{10}$ of being the next digit.

Dígito aleatorio (pág. 558) Dígito escogido de 0, 1, 2, 3, 4, 5, 6, 7, 8, 9, de modo que cada uno tenga la misma probabilidad de elegirse que cualquier otro (tiene probabilidad $\frac{1}{10}$); los dígitos aleatorios consecutivos deben ser independientes, o sea, si conoces los dígitos aleatorios ya escogidos, cada dígito de 0 a 9 aún tiene $\frac{1}{10}$ de probabilidad de escogerse como el dígito siguiente.

Range (p. 77) A measure of spread; the difference between the largest value and the smallest value in a data set.

Rango (pág. 77) Medida de dispersión; diferencia entre los valores máximo y mínimo de un conjunto de datos.

Rate of change (p. 155) The ratio of change in value of a dependent variable to change in value of a corresponding independent variable.

Tasa de cambio (pág. 155) La razón de cambio en valor de una variable dependiente al cambio en valor de la variable independiente correspondiente.

Rectangle (p. 365) A quadrilateral with opposite sides of equal length and one right angle.

Rectángulo (pág. 365) Un cuadrilátero con los lados opuestos de la longitud igual y de un ángulo recto.

Rectangular distribution (uniform) (p. 96) A distribution where all values in intervals of equal length are equally likely to occur.

Distribución rectangular (uniforme) (pág. 96) Distribución en que todos los valores en intervalos de la misma longitud son equiprobables.

Reflection symmetry (p. 401) In two dimensions, a figure has reflection symmetry if there is a line (called the *line of symmetry*) that divides the figure into mirror-image halves. Also called *mirror symmetry*. In three dimensions, a figure has reflection symmetry if there is a plane (called the *symmetry plane*) that divides the figure into mirror-image halves. Also called *plane symmetry*.

Simetría de reflexión (pág. 401) En dos dimensiones, una figura posee simetría de reflexión si hay una recta (el *eje de simetría*) que la divide en mitades especulares. También llamada *simetría especular*. En tres dimensiones, una figura posee simetría de reflexión si hay un plano (el *plano de simetría*) que la divide en mitades especulares. También llamada *simetría con respecto a un plano*.

English	Español
Regular dodecahedron (p. 441) A regular polyhedron with twelve congruent, regular pentagonal faces.	**Dodecaedro regular** (pág. 441) Poliedro regular con doce caras pentagonales regulares congruentes.
Regular hexahedron *see* **cube**	**Hexaedro regular** *véase* **cubo**
Regular icosahedron (p. 441) A regular polyhedron with twenty congruent, equilateral triangular faces.	**Icosaedro regular** (pág. 441) Poliedro regular con veinte caras triangulares equiláteras congruentes.
Regular octahedron (p. 441) A regular polyhedron with eight congruent, equilateral triangular faces.	**Octaedro regular** (pág. 441) Poliedro regular con ocho caras triangulares equiláteras congruentes.
Regular polygon (p. 400) A polygon in which all sides are congruent and all angles are congruent.	**Polígono regular** (pág. 400) Polígono cuyos lados son todos congruentes y cuyos ángulos son todos congruentes.
Regular polyhedron (platonic solid) (p. 439) A polyhedron in which all faces are congruent, regular polygons, and the arrangement of faces and edges is the same at each vertex.	**Poliedro regular (sólido platónico)** (pág. 439) Poliedro cuyas caras son todas polígonos regulares congruentes y la disposición de caras y aristas en cada vértice es la misma.
Regular tessellation (p. 409) A tessellation that consists of repeated copies of a single regular polygon.	**Teselado regular** (pág. 409) Teselado que consta de copias de un solo polígono regular.
Regular tetrahedron (equilateral triangular pyramid) (p. 441) A regular polyhedron with four congruent, equilateral triangular faces.	**Tetraedro regular (pirámide triangular equilátera)** (pág. 441) Poliedro regular con cuatro caras triangulares equiláteras congruentes.
Relative frequency histogram (p. 79) A histogram that shows the proportion or percentage that fall into the interval covered by each bar, rather than the frequency or count.	**Histograma de frecuencias relativas** (pág. 79) Histograma que muestra la proporción o porcentaje que cae en el intervalo cubierto por cada barra, en vez de la frecuencia o cuenta.
Relative frequency table (p. 100) A summary table for numerical data, where typically the column on the left gives the different data values and the column on the right gives the proportion (*relative frequency*) of measurements that have that value.	**Tabla de frecuencias relativas** (pág. 100) Tabla sumarial de datos numéricos, donde la columna de la izquierda lleva típicamente los diversos valores de los datos y la columna de la derecha lleva la proporción (*frecuencia relativa*) de las medidas que tienen dicho valor.
Resistant to outliers (less sensitive to outliers) (p. 86) Condition where a summary statistic does not change much when an outlier is removed from a set of data.	**Resistencia a los valores atípicos (menos susceptible a los valores atípicos)** (pág. 86) Condición en que una estadística sumarial no cambia mucho cuando se elimina un valor atípico de un conjunto de datos.
Rhombus (p. 366) A quadrilateral with all four sides of equal length.	**Rombo** (pág. 366) Cuadrilátero con cuatro lados que son de longitudes iguales.
Right triangle (p. 45) A triangle with a *right angle* (an angle with measure of 90°). The side opposite the right angle is the *hypotenuse*. The other two sides are the *legs*.	**Triángulo rectángulo** (pág. 45) Triángulo con un *ángulo recto* (ángulo que mide 90°). El lado opuesto al ángulo recto se llama *hipotenusa* y los otros dos lados son los *catetos*.

Glossary/Glosario

Rigid shapes (p. 366) Shapes that cannot flex when pressure is applied.

Formas rígidas (pág. 366) Formas que no se pueden doblar al aplicárseles presión.

Rotational symmetry (p. 401) In two dimensions, a figure has rotational symmetry if there is a point (called the *center of rotation*) about which the figure can be turned less than 360° in such a way that the rotated figure appears in exactly the same position as the original figure. In three dimensions, a figure has rotational symmetry if there is a line (called the *axis of symmetry*) about which the figure can be turned less than 360° in such a way that the rotated figure appears in exactly the same position as the original figure.

Simetría de rotación (pág. 401) En dos dimensiones, una figura posee simetría de rotación si hay un punto (el *centro de la rotación*) alrededor del cual la figura puede girar en menos de 360° de modo que la figura girada aparece en la misma posición que la figura original. En tres dimensiones, una figura posee simetría de rotación si hay una recta (el *eje de simetría*) alrededor de la cual la figura puede girar en menos de 360° de modo que la figura girada aparece en la misma posición que la figura original.

Row sum of a matrix (p. 249) The sum of the numbers in a row of a matrix.

Suma de fila de una matriz (pág. 249) Suma de las entradas de la fila de una matriz.

Run (trial) (p. 553) One repetition of a simulation.

Prueba (pág. 553) Una repetición de un simulacro.

S

Sample space (p. 534) The set of all possible outcomes of a chance situation.

Espacio muestral (pág. 534) Conjunto de todos los resultados posibles de una situación probabilística.

Scalene triangle (p. 68) A triangle with no two sides of equal length.

Triángulo escaleno (pág. 68) Triángulo sin ningún par de lados de la misma longitud.

Scatterplot (p. 5) A plot on a coordinate grid of the points whose (x, y) coordinates correspond to related data values of two variables.

Gráfica de dispersión (pág. 5) Gráfica en un cuadriculado de los puntos (x, y) cuyas coordenadas corresponden a datos relacionados de dos variables.

Schlegel diagram (p. 449) A vertex-edge graph resulting from "compressing" a three-dimensional object down into two dimensions.

Diagrama de Schlegel (pág. 449) Grafo que resulta de "comprimir" a dos dimensiones un objeto tridimensional.

Semiregular polyhedron (p. 442) A polyhedron whose faces are congruent copies of two or more different regular polygons and whose faces and edges have the same arrangement at each vertex.

Poliedro semirregular (pág. 442) Poliedro cuyas caras son copias congruentes de dos o más polígonos regulares distintos y cuyas caras y aristas poseen la misma disposición en cada vértice.

Semiregular tessellation (p. 410) A tessellation of two or more regular polygons that has the same arrangement of polygons at each vertex.

Teselado semirregular (pág. 410) Teselado de dos o más polígonos regulares que posee la misma disposición de polígonos en cada vértice.

Sensitive to outliers (p. 86) Condition where a summary statistic changes quite a bit when an outlier is removed from a set of data.

Susceptible a los valores atípicos (pág. 86) Condición en que una estadística sumarial cambia bastante cuando se elimina un valor atípico de un conjunto de datos.

Simulation (p. 553) Creating a mathematical model that copies (simulates) a real-life situation's essential characteristics.

Simulacro (pág. 553) Modelo matemático que copia (simula) las características esenciales de una situación concreta.

Glossary/Glosario

English

Single-variable data (p. 75) Data where a single measurement or count is taken on each object of study, such as height of each person or age of each person.

Skewed distribution (p. 77) A distribution that has a *tail* stretched either towards the larger values (*skewed right*) or towards the smaller values (*skewed left*).

Slope-intercept form (p. 160) A linear function with rule in the form $y = mx + b$ is said to be written in slope-intercept form because the value of m indicates the slope of the graph and the value of b indicates the y-intercept of the graph.

Slope of a line (p. 155) Ratio of change in y-coordinates to change in x-coordinates between any two points on the line; $\frac{\text{change in } y}{\text{change in } x}$ or $\frac{\Delta y}{\Delta x}$; indicates the direction and steepness of a line.

Solve (an equation, inequality, or system of equations) (p. 189) To find values of the variable(s) that make the statement(s) true.

Speed (p. 11) When a person or object moves a distance d in a time t, the quotient $\frac{d}{t}$ gives the average speed of the motion. The units of speed are given as "distance per unit of time."

Spreadsheet (p. 32) A spreadsheet is a two-dimensional grid of cells in which numerical data or words can be stored. Numerical values in the cells of a spreadsheet can be related by formulas, so that the entry in one cell can be calculated from values in other cells.

Square (p. 366) A rhombus with one right angle.

Square root (p. 335) If r is a number for which $r^2 = n$, then r is called a square root of n. Every positive number n has two square roots, denoted with the radical forms $\sqrt{n}$ and $-\sqrt{n}$.

Standard deviation (*s*) (p. 116) A useful measure of spread; based on the sum of the squared deviations from the mean; in a normal distribution about 68% of the values lie no more than one standard deviation from the mean.

Español

Datos de una sola variable (pág. 75) Datos en que se ejecuta una sola medida o cuenta en cada objeto de estudio, como la estatura o la edad de una persona.

Distribución asimétrica (pág. 77) Distribución que posee una *cola* extendida ya sea hacia los valores más grandes (*asimétrica derecha*) o hacia los valores más pequeños (*asimétrica izquierda*).

Forma pendiente-intersección (pág. 160) Una función lineal de la forma $y = mx + b$ se dice que está escrita en la forma pendiente-intersección porque m es la pendiente de la gráfica y b es la intersección y de la misma.

Pendiente de una recta (pág. 155) Razón del cambio en las coordenadas y al cambio en las coordenadas x entre dos puntos de una recta; $\frac{\text{cambio en } y}{\text{cambio en } x}$ o $\frac{\Delta y}{\Delta x}$; indica la dirección e inclinación de la recta.

Solución (de una ecuación, desigualdad o sistema de ecuaciones) (pág. 189) Calcular valores de la variable o variables que las satisfagan.

Rapidez (pág. 11) Cuando una persona o un cuerpo se desplaza una distancia d en un tiempo t, el cociente $\frac{d}{t}$ da la rapidez media del movimiento. Las unidades de rapidez son "distancia por unidad de tiempo."

Hojas de cálculos (pág. 32) Cuadriculado bidimensional de celdas en que pueden almacenarse datos numéricos o palabras. Los valores numéricos en una hoja de cálculos pueden estar relacionados por fórmulas, de modo que la entrada en una celda puede calcularse de los valores en otras celdas.

Cuadrado (pág. 366) Un rombo con un ángulo recto.

Raíz cuadrada (pág. 335) Si r es un número que cumple $r^2 = n$, r se llama una raíz cuadrada de n. Todo número positivo r posee dos raíces cuadradas, designadas por los radicales $\sqrt{n}$ y $-\sqrt{n}$.

Desviación estándar (pág. 116) Medida útil de dispersión; se basa en la suma de las desviaciones al cuadrado de la media; en una distribución normal, cerca del 68% de los valores yacen a no más de una desviación estándar de la media.

Glossary/Glosario

Stemplot (stem-and-leaf plot) (p. 97) A statistical display using certain digits (such as the tens place) as the "stem" and the remaining digit or digits (such as the ones place) as "leaves."

Diagrama de tallos (diagrama de tallo y hojas) (pág. 97) Presentación estadística en que se usan ciertos dígitos (las decenas, por ejemplo) como los "tallos" y el dígito o dígitos restantes (las unidades, por ejemplo) como "las hojas."

Summary statistic (p. 77) A numerical summary of the values in a distribution. For example, the mean, median, or range.

Estadística sumarial (pág. 77) Resumen numérico de los valores de una distribución. Por ejemplo, la media, la mediana o el rango.

Symmetry plane (mirror plane) *see* **reflection symmetry**

Plano de simetría (plano especular) *véase* **simetría de reflexión**

System of equations (p. 199) Two or more equations. The *solution of a system* is the set of solutions that satisfy each equation in the system.

Sistema de ecuaciones (pág. 199) Dos o más ecuaciones. La *solución de un sistema* es el conjunto de soluciones que satisfacen cada ecuación del sistema.

T

Tessellation (tiling) (p. 408) Repeated copies of one or more shapes so as to completely cover a planar region without overlaps or gaps.

Teselado (embaldosado) (pág. 408) Copias repetidas de una o más formas que cubren una región plana completamente sin traslapos o espacios.

Translation symmetry (p. 408) A pattern has translation symmetry if it coincides with itself under some translation (slide).

Simetría de traslación (pág. 408) Un patrón posee simetría de traslación si coincide consigo mismo bajo alguna traslación (deslizamiento).

Triangle (p. 389) A polygon with three sides. A *base* of a triangle is the side of the triangle that is perpendicular to an altitude.

Triángulo (pág. 389) Polígono con tres lados. Una *base* de un triángulo es un lado del mismo el cual es perpendicular a una altura.

Triangulate (p. 366) To divide a polygon into a set of nonoverlapping triangles where the vertices of the triangles are the vertices of the polygon.

Triangulación (pág. 366) División de un polígono en un conjunto de triángulos que no se traslapan y en que los vértices de los triángulos son los del polígono.

V

Variability (p. 103) The spread in the values in a distribution. (*see* **measure of spread**)

Variabilidad (pág. 103) Dispersión de los valores de una distribución. (*véase* **medida de dispersión**)

Variable (p. 389) A quantity that changes. Variables are commonly represented by letters like x, y, z, s, or t. (*see* **dependent variable** and **independent variable**)

Variable (pág. 389) Cantidad que cambia. Se representan en general por letras como x, y, z, s o t. (*véanse* **variable dependiente** y **variable independiente**)

Vertex (*plural:* vertices) (p. 241) A point where edges of a vertex-edge graph meet. Also, a point where two sides of a polygon meet.

Vértice (pág. 241) Punto al que concurren aristas de un grafo. También, punto al que concurren dos lados de un polígono.

Vertex angle defect (p. 437) In a convex polyhedron, the vertex angle defect is the positive difference between the sum of the measures of the *face angles* (the angle formed by two edges of a polyhedral angle) at that vertex and 360°.

Defecto del ángulo de un vértice (pág. 437) En un polígono convexo, la diferencia positiva entre la suma de las medidas de los *ángulos de cara* (el ángulo formado por dos aristas de un ángulo poliedro) en ese vértice y 360°.

Glossary/Glosario

English	Español

Vertex coloring *see* **color a graph**

Coloración de vértices *véase* **colorear un grafo**

Vertex-edge graph (graph) (p. 241) A diagram consisting of a set of points (called *vertices*) along with segments or arcs (called *edges*) joining some of the points.

Grafo (pág. 241) Diagrama que consta de un conjunto de puntos (los vértices) junto con segmentos o arcos (las aristas) que unen algunos de los puntos.

Vertical angles (p. 376) Two angles whose sides form two pairs of opposite rays.

Ángulos opuestos por el vértice (pág. 376) Dos ángulos cuyos lados forman dos pares de rayos opuestos.

Venn diagram (p. 540) A diagram involving circles that depicts collections of objects and the relationships between them.

Diagrama de Venn (pág. 540) Diagrama que consta de círculos que exhiben colecciones de objetos y las relaciones entre ellos.

W

With replacement (p. 562) Selecting a sample from a set so that each selection is replaced before selecting the next; thus, each member of the set can be selected more than once.

Con devolución (pág. 562) Selección de una muestra de un conjunto de modo que cada selección se devuelve antes de elegir la siguiente; así, cada miembro del conjunto puede escogerse más de una vez.

Without replacement (p. 562) Selecting a sample from a set so that each selection is not replaced before selecting the next; each member of the set cannot be selected more than once.

Sin devolución (pág. 562) Selección de una muestra de un conjunto de modo que cada selección no se devuelve antes de elegir la siguiente; así, cada miembro del conjunto no puede escogerse más de una vez.

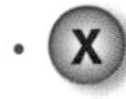

***x*-intercept of a graph** (p. 477) The point(s) where the graph intersects the x-axis.

Intersección *x* de una gráfica (pág. 477) El punto o los puntos en que una gráfica interseca el eje x.

Y

***y*-intercept of a graph** (p. 155) The point(s) where the graph intersects the y-axis.

Intersección *y* de una gráfica (pág. 155) El punto o los puntos en que una gráfica interseca el eje y.

Index of Mathematical Topics

A

B

C

D

E

F

Index of Mathematical Topics *(continued)*

Index of Contexts

Photo Credits

Cover (tl tr bl)Getty Images, (br)Japack Company/CORBIS; **xvi** (t)Dennis MacDonald/Alamy Images, (b)The Image Bank/Getty Images; **xvi** John Gilroy and John Lacko; **1** Anne-Marie Weber/Taxi/Getty Images; **2** David R. Frazier; **4** Tim Fuller; **5** Tim Fuller; **6** Michael Newman/PhotoEdit; **8** Index Stock; **10** (l)Getty Images, (c)The McGraw-Hill Companies, (r)CORBIS; **11** AP/Wide World Photos; **12** Kimberly White/Reuters/CORBIS; **13** Kevin Horan/Getty Images; **14** Getty Images; **16** S. Carmona/CORBIS; **17** Sergio Dorantes/CORBIS; **20** Alamy Images; **22** AP/Wide World Photos; **23** Ole Graf/Zefa/CORBIS; **24** Index Stock; **26** Bob Krist/CORBIS; **29** Tom Brakefield/Digital Vision/Getty Images; **31** Printed by permission of Norman Rockwell Family Agency Copyright © 1940 the Norman Rockwell Family Entities; **34** Tim Pannell/CORBIS; **35** (t)Robert Essel NYC/CORBIS, (b)Steve Terrill; **36** Panorama Stock/Alamy Images; **37** Alan and Sandy Carey/CORBIS; **40** Jeff Greenberg/PhotoEdit; **43** Travel Ink Photo/Index Stock; **44** Douglas Faulkner/Photo Researchers; **45** CORBIS; **49** Kim Karpeles; **56** Getty Images; **59** Stockbyte/Punchstock; **66** Brent Smith/Reuters/CORBIS; **69** Anne-Marie Weber/Taxi/Getty Images; **71** Alan Oliver/Alamy; **73** NOAA; **74** Diane Moore; **76** Kennan Ward/CORBIS; **77** Basler Turbo; **82** Lois Ellen Frank/CORBIS; **86** Tim Fuller; **87** Owen Humphreys/AP/Wide World Photos; **89** Jeff Greenberg/PhotoEdit; **90** Justin Guariglia/CORBIS; **91** Martin B. Withers/Frank Lane Picture Agency/CORBIS; **92** Scott Neff/Cinematour.com; **96** Brian L. Joiner; **97** CORBIS; **98** Reed Saxon/AP/Wide World Photos; **99** Entrepreneur.com; **101** Alamy Images; **103** Michel Tcherevkoff/The Image Bank/Getty Images; **104** Michael A. Keller/zefa/CORBIS; **106** Brand X/Superstock; **107** Jeff Greenberg/Alamy Images; **109** Carey Wolinsky/Stock Boston; **111** CORBIS; **112** DK Limited/CORBIS; **119** Ron Kuntz/Reuters/CORBIS; **120** Tim Fuller; **123** Bill Ross/CORBIS; **128** H.G. Rossi/zefa/CORBIS; **129** Mary Langenfeld Photo; **133** Ausloeser/zefa/CORBIS; **138** Dennis MacDonald/Alamy Images; **141** Tim Fuller; **142** Aaron Haupt/Stock Boston; **144** Michael Dwyer/Alamy Images; **149** Paul Freytag/CORBIS; **150** Jessica Tucker; **152** CORBIS; **154** Chuck Savage/CORBIS; **156** Arizona Republic; **157** Kayte M. DeiomaPhotoEdit; **158** Tony Freeman/PhotoEdit; **159** (t)A.E. Cook, (b)Justin Sullivan/Getty Images; **162** NASA; **165** Don Mason/Getty Images; **166** Jump Run Productions/Getty Images; **170** (t)The Kirby Company, (b)Richard J. Stoner II/AgriHouse, Inc.; **171** Tom Sanders/Aerial Focus; **172** Townsend P. Dickinson/The Image Works; **173** A.E. Cook, Ariel Skelley/CORBIS; **175** CORBIS; **181** Stuart Hannagan/Getty Images; **182** Andy Lyons/Getty Images; **184** Digital Vision/Getty; **186** Susan Van Etten/PhotoEdit; **188** Brand X Pictures/Alamy; **190** Jose Luis Pelaez/CORBIS; **191** Tom Pantages; **192** Journalist 2nd class Tim Walsh/U.S. Navy; **194** Kazuhiro/AFP/Getty Images; **196** Michael Keller/CORBIS; **200** Doug Wilson/Alamy; **201** (t)Profusion Picture Library/Alamy, (b)William Manning/CORBIS; **202** Getty Images; **203** A.E. Cook; **204** Tom Pantages; **207** AP/Wide World Photos; **214** Comstock Images/jupiterimages; **217** Adam Martin/CinemaTour.com; **218** John Gress/Reuters/CORBIS; **224** Scott Barbour/Getty Images; **229** Getty Images; **232** Bettmann/CORBIS; **235** Lawrence Migdale; **237** Stuart Ramsden; **238** London Transport Museum; **243** Michael Newman/PhotoEdit; **247** Laszlo Balogh/Reuters/CORBIS; **253** The Image Bank/Getty Images; **257** London Transport Museum; **261** Leonard Lessin/Photo Researchers; **264** Steve Hamblin/Alamy; **266** Philip James Corwin/CORBIS; **271** Michael Newman/PhotoEdit; **276** Zoran Milich/Masterfile; **277** Gerald French/CORBIS; **281** Courtesy Douglas Prince/University of New Hampshire Photographic Services; **283** Diana Venters; **286** Peter Southwick/Stock Boston; **289** Tom Brakefield/CORBIS; **290** ThinkStock/SuperStock; **293** Mary Langenfeld; **295** CNRI/SPL/Photo Researchers; **297** Scimat/Photo Researchers; **298** (t)Science Source/Photo Researchers, (b)Photonica/Getty Images; **301** James Marshall/CORBIS; **302** Jeffrey LePore/Photo Researchers; **307** Peaktorial/Alamy; **310** Michael Nagle/Getty Images; **313** Jeff Greenberg/PhotoEdit; **316** Phanie/Photo Researchers; **322** AP/Wide World Photos; **323** Henry Groskinsky/Peter Arnold; **324** Tim Fuller; **326** CORBIS; **329** Comstock SuperStock; **332** Henry Groskinsky/Peter Arnold; **338** John Gilroy and John Lacko; **339** Alamy Images; **340** David R. Frazier Photolibrary; **341** Arco Images/Alamy; **347** blickwinkel/Alamy; **348** Getty Images; **350** Tony Freeman/PhotoEdit; **352** Tom Pantages; **355** (t)Tom Brakefield/CORBIS, (b)A.E. Cook; **356** Richard Heathcote/Getty Images; **357** Roger Ressmeyer; **361** Jon Arnold Images/Alamy; **362** CORBIS; **363** Brand X Pictures/AGE Fotostock; **365** Jim Laser; **368** Diane Moore; **369** Teri Ziebarth; **376 378** Matt Meadows; **381** Todd Stailey/Tennessee Aquarium; **385** Matt Meadows; **388** Peter Casolino/Alamy; **390** Photo courtesy of Hurd Windows & Doors; **398** (tr)Jim Laser, (tl)Royalty-Free/CORBIS, (b)David Parker/Photo Researchers, **399** (r)Doug Martin; (others)StudiOhio; **400 401** Jim Laser; **407** (t)Brand X Pictures/Getty Images, (b)Hilary Bullock/Carleton College; **409** Bill Varie/Alamy; **410** Mel Curtis/Getty Images; **413** (t)UNI-GROUP U.S.A., (b)Manfred Kage/Peter Arnold, Inc.; **414** George H.H. Huey/CORBIS; **415** (b)M.C. Escher's "Symmetry Drawing E105" The M.C. Escher Company-Netherlands. All rights reserved. www.mcescher.com, (t)David Parker/Photo Researchers; **421** Jim Laser; **424** Getty Images; **426** Matt Meadows; **428** (l)Neil Emmerson/Robert Harding World Library/Getty Images, (r)Kathleen Lang/Art a GoGo; **429** Pacific Domes; **431** Geoff Butler; **435** Philippa Lewis/Edifice/CORBIS; **436** Jim Laser; **439** StudiOhio; **442** PhotoLink/Getty Images; **443** Andrew Ward/Life File/Getty Images; **444** Susan Van Etten; **446** Will & Deni McIntyre/Photo Researchers; **452** Emely/zefa/CORBIS; **456** (t)Jon Arnold Images/Alamy, (b)Kalamazoo Gazette; **457** David R. Frazier; **459** Courtesy of Marshfield Area Habitat for Humanity.; **461** Mike Powell/Getty Images; **462** John Larew; **464** courtesy Susie Flentie; **465** David S. Holloway/Getty Images; **468** Martin Mills/Getty Images; **469** Martin Child/Getty Images; **470** Ronald Siemoneit/Sygma/CORBIS; **472** Gerolf Kalt/zefa/CORBIS; **473** Digital Vision/Getty Images; **480** Anthony Reynolds/Cordaiy Photo Library/CORBIS; **481** Eberhard Streichan/zefa/CORBIS; **482** Jim Laser; **485** Shotfile/Alamy; **486** Jamie Squire/Getty Images; **489** Reuters/CORBIS; **491** Mario Tama/Getty Images; **493** Tony Freeman/PhotoEdit; **499** Getty Images; **500** Galen Rowell/CORBIS; **510** Jonathan Ferrey/Getty Images; **518** Roger Hill; **519** APWide World Photos; **522** Peter Barrett/Masterfile; **526** Image 100/Alamy; **528** Ambient Images/Alamy; **531** Courtesy of Didax Educational Resources; **532** Comstock/SuperStock; **533** Creatas/SuperStock; **535** Doug Martin; **538** White Packert/Getty Images; **542** Aaron Haupt; **543** Bettmann/CORBIS; **546** Creatas/SuperStock; **548** Getty Images; **551** Liu Liqun/CORBIS; **553** (t)Getty Images, (b)CORBIS; **554** SuperStock; **557** John Wickline/The Wheeling (W.Va.) News-Register/AP/Wide World Photos; **561** Tim Kiusalaas/Masterfile; **562** The McGraw Hill Companies; **567** Gen Nishino/Getty Images; **568** David Young-Wolff/PhotoEdit; **570** Courtesy The Splendid Peasant; **571** Bob Krist/CORBIS; **573** John G. Mabanglo/CORBIS; **574** Chris O'Meara/AP/Wide World Photos; **576 581** The McGraw Hill Companies; **582** NBAE/Getty Images; **586** Mark Steinmetz; **587** Michael Kelley/Getty Images; **588** CORBIS